# PESTICIDES IN THE SOIL ENVIRONMENT: PROCESSES, IMPACTS, AND MODELING

## Soil Science Society of America Book Series

Books in the series are available from the Soil Science Society of America, 677 South Segoe Road, Madison, WI 53711 USA.

1. MINERALS IN SOIL ENVIRONMENTS. Second Edition. 1989.
   J. B. Dixon and S. B. Weed, *editors* R. C. Dinauer, *managing editor*
2. PESTICIDES IN THE SOIL ENVIRONMENT: PROCESSES, IMPACTS, AND MODELING. 1990.
   H. H. Cheng, *editor* S. H. Mickelson, *managing editor*

# Pesticides in the Soil Environment: Processes, Impacts, and Modeling

**Editor: H. H. Cheng**

*Editorial Committee:* H. H. Cheng
G. W. Bailey
R. E. Green
W. F. Spencer

*Managing Editor:* S. H. Mickelson
*Associate Editor-at-Large:* Annette J. Tallard
*Editor-in-Chief SSSA:* David E. Kissel

**Number 2 in the Soil Science Society of America Book Series**

**Published by: Soil Science Society of America, Inc.**
**Madison, Wisconsin, USA**

**1990**

Soil Science Society of America, Inc.
677 South Segoe Road, Madison, Wisconsin 53711 USA

**Library of Congress Cataloging-in-Publication Data**

Pesticides in the soil environment : processes, impacts, and modeling / editor, H.H. Cheng.
p. cm. — (Soil Science Society of America book series : no. 2)
Includes bibliographical references and indexes.
ISBN 0-89118-791-X
1. Pesticides—Environmental aspects. 2. Soil pollution. 3. Soils—Pesticide content. I. Cheng, H. H. (Hwei-Hsien), 1936– II. Series.
TD879.P37P47 1990
628.5 '5—dc20
90-10070
CIP

Printed in the United States of America

# CONTENTS

# FOREWORD

Few environmental concerns now occupying public attention are as fraught with emotion as is the pesticide issue. Human exposure to pesticides may come directly through consumption of foods containing pesticide residue. This pathway is relatively straightforward and not too difficult to trace and assess. Pesticides may also find their way into the food chain or into drinking water, also leading to potential human exposure.

This second pathway is much more complex and difficult to quantify. Following a now well-established publication policy of the Soil Science Society of America, this volume seeks to bring to scientists, engineers, and regulators the best and most up-to-date scientific information. It has been wisely said: ''Dogmatic assertion is an inadequate substitute for truth.'' While science never achieves final and ultimate truth, we hope that judicious use of the information in this volume will help reduce the emotional debate over pesticide exposure to rational studies and reasoned debate. Through such an approach lies our best hope of ensuring both human health and an adequate food supply.

W. R. Gardner, *president*
*Soil Science Society of America*

# PREFACE

Soil scientists have been interested in the fate of pesticides in the soil environment ever since synthetic organic pesticides first came to use. Beginning in the 1960s, many symposia and special sessions have been held at the annual meetings of the American Society of Agronomy and Soil Science Society of America. An early symposium publication was entitled "Pesticides and Their Effects on Soils and Water," which was published in 1966 as ASA Special Publication Number 8. Subsequently, a comprehensive review of the principles controlling the environmental effects of pesticide-soil-water interactions was published by the Soil Science Society of America in 1974 as a book entitled *Pesticides in Soil and Water,* edited by an editorial committee chaired by W.D. Guenzi.

Early in 1981, the Executive Committee of SSSA appointed an Ad Hoc Committee to evaluate the need for a revision of the 1974 book. After surveying widely the opinions of scientists who have been active in pesticide research, the committee concluded that the basic principles described in the 1974 book were valid, although the coverage could be extended to new subject areas. Thus, a revision of the earlier book would not fulfill the purpose. A new book was recommended to present not only up-to-date information but also new concepts and approaches to the understanding of the subject matter. The committee report states that "The Soil Science Society of America has a corporate responsibility in providing the public authoritative and scholarly publications on subject matter pertinent to its field. Pesticides in soil and water constitute legitimate concerns for soil scientists."

In 1982, an Organizing Committee was then appointed to design an outline for the book. From the start, it was clear that the book would emphasize the application of soil science principles to the understanding of the transport and transformation processes that affect the behavior and fate of pesticides and related organics in the environment. It would concentrate on presentations of fundamental knowledge on processes, mechanisms, kinetics, and assessment techniques rather than a comprehensive citation of existing literature. This committee laid the foundation and the framework to construct the chapters in a sequence from the entry of the pesticides into the environment to its eventual fate in the soil so that the environment could be meaningfully assessed. The Editorial Committee has faithfully carried out the intent of the book established by the Organizing Committee.

Numerous scientists have contributed to this book project during the last 8 yr. Serving on the initial Evaluation and Organizing Committees were: H.H. Cheng, C.A.I. Goring, R.E. Green, W.D. Guenzi, W.F. Spencer, A.W. Taylor, and J.B. Weber. Others who have contributed significantly during these deliberations included: G.W. Bailey, D.A. Laskowski, G.H. Willis, and colleagues on the Western Regional Research Committee W-82. During the writing stage, the authors as well as the Editorial Committee had consulted with so many colleagues that it would not be feasible to list all of them here. All the chapters were reviewed in several rounds by the Editorial Committee, by authors of the other chapters, and by selected peer reviewers, including: S.A. Boyd, R.F. Carsel, C.E. Castro, E.P. Caswell, W.J. Farmer, D.E. Glotfelty, J.B. Harsh, P.G. Hunt, Edwin Johnson, W.G. Knisel Jr., K.M. Loague, D.J. Mulla, A.N. Sharpley, C.N. Smith, and G.H. Willis. To all those named and unnamed who have contributed to this book, the Editorial Committee wishes to express our profound thanks and appreciation for their time and effort.

During the past 8 yr, there has been a great deal of research activity on pesticide behavior and fate in the environment. Thus, the preparation of the book was at the same time a learning experience for all of us, as concepts were matured and methodology and techniques improved. The exercise of putting together a book has also proven to be a test for persistence and perserverance. Most of the authors were conscientious and cooperative in meeting deadlines and commitment. We had hoped to include a chapter on the kinetics of pesticide biotransformation. However, this planned chapter was not realized. Since the rates of biotransformation under natural conditions is one of the most difficult to characterize, the absence of this coverage could serve as a challenge to future research efforts.

Editorial Committee
H. H. Cheng, *chair*
G. W. Bailey
R. E. Green
W. F. Spencer

# CONTRIBUTORS


| | |
|---|---|
| **George W. Bailey** | Research Soil Physical Chemist, USEPA, Environmental Research Laboratory, Athens, GA 30613-7799 |
| **Gary L. Ballard** | Economist, USEPA, Office of Pesticide Programs, Washington, DC 20460 |
| **Jean-Marc Bollag** | Professor of Soil Microbiology, Laboratory of Soil Biochemistry, The Pennsylvania State University, University Park, PA 16802 |
| **H.H. Cheng** | Formerly, Professor, Department of Agronomy and Soils, Washington State University, Pullman, WA 99164; currently, Professor and Head, Department of Soil Science, University of Minnesota, St. Paul, MN 55108 |
| **Carl G. Enfield** | Soil Scientist, USEPA, Robert S. Kerr Environmental Research Laboratory, Ada, OK 74820 |
| **Virgil H. Freed** | Emeritus Professor, Department of Agricultural Chemistry, Oregon State University, Corvallis, OR 97331 |
| **Richard E. Green** | Professor of Soil Science, Department of Agronomy and Soil Science, University of Hawaii, Honolulu, HI 96822 |
| **Sidney S. Harper** | Formerly, Soil Scientist, USDA-ARS, Southern Weed Science Laboratory; currently, Research Chemist, Tennessee Valley Authority, Muscle Shoals, AL 35660 |
| **Chester M. Himel** | Research Professor (retired), Department of Entomology, University of Georgia, Athens, GA 30602 |
| **Samuel W. Karickhoff** | Research Scientist, USEPA, Environmental Research Laboratory, Athens, GA 30613 |
| **William C. Koskinen** | Soil Scientist, USDA-ARS, Soil Science Department, University of Minnesota, St. Paul, MN 55108 |
| **Minze Leistra** | Research Scientist, The Winand Staring Centre for Integrated Land, Soil and Water Research, Wageningen, The Netherlands |
| **R.A. Leonard** | Supervisory Soil Scientist, USDA-ARS, Southeast Watershed Research Laboratory, Tifton, GA 31793 |
| **Shu-Yen Liu** | Research Associate, Laboratory of Soil Biochemistry, The Pennsylvania State University, University Park, PA 16802 |
| **Harry L. Loats** | President, Loats Associates, Inc., Westminster, MD 21157 |
| **Yousef A. Madhun** | Environmental Research Chemist, Department of Agricultural Chemistry, Oregon State University, Corvallis, OR 97331 |
| **Glenn C. Miller** | Professor, Department of Biochemistry, University of Nevada, Reno, NV 89557 |
| **U. Mingelgrin** | Professor, Institute of Soils and Water, The Volcani Center, ARO, Bet-Dagan, Israel |
| **P.S.C. Rao** | Professor, Soil Science Department, University of Florida, Gainesville, FL 32611-0151 |
| **David J. Severn** | Formerly, Clement Associates, Inc., Fairfax, VA 22031-1207; currently, Manager for Chemistry, Jellinek, Schwartz, Connolly & Freshman, Inc., Washington, DC 20005 |

| | |
|---|---|
| **William F. Spencer** | Soil Scientist, USDA-ARS, Department of Soil and Environmental Sciences, University of California, Riverside, CA 92521 |
| **Alan W. Taylor** | Consultant in Environmental Sciences, Agricultural Experiment Station, University of Maryland, College Park, MD 20742 |
| **R.J. Wagenet** | Professor of Soil Physics, Department of Agronomy, Cornell University, Ithaca, NY 14853 |
| **N. Lee Wolfe** | Research Chemist, USEPA, Environmental Research Laboratory, Athens, GA 30613-7799 |
| **Scott R. Yates** | Soil Scientist, U.S. Salinity Laboratory, University of California, Riverside, CA 92521 |

# Conversion Factors for SI and non-SI Units

# Conversion Factors for SI and non-SI Units

| To convert Column 1 into Column 2, multiply by | Column 1 SI Unit | Column 2 non-SI Unit | To convert Column 2 into Column 1, multiply by |
|---|---|---|---|
| | | **Length** | |
| 0.621 | kilometer, km ($10^3$ m) | mile, mi | 1.609 |
| 1.094 | meter, m | yard, yd | 0.914 |
| 3.28 | meter, m | foot, ft | 0.304 |
| 1.0 | micrometer, $\mu$m ($10^{-6}$ m) | micron, $\mu$ | 1.0 |
| $3.94 \times 10^{-2}$ | millimeter, mm ($10^{-3}$ m) | inch, in | 25.4 |
| 10 | nanometer, nm ($10^{-9}$ m) | Angstrom, Å | 0.1 |
| | | **Area** | |
| 2.47 | hectare, ha | acre | 0.405 |
| 247 | square kilometer, $km^2$ $(10^3\ m)^2$ | acre | $4.05 \times 10^{-3}$ |
| 0.386 | square kilometer, $km^2$ $(10^3\ m)^2$ | square mile, $mi^2$ | 2.590 |
| $2.47 \times 10^{-4}$ | square meter, $m^2$ | acre | $4.05 \times 10^3$ |
| 10.76 | square meter, $m^2$ | square foot, $ft^2$ | $9.29 \times 10^{-2}$ |
| $1.55 \times 10^{-3}$ | square millimeter, $mm^2$ $(10^{-3}\ m)^2$ | square inch, $in^2$ | 645 |
| | | **Volume** | |
| $9.73 \times 10^{-3}$ | cubic meter, $m^3$ | acre-inch | 102.8 |
| 35.3 | cubic meter, $m^3$ | cubic foot, $ft^3$ | $2.83 \times 10^{-2}$ |
| $6.10 \times 10^4$ | cubic meter, $m^3$ | cubic inch, $in^3$ | $1.64 \times 10^{-5}$ |
| $2.84 \times 10^{-2}$ | liter, L ($10^{-3}$ $m^3$) | bushel, bu | 35.24 |
| 1.057 | liter, L ($10^{-3}$ $m^3$) | quart (liquid), qt | 0.946 |
| $3.53 \times 10^{-2}$ | liter, L ($10^{-3}$ $m^3$) | cubic foot, $ft^3$ | 28.3 |
| 0.265 | liter, L ($10^{-3}$ $m^3$) | gallon | 3.78 |
| 33.78 | liter, L ($10^{-3}$ $m^3$) | ounce (fluid), oz | $2.96 \times 10^{-2}$ |
| 2.11 | liter, L ($10^{-3}$ $m^3$) | pint (fluid), pt | 0.473 |

| | | | |
|---|---|---|---|
| | | **Mass** | |
| $2.20 \times 10^{-3}$ | gram, g ($10^{-3}$ kg) | pound, lb | 454 |
| $3.52 \times 10^{-2}$ | gram, g ($10^{-3}$ kg) | ounce (avdp), oz | 28.4 |
| 2.205 | kilogram, kg | pound, lb | 0.454 |
| 0.01 | kilogram, kg | quintal (metric), q | 100 |
| $1.10 \times 10^{-3}$ | kilogram, kg | ton (2000 lb), ton | 907 |
| 1.102 | megagram, Mg (tonne) | ton (U.S.), ton | 0.907 |
| 1.102 | tonne, t | ton (U.S.), ton | 0.907 |
| | | **Yield and Rate** | |
| 0.893 | kilogram per hectare, kg $ha^{-1}$ | pound per acre, lb $acre^{-1}$ | 1.12 |
| $7.77 \times 10^{-2}$ | kilogram per cubic meter, kg $m^{-3}$ | pound per bushel, $bu^{-1}$ | 12.87 |
| $1.49 \times 10^{-2}$ | kilogram per hectare, kg $ha^{-1}$ | bushel per acre, 60 lb | 67.19 |
| $1.59 \times 10^{-2}$ | kilogram per hectare, kg $ha^{-1}$ | bushel per acre, 56 lb | 62.71 |
| $1.86 \times 10^{-2}$ | kilogram per hectare, kg $ha^{-1}$ | bushel per acre, 48 lb | 53.75 |
| 0.107 | liter per hectare, L $ha^{-1}$ | gallon per acre | 9.35 |
| 893 | tonnes per hectare, t $ha^{-1}$ | pound per acre, lb $acre^{-1}$ | $1.12 \times 10^{-3}$ |
| 893 | megagram per hectare, Mg $ha^{-1}$ | pound per acre, lb $acre^{-1}$ | $1.12 \times 10^{-3}$ |
| 0.446 | megagram per hectare, Mg $ha^{-1}$ | ton (2000 lb) per acre, ton $acre^{-1}$ | 2.24 |
| 2.24 | meter per second, m $s^{-1}$ | mile per hour | 0.447 |
| | | **Specific Surface** | |
| 10 | square meter per kilogram, $m^2$ $kg^{-1}$ | square centimeter per gram, $cm^2$ $g^{-1}$ | 0.1 |
| 1000 | square meter per kilogram, $m^2$ $kg^{-1}$ | square millimeter per gram, $mm^2$ $g^{-1}$ | 0.001 |
| | | **Pressure** | |
| 9.90 | megapascal, MPa ($10^6$ Pa) | atmosphere | 0.101 |
| 10 | megapascal, MPa ($10^6$ Pa) | bar | 0.1 |
| 1.00 | megagram per cubic meter, Mg $m^{-3}$ | gram per cubic centimeter, g $cm^{-3}$ | 1.00 |
| $2.09 \times 10^{-2}$ | pascal, Pa | pound per square foot, lb $ft^{-2}$ | 47.9 |
| $1.45 \times 10^{-4}$ | pascal, Pa | pound per square inch, lb $in^{-2}$ | $6.90 \times 10^3$ |

(continued on next page)

# Conversion Factors for SI and non-SI Units

| To convert Column 1 into Column 2, multiply by | Column 1 SI Unit | Column 2 non-SI Unit | To convert Column 2 into Column 1, multiply by |
|---|---|---|---|
| | | **Temperature** | |
| 1.00 (K − 273) | Kelvin, K | Celsius, °C | 1.00 (°C + 273) |
| (9/5 °C) + 32 | Celsius, °C | Fahrenheit, °F | 5/9 (°F − 32) |
| | | **Energy, Work, Quantity of Heat** | |
| $9.52 \times 10^{-4}$ | joule, J | British thermal unit, Btu | $1.05 \times 10^{3}$ |
| 0.239 | joule, J | calorie, cal | 4.19 |
| $10^{7}$ | joule, J | erg | $10^{-7}$ |
| 0.735 | joule, J | foot-pound | 1.36 |
| $2.387 \times 10^{-5}$ | joule per square meter, J $m^{-2}$ | calorie per square centimeter (langley) | $4.19 \times 10^{4}$ |
| $10^{5}$ | newton, N | dyne | $10^{-5}$ |
| $1.43 \times 10^{-3}$ | watt per square meter, W $m^{-2}$ | calorie per square centimeter minute (irradiance), cal $cm^{-2}$ $min^{-1}$ | 698 |
| | | **Transpiration and Photosynthesis** | |
| $3.60 \times 10^{-2}$ | milligram per square meter second, mg $m^{-2}$ $s^{-1}$ | gram per square decimeter hour, g $dm^{-2}$ $h^{-1}$ | 27.8 |
| $5.56 \times 10^{-3}$ | milligram ($H_2O$) per square meter second, mg $m^{-2}$ $s^{-1}$ | micromole ($H_2O$) per square centimeter second, μmol $cm^{-2}$ $s^{-1}$ | 180 |
| $10^{-4}$ | milligram per square meter second, mg $m^{-2}$ $s^{-1}$ | milligram per square centimeter second, mg $cm^{-2}$ $s^{-1}$ | $10^{4}$ |
| 35.97 | milligram per square meter second, mg $m^{-2}$ $s^{-1}$ | milligram per square decimeter hour, mg $dm^{-2}$ $h^{-1}$ | $2.78 \times 10^{-2}$ |
| | | **Plane Angle** | |
| 57.3 | radian, rad | degrees (angle), ° | $1.75 \times 10^{-2}$ |

| | | | |
|---|---|---|---|
| | **Electrical Conductivity, Electricity, and Magnetism** | | |
| 10 | siemen per meter, $S\ m^{-1}$ | millimho per centimeter, $mmho\ cm^{-1}$ | 0.1 |
| $10^4$ | tesla, T | gauss, G | $10^{-4}$ |
| | **Water Measurement** | | |
| $9.73 \times 10^{-3}$ | cubic meter, $m^3$ | acre-inches, acre-in | 102.8 |
| $9.81 \times 10^{-3}$ | cubic meter per hour, $m^3\ h^{-1}$ | cubic feet per second, $ft^3\ s^{-1}$ | 101.9 |
| 4.40 | cubic meter per hour, $m^3\ h^{-1}$ | U.S. gallons per minute, $gal\ min^{-1}$ | 0.227 |
| 8.11 | hectare-meters, ha-m | acre-feet, acre-ft | 0.123 |
| 97.28 | hectare-meters, ha-m | acre-inches, acre-in | $1.03 \times 10^{-2}$ |
| $8.1 \times 10^{-2}$ | hectare-centimeters, ha-cm | acre-feet, acre-ft | 12.33 |
| | **Concentrations** | | |
| 1 | centimole per kilogram, $cmol\ kg^{-1}$ (ion exchange capacity) | milliequivalents per 100 grams, $meq\ 100\ g^{-1}$ | 1 |
| 0.1 | gram per kilogram, $g\ kg^{-1}$ | percent, % | 10 |
| 1 | milligram per kilogram, $mg\ kg^{-1}$ | parts per million, ppm | 1 |
| | **Radioactivity** | | |
| $2.7 \times 10^{-11}$ | becquerel, Bq | curie, Ci | $3.7 \times 10^{10}$ |
| $2.7 \times 10^{-2}$ | becquerel per kilogram, $Bq\ kg^{-1}$ | picocurie per gram, $pCi\ g^{-1}$ | 37 |
| 100 | gray, Gy (absorbed dose) | rad, rd | 0.01 |
| 100 | sievert, Sv (equivalent dose) | rem (roentgen equivalent man) | 0.01 |
| | **Plant Nutrient Conversion** | | |
| | ***Elemental*** | ***Oxide*** | |
| 2.29 | P | $P_2O_5$ | 0.437 |
| 1.20 | K | $K_2O$ | 0.830 |
| 1.39 | Ca | CaO | 0.715 |
| 1.66 | Mg | MgO | 0.602 |

# Chapter 1

# Pesticides in the Soil Environment—An Overview

**H. H. CHENG,** *Washington State University, Pullman, Washington*

The discovery of the potent power of certain synthetic organic chemicals for controlling unwanted organisms has led to the rapid rise in their use as pesticides. In less than 40 yr, synthetic organic pesticides have become a major element in modern agriculture production practices. The advent of pesticide use has coincided with the tremendous increase in agricultural productivity. Together with the adoption of improved varieties, the use of synthetic fertilizers to increase nutrient supplies, improved irrigation practices for water supplies, and more efficient machineries, synthetic organic pesticide use has been credited as one of the major contributors that modernized agricultural production. These innovations led to dramatic improvement in crop yields and nutritional quality of the products as well as efficiency in production management. Today, <2% of the population of the USA is producing a food surplus for domestic consumption.

The importance of pesticides to modern agricultural production practices is now well-recognized. Synthetic organic chemicals have essentially replaced inorganic chemicals and many tillage and cultural practices as the tool of choice for pest control. The recent trend toward conservation-tillage systems has also meant an increased reliance on chemical pesticide use for insect, weed, and disease control, although other means of pest control are constantly being sought, such as the integrated pest management approach combining nonchemical means with the chemical use for pest control. In this book, the term *pesticides* will refer to the synthetic organic chemicals now in general use. The term will be used in the most inclusive sense in that any synthetic organic chemicals that are manufactured for use in agricultural production to prevent or reduce adverse effects of pests, whether they are harmful insects, deleterious microorganisms, or unwanted plants, would qualify under this definition. Thus, the term includes all insecticides, fungicides, herbicides, fumigants, and other organic chemicals used for related functions. The advantages of these modern pesticides over other means of pest control include their effectiveness in controlling pests even when the chemicals are applied at such low levels as a few milligrams per hectare concentration. When pesticides are applied under appropriate soil and environ-

 *Pesticides in the Soil Environment*—SSSA Book Series, no. 2.

mental conditions in prescribed amount using specified procedures, they can be proven to be effective in pest control with little adverse effects on the surrounding environment.

Growing evidence indicates, however, that trace amounts of pesticides are present on nonagricultural land, in the atmosphere, and in water, both in surface bodies and underground, far from the sites of pesticide applications. Since pesticides are toxic by design, there is a natural concern on the impact of their presence in the environment on human health and environmental quality. In recent years, the adverse effects of certain pesticides to human and environmental health have become known. Although sporadic information on these adverse effects can be found in the literature since the early days of pesticide use, the impetus for the current awareness and concern for the adverse effects of pesticides probably had its beginning in the 1960s. Today, in addition to concern for the acute and chronic toxicity of pesticide chemicals, their potential as carcinogens, teratogens, and mutagens, and their presence in all corners of the Earth, from the arctic icecap to the groundwaters used for drinking purposes, have led to questions about the wisdom of continued pesticide use. These concerns on the potential threats of pesticides on human health and environmental quality must be viewed as serious. Pesticide use may have to be restricted to certain geographic locations or under certain agricultural practices. On the other hand, before any decisions are made to eliminate pesticide use for agricultural production purposes, it must be established that pesticides in unwanted locations is from pesticide use in agricultural production and not the result of other means of transport such as improper disposal of pesticides.

The cause-and-effect relationship between pesticide use for agricultural production at a locale and observation of pesticide contamination of the surrounding environment, causing adverse effect on environmental health, cannot be readily established in many cases. Reported incidents on pesticide contamination of the environment or on adverse environmental effects from pesticide use can often be traced to improper application or inappropriate practices. Either a lack of knowledge or a disregard for the sensitive nature of the environment has been the root of many pesticide contamination problems. For instance, common sense would predict that injudicious application of highly water-soluble pesticides to irrigated crops in sandy soils would likely result in the presence of these pesticides in shallow aquifers under the irrigated fields. The basic problem for most cases has been a lack of understanding of the processes affecting the behavior and fate of pesticides from the point they enter the environment to the point at which they would affect the target organisms.

Since concerns are mostly associated with the presence of pesticides in the soil environment, it is essential that the processes affecting the transport of pesticides in the soil be understood before any cause-and-effect relationship can be established. The purpose of this book is to provide a perspective on the fate and transport of pesticides in the soil environment that would help evaluate the effectiveness of pesticides for pest control and the impact of pesticide use on environmental health. Emphasis will be placed on the delineation of the principal mechanisms and processes governing the fate of pesticides in the environment and factors affecting these processes. No attempt will be made to provide a comprehensive description of all the processes affecting the fate of all pesticides in the soil environment. Specific chemicals will be mentioned only to illustrate the processes and their interactions.

The fate of a pesticide in the environment is governed by the retention, transformation, and transport processes, and the interaction of these processes. The interrelationship among the processes is depicted diagrammatically in Fig. 1-1. Retention is the consequence of interaction between the pesticide chemical and the soil particle surface or soil components thereon. It manifests the extent of their affinity. The retention processes are frequently described as adsorption or simply as sorption. They may be reversible or irreversible; they can retard or prevent the pesticide movement and affect its availability for plant or microbial uptake or for biotic or abiotic transformation. Whereas retention is mainly considered to be a physical process, transformation is characteristically a change in the chemical nature of the molecule, although the demarcation between a physical and a chemical process is not always easily differentiated. The transformation processes may be purely chemical in nature; they may be catalyzed by soil constituents or induced photochemically. Most pesticides are, however, transformed predominantly by biochemical means, such as by soil microorganisms. Biotic transformations of a pesticide generally result in degrading the molecular structure into simpler forms. Degradation tends to decrease the chemical's toxicity although occasionally the metabolic products could be even more toxic than the parent compound. Just as the transformation processes dictate *whether* and *how long* pesticides may be present in an environment, the transport processes determine *where* the pesticides may be present. Volatilization leads to the

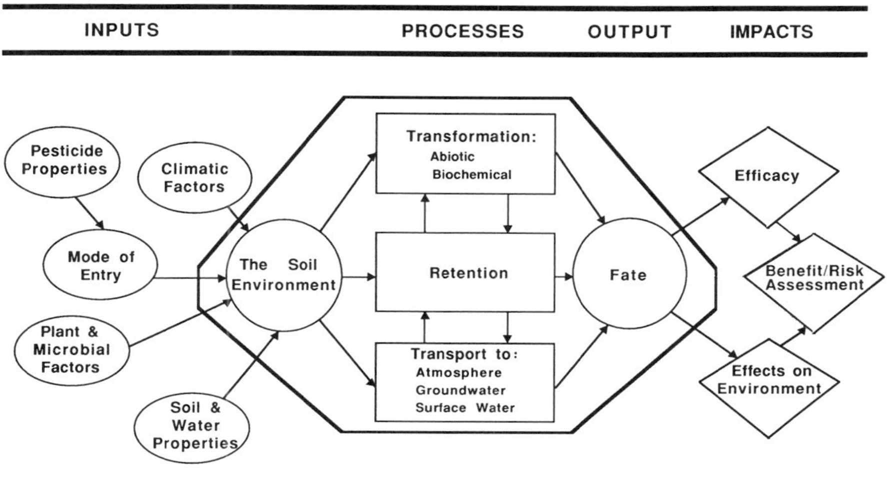


Fig. 1-1. A conceptual framework depicting the factors and processes that govern the fate of pesticides in the soil environment and how pesticide fate affects their efficacy and their impact on the environment.

distribution of pesticides from the soil to the atmosphere; leaching leads to the movement of the pesticides toward the groundwaters; and overland flows move the pesticides into surface waters. To assess the fate of a pesticide in the soil environment, one must assess the kinetics of the individual processes as well as the combined effects of all the processes.

In addition to the variety of processes involved in determining pesticide fate, many factors can affect the kinetics of the processes. Differences in chemical structures and properties, soil properties and conditions, the climatic status, presence or absence of plants and microorganisms, and soil management practices can all affect the kinetics of the processes and the eventual fate of a pesticide in the soil environment, adding complexities to the interactions. The chemical structure of the pesticide governs the reactivity of the chemical, such as its efficacy in controlling pests and its mobility and degradability. Properties of the chemical, such as its solubility in water, vapor pressure, and polarity, affect its behavior in the environment. Likewise, soil properties such as its organic matter and clay contents, pH, Eh, and ion exchange capacity affect the behavior of the chemical in the environment. Moreover, the condition of the soil, such as its moisture or oxidation state, and the location of the soil on the landscape can also affect the fate of the pesticide under natural field conditions. Soil conditions are affected by the climatic conditions. Temperature, precipitation, wind, and radiation can all affect the fate of the chemical. Similarly, the variety of plant species and their growth stages and the presence of multivarious types of microorganisms are additional variables affecting the fate of a chemical in the environment.

Thus, an assessment of the fate of a pesticide in the soil environment must include an assessment of all the processes that could potentially affect the chemical's fate and all the factors influencing the various processes in a specific setting. Because the processes and their interactions are complex and difficult to characterize experimentally, mathematical modeling using computers has proved to be a valuable tool not only to simulate the processes occurring under a particular set of conditions, but also to depict the entire system of interactions. As the modeling techniques become more mechanistic in describing the processes, the models are also becoming more useful for prediction and management purposes. By using a systems approach to understand the fate of a pesticide in the soil environment, a means is now accessible to evaluate the availability of the pesticide for pest control as well as the impact of the pesticide on the various components in different compartments of the environment. The availability of a pesticide at any given time and locale determines its efficacy for pest control and defines the benefit of its use. An assessment of the fate of a pesticide will also determine how much the environment is exposed to the pesticide. Combining the exposure with a hazard assessment based on pesticide toxicity, the risk of the presence of a pesticide can be determined. When both the benefits and risks of the presence of a pesticide in the soil environment are weighed, approaches to manage pesticide use and simultaneously to maintain environmental quality can finally be realistically established.

The chapters of this book have been organized to follow the pathways depicted in Fig. 1-1. The presentations will follow the pathways of pesticides from their entry into the environment, through their progression in the various retention, transformation, and transport processes under a variety of factors and conditions, to their eventual fate at a specific time and locale in the environment. The fate of a pesticide will govern its availability and efficacy for pest control as well as its potential adverse effects on nontarget organisms and other components in the environment. The systems and modeling approach for examining the fate and impact of pesticides in the environment will enable us to resolve environmental quality problems and anticipate, plan, and prevent the occurrence of such problems. If pesticides are used, their benefits will be maximized. Thus, understanding the processes that affect the environmental fate of pesticides will become an integral requirement of the pesticide registration procedure.

# Chapter 2

# Pesticide Sources to the Soil and Principles of Spray Physics

**CHESTER M. HIMEL,** *University of Georgia, Athens, Georgia*

**HARRY LOATS,** *Loats Associates, Inc., Westminister, Maryland*

**GEORGE W. BAILEY,** *U.S. Environmental Protection Agency, Athens, Georgia*

In the past 50 yr, organic pesticides have greatly enhanced the production and quality of food, feed, and fiber, as well as the control of disease vectors and pests adversely affecting the health and welfare of the world. Specifically, pesticides have been used in: (i) agriculture to increase productivity and the quality and quantity of food, feed, and fiber; (ii) forestry for pest control in silviculture; (iii) the industrial, commercial, municpal, and military sector for rodent and weed/brush control around plant sites and in right-of-ways; (iv) medical vector control—for control of mosquitoes and rodents etc.; and (v) urban environment for termite control around structures and pest control in gardens. For example, more than 500 different formulations of pesticides are used in the urban environment and in excess of 11 363 631 kg (25 million lb) of active ingredients are used in the gardens, yards, and homes annually in the USA (USEPA, 1979). Agriculture represents the largest single market for pesticides.

The purpose of this chapter is to review the literature to: (i) characterize the pesticide source term as it determines pesticide transport to the soil surface that involves both primary and secondary pathways of transport to the soil; (ii) define factors influencing intra- and intercompartmental transport; (iii) define the source term from a modeling perspective; (iv) discuss spray physics principles (influence of application methodology, formulation on spray propagation, drift impingement, and metabolic efficacy in killing pests); and (v) describe predictive models of spray movement.

## 2-1 PESTICIDE ENTRY INTO THE ENVIRONMENT

Pesticides may be introduced directly into the environment in a liquid phase, as a dispersion or solution, or in the solid phase, e.g., as a powder, dust, microcapsule, or granule. Sprays are directed to foliage, the soil, or stoichastically applied to these same media as rainout (Fig. 2–1 and 2–2). Solids are applied directly to the soil surface or to foliage. Pesticides may also be incorporated directly into the soil, usually in the top few millimeters to exert a biological effect. A pesticide may also enter the environment as a result of accidental spill or malicious application for waste disposal purposes.

Environmental entry and transport processes can occur either between or during rainfall events. Those direct or primary transport processes to the soil surface that occur during a rainfall event include wet deposition and foliar washoff. Secondary transport processes to the soil surfaces that occur between rainfall events include spray drift and dry deposition (the source term is particulates emanating from the pesticide drift process and pesticide found on dust particles generated by wind pickup and atmospheric transport). Pesticides dissolved in atmospheric water vapor may be deposited either on the soil surface directly or indirectly to the foliar surface with subsequent removal during rainfall events via the process of foliar washoff. Therefore, pesticide sources to the soil surface include application events, atmospheric wet or dry deposition, foliar washoff, or accidental spills onto the soil surface or into the soil profile. Environmental sinks for pesticides include chemical, photochemical and biological transformation, volatilization losses, erosion and runoff, leaching, and harvest removal and storage.

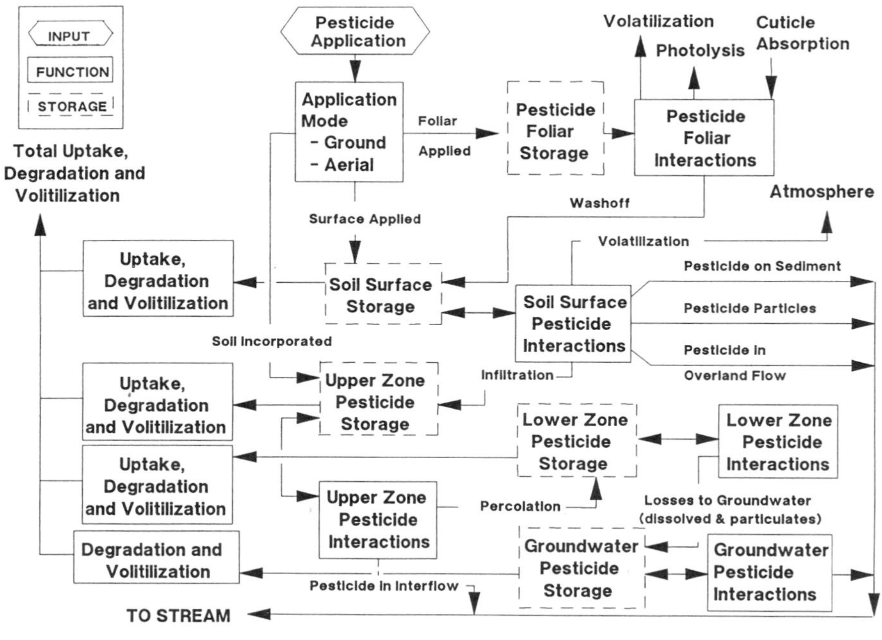


Fig. 2–1. Pesticide transport and transformation in the soil-plant environment and vadose zone. After Donigian and Crawford (1976).

The mass balance is given by:

Pesticide application, kg/ha = Σ *Pesticide sinks* [Chemical, Photochemical, Biological and Volatilization Losses + Erosion + Runoff + Leaching + Harvest + Storage (surface, upper and lower zone) (Bailey et al., 1985)].

Pesticides are directly applied to the soil as: (i) preplanting treatments; (ii) preemergence treatments; (iii) postemergence treatments; or (iv) soil

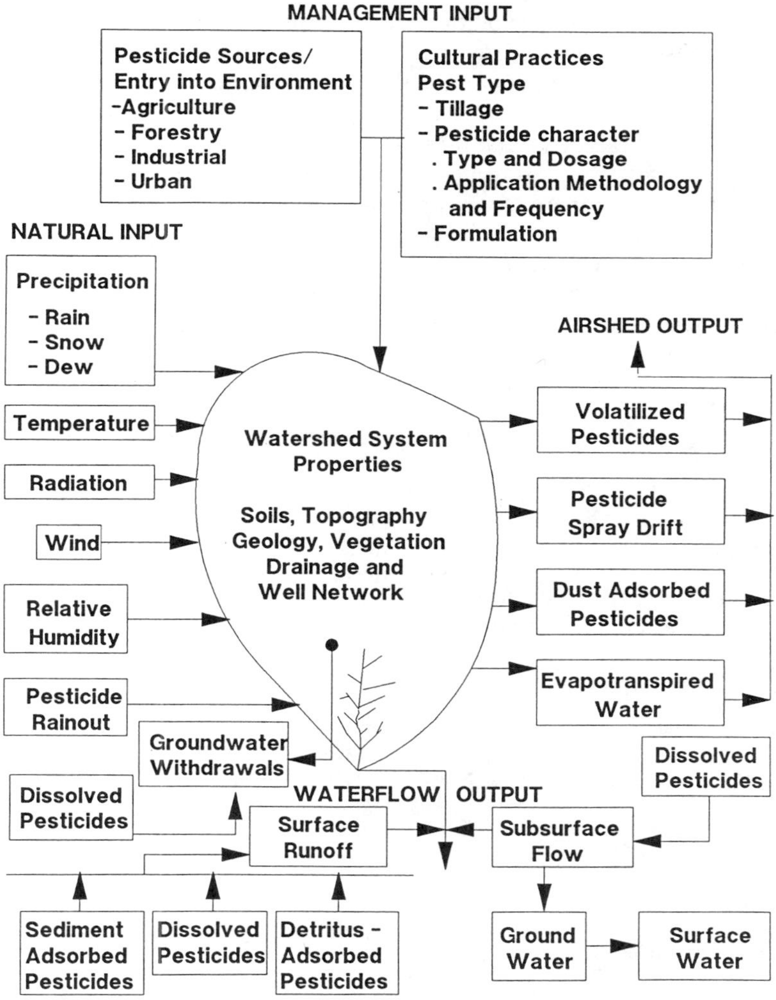


Fig. 2–2. Factors influencing the behavior and export of pesticides from a watershed. After Bailey et al. (1985).

sterilants. Several factors influence the final concentration of soil-applied pesticide including: (i) volatilization, (ii) photochemical degradation, (iii) chemical and biological transformation, (iv) leaching, and (v) sorption. The initial pesticide source location, i.e., foliage or soil surface on day of application, depends upon the pest being controlled, pesticide concentration and formulation, and mode of application. From a mass balance modeling standpoint, it is imperative to know at the time of application the pesticide concentration on the soil surface, plant canopy, or plant residue surface (or, if incorporated, the depth of incorporation). The knowledge of the initial concentration of pesticide in the applicator tank is not sufficient as up to 60% of the applied tank concentration (Smith et al., 1976) may be lost through spray drift and volatilization. Secondary transport processes include foliar washoff, foliar drip (occurring on application day), rainout, and volatilization from foliar and soil surfaces. The concentration of pesticide available for transport is determined by the time-varying processes/sinks including biotic/abiotic transformation, volatilization, plant uptake, plant harvest (root and foliar absorption), leaching, and foliar residue washoff. These sources and major transport process can be seen in Table 2-1. The major pesticide sources for offsite transport include canopy, pesticide residue bound on or in plant residue, soil surface, root or grain, foliage runoff (for dissolved pesticide), eroded soil particles (for sorbed pesticide), granules, dust or microcapsules, and leachates (for pesticide dissolved in solution or bound to colloidal particles).

Table 2-1. Pesticide sources and environmental exposure pathways.

| Pesticide source | Transport process | Exposure pathway | Receptor (Population at risk) |
|---|---|---|---|
| Canopy | Volatilization | Inhalation/skin contact | Human, animals |
| Crop residue | Volatilization | Inhalation/skin contact | Human, animals |
| Soil surface | Volatilization | Inhalation/skin contact | Human, animals |
| Root zone and below | Volatilization | Inhalation/skin contact | Human, animals |
| Grain/foliage | Manufacturing/ feeding† | Food/ingestion | Human, animals |
| Crop residue | Overland flow† | Surface water (potable water)/ingestion | Human, fish |
| Runoff | Overland flow† | Surface water (potable water and food chain) | Human, fish |
| Eroded soil particles and formulation—bound | Overland flow† | Ingestion surface water (food chain)/ingestion, contact | Fish |
| Leachate-dissolved-bound on colloidal particles | Leaching/percolation† | Groundwater (potable water)/ingestion | Human, animals |

† Environmental flux of concern to either on-site or off-site receptors. After Bailey et al. (1985).

## 2-2 PESTICIDE PROPERTIES

Traditionally, in a presentation on pesticide properties, an extensive discussion would focus on the major families and classes of pesticides, their structure and specific physical and chemical properties. However, because of a proliferation of classes of pesticides over the years, this is not feasible. The reader is referred to a variety of references on the structures and physical and chemical properties of specific pesticides—insecticides (Brown, 1951; Metcalf, 1955; Kubr & Poogh, 1976) herbicides, (Kearney & Kaufman, 1975, 1976; Klingman & Ashton, 1982; Weed Society of America, 1983) and fungicides (Torgeson, 1969; Luken, 1971; Marsh, 1972, Sharville, 1979).

Since we are interested in pesticide source characterization, their fate, transport, and behavior of pesticide to, within, and from soil, insight into important chemical characteristics is imperative. Chemical characteristics of pesticides that influence transport to and from soil include: (i) ionic state (cationic, anionic, basic, or acidic); (ii) water solubility; (iii) vapor pressure; (iv) hydrophobic/hydrophilic characters; (v) partition coefficient; and (vi) chemical, photochemical, and biological reactivity (Bailey et al., 1985).

Over the last 10–15 yr, there has been a drastic change in the general characteristics of pesticides. They have changed from those of extremely low water solubility, strongly sorbed (extremely hydrophobic and with a very high partition coefficient), and nonmobile to those that are more water-soluble, only slightly sorbed (more hydrophilic and lower partition coefficient value—a $K_d$ of 1 or less), and more mobile (Carsel & Smith, 1987). Part of this change is because of environmental problems (bioaccumulation and magnification) associated with past compounds and the increased pest resistance to chlorinated hydrocarbon pesticides.

These properties, combined with the amount of pesticide applied and entering the environment, determine how much pesticides can be transported along each route—specific routes can be seen in Fig. 2–1 and 2–2. Possible intracompartmental transport routes include foliar washoff to the soil surface, pesticide infiltration, percolation, and movement to lower storage zones and finally to groundwater. Possible intercompartmental transport routes include overland flow, interflow to subsurface water, and direct loss to the atmosphere from the soil, subsoil or foliar surfaces or losses because of movement of sediment-bound pesticide (both water and wind modes of transport) (Bailey et al., 1985).

Sorption processes determine the relative distribution between the amount in vapor phase, in solution phase and bound to environmental surfaces, and this dictates the primary or secondary transport routes.

Environmental persistence of pesticide in the soil is related to pesticide properties. A persistent compound is one that does not hydrolyze or biodegrade readily (first-order transformation rate constant is $<0.02\ d^{-1}$), has a low volatilization or Henry's law constant, has a high partition coefficient, and has a low potential to move to and contaminate groundwater (Carsel

& Smith, 1987). A nonpersistent pesticide is one that hydrolyzes or biodegrades readily (first-order degradation rate in soils at least 0.02 $d^{-1}$), has a high Henry's law constant, is highly water soluble, (partition coefficient $K_d < 1$); high potential to move to groundwater and has an appreciable mammalian or phytotoxicity value.

## 2-3 FORMULATION

Technical pesticides (90+ % active ingredient) are generally formulated prior to use except when they are liquids that can be sprayed neat (undiluted). Formulations include solutions, suspensions, powders, granules, or other forms designed to meet application needs and use requirements.

Van Valkenburg (1973) notes that several factors should be considered in formulation chemistry including: (i) importance of the solubility parameters of the hydrophobic and hydrophilic portion of surface active agent in selecting an emulsifier, (ii) effect of the phase inversion temperature on emulsion stability, (iii) the dehydrating effect of fertilizer salts and emulsifiers and the resultant stability or unstability of the emulsion, (iv) the all important flowability property of a mixture as it relates to the development of a dry pesticidal formulation, (v) catalytic effect of clays and other carriers on pesticide degradation, and (vi) the physical properties of formulations and how they affect the size of the particles in a spray and the movement of the particles in air.

The physical and chemical properties of the formulation and its resultant spray characteristics, as indicated by particle size and surface properties, (surface tension, spreading coefficient, and contact angles), dictates whether the pesticide will adhere to the plant surface and penetrate the lipid barrier of the cuticle and exert metabolic efficacy. Common pesticide formulation types are wettable powder (WP), emulsifiable concentrate (EC), water miscible concentrate (WMC), and flowables (F). These are designed to be diluted with varying ratios of water and applied by various types of spraying equipment. Important properties of the active ingredient that affect its alternate formulation includes physical states, water/oil solubility, miscibility, dissolution rate, melting point, density, and physical chemical stability to heat, oxygen, and moisture (Van Valkenburg, 1973). The pH both in the bulk and in solution/suspension of pesticide formulations is also important with mixtures of pesticidal ingredients or sometimes even with a single ingredient with additives to achieve desired physical properties, ease of application, and favorable metabolic efficacy. Formulation character affects stability of the active ingredient, ease of field mixing, bioavailability of active ingredient to the target organism, residuality of active ingredient, and adducts and potential side effects of pesticides. Certain aspects of pesticide formulation and use are subject to regulation by USEPA and other government agencies.

Pesticide formulations are available in many forms permitting the use of a variety of equipment types—ground vs. aerial application—to control target organisms effectively and successfully. Pesticides are formulated rather than used directly because of the necessity to: (i) dilute high-potency material to field concentration and (ii) apply them to a location where the target organism can be controlled while side-effects damage to humans, plants, wildlife, and the environment is minimized.

### 2-3.1 Emulsifiable Concentrates-Solution Systems

Most pesticides are dissolved in suitable oils or other organic solvents, with emulsifiers added and packaged as emulsifiable concentrates (EC). Typical EC formulations contain from 0.12 to 0.48 kg of active pesticide per liter (1-4 lb $gal^{-1}$). Solution systems without emulsifier are also used, particularly for low-volume (LV), or ultra-low-volume (ULV) sprays. Emulsifiable concentrates are the bulwark product for pesticide sprays. They can be diluted with water and are simple to use. Low surface tension increases ease of atomization and impingement probability.

### 2-3.2 Wettable Powders, Granular Products, and Flowables

Wettable powders are produced by adsorption of technical pesticides on finely ground clay-containing wetting agents. The resulting products are dispersible in water. Granular products with a high percentage of active ingredient have been formulated to alleviate some of the problems in handling wettable powders. Flowable products are finely ground, insoluble active ingredients suspended in oils with emulsifiers. They are nondusting and can be pumped as viscous liquids.

Granular pesticides are pellets or grains containing the pesticide diluted to the permissible concentration by mixing with other inert and functional ingredients. Granular carriers are inert, sorptive particles with a size range of 4 to 200 mesh, and are produced in two forms: disintegratable (regular volatile material, RVM) and nondisintegratable (low volatile material, LVM) in water. The major advantages of granular formulation over dusts, wettable powders, water-soluble powders and liquids are that it: (i) permits uniform application at desired rates, (ii) provides minimum dusting, (iii) delivers toxicant to desired location, (iv) has carriers either water disintegratable or nondisintegratable, (v) minimizes phytotoxicity, and (vi) regulates pesticide release so to act as a reservoir to extent pesticidal activity for a longer period. Granular pesticides applied to the soil surface can be picked up and moved in overland flow offsite as a result of a high-intensity rainfall event (Bailey et al., 1974).

Ground applicators for granular pesticides usually are attachments on row crop planters. They include capability for application in the seed furrow or a broadcast band over the row that remains either on the soil surface or incorporated by some mechanical means. Effective operation of a granular applicator requires that it meters, delivers, and then distributes the pesticide granules over the soil or in some cases incorporates them to the proper soil depths.

The granules are generally metered either through a gravity-fed metering orifice or through a positive metering device in which a constant granule volume is metered by a rotor. The granule is delivered to the target location by mean of gravity or pneumatic pressure.

## 2-4 PESTICIDE APPLICATION SYSTEMS

Most pesticides are applied by spray processes whose atomization (spray spectra) is dependent on the type of nozzle, pressure throughput, and other factors. Dusts are no longer widely used. High technology and engineered delivery systems need development, such as for controlled release (CR) of the active agent. In Australia, a new CR product emits a biodegradable, short-life insecticide within the root system of sugarcane (*Saccharum officinarum* L.) (McGuffog et al., 1984). It protects the first and successive ratoon crops up to 3 to 4 yr. This delivery system meets a critical plant protection need, is highly efficient in pesticide delivery, minimizes ecosystem effects, and limits the build-up of resistance.

### 2-4.1 High-Volume Dilute Spray Systems

High-volume dilute sprays in the range of 400 to 3000 L $ha^{-1}$ (100–800 gal $acre^{-1}$) are ineffecient and labor and energy intensive. They operate by attempting to saturate foliage, often using turbulent, high velocity air streams. Large losses occur to the soil and to drift.

### 2-4.2 Low-Volume and Ultra-Low Volume Sprays

Spray volumes in the range of 4 to 40 L $ha^{-1}$ (1–10 gal $acre^{-1}$) are often referred to as low volume (LV). Ultra-low volume (ULV) refers to sprays $<2$ L $ha^{-1}$ ($<0.5$ gal $acre^{-1}$). In these low volume spray clouds reduce foliar runoff (foliage saturation). No LV or ULV spray, however, can provide even coverage. Visual observations during spray application have no actual validity. Even coverage is a visual illusion. Measurement of the actual coverage of typical foliage systems is complex and time consuming. It requires use of tracer sampling and analysis systems that meet suitable probability standards. In the case of foliage systems that have substantial vertical dimensions, i.e., forests and dense crops, deposition within the canopy falls to a few droplets per cubic centimeter. Transport and impingement in foliage from LV and ULV sprays reflect turbulence within the canopy, the size and number of droplets in each size category, velocity of travel, geometry of the target, spray volume, foliage density and dimensions, and other factors. In typical cases, deposition is greatest on the tops of leaves or needles, on leaf hairs or insect setae and on rough surfaces including leaf veins and insect appendages. Deposition on the underside of leaves is low but increases as turbulence increases. Insects and disease vectors that are resident on the underside of leaves represent a major pest management problem. In dense foliage, pest management often requires an increase in numbers of effective droplets. Specialized formulations may be required to maximize adherence and control on waxy surfaces.

The ULV sprays are increasing in use. The concept of ULV was developed to meet spatial and temporal problems in control of the desert locust during its migrational flights in intercontinental space measured in thousands of cubic kilometers (Sayer, 1959). Numbers of droplets in a spray are an inverse cube function of size, hence, as spray volumes were reduced to meet logistic requirements, small size spray spectra were crucial. The original ULV concept involved a concomitant decrease in size of spray spectra and the volume of the spray; however, the significance of the size of the spray spectra was lost as ULV was adapted to agriculture using hydraulic nozzles.

As spray clouds diffuse, the numbers of droplets per cubic centimeter decreases, but lethal deposits can occur on insects when concentration of the pesticide in each droplet is high (Keathley, 1972; Himel, 1974). Volume-concentration factors in the design of insecticide spray programs are discussed by Spillman and Joyce (1978). Factors that control spray efficiency are discussed in sections 2–8 and 2–9.

Mosquito abatement groups have used ULV sprays for decades with typical applications of 60 to 120 mL of concentrated or neat insecticide per hectare (1–2 oz $acre^{-1}$). Mosquito sprayer use special nozzles (centrifugal and pneumatic) that are designed to limit the production of droplets larger than 40 to 50 $\mu$m. Size specifications for pesticide sprays to meet public health needs are not in controversy. The size of aerosol droplets impinging on mosquitoes has been reported by Lofgren et al. (1973) to center in the 5 to 10-$\mu$m range, conforming to impingement dynamics on a small, flying insect.

## 2–5 FOLIAR APPLICATION AND LOSS

Foliage or plant canopy may be the desired initial and final site for pesticide delivery. Plant canopy, as seen in Fig. 2–1 also may be the source of further pesticide transport to the soil or to the atmosphere. The latter may impact worker reentry and impose a human health risk. Characterization of this sink-source is necessary but has been lacking. Prediction of delivery and redistribution of pesticides within the plant canopy is the key to efficient use of pesticides. It requires a knowledge of pesticide spray physics, aerial transport phenomena, foliage intercept factors and storage; decay/transformation, plant uptake, and volatilization; and washoff behavior.

### 2–5.1 Initial Deposit of the Pesticide

Several factors influence the amount of initial deposit on the crop cover after application (Ripley & Edgington, 1983). The rate of application is the most obvious factor with increasing deposit occurring with increasing application rate. A limit is reached, however, beyond which no pesticide can adhere to the foliar surface. The amount of the deposit is slightly affected by the height of the spray boom above the canopy surface; a slightly lower amount of residue is found in the middle compared to the top of the plant canopy. The extent of groundcover is important but plant surface characteristics, morphology, and surface area/plant mass ratio of the crop intercepting the pesticide spray also can result in a differential deposition over the crop. Formulation also affects residue deposit and, in most cases, a lower amount of residue is found with wettable powders compared to emulsifiable concentrates, or flowable formulations. Formulations may facilitate greater penetration, reduce early physicochemical loss, and enhance foliar persistence. The amount of pesticide intercepted by the crop canopy is assumed to be proportional to the extent of groundcover at the time of pesticide application (Beyerlein & Donigian, 1979).

### 2–5.2 Pesticide Disappearance/Persistence

*Persistence,* in this case, is defined as the residence time of a pesticide in a defined compartment of the environment expressed in units of time relative to a bench mark (Greenhalgh, 1986). *Pesticide persistence* is the integrated effect or weighted summation of all the transformation and loss processes including chemical (e.g., hydrolysis), photochemical, and biological transformation; volatilization; cuticle adsorption; and washoff. Factors influencing persistence or the converse disappearance include crop factors (morphology, cuticle characteristics, stage of growth at treatment, and rate of growth), pesticide characteristics (Henry's law constant; water solubility; susceptibility to physical, chemical, and biological transformation processes), and environmental (rainfall intensity, duration and frequency; wind velocity; humidity; temperature; light-intensity; and percent cloud cover) conditions.

Disappearance curves after foliar application show an exponential decline for most pesticides. Pseudo first-order reaction kinetics may often be used to describe the decline in pesticide residue. One problem is that many such studies only treat a portion of the disappearance curve (time $>1$ d) or only 1 to 2 half-life periods (generally 5 half-lives or more are preferred). Several mixed-order rate equations may be operating simultaneously since several different processes are involved and bi- or trilinear curves have been reported (Ebeling, 1963). Many studies have failed to account for either the large initial residue loss (possibly because of volatilization, hydrolysis, or photolysis) in the early period after deposit (time $<1$ d) or the apparent long persistence of low concentrations of pesticide in the plant. The application of regression analysis using semilog first-order approach has not resulted in good correlation particularly where two or more apparent first or even second-order processes are involved.

For the persistance of certain pesticides in plants, Sirons et al. (1982) found a relation of $C = k\,(a * b * \log t)$ for the persistence of certain pesticides in plants, where $k$ = rate of dissappearance, $t$ = time after application, and $a$ and $b$ are coefficients.

### 2-5.3 Washoff—A Specific Loss Pathway

Attempts have been made to study and, in certain cases, to model the various dissipation or attenuation pathways. Washoff as an attenuation pathway will be discussed; little is known about the other pathways.

Several different approaches have been reported (Beyerlein & Donigian, 1979; Smith et al., 1981; Willis et al. 1982) for describing the foliar washoff process, for any rainfall event; the form generally is $PWO = FP * WOC * R$ where $PWO$ = pesticide washoff ($MT^{-1}$), $FP$ = foliage pesticide ($ML^{-2}$) i.e., kg $ha^{-1}$, $WOC$ = washoff coefficient ($L^{-1}$), $R$ = rainfall ($LT^{-1}$). Several models have been proposed to account for or predict foliar washoff. Beyerlein and Donigian (1979) proposed an equation that describes the removal of pesticides from crops by rainfall as an exponential function of the rainfall intensity. Such an approach necessitates the use of hourly rainfall data and a crop constant based on crop and pesticide properties to determine the washoff coefficient (WOC). A foliar washoff algorithm is contained in the CREAMS model (USDA, 1980) for predicting the proportion of pesticide lost from foliar surfaces. The residue is partitioned into dislodgeable (available for washoff) and nondislodgeable fractions. Washoff is calculated for each rainfall incident from the dislodgeable fraction, less the amount attenuated, after a plant washoff threshold has been reached. The washoff threshold (0.1–0.3 cm) is the amount of rainfall that a plant canopy can intercept and store prior to washoff. Smith and Carsel (1984) proposed a pesticide washoff model (FWOP) which requires daily rainfall amounts (cm $d^{-1}$) and a lumped first-order rate constant for foliar transformation processes ($k_f$). Specific formulation includes:

$$\frac{d[FP(t)]}{dt} = [-k_f FP(t) - PWO(t) + APP(t)] \qquad [1]$$

where

$$\frac{d[FP(t)]}{dt} = \text{rate of change of pesticide in foliage,}$$

$k_f$ = first-order degradation rate constant on the foliage ($T^{-1}$),
$k_f = k_p + k_r + k_b + k_v + k_h$,
$k_p$ = rate constant for photolysis ($T^{-1}$),
$k_r$ = rate constant for chemical reactions ($T^{-1}$),
$k_b$ = rate constant for biodegradation ($T^{-1}$),
$k_v$ = volatilization rate constant ($T^{-1}$),
$k_h$ = hydrolysis rate constant ($T^{-1}$),

and

$$PWO(t) = FP(t) * WOC * R(t) \qquad [2]$$

where

$PWO(t)$ = pesticide washoff ($MT^{-1}$),
$FP(t)$ = foliage pesticide ($L^{-1}$), and
WOC = washoff coefficient ($L^{-1}$)
$R$ = rainfall ($LF^1$), and
$APP(t)$ = application rate ($MT^{-1}$).

This equation cannot be solved analytically in a closed form because rainfall is stochastically variable in time. The equations, therefore, must be solved numerically as a part of a continuous simulation model. The FWOP model has been coupled with the Pesticide Runoff Simulator (PRS) (Computer Sciences Corporation, 1980) to make such calculations.

Willis et al. (1982) found that total storm losses from toxaphene on cotton (*Gossypium hirsutum* L.) under simulated rainfall were independent of intensity when 24 mm of simulated rainfall was applied at the rate of 13, 25, 51, or 100 mm $h^{-1}$ just 2 h after toxaphene had been applied at 2.24 kg $ha^{-1}$. Approximately 5 to 10% of the toxaphene was washed off depending upon the initial load. Linscott and Hagin (1968) reported that mean losses of sprinkler-applied 2,4-DB were 21, 60, and 93% of the amount on or in the plant when 2, 5, 13, and 51 mm of rain were applied, respectively. A paucity of information exists on the loss of pesticides on foliar surface because of photolysis, volatilization, chemical, and biological transformation processes. Research is urgently needed to define those processes and pathways both to improve model formulation and identify sources for pesticide availability in making risk assessment.

## 2–6 SOURCE-TERM DESCRIPTOR FOR MODELING

A quantitative definition of the source terms is needed for modeling pesticide exposure and for making risk assessment. A general approach described by Bailey et al. (1974) will be updated and used to quantify the pesticide source term. The initial consideration in the description of the source term is the physical state of the pesticide and its spatial distribution with respect to the soil surface. Figure 2–1 shows that it may be (i) on the foliar surface, (ii) on the soil surface, or (iii) incorporated into the soil and residing in the upper storage zone. The physical state of the pesticide is determined by the nature and properties of the formulation. Pesticide formulations as discussed earlier consist of five different physical types: solutions, wettable powders, emulsions, dust, and granular pellets. Their physical-chemical composition varies with the type and initial concentration of pesticide, solvent, surfactants/wetting agents, sticking agents, particle-size distribution of the dust or wettable powder, and rate of release from the granular pellet. These factors influence pesticide-soil interactions and equilibria in the various storage zones and affect directly the availability of the pesticide for surface and vertical transport as well as its susceptibility to various transformation processes.

The source definition requires that the initial pesticide phase and spatial distribution condition as dictated by the application methodologies be categorized and quantified. Pesticides are: (i) directly applied to the soil surface, (ii) incorporated into the soil, or (iii) purposefully or accidentally sprayed, dusted, or applied as granules/pellets to the canopy or to crop residues. Because of the restricted mobility of the pesticide, its initial placement is a crucial factor in mathematically describing the pesticide distribution on the foliage or pest surface, and on/or and throughout the soil profile, not only before and during the first rainfall event, but also between subsequent events. The initial placement also determines the relative importance of the four modes of transport: (i) mass transfer of the pesticide in the moving liquid boundary; (ii) pesticide desorption into or towards the moving liquid boundary; (iii) stationary dissolution and transfer to the moving liquid boundary; and (iv) moving dissolution in the moving liquid boundary—into runoff water. Initial placement determines whether the source is planar (initially surface-applied or subsequent foliar washoff onto the soil surface) or volumetric in nature (incorporated). Depth of incorporation and homogeneity of mixing (influenced by type of incorporation tool used) also influence the effective source concentration.

In the case of either directed or accidential foliar application, the soil source-term description is further complicated because of such processes as foliar interception, direct foliar drip, delayed foliar washoff and the various transformation processes—photolysis, chemical transformation (e.g., hydrolysis), and biological transformation—volatilization, and plant or pest uptake and metabolism—prior to reaching the soil surface. The role of atmosphere source characterization as a transport media is reviewed by Medved (1975). Pesticides released in the environment via spraying—either aerially or via ground sprayers—pass to the target plant, pest, or soil, but some are carried away by the wind both horizontally and vertically for varying distances. These offsite sprays may either settle upon the soil surface or be deposited in rainfall (Fig. 2–2—secondary transport processes).

Gohlich (1983) defines key features of the pesticide application distribution systems that need to be considered in reducing the amount of drift to include spray atomization and transport distance between atomizer and receptor area. Atomization is influenced by physical properties of the liquid, types of nozzle, and the atomization process. Transport is influenced by air turbulence, size of particle, evaporation behavior of the liquid particle, direction and force of the flow stream and particle motion, and forward speed of the spraying device. Drift and deposit are influenced by structure of the plant or receptor and its density and by turbulence within the canopy. The latter is related to external wind velocity.

## 2–7 PESTICIDE SOURCE CHARACTERIZATION AND RISK ASSESSMENT

Risk assessment is a tool used to evaluate the probability of an adverse impact that a pesticide, being a toxic or hazardous chemical, may have on

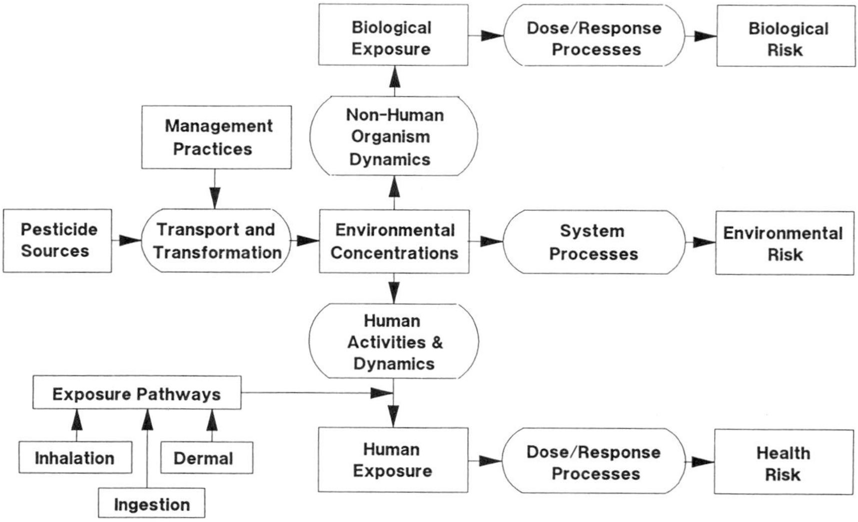


Fig. 2-3. Concept of pesticide exposure and risk assessment. After Bailey et al. (1985).

human health, or the environment. The key role that a knowledge of the physical state of the pesticide and its spatial distribution with respect to the soil surface can be seen in Fig. 2-1. The role of management practices on pesticide transport and transformation can be seen in Fig. 2-2 and 2-3. These interactions determine which environmental pathways are crucial to the estimation of both human and biological exposure and the calculations of environmental risk. This topic is discussed at greater length in chapter 13.

## 2-8 PRINCIPLES OF PESTICIDE SPRAY PHYSICS

The developing science of pesticide spray physics is concerned with the complex interactions of spray droplets in a spray cloud and during impingement. These interactions occur (i) during transport in the air (includes both turbulent transport and sedimentation); (ii) at the foliage-air interface (includes impingement probability and impingement filtration); and (iii) within the canopy (includes impingement probability, target geometry, turbulence within the canopy and impingement delivery).

Droplet impingement involves size of the droplet, velocity of travel, and geometry of the substrate. A droplet large enough to have a critical impingement velocity of zero (impingement probability = 1) will impinge on a target of any geometry at zero velocity. Penetration into a foliage canopy requires an impingement $P < 1$.

The hydraulic nozzle is used in most aerial and ground spray equipment because it is cheap, widely available, and because there has not been substantial pressures for change to more efficient delivery systems. Hydraulic nozzles atomize spray liquids into thin sheets and filaments and subsequently into a random, polydisperse spectrum of droplets whose sizes range from submicron to large. The distribution frequency approximates a bell-shaped, gaussian curve. A significant portion of the pesticide controversy stems from the fact that these large-droplet, broad spectrum sprays are grossly inefficient in transport of pesticides to target foliage. Its inefficiency and environmental costs are not widely recognized. Spray physics evaluates the transport and impingement charactersitics of each size class of droplets in a polydisperse spray cloud. For all practical purposes, spray spectra can be divided into two size classes:

1. Small droplets ($<$100–150 $\mu$m diam.).
2. Large droplets ($>$150 $\mu$m diam., ranging to 400–600+ $\mu$m).

The significance of these two size classes is that small droplets as a class are transported primarily on turbulent eddies and have an impingement $P < 1$. Their impingement on target systems depends on droplet size, travel velocity and the geometry of the target. On the other hand, the primary force acting on large droplets is gravity. They have an impingement probability $P = 1$ and impinge on the first solid or liquid substrate in their flight path. Typically, this means that they are delivered to peripheral foliage or the ground.

The objective of most pesticide spray applications is to deliver the spray cloud to a foliage target. Herbicide sprays are directed at any portion of the target plant whereas most insecticide and fungicide sprays are directed at microhabitats within the foliage canopy. Spray clouds are difficult to measure and portions of the spray are invisible. An empirical spray technology based on the use of polydisperse sprays was developed in the early and middle decades of the 20th century. It was "successful" and appeared to be "simple." It was buttressed by the visual illusion that spray clouds "fell" into the foliage. The actual interaction of spray clouds with the foliage canopy and the mechanism of pest control from sprays was unknown. This conventional spray technology was developed in an analytical void since nanogram sensitivity gas chromatography (GC) was not developed until 1959 (Coulson et al., 1959) and analytical methods for spray mass transport were not developed until the late 1960s. Spray physics became a part of pesticide controversies only in the decades since 1970.

In this part of chapter 2, we will discuss at length the role of mass transfer as the initial step by which pesticides are introduced into the environment. This is a different approach from those that relate to the study of pesticides *after* their original transfer. Controversies focus on the significance of droplet spectra in spray clouds, the mechanism of delivery of sprays, effects of a foliage environment, pesticide volatility, redistribution of pesticide, and the effect of spray drift.

Spray physics provides one simple, overriding guide to pesticide use—*efficient delivery of pest management sprays is size-limited.* This is because foliage systems present a narrow, size-limited "window of opportunity" for pesticide sprays. The upper size limit is determined by impingement on peripheral foliage and the lower size limit by impingement probability within the canopy. All broad spectrum sprays overlap that window and are therefore successful, whatever may be their economic and ecological limitations.

### 2–8.1 Spray Propagation

Mass transport and impingement can now be measured for all discrete spray droplets that travel as a spray cloud. This includes spray clouds that are electrostatically charged but excludes chemigation and high volume sprays which saturate and run-off from foliage. As spray clouds move away from their source, they expand as droplets within the cloud and are subjected to turbulence and gravitational forces. Turbulence is a key factor in the transport of small spray droplets. Large droplets and light refraction allow spray clouds to be seen. Under typical spray conditions turbulence and sedimentation causes the spray cloud to move in an expanding, usually downward pattern.

Although it is well known that the ground under a tree remains dry after a light rain, the light rain of a pesticide spray at the foliage-air interface is assumed to penetrate and to provide even coverage within the foliage canopy. Although nature had provided a clue in the size-related transport and impingement of air-borne pollen, and although the engineering physics of air filtration of particulates was well known, no interdisciplinary transfer of physical principles occurred and spray physics played no part in an empirical simplified, conventional spray technology which was accepted worldwide without any analytical basis.

To some significant extent this is because of spray drift. Spray drift is a fact. However, it typically constitutes <5% of the mass of most sprays and is dwarfed in potential ecological significance by losses (60–80%) of most pesticide sprays on peripheral foliage or the ground (Table 2–2). The direct effects of spray drift can be seen. The ecological and economic significance of pesticides wasted and transferred to the soil is difficult to measure or assess.

The typical assumption of spray drift is that it stems from downwind transport of small droplets in the spray cloud outside the spray area. Volatility, evaporation, and secondary transfer effects are not considered. The ubiquitous presence of small droplets in the foliage canopy was unknown or ignored. Analytical proof of the delivery of small droplets into foliage

Table 2–2. Major components of pesticides losses in spray processes.

1. Delivery losses (60–80% of most sprays)
   a. Directly to the soil
   b. Peripheral foliage
2. Primary volatilization losses to the air
   (3–10% of many sprays, but depends on volatility)
3. Primary particulate drift losses (in the air)
   (3–5% of many sprays)

canopies is the crux of the analytical data provided in the case studies in this chapter. These case studies require some knowledge of spray spectra. Spray spectra measurements include:

$D_{max}$ = the diameter of the largest droplet in the cloud, as estimated.
*MMD* (*VMD*) = the mass median diam. (volume median diam.). The diameter that separates the spray (mass or volume) into equal parts.
*NAD* = number average diam., the average size of all spray droplets measured.

Most pesticide sprays have a $D_{max}$ of 300 to 600 $\mu$m with a concomitant *MMD* in the range of 125 to 400 $\mu$m. Broad spectrum herbicide sprays have a $D_{max}$ of 1000 $\mu$m or more. Grover et al. (1972) and Barry (1984) have reported on the distribution of sizes in sprays.

Pesticide spray transport and impingement is controversial. Two theories (sedimentation theory and turbulence theory) have been used to explain the effects of pesticide sprays. The multibillion pound pesticide spray industry is largely guided by sedimentation theory. Turbulence theory was developed after 1970. It represents a basis for optimization of pest management and risk management of drift. The two theories diverge in their predictions as to the bioefficacy of large- and small-size portions of spray clouds. In this divergence, the theories are mutually exclusive; both cannot be correct. The case studies cited in this chapter provide an analytical basis for turbulence theory. A brief review of sedimentation and turbulence theories is given below.

### 2-8.1.1 Sedimentation Theory

Sedimentation theory is empirical and has no rigorous analytical basis. It evolved prior to development of analytical systems for mass transport of sprays. Its primary concept is Stokes' Law of Falling Spherical Bodies and a Reynolds number of $<0.1$. It assumes that gravitational forces are predominant even in a cross wind. Some of its concepts are depicted in Fig. 2-4. "Small droplets," having minimal mass, are assumed to travel "an in-

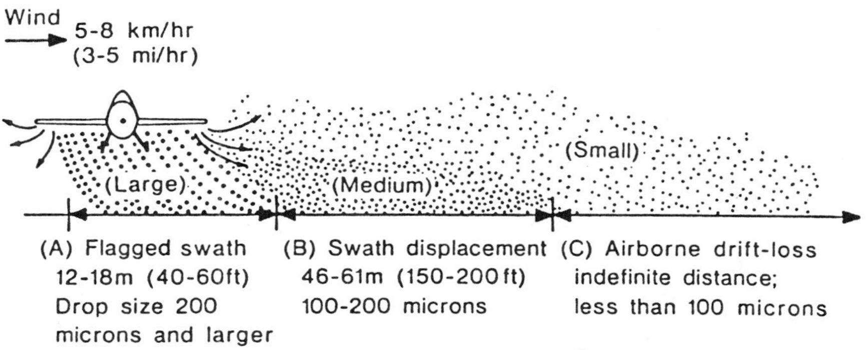


Fig. 2-4. Drift calculated by sedimentation reproduced from Akesson and Yates (1984). Reprinted with permission from ACS Symp. Ser. 238. Copyright, 1984.

finite distance." If small droplets "blow away," large droplets must then provide the pest management which is observed. Actually impingement probability limits spray droplet access in and on foliage systems and thus the validity of sedimentation theory.

The theory gained credibility because of the visual illusion of spray clouds falling into foliage systems. The drift problem reached a critical stage in the use of herbicide sprays and was solved by the use of ultra large spray spectra with a $D_{max}$ as much as 1000 $\mu$m without recognition of transport inefficiency and ecological effects. As an empirical theory, it has no analyical base and is an anachronism in a scientific century.

#### 2-8.1.2 Turbulence Theory

Turbulence theory stems from principles of spray physics and the mass transfer analytical data presented in the case studies in this chapter. The theory assumes that mass transport of spray droplets in a spray cloud is the vector resultant of turbulence and gravitational forces. As droplet size decreases, the contribution of turbulent forces increases. Transport and delivery patterns of spray droplets in spray clouds are determined by turbulent diffusion and sedimentation. Spray droplets move in envelopes of sizes determined by the extent of turbulent forces (Fig. 2-5). Spray efficiency and spray drift are inversely related dependent variables. Spray efficiency is a determinant in risk management of drift.

The key to efficient pest management lies in favorable turbulence and optimization of the size of the spray spectra used. Data for optimum spray droplet sizes are found in the case studies cited in this chapter. Computer-aided spray models such as those discussed in section 2-11 involve turbulent forces. Turbulence can be either favorable (average downward) or unfavorable (average upward). Favorable turbulence can maximize pest management and minimize drift.

Where the Reynolds number is $>0.1$, the terminal sedimentation velocities in Table 2-3 depict rates of gravitational fall only for large droplets. Sedimentation and turbulence calculations for spray cloud transport are shown in Fig. 2-5. The wide divergence in transport calculated by sedimentation and turbulence is illustrated. Under the conditions cited sedimentation predicts that a 40-$\mu$m droplet will travel 5000 m downwind while turbulence indicates a downwind travel in a Gaussian envelope centering at 300 m, not significantly different from the downwind travel of Gaussian envelopes containing droplets from 10 to 300 $\mu$m in diameter.

### 2-8.2 Estimation of Droplet-Size Distribution

Numerous investigations have attempted to relate liquid characteristics, nozzle type and size, and boom pressure to the prediction of spray droplet size distribution during the atomization process. Spray ejected into still air or air in streamline flow is not identical to that developed in turbulent air stream, which has additional shear effects (Miller, 1980).

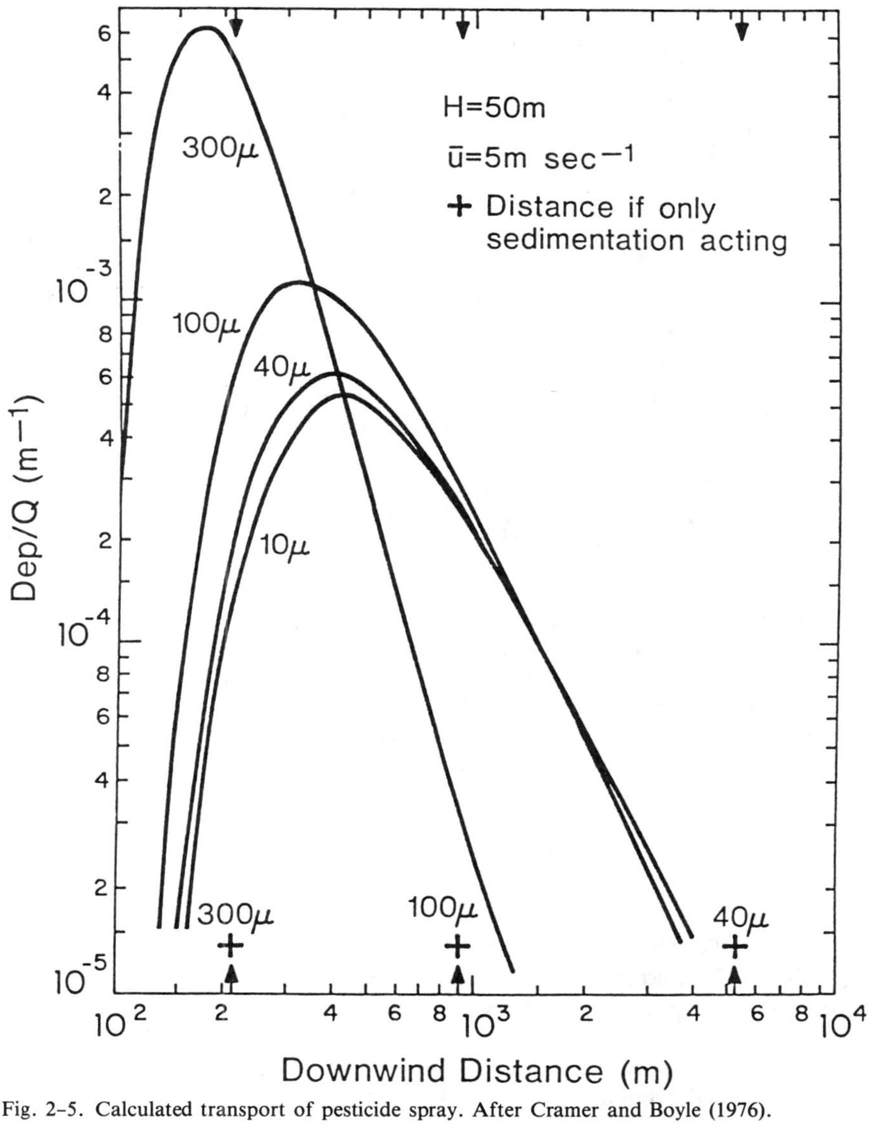


Fig. 2–5. Calculated transport of pesticide spray. After Cramer and Boyle (1976).

The following model can be used to estimate the size and number distribution of the spray from measures of the volume median diam. (*VMD*) and to estimate the spray cloud size-number distribution from spray recovery data. The computations follow the description of Saucier (1981) and Miller (1980) and are based on the hypothesis that the initial number of subclouds (sc), each contains a small range of droplet sizes. By making use of the droplet-size distribution, the range of droplet sizes contained in each subcloud is replaced by a single representative size, defined by the median size of the

Table 2–3. Terminal velocity of spheres (in still air) (Reynolds number <0.1). After Himel (1982).

| Droplet diameter, μm | Fall rate, (sp. gr. = 1) m s$^{-1}$ |
|---|---|
| 1 | 0.00003 |
| 10 | 0.003 |
| 20 | 0.012 |
| 50 | 0.075 |
| 100 | 0.275 |
| 200 | 0.721 |
| 500 | 2.139 |

droplets in the column. The number of droplets and the volume of spray contained in the subcloud is held fixed during this process. The mean volume droplet size is specified by the average diameter ($D_{sc}$)(μm) and is given by:

$$\overline{D_{sc}} = 2\left\{4/3\pi \int_{D_1}^{D_2} N[(D)\, dD]^{-1} \int_{D_1}^{D_2} dV\right\}^{1/3} \quad [3]$$

where $D_1$ and $D_2$ denote the lower and upper bounds, respectively, of the droplet sizes contained in the original subcloud.

If $V_f$ represents the total volume of spray (mL) per unit distance (cm) along the flight line in droplet form and $N$ represents the total number of droplets, then $N(D)$, the number distribution as a function of droplet diameter is given by:

$$N(D) = V_f(D)\left[4/3\pi \int_0^\infty (D/2)^3 f(D)\, dD\right]^{-1} \quad [4]$$

where $f(D)$ denotes the droplet size distribution of the spray cloud.

Figure 2–4, adapted from Akesson and Yates (1974), shows observed droplet-size distributions for a variety of spray types and nozzles with a logarithmic vertical drop-size scale. The near linearity of the plots illustrates the validity of log-normal distribution for modeling the size distribution. Using the log-normal assumptions, the droplet-size distribution is given by:

$$f(D) = \frac{1}{\sqrt{2\pi}\,\sigma_n\, D} \exp\left[-1/2\left(\frac{\ln D - a}{\sigma_n}\right)^2\right] \quad [5]$$

where

$a$ = ln(VMD) (volume median diam.)
$\sigma_n$ = $\ln(\sigma_o)$
$\sigma$ = standard deviation.

Substituting and integrating yields the number of droplets ($n_{sc}$) and the volume ($V_{sc}$) that lie in the subcloud (sc) range $D_1$ to $D_2$:

$$n_{sc} = \frac{3V}{\pi} \exp\left(-3a - \frac{9\sigma_n^2}{2}\right) \left\{ \text{erf}\left[\frac{\ln(D_2) - a}{\sigma_n\sqrt{2}}\right] - \text{erf}\left[\frac{\ln(D_1) - a}{\sigma_n\sqrt{2}}\right]\right\} \quad [6]$$

$$V_{sc} = \frac{V}{2}\left\{ \text{erf}\left[\frac{\ln(D_2) - (a + 3\sigma_n^2)}{\sigma_n\sqrt{2}}\right] - \text{erf}\left[\frac{\ln(D_1) - (a + 3\sigma_n^2)}{\sigma_n\sqrt{2}}\right]\right\}. \quad [7]$$

The mean volume droplet diameter of any subcloud can be computed from the following:

$$D_{sc} = 2\left(\frac{4\pi\ V_{sc}}{3n_{sc}}\right)^{1/3}. \quad [8]$$

### 2-8.2.1 Effects of Evaporation on Droplet Size

The evaporation of water-based droplets plays an important role in determining the size of spray droplets on impaction, or the final size of droplets in the droplet spectrum. Because of the substantial difference in rates of evaporation as a function of size, highly diluted sprays end with a skewed spray spectrum that increases spray drift.

An effective droplet size distribution can be computed based on the time to maximum deposition computed from the model and the thermodynamic relationships based on the relative humidity and droplet viscosity. The effective droplet distribution is directly related to the droplet distribution resulting from each specific nozzle configuration by partitioning the droplet into a number of discrete levels. A correction factor is applied separately to each level to derive an effective characteristic diameter for each level.

The effective droplet-size distribution by number is related to the droplet size distribution at the exit of the nozzle. Droplet size changes because of mass transport during the delivery. A serious drawback of dilute, water-based sprays is the potential for droplet size reduction by evaporation and the potential for skewing the droplet spectrum toward sizes that are not filtered out by foliage ($< \sim 5\ \mu m$). The time rate of change of a water droplet diameter is given by the following relationship adapted from the work of Trayford and Welsh (1977):

$$d = d_o\,[1 - (t/t_1)]^{1/2} \qquad [9]$$

where

$d$ = droplet diam. ($\mu$m) at the time $t$ after release,
$d_o$ = initial droplet diam., ($\mu$m)
$t$ = time after droplet release (s), and
$t_1$ = expected lifetime of droplet (s).

The expected lifetime of the droplet is given by:

$$t_1 = d_o^2/\lambda w_d \qquad [10]$$

where $w_d$ = wet bulb depression.
For water droplets,

$$\lambda = 84.76\,[1 + 0.24\mathrm{R_e}]^{1/2} - 10^{-12} \qquad [11]$$

where $\mathrm{R_e}$ = Reynolds number of the nonevaporated droplet.

$$t_1 \approx d_o^2/80\Delta\mathrm{T} \qquad [12]$$

where $\Delta$T = the difference in temperature (°C) between wet and dry bulb thermometers commonly called the *wet bulb depression*.

For example, at a wet bulb depression of 17 °C, a 600-$\mu$m droplet has an expected lifetime of 264 s (4.4 min.)

### 2-8.2.2 Extrapolation to Nonaqueous Formulations

Evaporation of spray formulations that contain no water can be calculated considering the basic thermodynamics in the evaporation processes. The rate of evaporation is inversely related to the amount of heat that must be transferred from the surrounding air to evaporate a unit volume of spray. Therefore, differences in the amount of heat required to vaporize a unit volume of nonwater spray material vis-a-vis that for water must be accounted for in computing the expected lifetime of the spray droplet based on water spray data and equations. In addition, the rate of evaporation is directly proportional to the difference between the saturation vapor pressure of the spray material, evaluated at the air temperature and the partial pressure of spray vapor already in the atmosphere.

For volatile sprays other than water, the following relationship gives the expected lifetime $\gamma_1$ (s) of a droplet:

$$\gamma_1 = (L/L_w)\,(P_{sw} - P_{air}/P_s)\gamma_{1w} \qquad [13]$$

where

$L$ = latent heat of evaporation of spray (J),
$L_w$ = latent heat of evaporation for water (J),
$P_{sw}$ = saturation vapor pressure of water evaluated at air temperature ($P_a$),
$P_{air}$ = partial pressure of water vapor in the air ($P_a$),
$P_s$ = saturation vapor pressure of spray evaluated at air temperature ($P_a$), and
$\gamma_w$ = expected lifetime of a water droplet (s).

### 2–8.3 Stokes' Law and Spray Cloud Sedimentation

Sedimentation velocities of pesticide spray droplets are usually calculated from Stokes' law (Table 2–3). Sedimentation calculations shown in Table 2–3 have no actual validity where droplets are small (<100 $\mu$m). They are usually transported on turbulent eddies, rather than by sedimentation, when Reynolds numbers are < 0.1, i.e., under most field spray conditions.

The Reynolds number and the terminal velocity of a droplet in free flight are related functions. The Reynolds number for any particular droplet size is computed from knowledge of the terminal velocity of the spray droplet as:

$$R_e = e_a VD/h \quad [14]$$

where

$R_e$ = Reynolds number,
$e_a$ = density of the air, (mg $m^{-3}$)
$V$ = terminal or settling velocity (m $s^{-1}$)
$D$ = droplet diam. ($\mu$m), and
$h$ = kinematic viscosity of the air (N s $m^{-2}$).

A droplet in free fall will accelerate downward until the gravitational force exerted on the droplet is counterbalanced by aerodynamic drag forces caused by the movement of the droplet through the viscous atmosphere. When these forces are balanced, the droplet then falls at a constant rate called the terminal or settling velocity of the droplet.

The terminal velocity is controlled by the coefficient of drag of the particles which, in general, is a nonlinear function of the Reynolds number. However, for the particle sizes common to aerial sprays, it is only appropriate to operate in the lower regime of Reynolds numbers in which the coefficient of drag of spherical particles is well represented by a linear approximation. In this regime, the settling velocity can be determined from Stokes' law. The settling velocity ($V$) given by Stokes' law is:

$$V = gD^2 e_d/18\ h \quad [15]$$

where

$V$ = settling velocity (m $s^{-1}$),
$g$ = gravitational acceleration = 9.8 m $s^{-2}$,
$D$ = droplet diam. ($\mu$m),
$e_d$ = specific gravity of the spray liquid (kg $m^{-3}$), and
$h$ = absolute viscosity of the air (181 N s $m^{-2}$).

The major factor determining the terminal or settling velocity of a droplet and thus its potential for turbulent transport is the droplet size. Table 2–3 presents the calculated terminal velocity for different size droplets (assuming laminar conditions and low Reynolds numbers).

In the case of liquids, the larger droplets may be deformed by aerodynamic forces that can reduce their terminal velocities below that shown.

#### 2–8.3.1 Turbulent Transport

Turbulent forces are significant in transport of spray clouds whose droplets are $<100$ $\mu$m (Glotfelty, 1978). Under favorable meteorological conditions, i.e., no thermal upwelling, turbulent transport carries the spray cloud downward into the foliage canopy. The greater the turbulence, the faster the transport and the greater the probability of reaching the foliage canopy and being filtered out within that canopy. This is a factor that underlies minimization of drift as cross wind speed increases.

The effect of turbulent transport on the drift and deposition of the spray cloud can be visualized by considering a sedimenting cloud as diffusing around its center of gravity and can be related to a line source spray cloud released in the lower 15 m of the atmosphere. Initially, the spray cloud follows a linear growth pattern primarily based on dispersion. Comparisons of the results of experiments using tracer distribution and independent comparison with spray data show good agreement with the assumed linear growth law and intensity of meteorologically derived atmospheric turbulence.

Models based on classical diffusion equations tend to underestimate the speed of growth of real spray clouds. Attempts to correct this effect by superimposing a pseudo-sedimentation velocity have been attempted but have not proven to be satisfactory. Bache and Sayer (1975) provided an empirically derived scaling factor to compensate for this effect. Empirical data, by Bache (1975, 1985), indicate that an appropriate scaling term for spray deposition was related to the location of the position of the spray cloud concentration maximum near the surface. This relationship is illustrated in Eq. [16]

$$(\sigma/\sigma_m) = g(x/x_m) \qquad [16]$$

where

$\sigma$ = the standard deviation of the cloud distribution,
$x$ = downwind distance (m),
$m$ = subscript for maximum, and
$g$ = acceleration because of gravity (9.8 m $s^{-2}$).

This result is supported by the work of Csanady (1973) and Jobson and Sayer (1970) related to the description of a diffusing cloud with sedimentary-type settling. For droplets that have relatively slow settling velocities, the concentration as a function of height and time is described by Eq. [17], which is related to standard Gaussian dispersion considerations for an elevated line source.

$$C = \frac{Q}{\sqrt{2\pi}} * \frac{2hz}{\sigma^2} * \exp\left[-\frac{h^2}{2\sigma^2}\right] \qquad [17]$$

where

$C$ = concentration (droplets, $mL^{-3}$),
$Q$ = source strength (droplets, $mL^{-3}$),
$h$ = height of release (m),
$z$ = height (m),
$\sigma^2$ = variance about center of cloud, and
$\sigma$ = standard deviation

Differentiating Eq. [17] and equating the result to zero, yields the following conditions for maximum concentration, $\sigma = h/\sqrt{3}$. Therefore, the maximum concentration, $c_m$, is:

$$C_m = Q\sqrt{(2/\pi)} * \frac{z}{h^2}\left[\frac{3}{e}\right]^{3/2}. \qquad [18]$$

A knowledge of the cloud growth is a key factor in the description of airborne concentration at or near the ground level. Generalizing the exact relationship between cloud growth and airborne concentration is described by the linear relationship,

$$cg = b * i * x, \qquad [19]$$

where

$cg$ = cloud growth, (m)
$b$ = a constant of proportionality which is empirically derived,
$i$ = the turbulent intensity, and
$x$ = downwind distance (m).

Therefore, the coefficient $b$ can be written as a function of $x_m$ (the downwind distance of maximum concentration), $h$ and the turbulent intensity $i$ as is given by

$$b = h/\sqrt{2}\, X_m\, i. \qquad [20]$$

Bean (1971) has developed an expression for turbulent intensity, $i$, which is related to wind speed, $w$, for neutral to unstable atmospheric conditions, and where $u$ is equal to wind speed at normalized height.

$$i = \sqrt{w^2}/u. \qquad [21]$$

## 2–8.4 Pesticide Spray Impingement

All spray droplets have finite impingement probabilities ranging from ~0 to 1 (from ~0 to 100%). The velocity required for impingement is termed the critical impingement velocity. Large spray droplets have a critical impingement velocity ~0 and an impingement $P = 1$ (100%). They impinge on peripheral foliage or fall to the ground. As droplet size decreases from the range of 100 to 150 μm, the probability of impingement decreases, i.e., becomes <1. As the probability of impingement decreases, critical impingement velocity increases and the actual point of impingement becomes a function of the droplet size, capture cross section, velocity, and turbulence and depends upon the size, geometry, and biology of the target. A significant number and mass of droplets <100 μm penetrate through the foliage barrier and enter insect and disease vector microhabitats.

Filtration is related to foliage density. Needles in conifer foliage are efficient filters. Filtration by broad-leaf foliage has not been quantified but appears to be related to density, shape, leaf hairs, veins, roughness, cuticle wax, and edge effects.

### 2–8.4.1 Spray Drift and Spray Efficiency/Efficacy

Spray drift is a complex scientific, social, economic, environmental, and legal problem. It involves principles of both spray physics and risk management. Aerial spray operators generally consider spray drift to be their most important problem. Early concepts of spray drift were reviewed by Akesson and Yates (1964). For our purposes, spray drift is defined as those air-borne components which move downwind and outside of the defined spray area. Since almost all pesticides are volatile, primary drift consists of particulate pesticide and pesticide vapor. It can be divided into near-field (short-range) and far-field (long-range) drift.

Secondary drift occurs post-spray and consist of vaporization from soil and foliage plus the aerial movement of soil-pesticide particulates (Fig. 2–6). All sprays, no matter how broad their spectrum, contain a driftable component. When aqueous sprays are highly dilute, evaporation occurs rapidly with droplets smaller than 150 μm (Spillman, 1984). This skews the resultant droplet spectrum, producing a wide range of small particulates that exacerbate drift problems.

The small droplet portion of the spray cloud is delivered by turbulent eddies. Its insecticide-fungicide bioefficacy depends on ability to circumvent impingement at the foliage barrier and a concomitant ability to deliver active agent to biological targets within the foliage system. Efficacy is limited to the portion of the spray mass that reaches biological targets. Drift constitutes vapor plus those droplets that do not fall to the ground or are not filtered out by foliage. Efficacy and drift are thus inversely related dependent variables. As spray efficacy is increased, the total amount of pesticide required is decreased with a concomitant decrease in spray drift and ecosystem burden. Spray drift is a probability problem, subject to risk management principles.

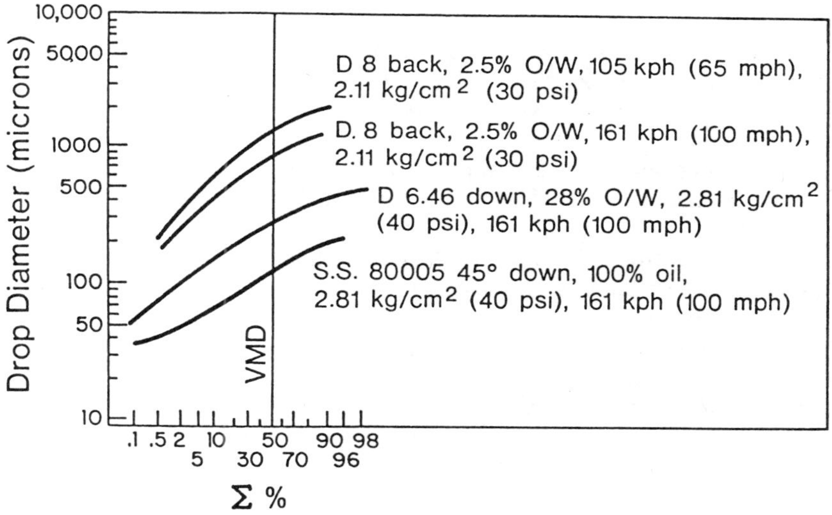


Fig. 2–6. Log normal distribution of drop diameters.

In the past, it was assumed that since small droplets have long calcuated sedimentation times they drift far downwind. This is aptly illustrated in Fig. 2–4 (Akesson & Yates, 1984). Based on sedimentation calculations, droplets $<100$ $\mu$m in diameter are labeled "drift-loss-indefinite distance". The spray concepts shown in Fig. 2–4 have no analytical basis but cast a wide shadow over the pesticide-use industry, acting to eliminate scientific progress in this worldwide industry.

Sedimentation is a widely used assumption in the spray efficacy literature. That literature is immense and largely without a quantitative basis. Spray spectra are usually unknown or unreported. Typically, bioassays of mortality are made, post-spray, without any knowledge of the actual sizes of droplets in the broad spectrum spray or what produced the mortality that was measured. Much of this literature ignores underlying principles of spray physics, is obsolete and a barrier to scientific progress.

The biological efficacy of pesticide sprays is determined by the efficiency of primary delivery and secondary redistribution. All spray droplets contain pesticide and, if delivered, can contribute to accumulation of a lethal dose by the target. Pesticide spray management is really a problem in delivery probability involving the use of polydisperse pesticide sprays in turbulent air streams.

The crux of pest management lies in recognition of the "window of opportunity" which is determined by probability of impingement and efficiency of delivery. Ekblad and Barry (1983) use this concept in optimization of forest sprays. When turbulence is low, droplets less than the range of 5 $\mu$m have a limited probability of impingement, i.e., they flow around the target. Velocity required for impingement defines the lower droplet size limit within the window of opportunity. The upper droplet size limit is determined by the efficiency of delivery. Efficiency of delivery into foliage microhabitats is size-limited to droplets < 100 to 150 μm. This is the upper range of efficiency for non-systemic insecticides and fungicides.

Herbicides are bioactive on any foliage, thus herbicide sprays are generally not size-limited, however their efficiency is size-limited because mass delivery to peripheral foliage appears to peak in the range of 250 μm. As the size of droplets of herbicide sprays is increased, the efficiency of the spray falls. Droplet size and mass transport relationships in herbicide sprays have never been quantified. Droplets > 150 μm in diameter have an impingement $P = 1$ (100%), impinging on anything in their flight path. They are filtered out at the foliage-air interface or fall to the ground. Large droplets, which contain most of the total mass in the spray, have a low delivery probability, and a low bioefficacy. Most fall directly to the ground. Actual analytical data on droplet transport and impingement are reported in Tables 2–4 to 2–7 and in Fig. 2–7.

The conifer forest can filter out spray droplets larger than ~5 μm (Crabbe et al., 1980a). They measured drift under these conditions to total <5% of a spray having a $D_{max}$ of 58 μm and showed that spray loss to drift (spray droplets <5 μm) along with volatilized pesticide can constitute the major portion of long range drift (far field drift). Increase in turbulence enhances delivery to forest foliage and decreases losses to the ground (Joyce and Spillman, 1978). As turbulence increases, delivery efficiency and foliage impaction increases and less pesticide is lost to the ground.

The efficiency of spray application is defined in Eq. [22]. Factors involved in spray losses are outlined in Table 2–2.

$$\text{Efficiency} = \frac{\text{Effective mass delivery}}{\text{Total spray mass}}. \qquad [22]$$

Effecient mass delivery in pesticide sprays is that which is delivered to targets. Present spray technology centers in the use of hydraulic nozzles in

Table 2–4. Comparison of low volume (LV) and ultra low volume (ULV) sprays in pine beauty moth control in Scotland, 1978.

| | LV† | ULV |
|---|---|---|
| Mean deposit, μg/20 needles | 23.6 | 41.2 |
| Mean deposit, μg/single bud | 17.5 | 18.8 |
| Mean deposit, μg/larvae | 4.4 | 22.4 |
| Lost outside target area, % | 20 | 1 |
| Lost to ground, % | 38 | 5 |
| Collected by target surface, % | 42 | 94 |
| Average size of droplet on needles | 50 μm | 50 μm |
| Maximum size droplet on needles | ~120 μm | ~120 μm |
| 90% of droplets found on ground | >150 μm | >150 μm |

† Both LV and ULV sprays were produced by a Micronaire nozzle under conditions to optimize the spray spectra available from the nozzle (VMP 120–150 μm). Pigment and the fluorescent particle spray droplet tracer methods were used. After Joyce and Beaumont (1978), Ruthven (1978), and Joyce et al. (1979).

Table 2-5. Number of droplets observed on conifer needles by size categories in a U.S. Forest Service field trial, 1975 with three different pesticides.†

| Drop size range, μm | Bt‡ | | Pesticide type: Carbaryl | | Trichlorfon | |
|---|---|---|---|---|---|---|
| | Number | Cum. % | Number | Cum. % | Number | Cum. % |
| <4 | 108 | 16 | 137 | 11 | 353 | 19 |
| 4-10 | 239 | 51 | 236 | 31 | 323 | 36 |
| 10-15 | 139 | 71 | 106 | 40 | 200 | 47 |
| 15-21 | 28 | 75 | 139 | 51 | 228 | 59 |
| 21-31 | 90 | 89 | 254 | 73 | 338 | 77 |
| 31-41 | 10 | 90 | 76 | 79 | 116 | 83 |
| 41-61 | 24 | 94 | 88 | 86 | 57 | 86 |
| 61-81 | 25 | 97 | 73 | 92 | 108 | 92 |
| 81-121 | 7 | 98 | 45 | 96 | 87 | 97 |
| 121-151 | 8 | 99 | 26 | 98 | 21 | 98 |
| 151-200 | 1 | 99 | 17 | 99 | 24 | 99 |
| >200 | 2 | 100 | 5 | 100 | 18 | 100 |

† After Barry (1984), Barry and Ekblad (1978), and Barry et al. (1978).
‡ Bt = *Bacillus thuringiensis.*

Table 2-6. Encapsulated droplets on conifer needles in a Canadian—U.S. Field Trial, 1982-1983.†

| Drop size | Avg. size | Number | Cum. | Volume | Cum. |
|---|---|---|---|---|---|
| μm | | % | | | |
| 2-5 | 3 | 3 | 3 | 0 | 0.0 |
| 6-10 | 8 | 14 | 17 | 0.3 | 0.3 |
| 11-20 | 15 | 47 | 64 | 7.8 | 8.14 |
| 21-30 | 25 | 24 | 88 | 24 | 31.6 |
| 31-40 | 35 | 8 | 96 | 24.5 | 56.1 |
| 41-50 | 45 | 2.3 | 98 | 13.8 | 70 |
| 51-75 | 63 | 0.9 | 99 | 9 | 79 |
| 76-100 | 88 | 0.7 | 99.8 | 15.5 | 94 |
| 101-125 | 100 | 0.1 | 100 | 5.9 | 100 |

† After Villaveces (1984) and Sundaram (1982).

Table 2-7. Insecticide spray droplet distribution on four insects from five different field studies.†

| | Droplet size range, μm: 21-30 | 31-40 | 41-50 | 51-60 | 61-120 | >120 |
|---|---|---|---|---|---|---|
| | Droplet % | | | | | |
| Spruce budworm‡ | 92 | 3.5 | 1.7 | 1.4 | -- | -- |
| Boll weevil§ | 97 | 2 | 0.4 | -- | -- | -- |
| Bollworm¶ | 98 | 1.3 | 0.2 | -- | -- | -- |
| Cabbage looper# | 97 | 1.9 | 0.6 | -- | -- | -- |

† After Himel and Moore (1967, 1969). Fluorescent particle tracer system used.
‡ *Choristoneura fumiferana.*
§ *Anthonomus grandis.*
¶ *Helothis zea.*
# *Trichoplusia ni.*

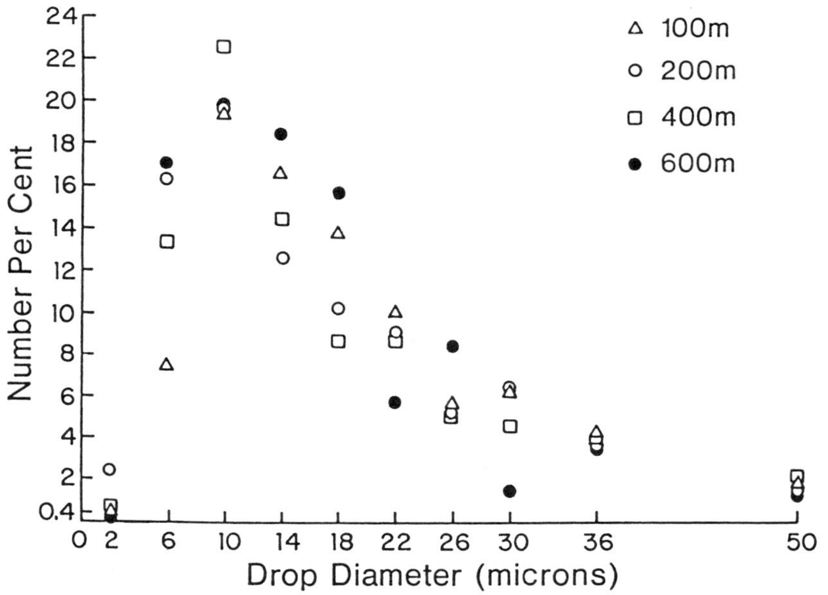


Fig. 2–7. Droplet spectra deposited on balsam fir needles under unstable conditions—100, 200, 400, and 600 m from the swath. Reproduced from Wiesner (1984). Reprinted with permission from ACS Symp. Ser. 238. Copyright, 1984.

which up to 75% or more of the spray has a low bioefficacy. Replacement of the hydraulic nozzle is a significant barrier to change because no alternative hardware has been widely accepted and no overriding pressures are acting to focus change.

Drift is also an impediment to change, for example, the use of "still" conditions during spray application has been a "safe refuge" in the "fight against drift." Analytical data that show an increase in drift under still conditions have a basis in spray physics but no standing in current pesticide regulation.

Three factors which affect pesticide losses during spray application are estimated in Table 2–2. Reduction in delivery losses constitute the major potential for effective change in pest management.

## 2–9 PROPERTIES OF THE CANOPY ENVIRONMENT ON SPRAY INTERACTION

The properties of the canopy and the microenvironment surrounding the canopy have a large influence on pesticide deposition and behavior.

### 2–9.1 Physical Properties of Canopy

The deposition of spray within the canopy is dependent upon the structure, including leaf (needle) area and geometry, orientation, veins, edges, hairs, and stem diameter. Plant or tree spacing, and canopy density are also significant factors. Needles, edges, veins, and hairs maximize impingement. The smaller the droplet size, the lower the collection efficiency for any given target. The smaller the target, the greater its collection efficiency (Eq. [22]). The greater the crop density and roughness, the larger the bulk removal of spray. The greater the turbulence, the greater is the foliage deposition. Far field drift is often composed of particulates $< \sim 5$ $\mu$m plus pesticide vapor. Still air increases drift (Crabbe et al., 1980a). Spray drift can be minimized as wind velocity increases (Joyce et al., 1979).

The turbulence regime within the canopy is different from that which occurs above the canopy. The collection efficiency of foliage or a crop canopy requires consideration of the probability that a droplet will hit an object (leaf or stem). Droplet collection efficiency within the canopy is directly related to the free stream air velocity, and the droplet size and inversely related to a representative foliar dimension:

$$E \propto \frac{eV_0 d^2}{18hl} \tag{23}$$

where

$E$ = collection efficiency,
$e$ = droplet density, (g cm$^{-3}$),
$V_o$ = free stream air velocity, (cm s$^{-1}$),
$d$ = droplet diameter, ($\mu$m),
$h$ = air viscosity (N s m$^{-2}$), and
$l$ = representative foliar dimension (cm).

### 2-9.2 Micrometeorological Factors

For spray droplet sizes where gravitational sedimentation is the predominant mechanism, the vertically falling droplets could fall on horizontal surfaces of the canopy, pass directly through the canopy, or bypass the canopy and fall to the ground. The smaller the droplet the greater is the effect of the horizontal components of airflow. Therefore, a droplet falling with a sedimentation velocity ($V_s$) affected by a horizontal wind ($V_h$) will approach a horizontal surface at the angle $= \tan^{-1} (V_s/V_h)$. This indicates that the effect of sedimentation is greatest for low wind speeds.

Since the wind is generally horizontal, droplets are more disposed to impact vertical surfaces. Furthermore, because the smaller droplets are more strongly influenced by the wind, they generally have a greater potential for inertial impact. Bache (1975) indicates that for very small droplets ($<10$ $\mu$m) inertial impact is the dominant factor over almost all windspeeds.

The wind velocity is rapidly attenuated within the crop canopy. At the top of the canopy, if the wind velocity is strong, the predominant deposition mechanism is by inertial impact on vertical surfaces. Down in the canopy where the wind velocity is low, the predominant deposition mechanism is

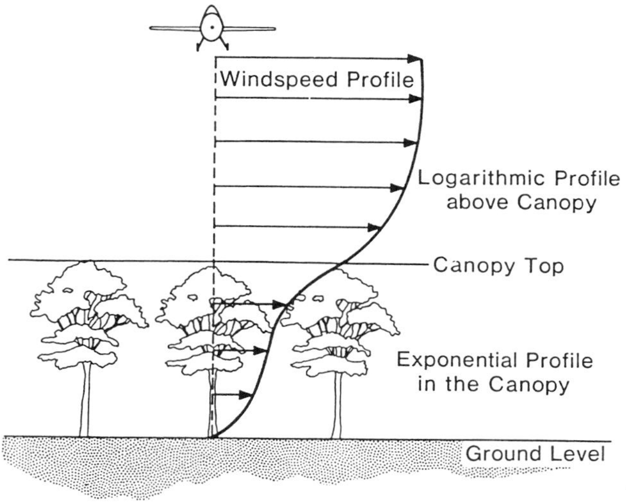


Fig. 2–8. Forest canopy influence on windspeed profile.

sedimentation on horizontal surfaces or by turbulent transport. Figure 2–8 illustrates the influence of a forest canopy upon the windspeed profile.

The average velocity required for prediction of spray drift and delivery is a function of the local canopy and can be estimated by extension from near-ground meteorological measurements. Extension requires knowledge of the wind profile as a function of height. The wind profile within a plant canopy can be approximated by the application of exponential scaling with height:

$$u = A \exp [\gamma(R - z)] \qquad [24]$$

where

$u$ = windspeed at normalized height (m $s^{-1}$),
$A$ = windspeed at the top of the canopy (m $s^{-1}$),
$R$ = normalized height ($Z$/canopy height, cm),
$z$ = height above the ground (m), and
$\gamma$ = extinction coefficient whose value increases with canopy density (ranges from about 1 to 5 for most naturally occurring vegetation canopies).

Above the canopy, the wind profile has logarithmic scaling and is approximated by:

$$u = \frac{u^*}{K} \ln(10R) \qquad [25]$$

where

$u$ = windspeed at normalized height $R > 1$,
$u^*$ = wind friction velocity (m $s^{-1}$),
$K$ = Von Karman constant = 0.4,
$R$ = Normalized height (Z/canopy ht.), and
$z$ = height aboveground (m).

The average velocity in the profile between the ground and the application height can be found by integrating the scaling equations with respect to $R$, the normalized height, over the applicable limits and dividing by the normalized application height, $R_{app}$, as follows:

$$\bar{u} = \frac{\lambda}{\gamma R_{app}} [1 - \exp(-r)] + 0.435A \left\{ 1.303 \left[ 1 - \frac{1}{R_{app}} \right] + \ln(R_{app}) \right\} . \quad [26]$$

The potential biological effect of each spray droplet in the spray cloud is dependent on both its transport pathways and its point of impingement. Impingement probability determines whether the spray droplet can penetrate the foliage barrier. Impingement probability also is a significant element in mathematical models being developed for computer modeling of pesticide sprays. Both validated mathematical models and the design of computer aided spray processes guide the recommended use of pesticide delivery systems.

The transport and impingement properties of pesticide spray droplets are predictable as a function of their size. Pesticide efficiency depends on both probability of transport to the target and probability of impingement on that target. Collection efficiency is represented in Eq. [23].

### 2–9.3 Spray Delivery to Canopy

Environmental and economic considerations make it important to manage the aerial application of pesticides for maximal pest control and minimum use of resources. Zones of potential damage from pests are both contiguous to and comingled with nontarget but sensitive areas. One method currently used for protecting sensitive areas is the establishment of *buffer zones*.

Pesticide spray released from aircraft or ground sprayers may, under various meteorolgic conditions, drift into areas of sensitive vegetative and animal species. Pesticide applications must be carried out in an optimal manner so that minimum quantities of active ingredients will produce the maximum beneficial results. Two parameter requirements needed during both planning and operations are: (i) data that determine the pesticide drift characteristics; and (ii) data that define or characterize the risks attendant upon pesticide encroachment upon nontarget areas.

Factors important to establishing drift characteristics include the following: (i) wind speed and direction; (ii) droplet density and size distribution; (iii) evaporation rate; (iv) aircraft altitude and swath characteristics; (v) volume and amount of pesticide applied; and (vi) atmospheric conditions (turbulent or stable). The parameters required to define the risk of impact upon the environment include the following: (i) sensitive areas/species (location and type); (ii) pesticide toxicities/safe exposure levels; and (iii) pesticide persistence/degradation rate.

## 2-10 CASE STUDIES ON PESTICIDE SPRAY TRANSPORT

Gas liquid chromatography for ecosystem analysis at the nanogram level was developed first by Coulson et al. (1959), followed by other chromatographic-mass spectral methods that increased detection and quantifiable sensitivity followed. These methods are widely used to study the fate of pesticides introduced into the environment.

### 2-10.1 Methods to Measure Pesticide Spray Transport

In the past half-century, artificial targets such as cards, slides, and plastic sheets have been used as tools in evaluation of spray management. Data produced through these methods are often misleading (Uk, 1977). Wiesner (1984) describes deposit cards as lacking diagnostic capabilities for interpretation of pesticide spray results and providing no insight into the reasons for success, variability, or failure of spray programs.

Impingement probability studies have been limited to deposition on artificial targets. Size-limited deposition means that these deposits have no quantitative or overall analytical significance. Further, such targets are usually placed in open areas and do not even mimic deposition on adjacent foliage.

Methods for analysis of spray transport in a foliage environment were developed by Himel et al. (1965) and Himel and Moore (1967). The quantitation of spray mass transfer in foliage systems is complex as well as time and labor intensive.

Spray droplets in target and nontarget areas can be determined by size and number on insects or natural foliage. Analytical systems that have had extensive field study include the fluorescent particle spray droplet system (Himel et al., 1965; Himel & Moore, 1967; Himel, 1969a, b); the dry liquids system, (Barry et al., 1977; Barry & Ekblad, 1978); the reversibly soluble pigment system, (Uk, 1977; Lawson & Uk, 1979); the fluorescent dye system, (Barry & Ekblad, 1978; Wiesner, 1984); the fluorescent pigment system, (Joyce & Beaumont, 1978); and the encapsulated droplet system (Himel, 1982; Sundaram, 1982; Villaveces, 1984).

### 2-10.2 Case Studies

Case studies cited below provide critical evidence for testing the validity of the turbulence theory.

#### 2-10.2.1 Scotland Forestry Service Case Study

Aqueous LV sprays (20 L $ha^{-1}$) were compared with nonaqueous sprays (1 L $ha^{-1}$) in a pesticide delivery study in Scotland. A Micronaire[1] nozzle provided a VMD of 120 to 160 μm. In the 1979 tests, wind speeds of 8 to 25 kg $h^{-1}$ (5–15 mi $h^{-1}$) were encountered. The high wind speeds increased deposition on the trees and drift was decreased. Analytical methodology included the use of fluorescent pigments and the fluorescent particle spray droplet tracer method (Joyce & Beaumont, 1978; Ruthven, 1978; Joyce et al., 1979).

#### 2-10.2.2 Canada Case Study

Downwind transfer of fenitrothion spray was measured using tracer systems in conjunction with meteorological measurements. The VMD of the spray was 58 μm (evaporated basis). In this research, droplet deposition (by number) on balsam fir needles peaked at 10μm. Of 5000 droplets analyzed on target needles, 90.4% (by number) were <60 μm (Wiesner, 1984).

#### 2-10.2.3 United States Case Study—1

Tracer studies involved BT, carbaryl, and trichlorfon sprays. More than 90% of the droplets (by number) found on needles were <60 μm (Barry, 1984; Barry & Ekblad, 1978; Barry et al., 1977).

#### 2-10.2.4 Canada—USA Case Study

The encapsulated droplet tracer system is a recently developed analytical system to identify deposition patterns. The method uses spray droplets that encapsulate during transit. The capsules are deposited on foliage or other surfaces as spheres or hemispheres that can be identified, measured, and counted. Research groups at the Forest Pest Management Institute (Canada), Canadian Forestry Service, U.S. Forest Service (CANUSA PROGRAM), and Univ. of Georgia are conducting encapsulated tracer studies (Villaveces, 1984).

#### 2-10.2.5 United States Case Study—2

The first analytical study of mass transfer by sprays using the fluorescent particle spray droplet tracer method was conducted from 1965 to 1969 (Himel & Moore, 1967, 1969b). Tests, carried out in Montana (spruce-fir

[1]Mention of a trade name does not imply endorsement by the authors and their institutes.

forest) and Georgia (cotton field) allowed comparison of foliage and habitat differences. The $D_{max}$ was 365 $\mu$m for the forest and 800 $\mu$m for the cotton field. Spray droplets on insects killed by the sprays were analyzed. Of 101 000 droplets analyzed, 98% (by number) were $<30$ $\mu$m in diameter. Delivery of spray droplets to the conifer forest insects was not different in kind from delivery to cotton insects. No spray droplets larger than 120 $\mu$m were found on insects killed by the spray (Himel & Moore, 1967, 1969b).

### 2–10.3 Case Study Summary

Since 1965, several mass transport studies of pesticide spraying have been carried out in the USA, Canada, and Great Britain by different research groups. The data from these studies are mutually consistent even though they represent different methodologies, test protocols, and field conditions. They confirm that the delivery and impingement of pest management sprays is governed by principles of spray physics. Only spray droplets $<100$ to 150 $\mu$m deliver pesticides into foliage environments; large droplets fall to peripheral foliage and the ground; and numbers of droplets, foliage density, impingement probability, and turbulence are also crucial factors. The spray application of insecticides, fungicides, and herbicides involve fundamental spray management principles. Mitigation of drift is a problem in risk management because drift is a concomitant result of all spraying techniques. Drift is exacerbated by evaporation and by unfavorable application conditions. Principles of spray physics are a basis for computer-aided spray models. It may take decades before sufficient analytical data becomes available for widespread acceptance of computer-aided spray models for general field use.

The analytical data collected in Tables 2–4 to 2–7 and Fig. 2–5 represent major interdisciplinary research studies with sprays. Each represents application of analytical tracer systems and literally thousands of hours of field and analytical time. It is important that these data be collected, presented, and reviewed because they are a basis for change in spray technology. Change is important because the spray application of pesticides is a multibillion-pound source of pesticides in the environment. The immensity of the problem of change in this industry can be illustrated by a composite of typical observations relative to current practices: "Yes I know that large droplets cannot enter foliage, but I have to use large droplet sprays to fight drift." Analytical data are given below as Tables 2–4 to 2–7 and illustrated in Fig. 2–7. Case studies cited provide critical evidence for testing the validity of the turbulence theory.

## 2–11 MODELING PESTICIDE SPRAY APPLICATION

Over the last 20 yr, the federal government has spent considerable effort in developing computer-based spray drift models in which theoretical approaches are represented by either diffusion or ballistic models.

Diffusion models are based on the assumption that the primary mechanism which drives the motion and deposition of the aerial spray particles is the mixing and interchange resulting from the spatial variation of particle concentration. Diffusion is a one-way process in which the total particle ensemble tends toward a state of equilibrium because areas having denser concentration with high-particle interaction tend to become less dense.

Since diffusion models consider particle ensembles, they are particularly useful for describing both spray drift and the environmental impact of the spray drift. There are two principal types of diffusion models: (i) those resulting from the mathematical solution of the diffusion equation, the so-called Gaussian assumptions and (ii) those based upon finite difference of the behavior of the pesticide spray. The latter is generally the more versatile because it permits more realism in incorporation of physical variables, for example, temperature gradients, humidity, wind, turbulence, and topography. The diffusion model is complex to operate but its benefits frequently justify the extra effort.

In the ballistic model, a spray droplet is considered as a moving spherical mass governed by Newton's laws of motion (force of gravity and air resistance). Averaging over the ensemble of particle size yields the overall time history and spatial distribution of the spray cloud. Ballistic models are used for studies of spray droplet having diameters $>100$ to 150 $\mu$m.

Complete description of polydisperse spray clouds whose particle diameters range from hundreds down to a few micrometers requires a combination of the diffusion and ballistic modeling approaches. In these combined models, the spray deposition is commonly modeled by ballistic assumptions, and drift is commonly approached on the basis of diffusion analogies.

Several aerial spray models are available:

### 2-11.1 FSCBG Model

The *F*orestry *S*ervice, *C*ramer, *B*arry, *G*rim (FSCBG) model is an extension of early work done for the U.S. Army at the Dugway Proving Ground by Dumbauld et al. (1975) and carried on by the H.E. Cramer Company by Dumbauld et al. (1977). The FSCGB model is a line source model that incorporates aircraft wake settling, line source dispersion, droplet evaporation, and canopy penetration. Concentration, deposition, and dosage above and below a downwind forest canopy are calculated for multiple spray lines.

This model has been used for spray prediction in a variety of spray applications by the U.S. Forest Service, including in pine seed orchards in the southern USA and in spruce budworm spraying programs in Maine (Ekblad & Barry, 1984).

### 2-11.2 AGDISP Model

The *AG*riculture *DISP*ersion (AGDISP) model developed by Bilianin et al. (1981) is based on a description of the motion of discrete droplets. Integral dynamic equations for calculating droplet trajectories are provided. The model includes the influence of the spray delivery system, wake turbulence, atmospheric turbulence, gravity, and evaporation. The model has been principally applied to large droplet sizes.

### 2-11.3 Picot Model

Picot model for forest spray dispersion and deposition (Bask & Picot, 1984) employs a finite difference technique to solve the diffusion-advection equation. Atmospheric turbulence, gravitational settling, droplet evaporation, atomizer type, foliage, and limited wake effects are considered.

### 2-11.4 EPAMS Model

The *E*nvironmental *P*rototype *A*utomatic *M*eteorological *S*ystem (EPAMS) model is based on the Dugway Proving Ground line source dispersion model (Dumbauld & Bjorkland, 1977). The model incorporates complex terrain wind field effects in a Gaussian instantaneous line source model. Mixing layer depths and wind velocity are estimated based on stable atmospheric assumptions. Dispersion is computed from the Gaussian assumptions. Evaporation and micrometeorological effects are ignored.

### 2-11.5 Reid and Crabbe Model

The Reid-Crabbe models (Crabbe et al., 1980a, b; Reid & Crabbe, 1980) are finite difference gradient transfer simulation techniques that use a Lagrangian-Markov chain formulation. The models consider atmospheric turbulence, droplet evaporation, gravitational settling, and initial droplet-size spectrum.

### 2-11.6 FWG Model

The FWG (Frost & Huang, 1981) model accounts for aircraft wake parameters, spray nozzle exit parameters, and mean and statistical atmospheric properties including turbulence. A fourth-order Runge-Kutta scheme is used to solve the three differential equations of motion. An aircraft vortex wake model is currently being incorporated. Evaporation is not considered.

### 2-11.7 Bragg Spray Model

The Bragg Spray model (Bragg, 1981) icorporates aircraft vortex wake, droplet evaporation, and propeller effects. The vortex path subroutine evaluates wing and helicopter blade surface potential flow and computes the spanwise location and strength of wing-tip vertices. Wake properties are computed based on a viscous wake model. Drift and deposition are computed from the interactions between the droplets and the wake.

### 2–11.8 Miller Spray Model

The Miller model (Miller, 1980) examines the dispersal of agricultural sprays from airplanes, helicopters, and ground spray rigs. The model is two-dimensional and neglects the micrometeorology of the canopy and droplet evaporation. Nozzle exit size distribution functions are computed based on measured spray deposition pattern with a few simplifying assumptions concerning the distribution shape. The model is applicable to a narrow range of turbulence parameters.

### 2–11.9 Trayford and Welsh Model

The Trayford and Welsh model (Trayford & Welsh, 1977) considers wing tip vertices, evaporation, and propeller swirl and computes droplet trajectories. The model applicability is generally limited to the near field (a few wing-spans from the flight line).

### 2–11.10 The Cranfield Model

The Cranfield model (Bache, 1975; Bache & Sayer, 1975; Bache & Uk 1975) simulates the deposition from a cloud diffusing about its center of gravity and transported with the mean wind. The model was originally developed for the evaluation of agricultural spraying and neglects canopy factors and droplet evaporation. When combined with the canopy models also developed at the Cranfield Institute of Technology, realistic estimates of droplet transport within canopies are developed.

### 2–11.11 Loats/Animal and Plant Health Inspection Service Model

This model, developed by Loats Associates, Inc. for the USDA, is an extension of the Cranfield model (Loats & Lloyd, 1983). The model was developed to simulate the dispersion and deposition of pesticides and herbicides to agricultural crops. The model includes methods for accounting for the effects of droplet evaporation. It is available in several forms including versions that run on small handheld calculators.

## 2–12 RESEARCH NEEDS FOR IMPROVED SPRAY TECHNOLOGY

Well-engineered, efficient pesticide delivery systems are urgently needed. The spray application of pesticides has never developed beyond its empirical basis. Most of the spray methods used today are deeply entrenched and difficult to change. If change is to come to pesticide spray methods, it must be based on widely available analytical mass transfer data. Because all mass transfer of sprays are time and labor intensive, there is a need for automation of foliage deposit analysis to make wide-ranging data available to field agriculturists. Analysis of foliage samples must be adapted to digital video methods. Wide availability of automated methods and their data is a key step to implement new methodologies in spray transfer and pest management.

There is a critical need to test and evaluate existing or improved mathematical models so that computer-directed field spray programs can be developed and made readily available. The available models need extensive field testing in conjunction with analytical systems to justify widespread adaptation to computer programs aimed at pesticide efficiency, optimized pest management and risk management of drift.

## 2–13 CONCLUSIONS

From the above discussion on spray physics we conclude that:

1. Delivery efficiency for pesticide sprays is size-limited. The ultimate question is whether investment in large droplets in sprays is economically sound and biologically relevant in treating foliage.

2. Present spray technology does not conform to underlying principles of spray physics, thus it needs to undergo major change in methods and hardware. That change will, however, be slow because present methods appear to be rational and are deeply rooted. Principles of spray physics are a valid basis for computerized spray models.

3. Most current pesticide spray methods are inefficient. New delivery technology must be based on efficient delivery to the target.

4. Broad-spectrum, polydisperse pesticide sprays can be evaluated by division into two discrete size segments—$>150$ $\mu$m having an impingement $P = 1$ and sizes $<100$ to 150 $\mu$m having an impingement $P < 1$. This is a major factor in determining the bioefficacy of sprays.

5. A foliage barrier exists for spray droplets $>100$ to 150 $\mu$m. They are filtered out in the first few millimeters of foliage.

6. Most insects and disease vectors live within a protective microhabitat. Effective management requires delivery of the pest control agent into and on that microhabitat.

7. Mass and droplet transfer in herbicide sprays has never been quantified.

8. Spray drift is the overriding factor in the use of herbicides. Control of spray drift using ultra-large, polydisperse sprays is not an ultimate solution.

9. Most pesticides are subject to volatilization from sprays, from foliage and from the ground. This is an environmental management problem whose significance has not been properly evaluated.

10. Efficient pesticide delivery systems must be developed.

## REFERENCES

Akesson, N.B., and W.E. Yates. 1964. Problems relating to application of agricultural chemicals and resulting drift residues. Annu. Rev. Entomol. 9:285–318.

Akesson, N.B., and W.E. Yates. 1974. The use of aircraft in agriculture. FAO Agric. Develop. Paper 9. FAO, Rome.

Akesson, N.B., and W.E. Yates. 1984. Physical parameters affecting aircraft spray application. p. 95–115. *In* W.Y. Garner and J. Harvey, Jr. (ed.) Chemical and biological controls in forestry. ACS Symp. Ser. 238. Am. Chem. Soc., Washington, DC.

Amaden, R.C. 1962. Reducing the evaporation of sprays. Agric.Aviat. 4:88–93.

Bache, D.H. 1975. Transport of aerial spray. III. Influence of microclimate on crop spraying. Agric. Meteorol. 15:379–383.

Bache, D.H. 1985. Prediction and analysis of spray penetration into plant canopies. Br. Crop Prot. Counc. Monogr. 28:183–190.

Bache, D.H., and W.J.D. Sayer. 1975. Transport of aerial spray. I. A model of aerial dispersion. Agric. Meteorol. 15:257–271.

Bache, D.H., and S. Uk. 1975. Transport of aerial spray. II. Transport within a crop canopy. Agric. Meteorol. 15:371–377.

Bailey, G.W., L.A. Mulkey, and R.R. Swank, Jr. 1985. Environmental implication of conservation tillage: A systems approach. p. 240–265. *In* F.M. D'Itri (ed.) A systems approach to conservation tillage. Lewis Publ., Chelsea, MI.

Bailey, G.W., R.R. Swank, Jr., and H.P. Nicholson. 1974. Predicting pesticide runoff from agricultural land: A conceptual model. J. Environ. Qual. 3:95–102.

Barry, J.W. 1984. Deposition of chemical and biological agents in conifers. p. 117–137. *In* W.Y. Garner and J. Harvey, Jr. (ed.) Chemical and biological controls in forestry. ACS Symp. Ser. 238. Am. Chem. Soc., Washington, DC.

Barry, J.W., W.M. Ciesla, M. Tysowsky, Jr., and R.B. Ekblad. 1977. Impaction of insecticide particles on western spruce budworm larvae and douglas-fir needles. J. Econ. Entomol. 70:387.

Barry, J.W., and R.B. Ekblad. 1978. Deposition of insecticide drops on coniferous foliage. Trans. ASAE 21:438–441.

Basak, N., and J.C.C. Picot. 1984. Langrangian simulator of forest pesticide spray dispersion and deposition. p. 373–398. *In* P.W. Volsey (ed.) Proc. Symp. Agric. and Forestry Aviation. Natl. Res. Counc., Ottawa, Canada.

Beyerlein, D.C., and A.S. Donigian Jr. 1979. Effectiveness of soil and water conservation practices for pollution control. p. 385–473. *In* USEPA, EPA-600/3-74-106. U.S. Gov. Print. Office, Washington, DC.

Bean, G.A. 1971. The variations of the statistics of wind temperature and humidity. Fluctuations with stability. Boundary-Layer Meteorol. 1:438–457.

Bilanin, A.J., M.E. Teske, and D.J. Morris. 1981. Predicting aerially applied particle deposition by computer. Soc. of Automotive Eng., Kansas City, KS.

Bragg, M.B. 1981. A numerical simulation of the dispersal of aerial sprays. NASA Contr. Rep. 165816. Ohio State Univ., Columbus.

Brown, A.W.A. 1951. Insect control by chemicals. John Wiley and Sons, New York.

Carsel, R.F., and C.N. Smith. 1987. Impact of pesticides on ground water contamination. p. 71–83. *In* G.J. Marco et al. (ed.) Silent spring revisited. Am. Chem. Soc., Washington, DC.

Computer Sciences Corporation. 1980. Pesticide runoff simulator (PRS) user's manual. USEPA, Washington, DC.

Coulson, D.M., L.D. Cavanagh, and J. Stuart. 1959. Gas chromatography of pesticides, J. Agric. Food Chem. 7:250–251.

Crabbe, R., L. Elias, M. Krzymien, and S. Davie. 1980a. New Brunswick forestry spray operations: Field study of the effect of atmospheric stability on long range pesticide drift. Natl. Aeronaut. Establ. LTR-UA-52. Nat. Res. Counc. of Canada, Ottawa, ON.

Crabbe, R., L. Elias, M. Krzymien, and S. Davie. 1980b. Field study of effect of atmospheric stability on target deposition and effective swath widths for aerial forest sprays in New Brunswick. Natl. Aeronaut. Estab. LTR-UA-61. Nat. Res. Counc. of Canada, Ottawa, ON.

Crabbe, R., M. Kzymien, L. Elias, and S. Davie. 1980c. New Brunswick forestry spray operations: Measurement of atmospheric fenitrothion concentrations near the spray area. Natl. Aeronaut. Establ. LTR-UA-56. Nat. Res. Counc. of Canada, Ottawa, ON.

Cramer, H.E., and G. Boyle. 1976. The meteorology and physics of spray particle behavior. Pesticide spray application and assessment: Workshop proceedings. USDA Forest Serv. Rep. PSW 16. U.S. Gov. Print. Office, Washington, DC.

Csanady, G.T. 1973. Turbulent diffusion in the environment. Reidel Publ., Boston.

Donnigan, A.S., Jr., and N.H. Crawford. 1976. Modeling pesticides and nutrients on agricultural land. USEPA, EPA 600/12-76-043. U.S. Gov. Print. Office, Washington, DC.

Dumbauld, R.K., and J.R. Bjorklund. 1977. Mixing-layer analysis routine and transport/diffusion aplication routine for EPAMS. Atmospheric Sciences Lab., U.S. Army Electronics Command, White Sands Missile Range, New Mexico, AD AO38399. U.S. Gov. Print. Office, Washington, DC.

Dumbauld, R.K., H.E. Cramer, and J.W. Barry. 1975. Application of meteorological prediction models to forest spray problems. Rep.TECOM 5-CO-403-000-051, DPG Doc.DPG-TR-M935P. U.S. Army Dugway Proving Ground. U.S. Gov. Print. Office, Washington, DC.

Ebeling, W. 1963. Analysis of the basic processes involved in the deposition, degradation, persistence and effectiveness of pesticides. Residue Rev. 33:35–163.

Ekblad, R.B., and J.W. Barry. 1983. A review of progress in technology of aerial application of pesticides. USDA Forest Serv. 3400-Forest Pest Manage. 8334 2803. U.S. Gov. Print. Office, Washington, DC.

Ekblad, R.B., and J.W. Barry. 1984. Technical progress in aerial application of pesticides. p. 79–94. *In* W.Y. Garner and J. Harvery, Jr. (ed.) Chemical and biological controls in forestry. ACS Symp. Ser. 238. Am. Chem. Soc., Washington, DC.

Frost, W., and K.H. Huang. 1981. Monte Carlo model for aircraft applications of pesticides. Paper 81-1507. ASAE, Chicago, IL.

Gohlich, H. 1983. Formation of drift and basic considerations for its reduction. p. 271–288. *In* R. Greenhalgh and N. Drescher (ed.) Pesticide chemistry: Human welfare and the environment. Permagon Press, New York.

Glotfelty, D.E. 1978. The atmosphere as a sink for applied pesticides. J. Air Pollut. Control Assoc. 28(a):917–921.

Glotfelty, D.E., and J.H. Caro. 1976. Pesticide volatilization. *In* V.R. Dietz (ed.) Removal of trace contaminants from the air. ACS Symp. Ser. 17, ACS, Washington, DC.

Greenhalgh, R. 1986. Definition of persistence in pesticide chemistry. Pure Appl. Chem. 52:2565–2566.

Grover, R., J. Maybank, and K. Yosida. 1972. Droplet and vapor from butylester and dimethylamine salt of 2,4-D. Weed Sci. 20(4):320–324.

Himel, C.M. 1969a. The fluorescent particle spray droplet tracer method. J. Econ. Entomol. 62:912–916.

Himel, C.M. 1969b. The optimum size for insecticide spray droplets. J. Econ. Entomol. 62:919–925.

Himel, C.M. 1974. Analytical methodology in ULV. Br. Crop Prot. Counc. Monogr. 11:113–119.

Himel, C.M. 1982. Analytical systems for pesticide spray transport and impingement. Paper 82-1001. ASAE, St. Joseph, MI.

Himel, C.M., and A.D. Moore. 1967. Spruce budworm mortality as a function of aerial spray droplet size. Science 156:1250–1251.

Himel, C.M.,L. Vaughn, R.P. Miskus, and A.D. Moore. 1965. A new method for spray deposit assessment. U.S. Forest Service Pacific Southwest Forest and Range Exp.Stn. PSW-87. U.S. Gov. Print. Office, Washington, DC.

Jobson, H.E., and W.W. Sayer. 1970. Predicting concentration profiles in open channel flows. Proc. ASCE J. Hydraul. Div. 96:1983–1996.

Joyce, R.J.V., and J. Beaumont. 1978. Collection of spray droplets and chemical by larvae, foliage and ground deposition. p. 63–80. *In* A.V. Holden and D. Bevan (ed.) Control of the pine beauty moth by fenitrothion in Scotland. Forestry Commission, Scotland.

Joyce, R.J.V., G.W. Schaefer, and K. Allsopp. 1979. Distribution of spray and assessments of larval mortality at Annabaglish. p. 15–46. *In* A.V. Holden and D. Bevan (ed.) Aerial application of insecticide against pine beauty moth. Forestry Commission, Scotland.

Joyce, R.J.V., and J.J. Spillman. 1978. Discussion of aerial spraying techniques. p. 13–24. *In* A.V. Holden and D. Bevan (ed.) Control of the pine beauty both by fenitrothion in Scotland. Forestry Commission, Scotland.

Kearney, P.C., and D.D. Kaufman. 1975. Herbicides: Chemistry, degradation and mode of action. Vol. I. Marcel Dekker, New York.

Kearney, P.C., and D.D. Kaufman. 1976. Herbicides: Chemistry, degradation and mode of action. Vol. II. Marcel Dekker, New York.

Keathley, J.P. 1972. Distribution of insecticide spray and efficiency in application. Ph.D. diss. Univ. of Georgia, Athens (Diss. Abstr. 73-5722).

Klingman, G.C., and F.M. Ashton. 1982. Weed science: Principles and practices. 2nd ed. John Wiley and Sons, New York.

Kubr, R.J., and H.W. Poogh. 1976. Carbamate insecticides: Chemistry, biochemistry, and toxicology. CRC Press, Cleveland.

Lawson, T.J., and S. Uk. 1979. The influence of wind turbulence, crop characteristics and flying height on the dispersal of aerial sprays. Atmos. Environ. 13:711–715.

Linscott, D.L., and R.D. Hagin. 1968. Interaction of EPTC and DNBP on seedlings of alfalfa and birdsfoot trefoil. Weed Sci. 16(2):182–184.

Loats, H.L., and D.G. Lloyd. 1983. Final report for "pesticide aerial drift/risk management model development." Contr. 53-3294-1-24. USDA-APHIS, Hyattsville, MD.

Lofgren, C.S., D.W. Anthony, and G.A. Mount. 1973. Size of aerosol droplets impinging on mosquitoes as determined with a scanning electron microscope. J. Econ. Entomol. 66:1085–1088.

Luken, R.J. 1971. Chemistry of fungicidal action. Springer-Verlag, New York, New York.

Marsh, R.W. (ed.). 1972. Systemic fungicides. Longman, London.

McGuffog, D.R., N. Plowman, and T.P. Anderson. 1984. Controlled release formulation of chlorpyrifos in thermoplastic granule matrix. p. 429–436. *In* Proc. 1984 Br. Crop Protection Conf.—Pests and diseases, Brigton, England. 19–22 Nov. Br. Crop Prot. Counc., Croydon, England.

Medved, L.E. 1975. Circulation of pesticides in the biosphere. p. 119–128. *In* P. Varo (ed.) Pesticide chemistry Vol. 3. (Helsinki-1974). Butterworth, London.

Metcalf, R.L. 1955. Organic insecticides: Their chemistry and mode of action. Interscience Publ., New York.

Miller, C.O. 1980. A mathematical model of aerial deposition of pesticides from aircraft. Environ. Sci. Tech. 14(7):824–831.

Reid, J.D., and R.S. Crabbe. 1980. Two models for the long-range drift of forest pesticide aerial spray. Atmos. Environ. 14:1017–1025.

Ripley, B.D., and L.V. Edington. 1983. Internal and external plant residues and relationships to activity of pesticides. p. 545–553. *In* Proc. of the 10th Int. Congr. Plant Protection for Human Welfare, Brigton, England. 20–25 Nov. Br. Crop Prot. Counc., Croydon, England.

Ruthven, A.D. 1978. Deposition of fenitrothion at ground level. p. 95–102. *In* A.V. Holden and D. Bevan (ed.) Control of pine beauty moth by fenitrothion in Scotland. Forestry Commission, Scotland.

Saucier, R. 1981. A mathematical model for the vapor and mass distribution from a falling evaporating aerosol cloud. U.S. Army Armament Res. and Development Command, Chemical Systems Lab., Aberdeen Proving Ground. ARCSL-TR-81007. U.S. Gov. Print. Office, Washington, DC.

Sayer, H.J. 1959. An ultra-low-volume spraying technique for the control of the desert locust. Bull. Entomol. Res. 50:371–386.

Sharville, E.G. 1979. Plant disease control. AVI Publ. Co., Westport, CT.

Sirons, G.J., G.W. Anderson, R. Frank, and B.D. Ripley. 1982. Persistence of hormone-type herbicide residue in tissue of susceptible crop plants. Weed Sci. 30:572–578.

Smith, C.N., G.W. Bailey, R.A. Leonard, and G.W. Langdale. 1978. Transport of agricultural chemicals from small upland piedmont watersheds. USEPA, EPA-600/3-78-056. U.S. Gov. Print. Office, Washington, DC.

Smith, C.N., and R.F. Carsel. 1984. Foliar washoff of pesticides (FWOP) model: Development and evaluation. J. Environ. Sci. Health. B19:323–342.

Smith, C.H., W.P. Payne, Jr., L.A. Mulkey, J.E. Benner, R.S. Parrish, and M.C. Smith. 1981. The persistence and disappearance by washoff and dry fall of methoxychlor from soybean foliage: A preliminary study. J. Environ. Sci. Health B16:774–794.

Spillman, J.J. 1984. Evaporation from freely falling droplets. J. R. Aero. Soc. 42:181–185.

Spillman, J.J., and R.J.V. Joyce. 1978. Low-volume and ultra-low-volume spray trials from aircraft over Thetford Forest in Control of pine beauty moth by fenitrothion in Scotland. p. 31–51. *In* A.V. Holden and D. Bevan (ed.) Control of the pine beauty moth by fenitrothion in Scotland. Forestry Commission, Scotland.

Sundaram, A. 1982. Foliar impaction and distribution of aerially sprayed encapsulated droplets. Paper 82-1006. ASAE, St. Joseph, MI.

Torgeson, D.C. (ed.). 1969. Fungicides. Vol. 2. Chemistry and physiology. Academic Press, New York.

Trayford, R.S., and L.W. Welsh. 1977. Aerial spraying: A simulation of factors influencing the distribution and recovery of liquid droplets. J. Agric. Eng. Res. 22:183–196.

Uk, S. 1977. Tracing insecticide spray droplets by sizes on natural surfaces. The state of the art and its value. Pestic. Sci. 8:501–509.

U.S. Department of Agriculture, Science and Education Administration. 1980. CREAMS, A field scale model for chemicals runoff and erosion for agricultural management systems. *In* W.G. Kniesel (ed.) Conserv. Res. Rep. 26. USDA-SEA, Washington, DC.

U.S. Environmental Protection Agency. 1979. National household pesticide usage study, 1976–1977. USEPA EPA-540/9-80-002.126. U.S. Gov. Print. Office, Washington, DC.

Van Valkenburg, J.W. 1973. The stability of emulsions. p. 93–112. *In* J.W. Van Valkenburg (ed.) Pesticide formulations. Marcel Dekker, New York.

Villaveces, A. 1984. Spray droplet transfer in forest pest management. M.S. thesis. Dep. of Entomology, Univ. of Georgia, Athens.

Weed Science Society of America. 1983. Herbicide handbook. 5th ed. Weed Sci. Soc. of Am., Champaign, IL.

Wiesner, C.J. 1984. Droplet deposition and drift in forest spraying. p. 140–147. *In* W.Y. Garner and J. Harvey, Jr. (ed.) Chemical and biological control in forestry. ACS Symp. Ser. 238. Am. Chem. Soc., Washington, DC.

Willis, G.H., L.L. McDowell, L.D. Meyer, and L.M. Southwick. 1982. Toxaphene washoff from cotton plants by simulated rainfall. Trans. ASAE 25:642–646.

# Chapter 3

# The Retention Process: Mechanisms

**WILLIAM C. KOSKINEN,** *USDA-ARS, St. Paul, Minnesota*

**SIDNEY S. HARPER,** *USDA-ARS, Stoneville, Mississippi*

Retention is one of the key processes affecting the fate of organic chemicals in the soil-water environment. *Retention* refers to the ability of the soil to hold a pesticide or other organic molecule and to prevent the molecule from moving either within or outside of the soil matrix. As such, retention refers primarily to the adsorption process, but also includes absorption into the soil matrix and soil organisms, both plants and microorganisms. Retention controls, and is subsequently controlled, by chemical and biological transformation processes. Retention strongly influences chemical transport to the atmosphere, groundwater, and surface waters. Not surprisingly, retention is a primary factor influencing the efficacy of soil-applied pesticides. The literature abounds with references on the retention of pesticides in soils (e.g., Bailey & White, 1964; Hamaker & Thompson, 1972; Green, 1974; Weed & Weber, 1974; Calvet, 1980).

*Adsorption* is defined as the accumulation of a pesticide or other organic molecule at either the soil-water or the soil-air interface. Adsorption is often used to refer to a reversible process involving the attraction of a chemical to the soil particle surface and retention of the chemical on the surface for a time that depends on the affinity of the chemical for the surface. The distinction between true adsorption in which molecular layers form on a soil particle surface, precipitation in which either a separate solid phase forms on solid surfaces or covalent bonding with the soil particle surface occurs, and absorption into soil particles and organisms is difficult. In practice, adsorption is usually determined only by chemical loss from solution, thus adsorption is often replaced by the more general term, *sorption*. Sorption refers to a general retention process with no distinction between the specific processes of adsorption, absorption, and precipitation.

The individual retention processes are highly complex. This complexity is primarily the result of soil heterogeneity and the soil's contiguity with biological, atmospheric, and water systems. Therefore, one of the keys to understanding the mechanisms of the retention process is the composition of the soil matrix.

## 3-1 NATURE OF THE SOIL MATRIX

Soil consists of three phases—a solid phase (about 50%) comprised of both organic and inorganic solids, a liquid phase (about 25%), and a gas phase (about 25%). The liquid and gas phases serve as the primary modes of transport for soluble and volatile chemicals in the soil. The solid phase is the primary site for chemical accumulation and transformations. Both organic and inorganic solids can serve as adsorption surfaces.

The adsorptive reactivity of soil organic and inorganic surfaces to organic molecules is dependent on the number and type of functional groups at accessible surfaces. When an organic molecule reacts with the surface functional groups either an inner- or outer-sphere surface complex is formed. An inner-sphere complex is a complex between the organic molecule and the surface functional group with no solvent (i.e., water) between the two. An outer-sphere complex has at least one solvent between the surface group and the organic molecule bound to it.

In addition to the functional group's inherent reactivity, the reactivity of any functional group also depends on both the type and the nearness of other functional groups. For instance, ionization of carboxyl or hydroxyl functional groups is enhanced by a nearby electron withdrawing substituent. The presence of a nearby halogen will stabilize the anion of a carboxyl or hydroxyl compound. The presence of an electron-donating substituent such as an alkyl group will reduce acidity (Table 3-1). This effect on dissociation tends to disappear when functional groups are separated by more than four C atoms. Steric hindrance caused by a large neighboring substituent or absorbate may negate reactivity of a functional group.

While functional groups account for much of the reactivity of soil to pesticide retention, accessibility of the functional groups to adsorbates is also an important factor. The intimate association between different soil minerals and between soil minerals and organic matter, makes many functional groups inaccessible to adsorbate molecules (Fig. 3-1). Some functional groups are accessible only to molecules that move through tiny soil pores, clay interlayers, or the polymeric soil organic matrix.

### 3-1.1 Inorganic Surfaces

The inorganic solids are composed of crystalline and noncrystalline, amorphous minerals, including primary (e.g., quartz and feldspars) and secondary minerals (e.g., phyllosilicates, carbonate and S minerals, and oxides). The structure and chemistry of the important alumino-silicate clay minerals in soils are described in detail elsewhere (Marshall, 1964; Dixon & Weed, 1977). Green (1974) and Sposito (1984) give a brief description of the key features of clay minerals in relation to clays as adsorbents for pesticides.

The major functional groups on inorganic surfaces contributing to the adsorptive reactivity are siloxane ditrigonal cavities in phyllosilicate clays and inorganic hydroxyl groups generally associated with metal (hydrous) oxides (Sposito, 1984). Siloxane ditrigonal cavities are Lewis bases whose basic

Table 3–1. Examples of the effects of chlorine substitution on acidity of organic acids.

| Compound | pKa |
|---|---|
| $CH_3COOH$ | 4.75 |
| $CH_2ClCOOH$ | 2.85 |
| $CH_3CH_2CH_2COOH$ | 4.81 |
| $CH_3CH_2CHCl\text{-}COOH$ | 2.86 |
| $CH_3CHClCH_2\text{-}COOH$ | 4.05 |
| $CH_2ClCH_2CH_2COOH$ | 4.52 |

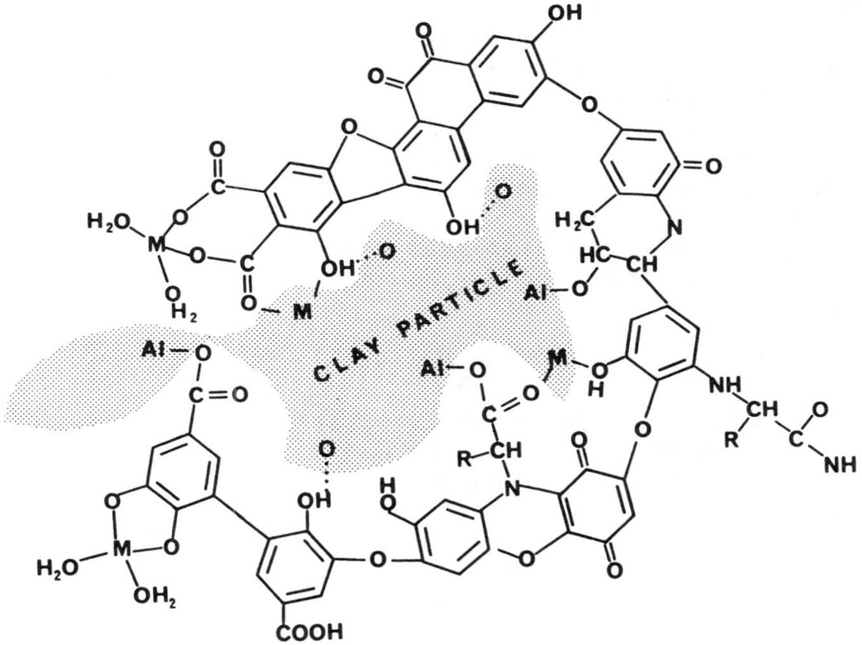


Fig. 3–1. Interaction of a clay particle and an organic molecule.

character can be enhanced by isomorphic cation substitutions in phyllosilicates that create a net negative charge. Unsubstituted cavities tend to weakly complex neutral dipolar molecules. Isomorphic substitutions of $Al^{3+}$ by $Fe^{2+}$ or $Mg^{2+}$ in the octahedral layer increases the Lewis base character resulting in the ability of the phyllosilicates to complex cations and dipolar molecules strongly. Isomorphic substitution of $Si^{4+}$ by $Al^{3+}$ in the tetrahedral layer can also result in strong complexes with cations and dipolar molecules because of the localization of the negative charge.

Inorganic hydroxyl groups are the most abundant and reactive functional groups on soil clays, particularly since they are associated with the surfaces of the clay minerals. Hydroxyl groups are found on phyllosilicates, amorphous silicate minerals, and metal oxides, oxyhydroxides, and hydroxides. Their reactivity varies depending on the number and type of coordination to metal ions.

### 3-1.2 Organic Components

Organic components of the solid phase include polymeric organic solids, decomposing plant and animal residues, and soil organisms. A variety of organic functional groups are present in the humic substances of soil. The exact structure of humic substances is largely unknown, but it has been theorized that they are large aromatic polymers made up of N heterocycles, quinones, phenols, and benzoic acids (Fig. 3-1) that occur as micelles in nature (Stevenson, 1972, 1982). Recent studies combining chemical analyses, infrared (IR), and nuclear magnetic resonance (NMR) have shown that humic substances contain a larger proportion of aliphatic material than previous studies using only elemental and functional group chemical analyses (Hatcher et al., 1981; Sciacovelli et al., 1977; Wilson et al., 1987).

The functional groups of humic substances are known to include carboxyl, carbonyl, phenylhydroxyl, amino, imidazole, sulfhydryl, and sulfonic groups. Soil humic substances also contain a relatively high concentration of stable-free radicals (Steelink & Tollin, 1967). The variety of functional groups in soil organic matter and the steric interactions between functional groups leads to a continuous range of reactivities in soil organic matter.

### 3-1.3 Soil Water

Water has an important role in the retention of pesticides by soil in that it is both a solvent for the pesticide and a solute that can compete for adsorption sites. Water is commonly and simplistically represented as having an ice-like tetrahedral structure in which the water molecules are associated by H bonds. In liquid water, however, this structure only exists in small groups of molecules over short distances. Water molecules outside the shell of nearest neighbors may have broken or distorted H bonds resulting in voids in the ice-like structure. Ionic solutes, polar and nonpolar organic molecules, and charged surfaces such as clays, can also break down the structure of liquid water.

Water molecules, having a dipole, will be attracted to cations. Monovalent cations contain between three and six water molecules in its primary solution shell while divalent cations contain between six and eight. Bivalent cations also contain about 15 water molecules in a secondary solvation shell. Water molecules also solvate cations on the mineral surface.

Polar sorptive solute molecules disrupt the H bonds responsible for the structure of liquid water and form H bonds between themselves and water molecules. In some cases, highly polar solute molecules may form stronger H bonds with water molecules than the water-water H bond. Nonpolar organic molecules are thought to fill the voids in the water structure when they dissolve.

The soil solution from which pesticides are adsorbed is composed of water and some dissolved components—inorganic salts, organic compounds, and gases. The predominant inorganic anions are nitrate ($NO_3$) and bicar-

Table 3-2. Elemental composition and acidic functional groups of humic and fulvic acids and soluble soil organic matter (SSOM). From Lineham (1977).

| | C | H | O | N | Acidity Total | COOH | OH |
|---|---|---|---|---|---|---|---|
| | % | | | | cmol $kg^{-1}$ | | |
| Humic acid | 52 | 4.8 | 41 | 2.2 | 10.0 | 2.5 | 6.5 |
| Fulvic acid | 43 | 3.4 | 52 | 1.8 | 14.6 | 7.0 | 7.6 |
| SSOM | 42 | 3.3 | 53 | 1.8 | 15.1 | 5.1 | 10.0 |

bonate ($HCO_3$) with Cl and sulfate ($SO_4$) becoming important in arid areas. The predominant cations are Ca, Mg, Na, K, and ammonium ($NH_4$). A small portion of the soil organic matter is also soluble. The composition of soluble soil organic matter (SSOM) is similar in composition to humic and fulvic acids (Table 3-2).

While the soil water serves primarily for chemical transport within the soil matrix, adsorption of a compound is affected by water as a solvent and solute and by other solutes contained in the water. An adsorbing solute, such as pesticides, must compete with water molecules, anions, or hydrophobic solutes for adsorption sites. The affinities of dissolved solutes for soil surfaces and for water in some cases will control the degree of adsorption and the types of adsorption sites available. In general, this means a polar surface will adsorb the more polar component of a solution and a nonpolar surface will adsorb the more nonpolar component (Stumm & Morgan, 1981). In the soil system, hydrophobic organic compounds and less hydrated ions will tend to have greater affinity for the surface than the solution.

The soil water also plays a direct role in many of the adsorption mechanisms such as water bridging and ligand exchange (section 3-3.1). Soil water also carries other solutes such as Fe and Al that may be involved in cation-bridging adsorption mechanisms. The solution pH will determine the dissociation and, therefore, the ionic character of the organic solutes. Hydrolysis of soil surfaces may change the types of adsorption sites that are available.

## 3-2 CHEMICAL NATURE OF PESTICIDES AND OTHER ORGANIC CHEMICALS

The chemical characteristics of pesticides and other toxic organics in soil are largely responsible for their behavior in soil. Observed differences in adsorption between organic compounds in the same soil are because of differences in the chemical characteristics of the compounds. As with the soil surfaces, functional groups on an organic molecule influence the strength and mechanism of chemical retention. For example, substitution in the phenyl ring with a halogen ($Cl^-$ or $Br^-$) or chlorophenoxy group, increasing the chain length of the dialkyls, or substituting the dialkyl with the corresponding alkyloxy derivative (Hance, 1965; Grover, 1975) increases the adsorption of phenylurea herbicides. The type (e.g., hydroxyl, methyl, halogen, or nitro), number, and placement of the functional groups determine the

Table 3-3. Effect of functional groups on pKa of substituted phenols.

| Compound | pKa | Solubility (mg/kg water at 25 °C) |
|---|---|---|
| phenol | 9.89 | 930 |
| *o*-chlorophenol | 8.49 | 280 |
| *m*-chlorophenol | 8.85 | 260 |
| *p*-chlorophenol | 9.18 | 270 |
| *o*-methylphenol | 10.20 | 250 |
| *m*-methylphenol | 10.01 | 260 |
| *p*-methylphenol | 10.17 | 230 |
| *o*-nitrophenol | 7.17 | 20 |
| *m*-nitrophenol | 8.28 | 140 |
| *p*-nitrophenol | 7.15 | 170 |

strength of bonding as well as the availability for bonding (Isaacson, 1985; Boyd, 1982) (Table 3-3). Using phenols as an example, an electron-donating group such as a methyl group tends to increase the pKa (decrease the acidity) of the hydroxyl, i.e., pKa of 10.2 for *o*-methylphenol and 9.89 for phenol. The pKa of the hydroxyl, however, decreases with the substitution of an electron-withdrawing group such as a halogen or nitro group, i.e., pKa of 7.17 for *o*-nitrophenol and 8.49 for *o*-chlorophenol.

If molecular structure only affected the pKa of a molecule, predicting sorption onto soil surfaces would be relatively easy. Besides determining the relative acidity or basicity of a molecule, the position, number, and type of functional groups also affect the ability of the molecule to form intramolecular bonds that affect the water solubility of the molecule and the ability of the molecule to form bonds with other molecules (Pauling, 1960). Again using phenol as an example, *o*-nitrophenol has a solubility of 20 mg/kg of water as opposed to 170 mg/kg of water for *p*-nitrophenol (Table 3-3). The lower solubility of *o*-nitrophenol is primarily due to its intramolecular bonding. Increased intramolecular bonding decreases intramolecular bonding with water (or a soil surface), but increases partitioning into nonpolar substances.

Several workers have reported a significant relationship between physical or chemical characteristics such as solubility and the sorption of pesticides or other organic compounds within a chemical category (Bailey et al., 1968; Lambert, 1967; Briggs, 1969; Chiou et al., 1979). There has been little success correlating physical and chemical characteristics to sorption over a broader range of chemicals or classes of chemicals (Harris & Warren, 1964).

## 3-3 ADSORPTION-DESORPTION PROCESS

### 3-3.1 Mechanisms/Forces

*Adsorption-desorption* is a dynamic process in which molecules are continually transferred between the bulk liquid and solid surface. The different intramolecular forces that can attract molecules to the interface and subsequently retain them on the surface have been classified as to mechanisms

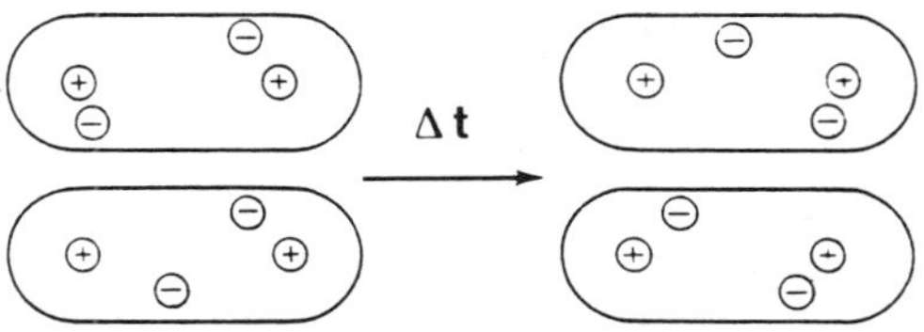

Fig. 3–2. van der Waals attraction. Electronic motion produces electrostatic attraction where $\Delta t$ is elapsed time (Streitwieser & Heathcock, 1976).

(Bailey & White, 1970; Burchill et al., 1981; Mortland, 1970, 1986; Sposito, 1984; Stevenson, 1982). Organic compounds can be sorbed by physical/chemical bonding such as van der Waals forces, H bonding, dipole-dipole interactions, ion exchange, covalent bonding, protonation, ligand exchange, cation bridging, and water bridging with varying degrees of strengths of interactions. Adsorption has also been described as a *hydrophobic partitioning process* between soil water and a soil organic matter phase for the sorption of hydrophobic (nonpolar) compounds (Chiou et al., 1979), although the extent and importance of this mechanism is uncertain (Mingelgrin & Gerstl, 1983). For any given compound, there is likely a continuum of mechanisms responsible for sorption onto soil. For example, an organic molecule may be sorbed initially by sites that provide the strongest mechanism, followed then by progressively weaker sites as the stronger adsorption sites become filled.

### 3–3.1.1 London-van der Waals

London-van der Waals forces are short-range bonds that are a result of dispersion forces. Induced dipole-induced dipole interactions or dispersion forces result from correlations in the electron movement between the molecules. In these interactions, the electronic motion in each of two adjacent molecules is mutually correlated to produce a small net electrostatic attraction at all times (Fig. 3–2). The strength of these interactions is relatively weak (2–4 kJ/mol) and decays rapidly with distance, varying as $1/r^6$ (Streitwieser & Heathcock, 1976). These interactions are additive, however, increasing with the area of contact, and can be especially important in sorption involving neutral polymeric or other high molecular weight materials. With large molecules such as these, several molecular elements may be involved in van der Waals bonding leading to the overall production of strong, short-range bonds (Sposito, 1984; Streitwieser & Heathcock, 1976).

### 3–3.1.2 Hydrogen Bonds

Hydrogen bonds are stronger dipole-dipole interactions than van der Waals interactions and are produced from the electrostatic attraction between an electropositive H nucleus and exposed electron pairs on electronegative atoms. Typically, this involves bonds such as –OH. . .O—, —OH. . .N—, as well as, —NH. . .O—, —NH. . .N— (Stevenson, 1982; Mortland, 1986; Burchill et al., 1981). Hydrogen bonds can occur both intra- and intermolecularly with energies that can vary from 2 to 4 kJ/mol in weak-bonding cases to 60 kJ/mol in strong bonding cases. Hydrogen bonding is apparently much more important in the bonding of organic compounds to organic surfaces than to inorganic O atoms and surface hydroxyls because of the prevalence of appropriate functional groups (Burchill et al., 1981; Sposito, 1984). Hayes (1970) has stressed the importance of H bonding in the adsorption of triazines. Li and Felbeck (1972) calculated the heat of formation of a humic and atrazine complex as 32 to 52 kJ $mol^{-1}$ that is in the range of H bonding. Hydrogen bonding has also been proposed as the major bonding mechanism for phenolic compounds (Boyd, 1982; Isaacson, 1985).

### 3-3.1.3 Cation and Water Bridging

Another weak adsorption mechanism is cation bridging or salt linkage. Generally, it involves the formation of an inner-sphere complex (i.e., a complex with no intervening molecules) between an exchangeable cation and an anionic or polar organic functional group. Because cations are normally surrounded by hydrating water molecules, the organic functional group must be able to either displace the water of hydration or it must react in the presence of a dry surface to form an inner-sphere complex. When the organic functional group is unable to displace the solvating water molecules, water bridging occurs. Water bridging is an outer-sphere interaction between a proton in a hydrating water molecule of an exchangeable cation and an organic functional group (Sposito, 1984). In general, the ability of an organic functional group to displace a hydrating water molecule around an exchangeable cation will depend on the ionic size and the heat of hydration of the cation. As the ionic size of the cation decreases, heat of hydration becomes more negative (Table 3-4). Therefore, displacement of the hydrating water molecules will

Table 3-4. Heats and entropies of ion hydration at 25 °C.

| Ion | Crystallographic ion radius | Heat of hydration, ΔH |
|---|---|---|
| | nm | kcal/mol |
| $H^+$ | -- | −258 |
| $Li^+$ | 0.060 | −121 |
| $Na^+$ | 0.095 | −95 |
| $K^+$ | 0.133 | −75 |
| $Rb^+$ | 0.148 | −69 |
| $Cs^+$ | 0.169 | −61 |
| $Be^{2+}$ | 0.031 | −591 |
| $Mg^{2+}$ | 0.065 | −456 |
| $Ca^{2+}$ | 0.099 | −377 |
| $Ba^{2+}$ | 0.135 | −308 |
| $Mn^{2+}$ | 0.080 | −438 |
| $Fe^{2+}$ | 0.076 | −456 |
| $Cd^{2+}$ | 0.097 | −428 |
| $Hg^{2+}$ | 0.110 | −430 |
| $Pb^{2+}$ | 0.120 | −350 |
| $Al^{3+}$ | 0.050 | −1109 |
| $Fe^{3+}$ | 0.064 | −1041 |
| $La^{3+}$ | 0.115 | −780 |

become more difficult causing water bridging to be favored over cation bridging. For instance, carboxylate groups would adsorb on montmorillonite through cation bridging and water bridging with monovalent and bivalent cations on the exchange sites, respectively. Cation and water bridging are important in the bonding of polar molecules to clay surfaces such as occur in the bonding of pyridine to montmorillonite (Farmer & Mortland, 1966). Water bridging would be especially important in natural systems where most of the metal cations are polyvalent and possess large solution energies.

### 3-3.1.4 Anion Exchange

Anion exchange mechanisms involve a stronger bonding mechanism. This is a nonspecific electrostatic attraction of an anion to a positively charged site on the soil surface, involving the exchange of one anion for another at the binding site. Anion exchange is not observed often because of the prevalence of negatively charged sites on soil surfaces. It is important in acidic soils containing significant amounts of pH dependent charge as in soils that are high in kaolinite or noncrystalline precipitates (e.g., aluminosilicates and iron oxides). Dissociated functional groups such as carboxylate will be the primary groups involved in anion exchange on organic molecules (Sposito, 1984; Mortland, 1986).

### 3-3.1.5 Ligand Exchange

Stronger than anion exchange or either bridging mechanisms is ligand exchange. This is an inner-sphere complex that forms when an organic functional group such as a carboxylate or hydroxyl displaces an inorganic hydroxyl or water molecule of a metal ion (Fe or Al) at the surface of a soil mineral. Although ligand exchange has been observed for the adsorption of organic molecules onto inorganic surfaces in simple systems, the evidence for ligand exchange mechanisms in the sorption of pesticides onto soil is mostly indirect (Sposito, 1984). This mechanism has been proposed for organic weak acids and anions on hydrous oxide surfaces (Stumm et al., 1980). Ligand exchange sorption reactions have been shown for oxalic and benzoic acids (Parfitt et al., 1977a) and humic and fulvic acids (Parfitt et al., 1977b) on goethite.

### 3-3.1.6 Protonation

Protonation of a herbicide, or formation of charge transfer complexes, at a mineral surface occurs when an organic functional group forms a complex with a surface proton. This retention mechanism will be particularly important for basic functional groups at acidic mineral surfaces at low pH and low water contents, particularly in the presence of Al, H, or other metal cations. This mechanism is important for organic molecules containing functional groups such as amino and carbonyl groups. The protonated complex can be sufficiently stable to render adsorption practically irreversible (Mortland, 1986; Sposito, 1984; Bailey & White, 1970). This mechanism has been proposed for some *s*-triazines (Senesi & Testini, 1984), particularly methylthiotriazine (Hayes, 1970) on organic matter.

### 3-3.1.7 Cation Exchange

Cation exchange mechanisms are essentially the reverse of anion exchange. While cation exchange sites are more prevalent in soil because of the large proportion of negatively charged sites associated with clays and organic matter, it is seldom observed with pesticides because of the lack of positively charged pesticides. This retention mechanism involves the electrostatic attraction of cationic organic functional groups such as amines and heterocyclic N by a negatively charge site at the soil surface, usually a clay mineral surface, that is occupied by a metal cation (Mortland, 1986). Cation exchange can occur directly or by protonation. Paraquat and diquat are examples of cationic and weakly basic pesticides adsorbed by cation exchange mechanisms (Knight & Tomlinson, 1967; Weber & Weed, 1968). While the sorption or exchange of short, straight chain aliphatic alkylammonium ions on clay may be reversible, compounds with long chains may not be. van der Waals forces are also involved in the binding of alkylammonium cations by clays, particularly those with long aliphatic chains.

### 3-3.1.8 Covalent Bonding

For most pesticides or other organic chemicals in soil, initially a rapid, reversible equilibrium is established between the chemical in solution and the chemical adsorbed onto the soil surface. Once adsorbed, however, the chemical is subject to other processes that can affect retention. Some chemicals may further react to become covalently and irreversibly bound (i.e., become "bound residues"), while others may become physically trapped in the soil matrix.

Phenols, catechols, and anilines are subject to formation of covalent bonds. The bonding of phenols and catechols to soils occurs relatively quickly with catechols having stronger intramolecular associations upon adsorption than phenols (Harper, 1984). Phenols in soil have also been shown to subsequently bind to soil organic matter by oxidative coupling (Kazano et al., 1972) and by enzymatic polymerization of the phenol with humic monomers (Bollag et al., 1980). Both mechanisms decrease chemical extractability. Kloskowski and Fuhr (1985) found that only 62 to 72% of the applied hydroxy-simazine could be extracted 1 d after application with various solvents compared to >90% for simazine. The extractability decreased with time.

Aromatic amines can also form covalent bonds with soil organic matter. Both biological and nonbiological mechanisms seem to be involved including oxidative-coupling reactions of anilines to phenols, and formation of anilinoquinone linkages (Hsu & Bartha, 1974; Parris, 1980). Graveel et al. (1985) suggest that binding of benzidine, $\alpha$-naphthylamine, and *para*-toluidine to soils involves both rapid, reversible equilibrium with soil followed by nucleophilic addition to the phenolic components of humus.

Some covalently bound residues of pesticides are either difficult to extract or are nonextractable using various solvents. Plants and soil organisms, however, have been shown to be able to extract small amounts of these residues. Therefore, these residues are not necessarily irreversibly bound to soil. For instance, Fuhremann and Lichtenstein (1978) found that earthworms (*Lumbricus* spp.) and oat (*Avena sativa* L.) plants were able to release and incorporate some soil-bound $^{14}$C-ring-labeled methylparathion. Oat plants were found to release more chemical from soil than the earthworms. Helling and Krivonak (1978) reported soybean [*Glycine max* (L.) Merr.] plants released from soil and, subsequently, absorbed small amounts of soil-bound $^{14}$C-dinitroaniline herbicide residues. Also, 9 yr after the last atrazine application to soils, 2-hydroxy-4-amino-6-isopropylamino-*s*-triazine and 2-hydroxy-4-ethylamino-6-amino-*s*-triazine were identified in oat plants (Capriel & Haisch, 1983).

#### 3–3.1.9 Physical Trapping

Decreased extractability of organic chemicals with increased incubation time may at times be because of physical trapping of the chemical in the soil matrix. Decreased extractability of fluridone after equilibration for 7 or 28 d compared to 1 d was in part attributed to the herbicide penetrating deeper into the soil micropores (Weber et al., 1986). X-ray diffraction studies indicated that fluridone can be adsorbed on the interlayer surfaces of montmorillonite. The herbicide prometon has also been shown to adsorb on the interlayer of montmorillonite (Weber et al., 1965).

Hydrophobic interactions and trapping of molecules in a molecular sieve formed by humic materials has been hypothesized as a retention mechanism (Khan, 1982). Khan suggested the incorporation of $^{14}$C-prometryn into humus is because of molecular sieving and hydrophobic partitioning where the prometryn molecules become bound and trapped within the pores of the humic structure.

### 3–3.2 Organic Chemical Characteristics Related to Adsorption

As previously discussed, a variety of mechanisms or forces can attract organic chemicals to a soil surface and retain them there. For a given chemical, or family of chemicals, several of these mechanisms may operate in the bonding of the chemical to the soil. For any given chemical, an increase in polarity, number of functional groups, and ionic nature of the chemical will increase the number of potential adsorption mechanisms for the chemical. For instance, within triazine herbicides, it has been suggested that mechanisms involving van der Waals forces, charge transfer, hydrophobic bonds, cation exchange, and cation bridging are responsible for bonding to soil surfaces (Hayes, 1970).

#### 3–3.2.1 Ionic/Ionizable Organic Chemicals

Bipyridylium cations such as diquat and paraquat are the most obvious members of this group. Ionizable compounds such as the weakly basic compounds (triazines and pyridinones) and the weakly acidic compounds (carboxylic acids and phenols) can adsorb by ionic mechanisms when these compounds are ionized. Weakly basic compounds may adsorb by cation exchange while weakly acidic compounds may adsorb by anion exchange. For these chemicals ion exchange is not the sole adsorption mechanism. For instance, adsorption of bipyridylium cations is primarily due to cation exchange. Other physicochemical forces such as charge-transfer interactions (Khan, 1974), H bonding, and van der Waals forces (Burns et al., 1973), however, can also be involved in the adsorption process.

Triazines are weakly basic chemicals that can be easily protonated at low soil pH levels. The pKa values for triazines range from about 1.7 for atrazine to 4.3 for prometon. There is abundant evidence for cation exchange as the bonding mechanism for triazines to soil (i.e., Carringer et al., 1975; Senesi & Testini, 1980). On the other hand, at soil pH values greater than two pH units above the pKa, triazines are not protonated and other mechanisms become more important such as H bonding (Hayes, 1970), and hydrophobic attractions (Hance, 1969; Bouchard & Lavy, 1985). Bonding by van der Waals forces has not been proved or disproved.

Pyridinones such as fluridone are also weakly basic compounds. With a pKa of 1.7, fluridone sorption would involve cation exchange only in low pH soils. Sorption on soil at pH 5 to 6 is suggested to be by the same mechanisms for sorption on both soil organic matter and montmorillonite, i.e., charge-transfer bonds, H bonding, and van der Waals forces (Weber, 1980; Weber et al., 1986).

For some basic compounds, the pKa may serve as a poor indicator of the solution pH range where exchange predominates. Sorption of quinoline to soil surfaces at suspension pH levels more than two units greater than its pKa is attributed to surface protonation and ion exchange (Zachara et al., 1986). Enhanced protonation at the sorption surface has also been shown for other bases such as triazines (Bailey et al., 1968).

Depending on the pH of the system, weakly acidic organic chemicals (carboxylic acids and phenols) exist either as the undissociated molecule or the corresponding anion. While numerous studies have shown that the anion of such herbicides as 2,4-D is readily adsorbed by anion exchange resins, adsorption of organic anions by soils via anion exchange is not likely, as clays and organic matter are generally either noncharged or negatively charged.

Adsorption of weakly acidic organics probably involves physical adsorption of the undissociated molecule and is not site specific. Phenols can, however, form covalent bonds under the appropriate conditions (see section 3–4.1). Numerous studies have shown that adsorption increases with a decrease in pH if there is a decrease in the percentage of the dissociated acid (i.e., Cheng, 1971).

Other adsorption mechanisms for weakly acidic organics are also possible. Charge-transfer and H bonding were postulated as the adsorption mechanisms for the weak acid chlorsulfuron (Shea, 1986).

### 3–3.2.2 Nonionic Organic Chemicals

Soil organic matter is the principal adsorbent for many nonionic organic compounds (Hamaker & Thompson, 1972). It has been suggested that the retention mechanism of nonionic organic chemicals in soil is a partitioning of the chemical between the aqueous phase and the hydrophobic organic matter (Chiou et al., 1979). The mechanism may not be that simple (Mingelgrin & Gerstl, 1983). For example, some clays have hydrophobic sites and many nonionic organic chemicals adsorb extensively on the clay mineral fraction of soil (e.g., Yaron et al., 1967).

There has been some success in using molecular structure indices to predict the sorption of nonionic compounds to soil organic matter. Boyd (1982) used Hammett constants, which are based on the reactivity of aromatic substituent groups, as a relative measure of phenol sorption in soil. Molecular connectivity indices, based on the topology of an organic molecule, have been successfully used to predict soil sorption coefficients for nonionic compounds (Sabljic, 1984; Sabljic & Protic, 1982), but less successfully for polar compounds such as many pesticides (Gerstl & Helling, 1987). Adsorption of nonionic, nonpolar hydrophobic compounds is because of weak attractive interaction such as van der Waals forces. Net attraction is because of dispersion forces and the strength of these weak forces are about 4 to 8 kJ/mol. These reactions are also entropy driven. Nonionic, nonpolar hydrophobic compounds will not find the polar environment of a solvent such as water to be favorable since they must disrupt the strong water-water bonds to solubilize. Instead these molecules tend to collect at nonpolar interfaces. Electrostatic interactions can also be important, especially when a molecule is polar in nature. Attraction potential can develop between polar molecules and the heterogeneous soil surface that has ionic and polar sites, resulting in stronger bonding.

While most nonionic organic chemicals are subject to low-energy bonding mechanisms, adsorption of phenyl- and other substituted-urea herbicides to soil or soil components has been attributed to a variety of mechanisms depending on the adsorbent. The mechanisms have included hydrophobic bonding (Carringer, 1975), cation bridging (Hance, 1971; Gaillardon et al., 1980), van der Waals forces (Hance, 1969; Senesi & Testini, 1980), and charge-transfer (Hance, 1969; Senesi & Testini, 1980). Even cation exchange has been reported for substituted ureas (Weber, 1980).

### 3-3.3 Dynamics of Adsorption-Desorption

The state of the soil system at any time depends on both the thermodynamics and the kinetics of the system. Thermodynamics identifies possible reactions and reaction end-products at equilibrium from basic relationships between matter and energy. It predicts the direction of a reaction and the relative composition of the constituents, but cannot predict the reaction rate or the intermediate steps between reactants and products—the kinetics of a reaction. Both thermodynamics and kinetics provide information about the retention of pesticides on solid surfaces. Most information to date has been derived from thermodynamic data, however, recent studies have demonstrated the importance of kinetics in controlling the adsorption process in soil. It must be remembered that this is because of the relative ease in obtaining macroscopic, time-independent data. The prediction of spontaneity of a reaction by thermodynamics says nothing about the kinetics of the reaction, only the possibility of a reaction at equilibrium. Activation energy barriers may slow a spontaneous reaction considerably.

### 3-3.3.1 Equilibrium

One of the main assumptions in characterizing adsorption of a pesticide or other organic chemical on a soil surface is that an equilibrium is established between pesticide remaining in solution and that adsorbed on the soil surface. Although it is questionable whether equilibrium is even reached in soil and whether it relates to real soil systems, the use of an equilibrium approach establishes a common ground for comparing pesticide distributions in soils. The equilibrium is most commonly and simply expressed as:

$$C_e \rightleftharpoons C_s$$

where $C_e$ is the pesticide in solution and $C_s$ is pesticide adsorbed. The relationship between adsorbed pesticide and that in solution can be expressed as the distribution coefficient:

$$K_d = [C_s]/[C_e] \quad [1]$$

if $[C_s]$ depends linearly on $[C_e]$. Most often, though, adsorption is not a linear relationship so that the distribution coefficient is commonly expressed in terms of the empirical Freundlich relationship:

$$K_f = [C_s]/[C_e]^{1/n}. \quad [2]$$

Wauchope and Myers (1985) found their adsorption-desorption data fit a sequential-equilibria model (Moore & Pearson, 1981) which assumes that two reversible equilibria occur sequentially:

$$C_e \rightleftharpoons C_{s-1} \rightleftharpoons C_{s-r}$$

where $C_e$, $C_{s-1}$, and $C_{s-r}$ are pesticide in solution, on "labile" soil surface sites, and "restricted" soil sites, respectively. The first equilibrium is the same as discussed above. The second equilibrium is assumed to result from a much slower reversible interchange of pesticide between labile and restricted soil sites. This is illustrated in Fig. 3-3 where the points are experimental data and curves are calculated from model parameters. If the rate of the reaction going from $C_{s-r}$ to $C_{s-1}$ is slow compared to that for the reaction going from $C_{s-1}$ to $C_e$, then this is a possible explanation for the hysteresis discussed in section 3-3.2.2.

The solution phase can also affect the adsorption-desorption equilibrium. There is increasing evidence that dissolved organic C (DOC) in water can bind some pesticides. This might be expected since adsorption of many pesticides to soil is correlated to the organic C content. Several investigators

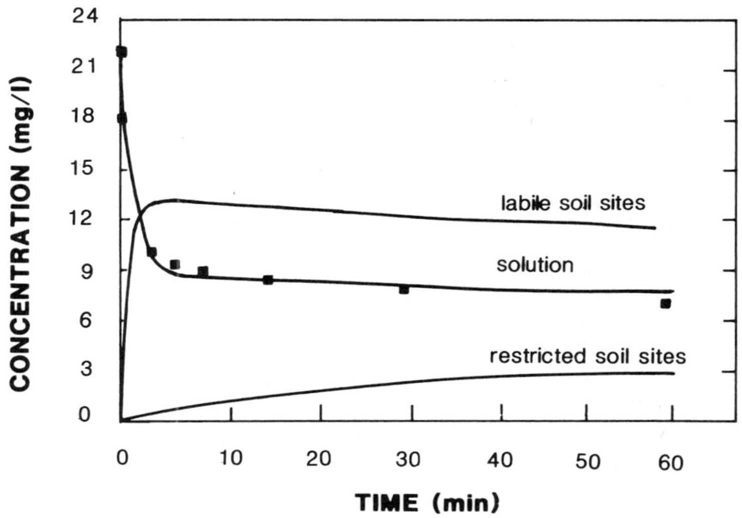


Fig. 3-3. Adsorption of atrazine on labile and restricted soil sites (Wauchope & Myers, 1985).

have reported that the apparent solubilities of strongly hydrophobic solutes increase in the presence of DOC (e.g., Wershaw et al., 1969). Carter and Suffet (1982) found that a significant fraction of the dissolved DDT found in natural waters may be bound to dissolved humic materials. Koskinen and Cheng (1983) hypothesized an equilibrium of:

$$C_{e-DOC} \rightleftharpoons C_e \rightleftharpoons C_s$$

to account for a shift to the adsorbed state for 2,4,5-T when the soluble organics were washed out of the soil.

Numerous additional interactions exist that could affect the equilibrium of different pesticides in various soils. For example, triazine herbicides can form relatively stable Cu(II) complexes in solution (Decock et al., 1985). It is beyond the scope of this chapter, however, to cover all possible interactions affecting the equilibrium for different compounds. Suffice to say that knowledge of the interactions involved in the adsorption process will aid in describing the equilibrium.

### 3-3.3.2 Reversibility

Because adsorption reactions do not always involve only weak, reversible bonding, hysteresis is often observed in adsorption-desorption studies in soil. In other words, the adsorption of a herbicide onto soil and its subsequent desorption is not a single-valued relationship and, therefore, a portion of the adsorbed pesticide is apparently resistant to desorption. While the causes of hysteresis are not understood, several factors have been proposed as contributing to hysteresis. One of the more obvious factors that could lead to a hysteretic effect would be nonattainment of equilibrium during the desorption process. Workers studying the effects of equilibration time on hysteresis, however, have found that the amount of hysteresis either increases as equilibration time increases (Koskinen et al., 1979) or remains the same (Di Toro & Horzempa, 1982) up to 48-h equilibration times.

The desorption methodology itself may have inherent factors that cause apparent hysteresis during desorption. In some cases, degradation has been found to account for at least part of the observed apparent hysteresis (Koskinen et al., 1979). Others have proposed that the actual adsorption-desorption equilibrations in the batch study may lead to changes in the composition of the slurry or to changes in the soil surfaces that cause apparent hysteresis (Koskinen & Cheng, 1983; Gschwend & Wu, 1985). Repeated centrifugation of the slurries in batch equilibration studies has been reported to be responsible in some way for hysteresis for some pesticides (Bowman & Sans, 1985). Perhaps centrifugation results in increased time to reestablish equilibrium. Another possibility is that in batch equilibration studies the supernatant removed for analysis contains some suspended sediment with its adsorbed pesticide and, therefore, is removed after each equilibration.

Some compounds that are strongly bound to soil surfaces such a diquat and catechol have little or no desorption, particularly at lower rates of chemical application (Corwin & Farmer, 1984; Cheng et al., 1983). These appear to be cases where the compound reacts irreversibly with the soil surface or else equilibrium is not established because the kinetics of the desorption are far slower than the adsorption. With sufficient time, even weakly adsorbed chemicals can react with the soil surface to become more strongly adsorbed or bound compared to when they were first applied. Desorption coefficients for cyanazine and metribuzin measured 56 and 121 d after application were two to three and six to eight times greater, respectively, than when measured 1 d after application (Boesten & van der Pas, 1983).

### 3-3.3.3 Thermodynamics

Thermodynamic parameters that can be related to bond energies may be used to assess the bonding mechanisms of organic chemicals to the soil matrix. Since the adsorption reaction is dependent on intensive properties such as temperature and pressure, methods have been developed for calculating thermodynamic values from the dependence of Freundlich isotherm data on temperature (Mills & Biggar, 1969; Biggar & Cheung, 1973; Wauchope & Koskinen, 1983) including standard free energies ($\Delta G$), entropies ($\Delta S$), and heats of adsorption or enthalpies ($\Delta H$). There are three basic problems with the general acceptance of Eq. [2] for determining thermodynamic parameters. First, the concentration of the adsorbed pesticide must be defined. Second, this equation is only applicable to chemicals where a rapid, reversible equilibrium is established and the adsorption is because of electrostatic interactions or other relatively weak forces. Third, it must be assumed that no reactions with competing species occur. Equation [2] overlooks other processes occurring in the pesticide-soil-water system such as competition with solvent and other solute molecules, and effects of pH, water content, and ionic strength.

The thermodynamic parameters can indicate the direction of the adsorption reaction, the strength of the bonds formed during adsorption of a chemical onto a solid matrix, and the degree of interaction of solute molecules with solvent and other solute molecules. A thermodynamic analysis has shown that the soil surface reactions with triazine, substituted urea, and uracil herbicides are similar and that there is little evidence of a specific bonding mechanism for any of them (Wauchope & Koskinen, 1983). In a related study, naphthalene molecules were determined to be "freer" on the soil surface than on the crystal naphthalene surface and apparently they must displace water molecules before they adsorb on the soil surface (Wauchope et al., 1983).

The various methods used to calculate the thermodynamic parameters all use the familiar relations:

$$\Delta G = -RT \ln K = \Delta H - T\Delta S$$

$$\ln K = \Delta S/R - \Delta H/RT$$

that allow the free energies of adsorption ($\Delta G$) to be broken down into their component enthalpy ($\Delta H$) and entropy ($\Delta S$) terms when the temperature ($T$) dependence of the equilibrium constant $K$ is measured. $R$ is the gas constant. The methods differ, however, in their approach to calculation of the constant $K$ for the equilibrium of chemical between solution and solid matrix. There is no standard method for defining and measuring the activity coefficient of the solute in the adsorbed phase.

The equilibrium constant $K$ in the calculation of $\Delta G$ is most commonly defined as $K = a_s/a_e = (\gamma C_s)/(\gamma C_e)$ (Biggar & Cheung, 1973), although the expression $K = a_e/a_o$ has also been used, where $a_s$ is the adsorbed solute activity, $a_e$ is the equilibrium solution activity, $\gamma$ is the activity coefficient, $C_s$ is solute concentration in the adsorbed phase, $C_e$ is the equilibrium solution concentration, and $a_o$ is the initial solution activity. Since the activity coefficient, $\gamma$, approaches one as the solute concentrations approaches zero, the equilibrium constant can be obtained by plotting ln $(C_s/C_e)$ vs. $C_s$ and extrapolating to $C_s = 0$. The units of $C_e$ used have ranged from microgram per milliliter to micromole per liter to mole fraction. The term $C_s$ has been defined in a variety of ways; either mass or moles of solute per mass soil or organic matter on the soil, or per layer of solvent in contact with the soil. Depending on the units of $C_s$, $\Delta G$ can vary by an order of magnitude.

One method to overcome the problem of adsorbed solute concentration is the use of microcalorimetry to provide a direct measure of the heat of adsorption. Microcalorimetric methods relate heat changes in a defined system to the intensity of a reaction such as adsorption. Enthalpies of adsorption are obtained from the energy associated with the reaction (i.e., the heat change resulting from addition of a reactant) in terms of kilojoule per mole. These enthalpies are then related to bond formation and can indicate bond strength with stronger bonds leading to higher enthalpies (Fig. 3–4). A few workers have used calorimetry to obtain enthalpies of adsorption for organic chemicals onto soil components (Hayes et al., 1972; Harter & Kill-

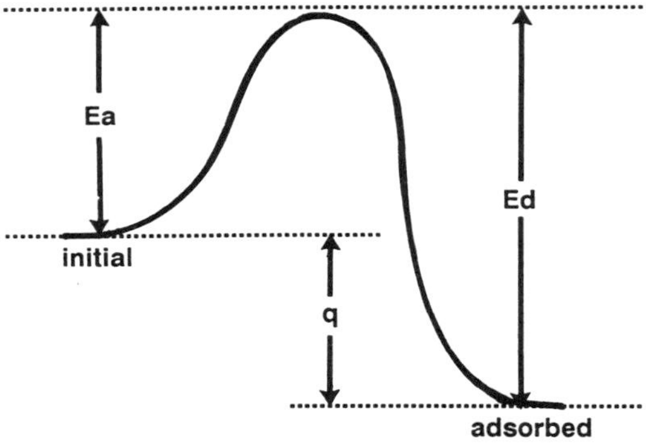


q = heat of adsorption
ΔEa = activation energy of adsorption
ΔEd = activation energy of desorption

Fig. 3-4. Relationship between activation energy of adsorption and heat of adsorption.

cullen, 1976; Burchill & Hayes, 1980; Harper, 1984), and found them comparable to values obtained by other methods. Usually a range of adsorption reactions are being measured because of the heterogeneity of soil and soil components. This means that any thermodynamic parameters obtained are actually average values for all sorption mechanisms occurring. Despite its infrequent use, microcalorimetry appears to be a promising technique to obtain enthalpy values that can be related to binding mechanisms.

Thermodynamic data and analysis have been used in an attempt to differentiate between partition and adsorption as the retention mechanism of certain organic chemicals and soil organic matter (Chiou et al., 1979). However, it has been argued that the thermodynamic analysis used to prove the partition mechnaism is too simplified and may lead to erroneous conclusions (Mingelgrin & Gerstl, 1983).

Calculation of thermodynamic parameters of adsorption has been used to assess bonding mechanisms of organic chemicals to soil. While thermodynamics can provide useful information about bonding, the data must be interpreted with caution. Assumptions used in the calculation and interpretation of the parameters may not be valid.

### 3-3.3.4 Kinetics

Adsorption in the soil is generally controlled by the rate of molecular diffusion into soil aggregates and the rate of reaction (rate of adsorption) at the soil-water interface. Diffusion has been found to be the rate-limiting step (van Genuchten & Wierenga, 1976; Leenheer & Alrichs, 1971; Khan, 1973; Wauchope & Myers, 1985) with solute movement from mobile pore water to the adsorbent surface surrounded by immobile pore water limiting the initial rate of adsorption and solute diffusion within a soil particle dominating the rate of adsorption as adsorption slows (van Genuchten &

Wierenga, 1976; Leenheer & Ahlrichs, 1971). The actual retention reactions tend to be relatively rapid, particularly the exchange-type reactions; however, Rao et al. (1979) proposed that two types of sorption sites may be involved that are controlled by the kinetics of the sorption process. Wauchope and Myers (1985) reported that adsorption and desorption of atrazine and linuron on sediments reached 75% of the equilibrium value within 3 to 6 min.

A pesticide may be retained initially by rapid low-energy bonding mechanisms and subsequently converted to more stable high-energy bonding mechanisms over time. This is particularly true of some of the pesticide metabolites such as phenols, catechols, and anilines that are believed to form covalent bonds with soil constituents over time (Martin et al., 1979; Cheng et al., 1983; Stott et al., 1983).

## 3-4 METHODS FOR DETERMINING RETENTION MECHANISMS

A number of mechanisms have been postulated to be involved in the retention of pesticides and related organic chemicals in soil. For most organic chemicals in soils, it is difficult to isolate a definitive retention mechanism. Most retention mechanisms are an interaction of a variety of forces and factors. Experimentally, few methods are available to characterize and differentiate the various mechanisms or the extent of each mechanism involved in the retention of organic residues in soil. Methods that have been used to determine mechanisms include infrared and x-ray spectroscopy, use of model sorbents including cation and anion resins, measurements of heats of adsorption, characterizing pH effects, and use of a variety of extraction techniques.

### 3-4.1 Spectroscopy

Spectroscopy, primarily x-ray and infrared, has been used with some success to determine retention mechanisms of various organic chemicals on selected well-defined sorbents. X-ray diffraction studies have been limited to studies of mechanisms of clay-organic chemical interactions. This method has shown that organic chemicals can be adsorbed in the interlayers of montmorillonite (section 3-3.1.9). X-ray analysis does not, however, permit ready analysis of adsorption mechanisms of organic chemicals on whole soils, soil organic matter, or other amorphous materials in soil.

Infrared spectroscopy has been limited primarily to the study of adsorption mechanisms of organic chemicals on clay surfaces. Based on infrared spectroscopy, cation-dipole interactions have been determined to be the binding mechanism between dehydrated montmorillonite saturated with different metal cations and both EPTC (Mortland & Meggit, 1966) and malathion (Bowman et al., 1970). Infrared studies have also provided evidence for bonding mechanisms involving coordination complexes (Russell et al., 1968) and water-bridging (Bowman et al., 1970) between pesticides and clays.

Infrared spectroscopy has been used in studies of soil- and organic matter-pesticide interactions with varying degrees of success. Mechanistic characterization involves interpretation of small shifts of absorption bands of different functional groups of the soil organic matter because of the bonding of the pesticide (or vice versa) (i.e., Sensei & Testini, 1980, 1982). Soil organic matter, however, has a complex spectra and any small shifts, if not masked, would be subject to questionable interpretation. The application of Fourier transform infrared spectroscopy is promising for future applications in sorption mechanism studies, however, the requirement of high pesticide concentrations is currently limiting its usefulness.

### 3–4.2 Model Adsorbents

Soil is a complex, heterogenous solid on which adsorption of organic chemicals takes place. In many cases, it is difficult to determine the mechanism and site of bonding of a chemical to soil. By studying the adsorption of a chemical on well-defined adsorbents, information on the types of bonding mechanisms possible for particular chemicals on soil can be determined. For instance, reversed-phase HPLC has been used to estimate and study hydrophobic bonding of aromatics (Opperhuizen, 1987). To measure the possible bonding mechanism of chlorsulfuron, Shea (1986) conducted adsorption studies on a variety of soils, montmorillonite, illite, kaolinite, $H^+$-saturated organic water from a peaty muck, activated charcoal, cellulose, strong- and weak-anion exchange resins, and nonionic polymeric adsorbent (XAD-2 resin), and acidic and basic aluminum oxide (AlO). Using the more well-defined adsorbents, chlorsulfuron was found to bond to anion exchange sites and to form weak physical bonds to organic substances. There is a problem of relating results such as these to real soil because of the interactions that take place between soil components, particularly organic matter and organic matter-clay complexes. These methods, however, are useful for studying a specific sorption mechanism without interference from other reactions.

Modification of adsorbents has also been used to study adsorption mechanisms. Cation saturation of organic matter and clays has been common. Selective removal of organic components from soil followed by adsorption studies has been done with varying degrees of success (Shin et al., 1970; Dunigan & McIntosh, 1971).

### 3–4.3 pH Effects

Effects of pH changes on adsorption of organic chemicals are commonly used to determine bonding mechanisms of acidic and basic organic chemicals. Weakly basic chemicals such as triazines and pyridinones exist as either the protonated cation or the neutral species depending on the pH of the system (section 3–3.2.1). Weakly acidic organic chemicals exist as either the undissociated molecule or the anion. Decreasing the pH of the system can increase the adsorption because of greater adsorption of the conjugate acid vs. the conjugate base.

Difficulties in interpretation of pH effects can result if the change in acidity affects the solution composition or the adsorbent. Rapid acidification of the solution may increase the concentration of soluble salts, and thereby, the ionic strength. Increases in ionic strength may affect adsorption of some herbicides through effects on solubility. Triazine solubilities are reduced at ionic strengths $>0.1\ \mu$ (Gaynor & Van Volk, 1981). Increasing the solution pH with NaOH can solubilize soil organic matter. Adsorption of pesticide onto the soluble soil organic material would decrease the amount adsorbed onto soil.

Decreasing the pH can affect the adsorbent. Low pH can cause hydrolysis of clays bringing $Al^{3+}$ and $Fe^{3+}$ ions to the surface subsequently forming hydroxides. These highly sorptive hydroxides may frequently be responsible for observed increases in adsorption at low pH. At low pH, soils dominated by variable charge clays can affect the adsorption of weak acid herbicides. Depending on the pH and ionic environment of the soil, these soils can have anion exchange capacities because of a net positive charge. Also, the protons at the soil surface can cause conformational and charge modifications of humic materials.

### 3–4.4 Solvent Extraction Techniques

The adsorption-desorption process is primarily characterized using the batch equilibration method. This method is limited for elucidating retention mechanisms. For instance, the energy resulting from shaking soil in an aqueous suspension in a desorption process is sufficient to only break weak bonds. Compounds that are adsorbed to soil by stronger forces would not be readily desorbable. On the other hand, extraction of adsorbed organic molecules by various solvents can provide information on bonding mechanisms.

Past approaches in devising solvent extraction procedures were based mostly on the affinity of the organic molecule to the solvent, such as preferential solubility or similarity in polarity. Little attention has been given to the need for breaking the bonds between the molecule and the surface before the chemical can be released from soil into the extraction solution. Strategies should either be to weaken the bonds holding the molecules to the surface or to select solvents that can effectively compete for sorption sites (Cheng, 1990).

Numerous solvents have been used for extraction of pesticides from soil. Aqueous solutions of acetone, methanol, or acetonitrile have been commonly used. Salt solution (i.e., 2 *M* KCl), acids (HCl), bases (NaOH), and metal complexing organics such as citrate and EDTA have also been used depending on the chemical bound to soil. Cheng (1971) found a decrease in the extractability of picloram by 2 *M* KCl with decreasing soil pH (Fig. 3–5). Adjusting the soil pH to 7 permitted quantitative recovery of picloram and picloram was equally extractable from freshly treated soil as from aged soil. The dependence on soil pH and reversibility of the adsorption process indicate the predominant mechanism is physical adsorption of the undissociated picloram molecules.

Studies have shown that catechols were gradually stabilized in soil by sorptive or binding mechanisms (Martin et al., 1979). Due to the nature of

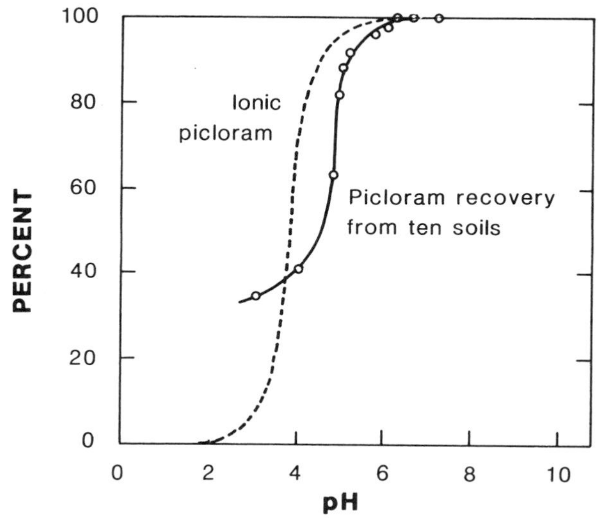


Fig. 3-5. Effect of soil pH on recovery of picloram from 10 soils using 2*M* KCl (Cheng, 1971).

the bonding, a number of individual organic solvents have been shown not to extract catechols from soil. Cheng et al. (1983) found that a mixed extractant of citric acid-ascorbic acid-acetone was most effective in extracting catechols from variously treated soil samples. The effectiveness of this mixture as an extractant could be associated with its power to complex metals (citric acid), to maintain the soil at a reduced state (ascorbic acid), and to reduce H bonding between chemicals in solution and the soil surface (acetone/water) thus maintaining the catechol in soil solutions and readily extractable.

Sequential extraction of chemicals adsorbed to soil using different solvents has been used to determine bonding mechanisms. For instance, Graveel et al. (1985) showed that benzidine, α-naphthylamine, and *para*-toluidine first bind to soil by electrostatic interactions, hydrophobic bonding, or reversible imine linkages between the amines and soil colloids (extractable with ethyl acetate/methanol and ammonium acetate). Subsequently, the amines form covalent bonds with the soil organic matter through 1,4-nucleophilic addition (extractable with NaOH).

## 3-5 CONCLUSION

Retention processes directly or indirectly control all the other processes in soil that affect organic chemicals. Retention processes dictate the amount of chemical that can move in the aqueous or gaseous phases, degrade, and be taken up by plants. To understand how the retention processes interact with other processes, the retention mechanisms must be known.

It is difficult to assess the relative importance of the various retention mechanisms for organic chemicals in soil. The heterogenous nature of the soil and the wide variety of organic chemicals preclude systematic determination of retention mechanisms. For example, adsorption processes probably involve a range of mechanisms with one mechanism predominating under a given set of conditions. To further confound the situation, the retention mechanism may change with time resulting ultimately in irreversible bonding of chemicals to soil surfaces and totally removing them from transport processes. Thus, any simplistic representation of adsorption such as depicting the processes as a partitioning between the aqueous phase and organic matter may not be indicative of the processes involved in the bonding of most organic chemicals to soil. A better understanding of retention mechanisms of organic chemicals in soil will help in developing a quantitative description of retention that can be used to model the fate of organic chemicals in soil and the environment.

## REFERENCES

Bailey, G.W., and J.L. White. 1964. Review of adsorption and desorption of organic pesticides by soil colloids, with implications concerning pesticide bioactivity. J. Agric. Food Chem. 12:324–332.

Bailey, G.W., and J.L. White. 1970. Factors influencing the adsorption, desorption, and movement of pesticides in soils. Residue Rev. 32:29–92.

Bailey, G.W., J.L. White, and T. Rothberg. 1968. Adsorption of organic herbicides by montmorillonite: Role of pH and chemical character of adsorbate. Soil Sci. Soc. Am. Proc. 32:222–234.

Biggar, J.W., and M.W. Cheung. 1973. Adsorption of picloram (4-amino-3,5,6-trichloropicolinic acid) on Panoche, Ephrata, and Palouse soils: A thermodynamic approach to the adsorption mechanism. Soil Sci. Soc. Am. Proc. 37:863–868.

Boesten, J.J.T.I., and L.J.T. van der Pas. 1983. Test of some aspects of a model for adsorption/desorption of herbicides in field soil. Aspects Appl. Biol. 4:495–501.

Bollag, J.-M., S.-Y. Lin, and R.D. Minard. 1980. Cross-coupling of phenolic humus constituents and 2,4-dichlorophenol. Soil Sci. Soc. Am. J. 44:52–56.

Bouchard, D.C., and T.L. Lavy. 1985. Hexazinone adsorption-desorption studies with soil and organic adsorbents. J. Environ. Qual. 14:181–186.

Bowman, B.T., R.S. Adams, Jr., and S.W. Fenton. 1970. Effect of water upon malathion adsorption onto five montmorillonite systems. J. Agric. Food Chem. 18:723–727.

Bowman, B.T., and W.W. Sans. 1985. Partitioning behavior of insecticides in soil-water systems: II. Desorption hysteresis effects. J. Environ. Qual. 14:270–273.

Boyd, S.A. 1982. Adsorption of substituted phenols by soils. Soil Sci. 134:337–343.

Briggs, G.G. 1969. Molecular structure of herbicides and their sorption by soils. Nature (London) 223:1288.

Burchill, S.A., and M.H.B. Hayes. 1980. Adsorption of poly(vinyl alcohol) by clay minerals. p. 109–121. *In* A. Banin and U. Kafkafi (ed.) Agrichemicals in soils. Pergamon Press, New York.

Burchill, S., M.H.B. Hayes, and D.J. Greenland. 1981. Adsorption. p. 221–400. *In* D.J. Greenland and M.H.B. Hayes (ed.) The chemistry of soil processes. John Wiley and Sons, Chichester, England.

Burns, J.G., M.H.B. Hayes, and M. Stacey. 1973. Some physico-chemical interactions of paraquat with soil organic materials and model compounds. II. Adsorption and desorption equilibria in aqueous suspensions. Weed Res. 13:67–78.

Calvet, R. 1980. Adsorption-desorption phenomena. p. 1–30. *In* R.J. Hance (ed.) Interactions between herbicides and the soil. Academic Press, London.

Capriel, P., and A. Haisch. 1983. Aufnahme von atrazin metaboliten durch hafer neun jahre nach der herbizid-applikation. Z. Pflanzenernaehr. Bodenk. 146:736–740.

Carringer, R.D., J.B. Weber, and T.J. Monoco. 1975. Adsorption-desorption of selected pesticides by organic matter and montmorillonite. J. Agric. Food Chem. 23:568–572.

Carter, C.W., and I.H. Suffet. 1982. Binding of DDT to dissolved humic materials. Environ. Sci. Technol. 16:735-740.

Cheng, H.H. 1971. Picloram in soil: Extraction and mechanism of adsorption. Bull. Environ. Contam. Toxicol. 6:28-33.

Cheng, H.H., K. Haider, and S.S. Harper. 1983. Catechol and chlorocatechols in soil: Degradation and extractability. Soil Biol. Biochem. 14:311-317.

Cheng, H.H. 1990. Organic residues in soils: Mechanisms of retention and extractability. Int. J. Environ. Anal. Chem. 39:165-171.

Chiou, C.T., L.J. Peters, and V.H. Freed. 1979. A physical concept of soil-water equilibria for nonionic organic compounds. Science 206:831-832.

Corwin, D.L., and W.J. Farmer. 1984. Nonsingle-valued adsorption-desorption of bromacil and diquat by fresh water sediments. Environ. Sci. Technol 18:507-415.

Decock, P., B. DuBois, J. Lerivrey, C. Gessa, J. Urbanska, and H. Kozlowski. 1985. Cu(II) binding by substituted 1,3,5-triazine herbicides. Inorg. Chim. Acta 107:63-66.

Di Toro, D.M., and L.M. Horzempa. 1982. Reversible and resistant components of PCB adsorption-desorption: Isotherms. Environ. Sci. Technol. 16:594-602.

Dixon, J.B., and S.B. Weed (ed.). 1977. Minerals in soil environments. SSSA, Madison, WI.

Dunigan, E.P., and T.H. McIntosh. 1971. Atrazine-soil organic matter interactions. Weed Sci. 19:279-282.

Farmer, V.C., and M.M. Mortland. 1966. An infrared study of the coordination of pyridine and water to exchangeable cations in montmorillonite and saponite. J. Chem. Soc. A:344-351.

Fuhremann, W., and E.P. Lichtenstein. 1978. Release of soil-bound methyl-$^{14}$C-parathion residues and their uptake by earthworms and oat plants. J. Agric. Food Chem. 26:605-610.

Gaillardon, P., R. Calvet, and J.C. Gaudry. 1980. Adsorption de quelques phenylurees herbicides par des acides humiques. Weed Res. 20:201-204.

Gaynor, J.D., and V. Van Volk. 1981. *s*-Triazine solubility in chloride salt solutions. J. Agric. Food Chem. 29:1143-1146.

Gerstl, Z., and C.S. Helling. 1987. Evaluation of molecular connectivity as a predictive method for the adsorption of pesticides by soils. J. Environ. Sci. Health B22:55-69.

Graveel, J.G., L.E. Sommers, and D.W. Nelson. 1985. Sites of benzidine, α-naphthylamine and *p*-toluidine retention in soils. Environ. Toxicol. Chem. 4:607-613.

Green, R.E. 1974. Pesticide-clay-water interactions. p. 3-37. *In* W.D. Guenzi (ed.) Pesticides in soil and water. SSSA, Madison, WI.

Grover, R. 1975. Adsorption and desorption of urea herbicides on soils. Can. J. Soil Sci. 55:127-135.

Gschwend, P.M., and S.-C. Wu. 1985. On the constancy of sediment-water partition coefficients of hydrophobic organic pollutants. Environ. Sci. Technol. 19:90-96.

Hamaker, J.W., and J.M. Thompson. 1972. Adsorption. p. 49-143. *In* C.A.I. Goring and J.W. Hamaker (ed.) Organic chemicals in the soil environment. Marcel Dekker, New York.

Hance, R.J. 1965. The adsorption of urea and some of its derivatives by a variety of soils. Weed Res. 5:98-107.

Hance, R.J. 1969. The adsorption of linuron, atrazine, and EPTC by model aliphatic adsorbents and soil organic preparations. Weed Res. 9:108-113.

Hance, R.J. 1971. Complex formation as an adsorption mechanism for linuron and atrazine. Weed Res. 11:106-110.

Harper, S.S. 1984. Evaluation of the mechanism of sorption of catechol and its chlorinated derivatives on a Palouse silt loam soil. Ph.D. diss. Washington State Univ., Pullman. (Diss. abstr. DEU85-10852).

Harris, C.I., and G.F. Warren. 1964. Adsorption and desorption of herbicides by soil. Weeds 12:120-126.

Harter, R.D., and B.M. Killcullen. 1976. Microcalorimeter adaption for measurement of heats of adsorption at solid-solution interfaces. Soil Sci. Soc. Am. J. 40:612-614.

Hatcher, P.G., M. Schnitzer, L.W. Dennis, and G.E. Maciel. 1981. Aromaticity of humic substances in soils. Soil Sci. Soc. Am. J. 45:1089-1094.

Hayes, M.H.B. 1970. Adsorption of triazine herbicides on soil organic matter, including a short review on soil organic matter chemistry. Residue Rev. 31:131-174.

Hayes, M.H.B., M.E. Pick, and B.A. Toms. 1972. Application of microcalorimetry to the study of interactions between organic chemicals and soil constituents. Sci. Tools 19:9-12.

Helling, C.S., and A.E. Krivonak. 1978. Biological characteristics of bound dinitroaniline herbicides in soils. J. Agric. Food Chem. 26:1164–1172.

Hsu, T.-S., and R. Bartha. 1974. Interactions of pesticide-derived chloroaniline residues with soil organic matter. Soil Sci. 116:444–452.

Isaacson, P.J. 1985. Sorption of phenol vapors and influence of ring substitution. Soil Sci. 140:189–193.

Kazano, H., P.C. Kearney, and D.D. Kaufman. 1972. Metabolism of methylcarbamate insecticides in soils. J. Agric. Food Chem. 20:975–979.

Khan, S.U. 1973. Equilibrium and kinetic studies of the adsorption of 2,4-D and picloram on humic acid. Can. J. Soil Sci. 53:429–434.

Khan, S.U. 1974. Adsorption of bipyridylium herbicides by humic acid. J. Environ. Qual. 3:202–206.

Khan, S.U. 1982. Distribution and characteristics of bound residues of prometryn in an organic soil. J. Agric. Food Chem. 30:175–179.

Kloskowski, R., and F. Fuhr. 1985. Behavior of [ring-$^{14}$C] hydroxy-simazine in a parabraun soil. Chemosphere 14:1913–1920.

Knight, B.A.G., and T.E. Tomlinson. 1967. The interaction of paraquat (1:1′-dimethyl 4:4′-dipyridylium dichloride) with mineral soils. J. Soil Sci. 18:233–243.

Koskinen, W.C., and H.H. Cheng. 1983. Effects of experimental variables on 2,4,5-T adsorption-desorption in soil. J. Environ. Qual. 12:325–330.

Koskinen, W.C., G.A. O'Connor, and H.H. Cheng. 1979. Characterization of hysteresis in the desorption of 2,4,5-T from soils. Soil Sci. Soc. Am. J. 43:871–874.

Lambert, S.M. 1967. Functional relationship between sorption in soil and chemical structure. J. Agric. Food Chem. 15:572–576.

Leenheer, J.A., and J.L. Ahlrichs. 1971. A kinetic and equilibrium study of the adsorption of carbaryl and parathion upon soil organic matter surfaces. Soil Sci. Soc. Am. Proc. 35:700–704.

Li, G.C., and G.R. Felbeck, Jr. 1972. The mechanism of adsorption of atrazine by humic acid from a muck soil. Soil Sci. 113:140–148.

Lineham, D.J. 1977. A comparison of the polycarboxylic acids extracted by water from an agricultural topsoil with those extracted by alkali. J. Soil Sci. 28:369–378.

Marshall, C.C. 1964. The physical chemistry and mineralogy of soils. John Wiley and Sons, New York.

Martin, J.P., K. Haider, and L.F. Linhares. 1979. Decomposition and stabilization of ring-$^{14}$C-labeled catechol in soil. Soil Sci. Soc. Am. J. 43:100–104.

Mills, A.C., and J.W. Biggar. 1969. Adsorption of 1,2,3,4,5,6-hexachlorocyclohexane from solution: The differential heat of adsorption applied to adsorption from dilute solutions on organic and inorganic surfaces. J. Colloid Interface Sci. 29:720–731.

Mingelgrin, U., and Z. Gerstl. 1983. Reevaluation of partition as a mechanism of nonionic chemicals adsorption in soils. J. Environ. Qual. 12:1–11.

Moore, J.W., and R.G. Pearson. 1981. Complex reactions. p. 296–300. *In* Kinetics and mechanism. 3rd ed. John Wiley and Sons, New York.

Mortland, M.M. 1970. Clay-organic complexes and interactions. Adv. Agron. 22:75–117.

Mortland, M.M. 1986. Mechanisms of adsorption of nonhumic organic species by clays. p. 59–76. *In* P.M. Huang and M. Schnitzer (ed.) Interactions of soil minerals with natural organics and microbes. SSSA Spec. Publ. 17. SSSA, Madison, WI.

Mortland, M.M., and W.F. Meggit. 1966. Interaction of ethyl-*N,N*-di-*n*-propyl-thiocarbamate (EPTC) with montmorillonite. J. Agric. Food Chem. 14:126–129.

Opperhuizen, A. 1987. Relationships between octan-1-ol/water partition coefficients, aqueous activity coefficients and reversed phase HPLC capacity factors of alkylbenzenes, chlorobenzenes, chloronaphthalenes and chlorobiphenyls. Toxicol. Environ. Chem. 15:249–264.

Parfitt, R.L., V.C. Farmer, and J.D. Russell. 1977a. Adsorption on hydrous oxides. I. Oxalate and benzoate on goethite. J. Soil Sci. 28:29–39.

Parfitt, R.L., A.R. Fraser, and V.C. Farmer. 1977b. Adsorption on hydrous oxides. III. Fulvic acid and humic acid in geothite, gibbsite and imogolite. J. Soil Sci. 28:289–296.

Parris, G.E. 1980. Environmental and metabolic transformations of primary aromatic amines and related compounds. Residue Rev. 76:1–30.

Pauling, L. 1960. The nature of the chemical bond. 3rd ed. Cornell Univ. Press, Ithaca, NY.

Rao, P.S.C., J.M. Davidson, R.E. Jessup, and H.M. Selim. 1979. Evaluation of conceptual models for describing nonequilibrium adsorption-desorption of pesticides during steady-flow in soils. Soil Sci. Soc. Am. J. 43:22–28.

Russell, J.D., M.I. Cruz, and J.L. White. 1968. The adsorption of 3-aminotriazole by montmorillonite. J. Agric. Food Chem. 16:21–24.

Sabljic, A. 1984. Predictions of the nature and strength of soil sorption of organic pollutants by molecular topology. J. Agric. Food Chem. 32:243–246.

Sabljic, A., and M. Protic. 1982. Relationship between molecular connectivity indices and soil sorption coefficients of polycyclic aromatic hydrocarbons. Bull. Environ. Contam. Toxicol. 28:162–165.

Sciacovelli, O., N. Senesi, V. Solinas, and C. Testini. 1977. Spectroscopic studies on soil organic fractions I. IR and NMR spectra. Soil Biol. Biochem. 9:287–293.

Senesi, N., and C. Testini. 1980. Adsorption of some nitrogenated herbicides by soil humic acids. Soil Sci. 130:314–320.

Senesi, N., and C. Testini. 1982. Physico-chemical investigations of interaction mechanisms between *s*-triazine herbicides and soil humic acids. Geoderma 28:129–146.

Senesi, N., and C. Testini. 1984. Theoretical aspects and experimental evidence of the capacity of humic substances to bind herbicides by charge-transfer mechanisms (electron donor-acceptor processes). Chemosphere 13:461–468.

Shea, P.J. 1986. Chlorsulfuron dissociation and adsorption on selected adsorbents and soils. Weed Sci. 34:474–478.

Shin, Y.-O., J.J. Chodan, and A.R. Wolcott. 1970. Adsorption of DDT by soils, soil fractions, and biological materials, J. Agric. Food Chem. 18:1129–1133.

Sposito, G. 1984. The surface chemistry of soils. Oxford Univ. Press, New York.

Steelink, C., and G. Tollin. 1967. Free radicals in soil. p. 147–169. *In* A.D. McLaren and G.H. Peterson (ed.) Soil biochemistry. Marcel Dekker, New York.

Stevenson, F.J. 1972. Role and function of humus in soil with emphasis on adsorption of herbicides and chelation of microorganisms. Bioscience 22:643–650.

Stevenson, F.J. 1982. Humus chemistry: Genesis, composition, reactions. Wiley-Interscience, New York.

Stott, D.E., J.P. Martin, D.D. Focht, and K. Haider. 1983. Biodegradation, stabilization in humus and incorporation into soil biomass of 2,4-D and chlorocatechol carbons. Soil Sci. Soc. Am. J. 47:66–70.

Streitwieser, A. Jr., and C.H. Heathcock. 1976. Introduction to organic chemistry. Macmillan Publ. Co., New York.

Stumm, W., R. Kummert, and L. Sigg. 1980. A ligand exchange model for the adsorption of inorganic and organic ligand at hydrous oxide interfaces. Croat. Chem. Acta 52:291–312.

Stumm, W., and J.J. Morgan. 1981. Aquatic chemistry. John Wiley and Sons, New York.

van Genuchten, M. Th., and P.J. Wierenga. 1976. Mass transfer studies in sorbing porous media. I. Analytical solution. Soil Sci. Soc. Am. Proc. 40:473–480.

Wauchope, R.D., and W.C. Koskinen. 1983. Adsorption-desorption equilibria of herbicides in soil: A thermodynamic perspective. Weed Sci. 31:504–512.

Wauchope, R.D., and R.S. Myers. 1985. Adsorption-desorption kinetics of atrazine and linuron in fresh water-sediment aqueous slurries. J. Environ. Qual. 14:132–136.

Wauchope, R.D., K.E. Savage, and W.C. Koskinen. 1983. Adsorption-desorption equilibria of herbicides in soil: Naphthalene as a model compound for entropy-enthalpy effects. Weed Sci. 31:744–751.

Weber, J.B. 1980. Adsorption of buthidazole, VEL 3510, tebuthiuron, and fluridone by organic matter, montmorillonite clay, exchange resins, and a sandy loam soil. Weed Sci. 28:478–483.

Weber, J.B., P.W. Perry, and R.P. Upchurch. 1965. The influence of temperature and time of the adsorption of paraquat, 2,4-D and prometone by clays, charcoal, and an ion-exchange resin. Soil Sci. Soc. Am. Proc. 29:678–688.

Weber, J.B., P.H. Shea, and S.B. Weed. 1986. Fluridone retention and release in soils. Soil Sci. Soc. Am. J. 50:582–588.

Weber, J.B., and S.B. Weed. 1968. Adsorption and desorption of diquat, paraquat, and prometone by montmorillonite and kaolinite clay minerals. Soil Sci. Soc. Am. Proc. 32:485–487.

Weed, S.B., and J.B. Weber. 1974. Pesticide-organic matter interactions. p. 39–66. *In* W.D. Guenzi (ed.) Pesticides in soil and water. SSSA, Madison, WI.

Wershaw, R.L., P.J. Burcan, and M.C. Goldberg. 1969. Interaction of pesticides with natural organic materials. Environ. Sci. Technol. 3:271–273.

Wilson, M.A., A.M. Vassallo, E.M. Perdue, and J.H. Reuter. 1987. Compositional and solid-waste nuclear magnetic resonance study of humic and fulvic acid fractions of soil organic matter. Anal. Chem. 59:551–558.

Yaron, B., A.R. Swoboda, and G.W. Thomas. 1967. Aldrin adsorption by soils and clays. J. Agric. Food Chem. 15:671–675.

Zachara, J.M., C.C. Ainsworth, L.J. Felice, and C.T. Resch. 1986. Quinoline sorption to subsurface materials: Role of pH and retention of the organic cation. Environ. Sci. Technol. 20:620–627.

# Chapter 4

# Sorption Estimates for Modeling

**R. E. GREEN,** *University of Hawaii, Honolulu, Hawaii*

**S. W. KARICKHOFF,** *U.S. Environmental Protection Agency, Athens, Georgia*

Sorption of pesticides by soil colloids is one of the principal processes determining the partitioning of a chemical between the solid, liquid, and gaseous phases. Mechanisms of sorption are discussed in chapter 3 in this book. The quest for knowledge of fundamental forces and mechanisms responsible for sorption in a variety of soil-pesticide systems is ongoing, but progress is slow. However, the need to account for the impact of sorption on the behavior of pesticides in soil-water systems has increased rapidly in recent years with the recognition that pesticide contamination of surface and ground waters is a common occurrence.

Assessment of the potential impact of pesticide use in agriculture on environmental quality requires quantification of sorption. The rapid development and acceptance of mathematical models for such assessments has intensified demand for quantitative measures of sorption. Thus, the deliberate efforts of researchers to provide a better understanding of sorption mechanisms and the urgent demand of environmental modelers for sorption distribution coefficients or functions may appear to be in conflict. In fact, both pursuits are complementary and should be encouraged, to provide reliable assessment methodologies required by federal and state regulatory agencies.

This chapter focuses on quantifying sorption from solution in the context of environmental modeling; sorption estimates can be based on measurements in the laboratory or on a variety of computation methods. The accuracy of sorption estimates varies widely with the method of estimation. In most environmental modeling endeavors there are numerous inputs required, each of which is estimated with some uncertainty. Characterization of the approximate uncertainty of each input parameter and the extent to which errors propagate through model calculations allows one to evaluate errors in sorption estimation in proper perspective. Thus, it may be acceptable in some cases to use sorption parameter estimates that are known to be

 *Pesticides in the Soil Environment*—SSSA Book Series, no. 2

accurate only within one order of magnitude. Improvements in understanding and description of transport mechanisms and associated modeling may warrant increasingly rigorous requirements for accuracy. Now, the best strategy is to (i) use the most appropriate methodology available, based on accuracy requirements and cost/benefit considerations and (ii) characterize the uncertainty in sorption estimates.

The approach in this chapter is to focus on estimates of sorption from solution at equilibrium as a first step in defining sorption quantitatively. Kinetics of sorption are discussed briefly. Only the simplest of equilibrium models are discussed, in keeping with the lack of rigor common to the estimation methods presented. Sorption measurement methods are discussed, but greater attention is given to indirect estimation procedures, particularly those which produce sorption distribution coefficients referenced to soil organic C content ($K_{oc}$) and to soil specific surface.

## 4-1 MECHANISMS IN RELATION TO SORPTION ESTIMATES

The complexity of the solid matrix of soils prevents clear definition of sorption mechanisms for most soil-pesticide combinations. Even so, enough is known about soil colloids and the physical-chemical properties of most registered pesticides to allow some intelligent speculation about predominant mechanisms. Koskinen and Harper (see chapter 3 in this book) identify various sorption forces that are thought to operate between functional groups of pesticides and constituents of soils. Identifying such interactions is easier in pesticide-clay-water systems (Green, 1974) than in soils, which contain both clays and organic matter. The dominant role of organic matter in most agricultural soils (with the exception of subsoils and low organic matter soils in arid regions) diminishes somewhat the practical value of our knowledge about sorption of pesticides on pure mineral components of soils. Organic matter is complex in chemical structure, varies substantially in space, and may change over time. Thus it is difficult, if not impossible, to sort out what mechanisms are responsible for the sorption of a given pesticide on organic matter-rich soils. Certain generalizations about pesticide sorption are possible. Cationic pesticides and weak bases, protonated at low pH, are highly sorbed on soil colloids with high negative charge, whether the colloid is organic or inorganic. Sorption of weak acids on the same colloids is generally greater at low pH, when the molecular form of the pesticide dominates. Sorption of most uncharged, low-solubility pesticides in surface soils is generally dominated by so-called *hydrophobic* interaction with soil organic matter. There is evidence that the sorption potential of mineral surfaces in natural surface soils is blocked by organic matter (Hance, 1969; Walker & Crawford, 1968; Karickhoff, 1984). The extent to which clay minerals contribute to sorption depends on both the ratio of clay mineral to organic C fractions ($f_{cm}/f_{oc}$) of the soil or sediment and on the nature of the organic sorbate. The type of soil clay becomes increasingly important when soil organic C contents are low. The high surface area and presence of exchangeable cations on smectite clays, for example, can be responsible for substantial sorption of phosphorothioate compounds by H bonding via a water bridge to the cations (Bowman, 1973). This mechanism would not contribute to sorption as much on kaolinite or iron oxide colloids.

The above ideas are reflected in the tentative procedure for estimating sorption coefficients presented by Green and Karickhoff (1990) and summarized later in this chapter. Better estimation methods will be possible when the complex interactions of soil clays and organic matter are better understood. In the meantime, we should attempt to use what is known about sorption mechanisms to select the most appropriate estimation method for a given soil-pesticide combination.

## 4-2 KINETIC CONSIDERATIONS

The complex nature of colloidal surfaces in soils, which defies exact identification of mechanisms, undoubtedly contributes to a suite of sorption rates for each soil, depending upon the specific interaction of the sorbate with each sorbent component. If sorption were truly *adsorption* (i.e., reactions on external surfaces) then sorption rates would likely be nearly instantaneous. But this is seldom the case in soils. Soil organic matter tends to dominate the sorption process for most pesticides in most soils, and sorption of nonionic pesticides on organic matter generally requires hours to days to achieve apparent equilibrium. Thus, it is doubtful that true equilibrium is reached in a few hours for sorbents high in organic C, especially if sorption is high. Karickhoff (1980) suggests that the sorption process has a rapid component and a slower component, the rate of the latter depending largely on movement of solute to less-accessible sorption sites. His analysis on one sediment indicates that the rate coefficient for the slow component varied inversely as the sorption partition coefficient.

For most practical applications, conventional batch equilibrations for 24 h probably produce sorption coefficients that represent macroscale equilibrium quite well, even though equilibrium may not be reached at internal sorbent sites. The question to be answered, then, is whether or not an equilibrium sorption coefficient measured in the laboratory is appropriate for use in pesticide transport models to represent partition of the pesticide between the aqueous and sorbed states. This question is difficult to answer in a general sense because field systems being simulated may or may not be near equilibrium. In near-static soil-water systems, such as a soil profile at field capacity without evapotranspiration, equilibrium is likely approached, and an equilibrium sorption coefficient may provide a good estimate of pesticide distribution in the soil. Laboratory results of Green and Obien (1969) for equilibration of atrazine in pots of soil at various water contents encourage such a conclusion. On the other hand, during infiltration and redistribution of water in soils, the equilibrium sorption coefficient appears to over-predict pesticide retardation in both laboratory columns and the field. Bilkert and Rao (1985) found that for two nematicides the equilibrium distribution coefficient, $K_d$, for the linear sorption equation had to be multiplied by a factor of about 0.4 to successfully predict the position of a pesticide peak in soil columns or in a field soil profile. Thus, assumption of equilibrium conditions for dynamic systems may result in calculating too much pesticide in the sorbed state during the sorption process and too little in the sorbed state during the desorption process. A realistic modeling description of sorption-desorption for newly applied pesticides in a dynamic flow system probably requires a kinetic formulation of the sorption process.

Wauchope and Myers (1985) successfully applied a kinetic model to the sorption of atrazine and linuron on a number of sediments during a 2-h period. The analysis was based on the assumption of sequential equilibria, first sorption from solution on labile sites, then sorption of the labile sorbate on restricted sites. The reaction kinetics were assumed to be first order. It is doubtful that this approach would be successful in describing long-term (weeks to months) sorption-desorption of persistent pesticides on soils with organic C contents exceeding 2% (for example), because of the variety of retention mechanisms that would be operative. A single reaction rate for restricted sites probably would not be adequate to describe long-term sorption-desorption. The hysteresis during desorption observed by some investigators is probably real for systems having extended equilibration periods (days to weeks), especially with soils or sediments having high organic C contents. Koshinen and Harper (see chapter 3 in this book) discussed possible reasons for apparent hysteresis that may not be real.

Progress in incorporating sorption kinetics into chemical-transport modeling has been slow. Hamaker and Thompson (1972) emphasized in their insightful discussion of pesticide leaching the need for additional research on sorption kinetics:

> As we attempt to look down the corridors of the future with respect to research in this area, we see two chief directions of opportunity and need: (a) achieving a better fundamental understanding of the nature of the soil adsorptive surface and (b) developing valid kinetics for adsorption by soil. These two areas, in particular, are needed for the application of sorption to the practical problems that attend the use of chemicals in soil.
>
> Upon these two advances in fundamental understanding, we can expect an increased ability to handle a number of applications of adsorption knowledge; and perhaps the most important of these is leaching. At the present time, the correlation between leaching and sorption is inadequate, largely because there is no distinction between short- and long-term adsorption. Therefore, we can predict only the leaching that would occur on freshly applied material, and it is not within our knowledge at present to deal with material that has been in the soil for some time, for which the leaching process is much less effective.

Recent work (Pignatello et al., 1987; Buxton & Green, 1987; Green et al., 1986) showing the resistance to desorption of long-term pesticide residues in soils, even for pesticides such as EDB and DBCP that are not highly sorbed, suggests that sorption modeling must be designed to accommodate sorption-desorption behavior in specific environmental systems. Recently applied pesticides may be modeled in one way, while residues will need to be modeled in another way. A kinetic approach may be most appropriate in some cases and equilibrium sorption may be adequate in others. The methods of measuring or estimating sorption parameters must be consistent with the intended modeling use. While some situations may be relatively easy to model, others are inherently difficult. Dynamic flow systems, such as surface soils subjected to periodic rainfall and irrigation, are difficult to model because of the range of water application fluxes and accompanying wide range of pore-water velocities over time. Even though a kinetic approach works well at low flow velocities, it may not be successful at high velocities (van Genuchten et al., 1974).

The focus of this chapter is on estimation of equilibrium sorption as a first approximation for modeling. The above discussion admits the limitations of equilibrium assumptions, thus equilibrium sorption estimates must be used with caution in modeling dynamic environmental systems.

## 4–3 MODELS OF SORPTION

While limiting ourselves to the equilibrium case simplifies sorption modeling considerably, further simplification is achieved by representing sorption isotherms by a power function known as the Freundlich equation,

$$S = K_f C^N \qquad [1]$$

in which $S$ is the sorbed concentration, expressed, for example, in units of micromoles of solute sorbed per kilogram of dry soil ($\mu$mol kg$^{-1}$), $C$ is the solution concentration ($\mu$mol L$^{-1}$), and $K_f$ and $N$ are empirical constants. The Freundlich equation has satisfactorily described experimental results of pesticide sorption in many cases. Alternative sorption equations, which are presented elsewhere (Boast, 1973; van Genuchten & Cleary, 1979; Voice & Weber, 1983), may describe sorption better in some instances, but lack the general utility and simplicity of the Freundlich expression. When $N = 1$ in Eq. [1], a linear equation results,

$$S = K_d C \qquad [2]$$

in which $K_d$ is the linear distribution coefficient (L$^3$ kg$^{-1}$). The value of $N$ in the Freundlich power function is usually $<1$, so that the resulting isotherm is curvilinear, with the sorbed quantity increasing rapidly as concentration is increased at low concentrations; sorption approaches a maximum slowly at high solution concentrations. This means that $K_d$ decreases as concentration increases. In some cases, however, $N$ exceeds one and the opposite is true. Compilations of the coefficients for many compounds (Hamaker & Thompson, 1972; Rao & Davidson, 1980) suggest that values of the coefficient $N$ in Eq. [1] commonly range between 0.75 and 0.95.

Rao and Davidson (1980) suggested that for many modeling purposes a linear isotherm can be used in lieu of the Freundlich isotherm with acceptable error. The error introduced by this approximation depends on the value

Table 4-1. Ratio of sorbed quantities calculated by the Freundlich and linear equations at different concentrations of pesticide in solution.†

| | Ratio: $\frac{S = K_f C^N}{S = K_d C}$ | | | | |
|---|---|---|---|---|---|
| | Solution concentration, C | | | | |
| $N$ | 0.01 | 0.1 | 1 | 10 | 100 |
| 1.0 | 1 | 1 | 1 | 1 | 1 |
| 0.9 | 1.6 | 1.3 | 1 | 0.79 | 0.63 |
| 0.8 | 2.5 | 1.6 | 1 | 0.63 | 0.40 |
| 0.7 | 4.0 | 2.0 | 1 | 0.50 | 0.25 |
| 0.6 | 6.3 | 2.5 | 1 | 0.40 | 0.16 |
| 0.5 | 10.0 | 3.2 | 1 | 0.32 | 0.10 |

† Summarized from Hamaker and Thompson (1972).

of $N$ and the solution concentration, as shown in Table 4-1 from calculations given by Hamaker and Thompson (1972). When $N = 1$ the two equations are equivalent, as shown by the ratio of 1 in the top row in the main body of the table. As the value of $N$ decreases, the linear approximation becomes less and less satisfactory, especially at high and low concentrations. Although $N \leq 1$ is the most common situation, the exceptions may be important in some cases.

The linear relationship of Eq. [2] is a desirable simplification for complex environmental models, as it allows simpler mathematical solutions. For many environmental contexts, e.g., runoff and stream flow, the linear approximation is probably acceptable considering errors associated with the estimation of water and sediment transport. Karickhoff (1981) cites evidence that for hydrophobic polycyclic aromatic compounds sorbed on natural sediments, the sorption isotherms are linear if the equilibrium concentration of solute in the aqueous phase is below $10^{-5}$ $M$ (1-3 mg/L) or less than one-half the solute solubility in water, whichever is lower. At the higher concentrations encountered in surface soils soon after application of a pesticide to the soil, the linear assumption may result in sorption estimates that are too high by two- to threefold or even more, depending on the extent of isotherm nonlinearity and the concentration of solute in solution.

If sorption isotherms are measured in the laboratory, the decision of whether to use Eq. [1] or [2] may be dictated by the model in which the sorption expression is to be used. If measured isotherms are not available, however, estimation procedures (section 4-5 and 4-6) may yield an acceptable estimate of $K_d$ in Eq. [2]. Such estimates are inherently approximate because the estimation procedures are not exact and because the assumption of isotherm linearity may be in error. If one requires accurate representation of equilibrium sorption in a modeling effort, it is desirable to actually measure a sorption isotherm for the situation of interest.

## 4-4 MEASURING SORPTION ONTO SOILS FROM SOLUTION

Pesticide sorption on soils or sediments can be easily measured in the laboratory if analytical methods are operational when and where they are required. Sorption measurement is considered an estimation procedure in this chapter because laboratory-derived sorption parameters constitute estimates of sorption in field soils. Such estimates, however, provide the basic information required for indirect estimation methods which will be discussed in subsequent sections.

Sorption methods can be divided into two principal categories: direct and indirect measurements. Both require equilibration of a soil sample with an aqueous pesticide solution of known initial concentration. Direct methods involve measurement of both solution concentration and sorbed concentration; after equilibration the sorbed pesticide is displaced from the soil by an appropriate solvent and measured directly. Indirect methods determine the quantity of sorbed pesticide by measuring the change in solution concentration resulting from sorption of pesticide from solution onto the soil; in this case, the quantity of sorbed solute is assumed to equal the quantity lost from solution. Green et al. (1980) discussed variations of these two approaches.

Most common is the indirect batch-suspension measurement, which consists of agitating soil and pesticide solution in a closed vessel for a sufficient time to achieve apparent equilibrium in the system. Experimental variables include the solution/soil ratio, temperature, antecedent condition of the soil, type of vessel, and nature of agitation. These variables have not all been standardized and can have a significant effect on the results (Koskinen & Cheng, 1983; Dao et al., 1982; Voice et al., 1983). McCall et al. (1981) presented a detailed discussion of the method.

One distinctive limitation of the batch-suspension method is poor precision when sorption is low (Green & Yamane, 1970). While sorption of most pesticides in surface soils is usually measurable with acceptable precision by the batch method, precision is often poor on subsoils or underlying materials. The current and future public and scientific interest in contamination of groundwater by pesticides will undoubtedly be accompanied by attempts to measure sorption of pesticides on low organic matter strata below the plow layer. Such determinations can be accomplished satisfactorily by direct measurement of the sorbed pesticide, such as in the flow-equilibration method (Green & Corey, 1971; Green et al., 1980). This method is more complicated and costly than the batch-suspension measurement, but precision is excellent even when sorption is extremely low.

A problem that plagues all sorption methods is the chemical or biological transformations of pesticides during the period of equilibration. Most modern pesticides are not highly stable in soil. The impact of transformation during sorption in a batch equilibration has been quantified experimentally for specific soil-pesticide combinations (Koskinen & Cheng, 1982) and simulated in a theoretical analysis based on first-order kinetics of transformation (Dekkers, 1978). A variety of procedures designed to attenuate microbial activity have been evaluated, but such treatments often have other adverse effects on the accuracy of sorption measurements (Dao et al., 1982). Use of initially air-dry soil, a brief equilibration period (e.g., $<2$ h) and low temperature have been suggested as possible ways to minimize transformation effects without resorting to autoclaving, radiation, or application of chemicals that are lethal to soil organisms (Dao et al., 1982). However, as mentioned previously, the antecedent soil conditions, equilibration period, and temperature are all experimental variables which themselves can affect sorption results. Thus, determination of what sorption method to use and what experimental protocol to follow requires careful consideration of the properties of the pesticide.

## 4–5 INDIRECT ESTIMATES OF SORPTION

### 4–5.1 Sorption Estimates Based on Soil Organic Carbon

When it is not possible to actually measure a sorption isotherm for a given pesticide-soil combination, a simplified estimation procedure must be used. The most attractive simplification is that which can use existing or easily developed data on pesticides and soils to provide sorption estimates that are sufficiently accurate for a given modeling objective. Currently, an approach which is widely accepted for management-oriented models is one which assumes that sorption of pesticides is related directly to the organic matter content of the soil, with the contribution of organic matter represented quantitatively by the organic C content of a soil (see, for example, Hamaker & Thompson, 1972; Chiou et al., 1979; Rao & Davidson, 1980; Kenaga & Goring, 1980; Karickhoff, 1981; McCall et al., 1981; Briggs, 1981a, b; Lyman, 1982; Chiou et al., 1983). An alternative approach is to express sorption on the basis of soil surface area rather than on the mass basis (Pionke & DeAngelis, 1980); however, the surface area calculation also weights sorption heavily toward organic C content. The organic C basis has gained the most acceptance because organic C content is commonly measured in soils, in contrast to surface area which is usually estimated from other soil properties. These two sorption estimation methods were discussed recently in a supporting document for the USDA-ARS SWAM Model (Green & Karickhoff, 1990); that discussion is presented here with only minor changes.

#### 4–5.1.1 Assumptions in $K_{oc}$ Approach

A number of simplifying assumptions are inherent in the organic C referenced method of representing sorption of pesticides ($K_{oc}$ approach). The errors in prediction of pesticide distribution that are caused by violation of the assumptions in actual field situations are difficult to assess. The impact and importance of such errors will depend on the objective of the modeling effort and the sensitivity of model output to variation in sorption parameters. The following discussion will, hopefully, aid the prospective modeler in understanding the advantages and limitations of the $K_{oc}$ approach.

The principal assumptions are (i) equilibrium in the sorption-desorption process, (ii) linearity of the sorption isotherm, (iii) singularity of the isotherm for sorption and desorption (sorption is reversible), (iv) sorption is limited to the organic component of the soil, and (v) soil organic C is invariant in its sorption capacity (i.e., organic C has the same sorption capacity per unit mass at different locations of a given soil or in different soils). In Eq. [2] (section 4–3), the sorbed concentration, $S$, represents pesticide sorption per unit mass of dry soil. $S$ can also be adjusted to represent the amount of pesticide sorbed per unit mass of organic C in the soil, i.e.,

$$S_{oc} = K_{oc}C \quad [3]$$

with $K_{oc}$ also having units of liter per kilogram when the mass of organic C is expressed in kilograms. If $f_{oc}$ = (mass organic C)/(mass dry soil or sediment), $K_{oc}$ is given by $K_{oc} = K_d f_{oc}$, and $S_{oc} = S/f_{oc}$. Thus Eq. [3] represents a linear sorption isotherm which is referenced to organic C content, and $K_{oc}$ is a distribution coefficient that describes the distribution of a pesticide between the aqueous and the soil organic phases. This approach is most appropriate for hydrophobic pesticides, that is non-ionic pesticides with a water solubility less than about $10^{-3}$ $M$, but may be practically suitable for many pesticides that are slightly polar and too water soluble to be considered hydrophobic.

The first three assumptions discussed above (equilibrium, linear, and reversible sorption) are inherent in the use of Eq. [2]. Adoption of Eq. [3] as a further simplification is attended by some additional assumptions, which if invalid, will introduce more error. It appears that the errors associated with the assumption of partitioning of pesticides on organic C are as large as those resulting from the first three assumptions. If the model user is to avoid unacceptably large errors which may lead to inappropriate conclusions from model calculations, an awareness of the limits of applicability of the partitioning assumption is necessary.

Clearly, none of these assumptions is valid in the strict sense, so that use of organic-C-referenced sorption will introduce errors, the magnitude of which depends on how severely the assumptions are violated in a given system. Opinions concerning the $K_{oc}$ approach for environmental assessments range from enthusiastic support (Chiou et al., 1979) to skepticism (Mingelgrin & Gerstl, 1983). The application of this concept has been proposed and also questioned repeatedly since its introduction relative to pesticides in the early 1960s. Acceptance of the approach has been encouraged by the increasing need for calculation procedures applicable to hundreds of organic chemicals in hazard assessments for a wide range of environmental conditions. The importance and urgency of developing a generalized procedure justifies some sacrifice in scientific rigor and accuracy. On the other hand, theoretical justification for the $K_{oc}$ approach has been provided by Karickhoff (1981, 1984), who identifies the organic-C/hydrophobic-chemical interaction as a first-order effect in contrast to the second-order effects of other factors of lesser importance. In practice, one must know when second-order effects are important. For example, Mingelgrin and Gerstl (1983) challenge the use of the partitioning concept (i.e., that hydrophobic organic chemicals are partitioned between water and the organic component of soils or sediments in a manner analogous to partitioning between two immiscible solvents); they contend that other mechanisms may play a dominant role, e.g., pesticide-clay-water interactions. These authors illustrate their point by showing the variability in $K_{om}$ (sorption coefficient referenced to organic matter rather than organic C) for 12 compounds sorbed on soils or sediments having a wide range of organic matter contents. The largest variation in $K_{om}$ was for parathion (factor of 50 between the largest and smallest value) and the smallest was for dieldrin (factor of 3). The variability that one might anticipate in $K_{oc}$ values for a given compound is indicated by the coefficient of variability calculated for 43 chemicals by Rao and Davidson (1980). Many important pesticides have polar functional groups that preclude strictly hydrophobic behavior, and several others are weak acids or bases, requiring that care be exercised in applying the $K_{oc}$ approach indiscriminantly to all pesticides. The scheme proposed in this chapter for determining a distribution coefficient for a given pesticide-soil system recognizes the need for alternative procedures for nonhydrophobic chemicals.

Karickhoff (1984) quantitatively addressed the role of clays in sorption of organic chemicals. His analysis of a few sets of experimental data suggests that swelling clays may have important effects on sorption of some hydrophobic organics in sediments having low organic C contents, but when the ratio of clay to organic C ($f_{cm}/f_{oc}$) is $<30$, contributions of mineral surfaces are masked regardless of the clay content. Estimates of organic C and mineral contributions to sorption are provided in Table 4–2 from Karickhoff (1984). The relative contributions of clay minerals and organic C to the sorption of three broad types of chemical compounds are indicated by the ratio of the clay mineral fraction ($f_{cm}$) to the organic C fraction ($f_{oc}$). The ratio $K_{oc}/K_{cm}$ indicates the relative preference of chemicals in each of the three groups for soil organic matter and soil clay minerals. The large ranges in $K_{oc}/K_{cm}$ and in the threshold for clay mineral contributions ($f_{cm}/f_{oc}$) for neutral organics with polar functional groups indicate the uncertainty that exists in such estimates for most pesticides. Thus, further scrutiny of the available data and continued research in this area will help to define applications in which the $K_{oc}$ approach is acceptable.

Table 4–2. Mineral and organic sorbent contributions to sorption as indicated by the ratio of organic sorbate partition coefficients for organic C and clay minerals ($K_{oc}/K_{cm}$) and the ratio of clay mineral fraction to organic C fraction ($f_{cm}/f_{oc}$) in soil or sediment.†

| | Compound type | | |
|---|---|---|---|
| Ratios | Neutral organics, polar functional groups | Nonpolar organics $<C_{10}$ | Large nonpolar organics $>C_{10}$ |
| $K_{oc}/K_{cm}$ | 10–50 | 50–100 | >100 |
| $f_{cm}/f_{oc}$‡ | 25–60 | >60 | Insignificant in sediments and surface soils |

† Adapted from Karickhoff (1984).
‡ "Threshold" ratio for mineral contribution to sorption; ratio of clay mineral fraction to organic C fraction in soil or sediment. If $f_{cm}/f_{oc}$ $<25$, then clay mineral contribution is insignificant for neutral organics.

#### 4–5.1.2 Sources of $K_{oc}$ Data

Once one has decided to use the $K_{oc}$ approach for a reasonably hydrophobic pesticide, the next step is to find the appropriate $K_{oc}$ value. Several tabulations of $K_{oc}$ values for a large number of pesticides are now available. A recent tabulation by Green and Karickhoff (1990) includes data summaries from Hamaker and Thompson (1972), Kenaga and Goring (1980), Rao and Davidson (1980), and Briggs (1981a). If a $K_{oc}$ value is not available for a given pesticide, it may be possible to calculate an approximate $K_{oc}$ from the octanol-water partition coefficient ($K_{ow}$), if one exists for the pesticide of interest. The $K_{ow}$-based prediction of $K_{oc}$ and other indirect methods of estimating $K_{oc}$ are presented in the next section. Additional error will usually be introduced by indirect approaches; these errors should be evaluated in the context of variability in $K_{oc}$ associated with variation of organic C content of soils in a watershed. More research is needed to quantify the relative magnitude of errors contributed by various sources of variation mentioned in this section.

#### 4–5.1.3 Calculation Methods for Obtaining $K_{oc}$

Estimating the sorption of uncharged organics, as discussed above, can often be reduced to estimating $K_{oc}$, which, when combined with organic C content, yields an estimate of the soil/water distribution coefficient, $K_d$.

The approach used in estimating $K_{oc}$ combines thermodynamic theory and empirical correlation to relate sorption parameters to widely measured properties of pesticides and soils. Theory is used to suggest the mathematical form relating the physical properties, and experimental data are used to empirically establish constants (by calibration) in the mathematical equation. This approach is far preferable to complete reliance on empirical correlation which precludes extrapolation out of the narrow range of substances or system properties upon which it is based.

$K_{oc}$ can be expressed as a ratio of solute activity coefficients in the aqueous phase, $\gamma^{w}$, and organic C phase, $\gamma^{oc}$, thus $K_{oc} = \gamma^{w}/\gamma^{oc}$. Activity coefficients quantify pesticide affinity for a given phase. In considering $K_{oc}$ for a series of uncharged pesticides, one would expect variations in $K_{oc}$ to be dominated by variations in aqueous phase affinity. This derives from the dissimilarity of the organic solute with the aqueous medium, with variations in $\gamma^{w}$ from solute to solute comparable to variations in aqueous solubilities. For uncharged solutes, one would expect much less variability in affinity for soil organic matter ($\gamma^{oc}$). Water outcompetes organic solutes for polar sites on soil surfaces, substantially reducing hydrophilic-bonding contributions for uncharged chemicals. Furthermore, lipophilic interactions with soil organic matter are not highly variable from solute to solute, and their differences in $K_{oc}$ are dominated by $\gamma^{w}$. Chiou et al. (1983) have addressed $\gamma^{oc}$ variability explicitly. The aqueous phase activity coefficient, $\gamma^{w}$, provides the thermodynamic link relating $K_{oc}$ to other solute properties, such as water solubility, molecular surface area, or octanol-water partition coefficient. Table 4–3 is a tabulation of semiempirical equations for estimating $K_{oc}$ from

Table 4-3. Semiempirical equations for estimating $K_{oc}$ from solute physical properties.

| | Calibration compounds† | $K_{oc}$ range | No. of compounds | $r^2$ | Reference |
|---|---|---|---|---|---|
| $\log K_{oc} = -0.68 \log S(\mu g/ml) + 4.273$ | pn, ha, aa, ch | $10$–$10^6$ | 23 | 0.93 | Hassett et al., 1980 |
| $\log K_{oc}$‡ $= -0.557 \log S(\mu mol/L) + 4.277$ | ch | $10$–$10^5$ | 15 | 0.99 | Chiou et al., 1979 |
| $\log K_{oc} = -0.83 \log S$(mole fraction)§ $- 0.01$ (mp $- 25\,°C$)¶ $- 0.93$ | pn, ch, ct, cb, op | $10^2$–$10^6$ | 47 | 0.93 | Karickhoff, 1984 |
| $\log K_{oc}$‡ $= -0.729 \log S$(molar) $- 0.0073$ (mp $- 25\,°C$) $+ 0.24$ | pn, ch | $10$–$10^5$ | 12 | 0.996 | Chiou et al., 1983 |
| $\log K_{oc}$‡ $= -0.808 \log [\bar{v}S$ (molar)]# $- 0.0081$ (mp $- 25\,°C$) $- 0.74$ | pn, ch | $10$–$10^5$ | 12 | 0.997 | Chiou et al., 1983 |
| $\log K_{oc}$‡ $= 0.904 \log K_{ow} - 0.539$ | pn, ch | $10$–$10^5$ | 12 | 0.989 | Chiou et al., 1983 |
| $\log K_{oc} = 1.029 \log K_{ow} - 0.18$ | ch, ct, cb, op, ur, pa | $10$–$10^5$ | 13 | 0.91 | Rao & Davidson, 1982 |
| $\log K_{oc} = \log K_{ow} - 0.317$ | pn, ha, aa, ch | $10$–$10^6$ | 23 | 0.98 | Hassett et al., 1980 |
| $\log K_{oc} = 0.72 \log K_{ow} + 0.49$ | cb, mb | $10^2$–$10^4$ | 13 | 0.95 | Schwarzenbach & Westall, 1981 |

† pn = polynuclear aromatic hydrocarbons; ha = heteronuclear aromatic hydrocarbons; aa = aromatic amines; ch = chlorinated hydrocarbons; ct = chloro-*s*-triazines; cb = carbamates; op = organophosphates; ur = uracils; pa = phenoxy acids; mb = methylated benzenes.

‡ Original data expressed as $K_{om}$; the factor 1.74 was used to convert $K_{om}$ to $K_{oc}$; $K_{oc} = 1.74\ K_{om}$.

§ For hydrophobic compounds, the solubility, $S$ (mole fraction) $\approx S$ (molar) $\times$ 18/1000.

¶ mp = melting point in degrees Celsius; reference temperature 25 °C; for liquids mp set at 25 °C, crystal term vanishes.

# $\bar{v}$ = pure solute molar volume (L/mol).

solute physical properties, such as solubility in water ($S$), melting point (mp), and octanol-water partition coefficient ($K_{ow}$). Briggs (1973) introduced the application of this approach to pesticide sorption on soils. Subsequently, numerous empirical equations have been developed for a variety of soil-pesticide combinations. These equations are generally applicable for uncharged pesticides of limited water solubility ($< 10^{-3}$ $M$) that are not susceptible to speciation changes or other complex formation in soil suspensions of interest. Soil organic matter dominance of sorption is presumed, so the ratio of clay/organic C must be $<40$; otherwise, mineral contributions should be addressed. The reader is cautioned that the generalized equations in Table 4-3 are expected to give only approximate estimates of $K_{oc}$. An equation developed for a broad range of chemical structures will likely give questionable results for pesticides that are not well represented by the chemicals used to develop the equation.

### 4-5.2 Sorption Estimates Based on Specific Surface of Soil

An alternative to the organic-C-referenced sorption approach may be desirable for some pesticide-soil combinations. For example, if the chemical is nonhydrophobic the relationship between $K_d$ and $f_{oc}$ may not be reliable. The $K_{oc}$ approach may also be inappropriate for soils or sediments having a clay/organic C ratio which is sufficiently high (perhaps $>40$) for the clay contribution to pesticide sorption to be a significant part of the total sorption.

Pionke and DeAngelis (1980) developed an estimation methodology which allows one to predict $K_d$ values for a large number of pesticides on the basis of the relationhsip of $K_d$ to either the organic C content or the specific surface of the soil or sediment. The specific surface (SS) was calculated from the equation:

$$SS = 100\ (\%\ OC) + 2.0\ (\%\ clay) + 0.4\ (\%\ silt) + 0.005\ (\%\ sand).$$

The coefficients for mineral fractions in the equation were taken from the work of Young and Onstad (1976), while the coefficient for percent organic C was determined from data Bailey and White (1964) presented, with the assumption that most clay in intact soils would exhibit the surface area of kaolinite (20 $m^2/g$). The Pionke-DeAngelis method is based on linear regressions between measured $K_d$ values for 35 pesticides and either percent organic C or SS; the SS calculation is heavily weighted toward organic C, i.e., percent organic C dominates except where percent clay is large and percent organic C is small. It was designed to give results similar to the $K_{oc}$ calculation (Pionke, 1984, personal communication). For a diverse group of pesticides the equation $K_d = mSS$ gave consistently higher coefficients of determination than did the equation $K_d = m\ (\%\ OC) + b$, in which m and b are fitted coefficients of linear regression. Thus, the authors recommended the surface area approach.

To extend the relationship between $K_d$ and soil surface area to other pesticides which were not included in the original data set, Pionke and

DeAngelis developed curves relating m values from the two equations to published $R_f$ values from soil-thin-layer chromatography (Helling, 1971). Since $R_f$ values were available for about 100 pesticides, the $R_f$-m relationship obtained for the original group of pesticides could then be used to predict m values for many other pesticides for which $R_f$ values had been measured. The procedure requires an assumption that the standard soil on which the $R_f$ values were measured (Hagerstown silty clay loam; fine, mixed, mesic Typic Hapludalfs) is reasonably representative of the soils for which the predictions of pesticide sorption are to be made. The authors present data of Helling (1971) that show that $R_f$ values for the Hagerstown soil were representative of the *average* mobility of 13 pesticides on a number of fine-textured soils. The Hagerstown data, however, did not adequately represent $R_f$ results for two coarse-textured soils. While the soil-thin-layer procedure is a good way to characterize relative mobilities of a large number of pesticides, it is risky to assume that the relationship between $R_f$ and m values can be applied to any soil of interest, even if the soil is fine textured. In the present context, we are interested in predicting a sorption coefficient for a specific pesticide on a specific soil, which may not be well represented by the *average* mobility of the pesticide on many soils. In spite of this principal limitation of the Pionke-DeAngelis method, however, the approach does provide an alternative for pesticide-soil combinations for which the $K_{oc}$ method appears inappropriate. The reader is encouraged to consult the paper by Pionke and DeAngelis for details.

## 4–6 DETERMINING A SORPTION COEFFICIENT FOR A GIVEN PESTICIDE-SOIL COMBINATION

In the event that a published value of a distribution coefficient is not available and it is not possible or practical to measure one for a specific pesticide-soil combination, then an estimation procedure can be followed. The following decision sequence (Green & Karickhoff, 1990) provides a systematic approach to obtain an estimate of a sorption coefficient.

Required information:

1. The nature of the chemical. Five useful groupings of chemicals are (i) hydrophobic (solubility in water $< 10^{-3}$ $M$ or about 300 mg/L); (ii) sufficiently polar so that solubility in water exceeds $10^{-3}$ $M$, but not cationic or a weak acid or weak base; (iii) a weak base; (iv) a weak acid; and (v) a cation.

2. Soil or sediment properties. The nature of the pesticide and the associated procedure used to obtain an appropriate sorption coefficient will determine what soil or sediment properties are required. In general, the order of priority for required data will be (i) organic C content, (ii) clay, silt, and sand contents, (iii) pH. Organic C content is required for both the $K_{oc}$ approach and the surface area approach. Percentages of the various textural separates are required for the surface area approach and will aid in estimating the mineral contribution to sorption. The pH is needed only if the pesticide of interest is an acid or base.

A scheme for obtaining a sorption coefficient for a given pesticide in a given soil or sediment is shown in Fig. 4-1. Each decision or action step is numbered for easy reference in the following discussion. While an attempt has been made to provide a definitive procedure that can be followed by anyone with a basic understanding of the chemistry of pesticides and soils, there are a number of points of uncertainty that will be encountered in using the scheme. For example, the definition of hydrophobic is somewhat arbitrary, but the criterion of $<10^{-3}M$ solubility in water is a useful, though perhaps conservative, upper boundary for hydrophobic compounds. Other criteria for decisions, such as in steps 2 and 11 are based on experimental evidence but may not be appropriate for all soil-pesticide combinations. Experience in using the scheme in Fig. 4–1 will hopefully lead to procedural improvements.

Some brief explanations or directions associated with several steps in Fig. 4–1 are given below to assist in the use of the scheme.

### 4–6.1 Steps 1 and 8 through 20

The use of the $K_{oc}$ approach is most appropriate for organic chemicals of limited solubility in water. Karickhoff (1984) has suggested a solubility of $<10^{-3}$ $M$, or about 100 to 300 mg/L, for most pesticides. $K_{oc}$ values are available in the literature for many chemicals with much higher solubilities in water, but the reliability of the $K_{oc}$ approach is more questionable for these pesticides than for hydrophobic chemicals.

There are unique problems associated with estimating sorption coefficients for weak acids and weak bases due to significant changes in sorption mechanism and quantity sorbed with changes in soil pH. Two generally accepted assumptions for such pesticide-soil systems provide guidelines for estimating $K_d$ values for weak acids and weak bases. One assumption is that the pH at colloid surfaces is lower than in a bulk soil-water suspension, such as in the 1:1 soil/water slurry often used for pH measurement. A reasonable estimate of surface acidity ($pH_s$) is two pH units lower than the bulk suspension pH, i.e., $pH_s = pH - 2.0$. The other assumption is that in aqueous solutions the ratio of molecular to ionic species is 1:1 when $pH = pK_a$. Most weakly acidic pesticides are in predominantly ionic (negatively charged) form at the pH of most agricultural soils, while most weakly basic pesticides are in molecular form. In either case, as soil pH decreases toward a value equal to the $pK_a$ of the pesticide, sorption tends to increase because the acidic pesticide has more of the molecular species at low pH and the basic pesticide has more of the protonated (positive) ionic species at low pH. The colloidal surfaces of most agricultural soils have a net negative charge and, thus, have an affinity for positively charged species of bases, but not much affinity for negatively charged species of acids. Pionke and DeAngelis (1980) adopted the following approximations: 50% of the basic pesticide is positively charged when the measured soil suspension pH minus 1.0 equals the $pK_a$, and 50% of the acidic pesticide is negatively charged when soil suspension pH equals the $pK_a$. These workers suggest that if measured pH minus

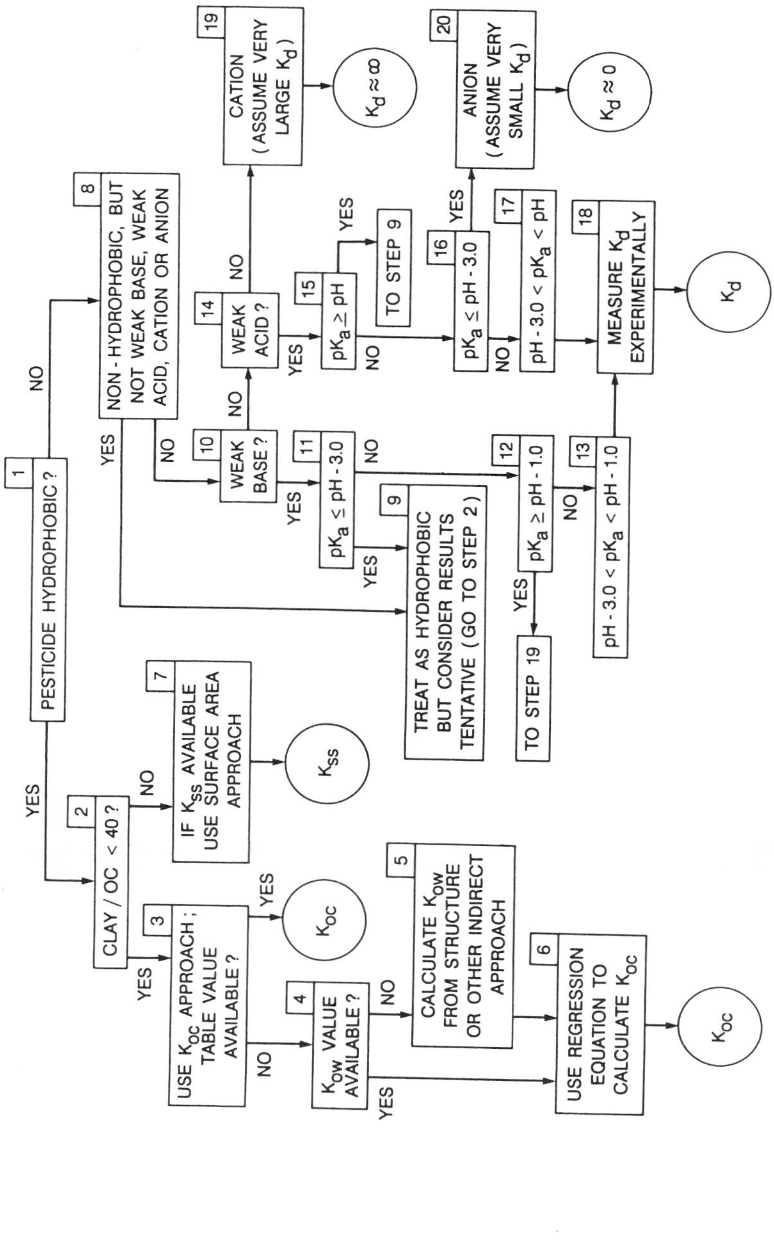


Fig. 4–1. Scheme for estimating sorption partition coefficients for pesticides and other toxic organics. From Green and Karickoff (1990).

2 is one or more pH units above the $pK_a$ of the pesticide, the effect of pH on $K_d$ can be considered insignificant. This limiting criterion can be summarized by $pK_a \leq$ (soil pH $-3$). In the absence of other more rational approaches, the above rule is adopted for weak bases in the scheme in Fig. 4–1 (see step 11).

Additional useful criteria have been suggested by Pionke (1984, personal communication) and are included in the sequence for weak bases and weak acids in Fig. 4–1. These criteria should be considered tentative since they have not been confirmed experimentally. The rationale for each step is given to aid the reader in evaluating the procedure. The relevant equilibrium for weak bases is

$$RH_2^+ \rightleftharpoons RH + H^+.$$

Then $K_a = \{[RH][H^+]\}/[RH_2^+]$ or, at the soil particle surface, $pK_a = pH_s + p[RH/RH_2^+]$, where the brackets indicate chemical activity. Assuming $pH_s = pH - 2.0$, we get

$$pK_a = pH - 2.0 + p[RH/RH_2^+].$$

If we require $[RH] \geq 10[RH_2^+]$, or $[RH]/[RH_2^+] \geq 10$, for a compound to be considered hydrophobic (in addition to the solubility criterion), then $pK_a \leq pH - 3.0$ is the appropriate criterion (step 11). Conversely, if $[RH]/[RH_2^+] < 0.1$, then the compound would be expected to behave like a cation because 90% or more is positively charged. Thus, for a weak base to be considered cationic, $pK_a \geq pH - 2.0 + p(0.1)$ or $pK_a \geq pH - 1.0$, as given in step 12. For pH/$pK_a$ relationships that are intermediate to the above extreme cases (i.e., the chemical cannot be considered either molecular or cationic, but rather a mixture of both species), a measured value of $K_d$ is required (step 18).

Many acidic pesticides are negatively charged at prevailing soil pH values and thus are not hydrophobic. However, a procedure analogous to that suggested for weak bases can be rationalized for weak acids.

$$RH \rightleftharpoons R^- + H^+$$

$K_a = \{[R^-][H^+]\}/[RH]$ or $pK_a = pH_s + p[R^-/RH]$ at the soil-particle surface. Again, with $pH_s = pH - 2.0$ as assumed earlier for the computation with weak bases, $pK_a = pH - 2.0 + p[R^-/RH]$. If we require $[R^-/RH] \geq 10$ (corresponding to 90% or more in the anion form) for the weak acid to be considered anionic, then the appropriate numerical criterion is given by

$$pK_a \leq pH - 2.0 + p[10]$$

or

$$pK_a \leq pH - 3.0$$

as shown in step 16. For the unusual case in which an acidic pesticide may be hydrophobic, a more restrictive criterion such as $[R^-/RH] < 0.01$ may be desirable to ensure a small fraction of the mobile anionic species. Thus, $pK_a \geq pH - 2.0 + p[0.01]$ or $pK_a \geq pH$, as given in step 15, is the appropriate criterion. Intermediate $pK_a$ values (step 17) would require that the $K_d$ be measured experimentally. Values of $pK_a$ for several basic pesticides are given by Pionke and DeAngelis (1980).

### 4–6.2 Steps 2 and 7

The $K_{oc}$ approach assumes no contribution of the mineral fraction of a soil or sediment to sorption of a hydrophobic chemical. However, when the organic C content ($f_{oc}$) of the sorbent is low relative to the clay mineral content ($f_{cm}$), the contribution of clay to sorption may become more evident (Minglegrin & Gerstl, 1983; Karickhoff, 1984). Karickhoff (1984) found a threshold for swelling-clay contribution to sorption at an $f_{cm}/f_{oc}$ ratio of about 30/1 for clay mineral fraction ($f_{cm}$) to organic C fraction ($f_{oc}$). This ratio can be expected to vary with clay type, swelling clays being more active in sorption than nonswelling clays. At the present time, there is no quantitative method of correcting for mineral contribution to sorption, but a threshold of about 40 for $f_{cm}/f_{oc}$ may be a reasonable value to adopt in view of the suggestion by Karickhoff of 25 to 60 for neutral organics with polar functional groups and >60 for nonpolar organics.

### 4–6.3 Steps 3 and 4

Green and Karickhoff (1990) summarized $K_{oc}$ and $K_{ow}$ data for a large number of pesticides. Such data will likely be provided by developers of new pesticides as part of registration requirements.

## 4–7 SORPTION ESTIMATION ERRORS IN THE MODELING CONTEXT

Throughout this chapter, we have referred to potential errors in sorption estimates that are likely to occur as a result of simplifying assumptions being invalid. The principal justification for adopting simplified approaches that are known to contribute additional uncertainty to modeling results is that a reasonable estimate of sorption is better than no estimate at all. Additionally, modelers may be inclined to rationalize that errors in sorption estimates are relatively small in comparison to parameter estimates for other processes involved in the modeling of pesticide fate in soils and water. For example, use of first-order kinetics of pesticide transformation and the associated estimation of first-order rate coefficients are procedures fraught with uncertainty. We have encouraged model users to evaluate for each application the impact of uncertainty arising from errors in sorption estimates. This can be accomplished qualitatively by observing model response to variations in input parameters, as in model sensitivity analyses. Such an analysis with a fertilizer N-movement model (Khan & Green, 1980) revealed that variation in the first-order rate coefficient for nitrification was more important than variability in any other input parameter. In this case, errors associated with estimation of sorption coefficients or dispersion coefficients would have much less impact than would errors in kinetic coefficients. One cannot, however, generalize such results from one chemical to another or for all scenarios to be modeled.

The impact of errors in model inputs can be evaluated quantitatively by first-order uncertainty analysis (Cornell, 1972). Such analyses have been used in hydrological modeling (e.g., Burges & Lettenmaier, 1975; Garen & Burges, 1981), but have yet to be employed much for contaminant transport modeling. Loague et al. (1989) conducted such an analysis to examine the impact of uncertainty in estimates of soil variables for Hawaii soils on pesticide retardation computed by a simple mobility index. Their analysis for one chemical (diuron) revealed that variation in estimated $K_{oc}$ values from the literature had a greater impact on the results than variation in relevant soil properties (organic C content, bulk density, and water content at 33 kPa). Other model inputs would likely have more impact than $K_{oc}$ for more comprehensive models that incorporate chemical transport and transformation, but we are not aware of any studies that address this question in a generic manner.

In the absence of a quantitative evaluation of error propagation in a given model, it is helpful to consider the magnitude and direction of potential errors associated with sorption estimates. In Table 4–4, we summarize the approximate errors associated with the various simplifying assumptions discussed earlier in this chapter. The potential error is expressed as a factor, with the plus or minus operator indicating the likely direction of the error.

Table 4–4. Approximate error in pesticide sorption estimates attributed to various simplifying assumptions.

| Simplifying assumption† | Likely error factor for sorption estimate‡ |
|---|---|
| Lab to field extension | |
| Lab measurement for field application | + 2X (recently applied pesticide); −10 to −1000X (aged residues) |
| General sorption estimation | |
| Equilibrium | No error for static systems; +1.5X to +5X for dynamic systems |
| Linear isotherm | ±2X |
| Singular isotherm§ | No error in many systems; perhaps −1.5X in some systems |
| Additional for $K_{oc}$ approach | |
| Sorption on organic matter only | Little error in soils with $f_{oc}$ >0.01; −2 to −10X on soils with $f_{oc}$ <0.002 |
| Uniformity of organic matter | ±2X |

† Discussed in previous sections of this chapter.

‡ Direction and approximate magnitude of error is indicated; overprediction due to simplifying assumption is indicated by a plus (+) factor, underprediction by a negative (−) factor. For example, +5X indicates the simplifying assumption produces a sorption estimate which is five times greater than is likely in field applications.

§ Desorption hysteresis ignored.

For example, the equilibrium assumption should result in little or no error for field systems which approach equilibrium, but for a dynamic system, the assumption of equilibrium may produce an estimate of sorption which is 1.5 to 5 times greater than that would occur in the field. The factors shown in Table 4–4 are highly subjective, but they at least indicate the order of magnitude and direction of potential errors.

If we consider the impact of all simplifying assumptions inherent in the use of the $K_{oc}$ approach, then all of the assumptions in Table 4–4 apply. It is difficult to assess the composite result of all assumptions since some assumptions lead to overestimation and others to underestimation. Also, the departure of modeling results from accurate field estimates depends on the process being modeled, so no generalizations are possible. Since errors are often additive, however, it appears that a $K_{oc}$ estimate could overpredict sorption of recently applied pesticide by several fold. On the other hand, the sorption of a pesticide residue in subsoils may actually be underpredicted, considering the assumptions of lab to field extension (aged residues) and sorption on organic matter only on soils with $f_{oc} < 0.002$. Clearly, the potential impact of errors in parameter estimation for modeling must be evaluated by the model user for each application. If $K_{oc}$ estimates are used only for ranking pesticides for relative mobility, then estimation errors may not be important. For example, $K_{oc}$ values provided by Kenaga and Goring (1980) for carbaryl and chlorpyrifos are 230 and 13 600, respectively. In this case, the difference in $K_{oc}$ is sufficiently large that errors due to simplifying assumptions associated with the $K_{oc}$ estimates would have little effect on the implied relative mobility of these chemicals. If one wanted to model the leaching of carbaryl at a given location, however, it would be desirable to know how much confidence one could have in the value, $K_{oc} = 230$. Other sources of $K_{oc}$ data give carbaryl $K_{oc}$ values of 298, 104, and 121 (Green & Karickhoff, 1990). Accurate predictions of pesticide fate at a given point in time and space require accurate parameter estimates. Thus, the model user must decide how much error can be tolerated.

## 4–8 CONCLUSION

Quantitative estimates of pesticide sorption on soils are now required routinely for pesticide regulation. Simple indices of pesticide mobility in soils, and also more complex dynamic models, require sorption coefficients. The choice of procedure to measure or estimate sorption coefficients is dependent upon the level of accuracy required for a given use. We have identified several potential sources of error that may result in unsatisfactory sorption estimates if stringent accuracy requirements are imposed. Presently, however, there are many environmental quality applications that require only approximate estimates of pesticide sorption, so that simplifying assumptions implicit in estimation procedures may be acceptable. On the other hand, regulators and modelers should be aware of the uncertainty associated with various types of sorption estimates. Determination of parameter error propagation in a model and its impact on model output is an important component of modeling in the regulatory context.

The estimation procedures discussed in this chapter are acceptable for many applications in which a reasonable estimate of sorption is better than no estimate at all. More site-specific regulation of pesticides in agriculture in the years ahead may require more accurate modeling of pesticide movement. Better estimates of sorption may then be required. Improved understanding of sorption mechanisms for a broad array of soil-pesticide combinations should lead to more accurate sorption estimation procedures.

## REFERENCES

Bailey, G.W., and J.L. White. 1964. Review of adsorption and desorption of organic pesticides by soil colloids, with implications concerning pesticide bioactivity. J. Agric. Food Chem. 12:324–332.

Bilkert, J.N., and P.S.C. Rao. 1985. Sorption and leaching of three nonfumigant nematicides in soils. J. Environ. Sci. Health B20(1):1–26.

Bowman, B.T. 1973. The effect of saturating cations on the adsorption of Dasanit® O,O-diethyl O-[*p*-(methyl sulfinyl) phenyl] phosphorothioate, by montmorillonite suspensions. Soil Sci. Soc. Am. Proc. 37:200–207.

Boast, C.W. 1973. Modeling the movement of chemicals in soils by water. Soil Sci. 115:224–230.

Briggs, G.G. 1973. A simple relationship between soil adsorption of organic chemicals and their octanol/water partition coefficients. p. 83–86. *In* Proc. 7th Br. Insectic. Fungic. Conf., 19–22 Nov. Br. Crop Protection Counc., Brighton, England.

Briggs, G.G. 1981a. Theoretical and experimental relationships between soil adsorption, octanol-water partition coefficients, water solubilities, bioconcentration factors, and the parachor. J. Agric. Food Chem. 29:1050–1059.

Briggs, G.G. 1981b. Adsorption of pesticides by some Australian soils. Aust. J. Soil Res. 19:61–68.

Burges, S.J., and D.P. Lettenmaier. 1975. Probabilistic methods in stream quality management. Water Resour. Bull. 11:115–130.

Buxton, D.S., and R.E. Green. 1987. Desorption and leachability of DBCP residues in soils. p. 167 *In* Agronomy abstracts. ASA, Madison, WI.

Chiou, C.T., L.J. Peters, and V.H. Freed. 1979. A physical concept of soil-water equilibria for nonionic organic compounds. Science 206:831–832.

Chiou, C.T., P.E. Porter, and D.W. Schmedding. 1983. Partition equilibria of nonionic organic compounds between soil organic matter and water. Environ. Sci. Technol. 17:227–231.

Cornell, C.A. 1972. First-order analysis of model and parameter uncertainty. p. 1245–1274. *In* C.C. Kisiel and L. Duckstein (ed.) Proc. Int. Symp. on Uncertainties in Hydrologic and Water Resource Systems, Tucson, AZ. 11–14 Dec. Vol. 3. Dep. Hydrology and Water Resourc., Univ. of Arizona, Tucson.

Dao, T.H., D.B. Marx, T.L. Lavy, and J. Dragun. 1982. Effect, and statistical evaluation, of soil sterilization on aniline and diuron adsorption isotherms. Soil Sci. Soc. Am. J. 46:963–969.

Dekkers, W.A. 1978. Computed effect of decomposition on the partition coefficient determined by the batch equilibration method. Weed Res. 18:341–345.

Garen, D.C., and S.J. Burges. 1981. Approximate error bounds for simulated hydrographs. J. Hydraul. Div., Am. Soc. Civ. Eng. 107(HY11):1519–1534.

Green, R.E. 1974. Pesticide-clay-water interactions. p. 3–37. *In* W.D. Guenzi (ed.) Pesticides in soil and water. SSSA, Madison, WI.

Green, R.E., and J.C. Corey. 1971. Pesticide adsorption measurement by flow equilibration and subsequent displacement. Soil Sci. Soc. Am. Pro. 35:561–565.

Green, R.E., and S.W. Karickhoff. 1990. Estimating pesticide sorption coefficients for soils and sediments. p. 1–18. *In* D.G. DeCoursey (ed.) Small watershed model (SWAM) for water, sediment, and chemical movement: Supporting documentation. ARS-80. USDA-ARS, Washington, DC.

Green, R.E., and S.R. Obien. 1969. Herbicide equilibrium in soils in relation to soil water content. Weed Sci. 17:514–519.

Green, R.E., and V.K. Yamane. 1970. Precision in pesticide adsorption measurements. Soil Sci. Soc. Am. Proc. 34:353–354.

Green, R.E., J.M. Davidson, and J.W. Biggar. 1980. An assessment of methods for determining adsorption-desorption of organic chemicals. p.73–82. *In* A. Banin and V. Kafkafi (ed.) Agrochemicals in soils. Pergamon Press, New York.

Green, R.E., C.C.K. Liu, and N. Tamrakar. 1986. Modeling pesticide movement in the unsaturated zone of Hawaii soils under agricultural use. p. 366–383. *In* W.Y. Garner et al. (ed.) Evaluation of pesticides in ground water. ACS Symp. Ser. 315. Am. Chem. Soc., Washington, DC.

Hamaker, J.W., and J.M. Thompson. 1972. Adsorption. p. 49–143. *In* C.A.I. Goring and J.W. Hamaker (ed.) Organic chemicals in the soil environment, Vol. 1. Marcel Dekker, New York.

Hance, R.J. 1969. Influence of pH, exchangeable cation and the presence of organic matter on the adsorption of some herbicides by montmorillonite. Can. J. Soil Sci. 49:357–364.

Helling, C.S. 1971. Pesticide mobility in soils. III. Influence of soil properties. Soil Sci. Soc. Am. Proc. 35:743–748.

Karickhoff, S.W. 1980. Sorption kinetics of hydrophobic pollutants in natural sediments. p. 193–205. *In* R.A. Baker (ed.) Contaminants and sediments, Vol. 2. Ann Arbor Sci. Publ., Ann Arbor, MI.

Karickhoff, S.W. 1981. Semi-empirical estimation of sorption of hydrophobic pollutants on natural sediments and soils. Chemosphere 10:833–846.

Karickhoff, S.W. 1984. Organic pollutant sorption in aquatic systems. J. Hydraul. Eng. 110:707–735.

Kenaga, E.E., and C.A.I. Goring. 1980. Relationship between water solubility, soil sorption, octanol-water partitioning and concentration of chemicals in biota. p. 78–115. *In* J.G.Eaton et al. (ed.) Aquatic toxicology. ASTM STP 707. Am. Soc. Test. and Materials, Philadelphia.

Khan, M.A., and R.E. Green. 1980. Simulation of processes relevant to N fertilization with irrigation. p. 132. *In* Agronomy abstracts. ASA, Madison, WI.

Koskinen, W.C., and H.H. Cheng. 1982. Elimination of aerobic degradation during characterization of pesticide adsorption-desorption in soil. Soil Sci. Soc. Am. J. 46:256–259.

Koskinen, W.C., and H.H. Cheng. 1983. Effects of experimental variables on 2,4,5-T adsorption-desorption in soil. J. Environ. Qual. 12:325–330.

Loague, K.M., R.S. Yost, R.E. Green, and T.C. Liang. 1989. Uncertainty in a pesticide leaching assessment for Hawaii. J. Contaminant Hydrol. 4:139–161.

Lyman, W.J. 1982. Adsorption coefficients for soils and sediments. p. 4-1 to 4-32. *In* W.J. Lyman et al. (ed.) Handbook of chemical property estimation methods. McGraw-Hill Book Co., New York.

McCall, P.J., D.A. Laskowski, R.L. Swann, and H.J. Dishburger. 1981. Measurement of sorption coefficients of organic chemicals and their use in environmental fate analysis. p. 89–109. *In* Test protocols for environmental fate and movement of toxicants. Proc. Symp. AOAC 94th Annu. Meet., Washington, DC. 21–29 Oct. 1980. Assoc. of Official Analytical Chem., Arlington, VA.

Mingelgrin, U., and Z. Gerstl. 1983. Reevaluation of partitioning as a mechanism of nonionic chemicals adsorption in soils. J. Environ. Qual. 12:1–11.

Pignatello, J.J., B.L. Sawhney, and C.R. Frink. 1987. EDB: Persistence in soil. Science 236:898.

Pionke, H.B., and R.J. DeAngelis. 1980. Method for distributing pesticide loss in field runoff between the solution and adsorbed phase. p. 607–643. *In* CREAMS, A Field Scale Model for Chemicals, Runoff, and Erosion from Agricultural Management Systems. USDA Conservation Res. Rep. 26. USDA, SEA, Washington, DC.

Rao, P.S.C., and J.M. Davidson. 1980. Estimation of pesticide retention and transformation parameters required in non-point source pollution models. p. 23–67. *In* M.R. Overcash and J.M. Davidson (ed.) Environmental impact of nonpoint source pollution. Ann Arbor Sci. Publ., Ann Arbor, MI.

Van Genuchten, M. Th., and R.W. Cleary. 1979. Movement of solutes in soil: Computer-simulated and laboratory results. p. 349–386. *In* G.H. Bolt (ed.) Soil chemistry B. Physico-chemical models. Elsevier Sci. Publ. Co., New York.

Van Genuchten, M. Th., J.M. Davidson, and P.J. Wierenga. 1974. An evaluation of kinetic and equilibrium equations for the prediction of pesticide movement in porous media. Soil Sci. Soc. Am. Proc. 38:29–35.

Voice, T.C., and W.J. Weber, Jr. 1983. Sorption of hydrophobic compounds by sediments, soils and suspended solids—I. theory and background. Water Res. 17:1433–1441.

Voice, T.C., C.P. Rice, and W.J. Weber. 1983. Effect of solids concentration on the sorptive partitioning of hydrophobic pollutants in aquatic systems. Environ. Sci. Technol. 17:513–518.

Walker, A., and D.V. Crawford. 1968. The role of organic matter in adsorption of the triazine herbicides by soils. p. 91–108. *In* Isotopes and radiation in soil organic-matter studies. Proc. Symp. IAEA, FAO and ISSS, Vienna. 15–19 July. Int. Atomic Energy Agency, Vienna, Austria.

Wauchope, R.D., and R.S. Myers. 1985. Adsorption-desorption kinetics of atrazine and linuron in fresh water-sediment aqueous slurries. J. Environ. Qual. 14:132–136.

Young, R.A., and C.A. Onstad. 1976. Predicting particle-size composition of eroded soil. Trans. ASAE 19:1071–1075.

# Chapter 5

# Abiotic Transformations in Water, Sediments, and Soil

**N. LEE WOLFE,** *U.S. Environmental Protection Agency, Athens, Georgia*

**URI MINGELGRIN,** *The Volcani Center, ARO, Bet-Dagan, Israel*

**GLENN C. MILLER,** *University of Nevada, Reno, Nevada*

The importance of abiotic transformations to the fate of pesticides in the environment became widely recognized only recently. Even in a medium in which intense biological activity takes place, such as soil, abiotic transformations can be important. Under some conditions, such as those that may occur below the root zone, abiotic transformations can dominate the fate of pesticides.

Abiotic transformations of pesticides have been reviewed by several investigators (e.g., Melnikov, 1971; Kearny & Kaufman, 1969; Morrill et al., 1982). In these publications, the pesticides were classified according to their chemical structure, demonstrating that transformations are based on functional groups of the pesticide. In the present chapter, the abiotic transformations will be described in terms of the specific environment in which they occur.

Numerous transformations occur in the homogeneous phases, especially in the liquid phase. Other transformations occur in the interface between phases. These include reactions that are heterogeneously catalyzed and those that occur in solution under the influence of the electric field of charged surfaces. Crosby (1970), in a review on abiotic transformations in the soil, presented some examples of both surface-enhanced reactions and reactions that take place in the bulk liquid phase.

It is often difficult to determine whether a pesticide undergoes abiotic or biotic transformations. In the case of many pesticides, significant biological and chemical degradation takes place simultaneously (Wolfe et al., 1980). For example, Deuel et al. (1985) reported such a situation for carbaryl degradation in flooded rice (*Oryza sativa* L.) fields. Barug and Vonk (1980) re-

 *Pesticides in the Soil Environment*—SSSA Book Series, no. 2

ported for bis(tributyltin) oxide degradation in soil. Biologically produced enzymes and other biochemical compounds in the soil can be involved in transformations of pesticides. Thus, sterilization, which destroys living organisms, also will affect the abiotic chemical reactions that are dependent on substances generated by biological processes. Sterilization processes, furthermore, can alter nonbiological constituents of the treated systems. Heat and radiation, for example, can affect the free-radical content of the soil. Finally, many degradative pathways include both biologically and chemically controlled steps.

## 5-1 ABIOTIC TRANSFORMATIONS IN NATURAL WATERS

### 5-1.1 Physical and Chemical Parameters that Influence the Reactions in Natural Waters

The intrinsic reactivity of the labile functional group of the pesticide is the first factor that should be considered when studying transformations of pesticides in natural waters. The reaction rate constant is a function of this reactivity. In fact, the rate constant often describes the velocity of the rate-determining step of the reaction in which bond clevage or bond formation takes place. The activity in natural waters of the most important species that may participate or enhance the transformation of pesticides and the major environmental conditions which affect these transformations are considered below.

#### 5-1.1.1 pH, Buffering, and General Acid Base Catalysis

Hydrogen ion activity affects transformation kinetics in two distinct ways. The most common is in acid-base mediated hydrolysis reactions. The rate of the reaction is directly proportional to the proton and hydroxide concentration (e.g., Wolfe et al., 1977). In the case of base-catalyzed hydrolysis, the rate equation is

$$(-dP/dt) = k[\mathrm{P}][\mathrm{OH}^-] \qquad [1]$$

where $k$ = second-order rate constant ($\mathrm{mol}^{-1}\ \mathrm{s}^{-1}$), [P] = pesticide concentration and $[\mathrm{OH}^-]$ = hydroxide ion concentration.

For compounds that have acidic or basic functional groups (e.g., Fig. 5-1), the pH of the water also governs the relative ratios of the associated and disassociated species. Because associated and disassociated species react differently, small changes in pH can effect large changes in half-lives (e.g., Bender & Silver, 1963).

The pH of natural waters varies over a wide range. In swamps and water bodies containing high concentrations of humic acids, the pH of the water can be as low as 3.5. On the upper end of the scale, eutrophic lakes and ponds have pH values as high as 9.5 (Hutchinson, 1957). The average pH of the ocean is estimated to be 8.2 (Turekian, 1969).

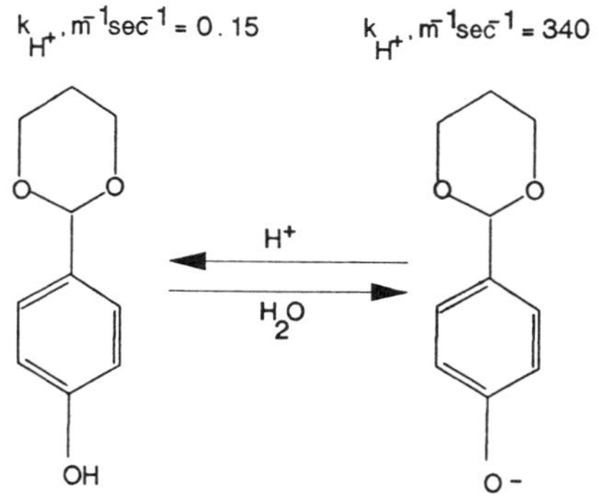


Fig. 5-1. Effect of dissociation on the second-order hydrolysis rate constants for the alkaline hydrolysis of a substituted 2-phenyl-1,3-dioxane (Bender & Silver, 1963).

Table 5-1. Concentration of selected organic and inorganic species in selected aquatic environments.

| Species | World avg. river | World avg. seawater | World interstitial water |
|---|---|---|---|
| | | mol/L | |
| Ammonia | -- | -- | $2.00 \times 10^{-4}$ |
| Carbonate | $9.78 \times 10^{-4}$ | $2.33 \times 10^{-3}$ | $9.09 \times 10^{-3}$ |
| Fulvic acid | $1.00 \times 10^{-4}$ | $5.00 \times 10^{-6}$ | $2.00 \times 10^{-4}$ |
| Phosphate | -- | $2.84 \times 10^{-6}$ | $1.80 \times 10^{-5}$ |
| Sulfide | -- | -- | $3.96 \times 10^{-7}$ |
| Silicate | $2.18 \times 10^{-4}$ | $1.03 \times 10^{-4}$ | -- |
| Borate | -- | $4.10 \times 10^{-4}$ | -- |

Acid-bases that are most likely to affect hydrolysis reactions in aerobic river waters are carbonate, fulvic acid, and silicate. The world average concentration (mol/L) of these species is $9.8 \times 10^{-4}$, $1.0 \times 10^{-4}$, and $2.2 \times 10^{-4}$, respectively (Table 5-1). Alternatively, in anaerobic interstitial waters, the concentration (mol/L) of the most important acid-bases: ammonia, carbonate, fulvic acid, phosphate, and sulfide are $2.0 \times 10^{-4}$, $9.1 \times 10^{-3}$, $2.0 \times 10^{-4}$, $1.8 \times 10^{-5}$, and $4.0 \times 10^{-7}$, respectively (Emerson, 1976).

### 5-1.1.2 Temperature

The rate of pesticide transformations increases with increasing temperature. The effect of the temperature on the rate is most often described in the environmental literature by the Arrhenius equation (e.g., Moore & Pearson, 1981):

$$k = A\, e^{-\mathrm{Ea}/\mathrm{RT}} \qquad [2]$$

where Ea is the activation energy for the reaction, R is the gas constant, T is the absolute temperature, and $A$ is a constant.

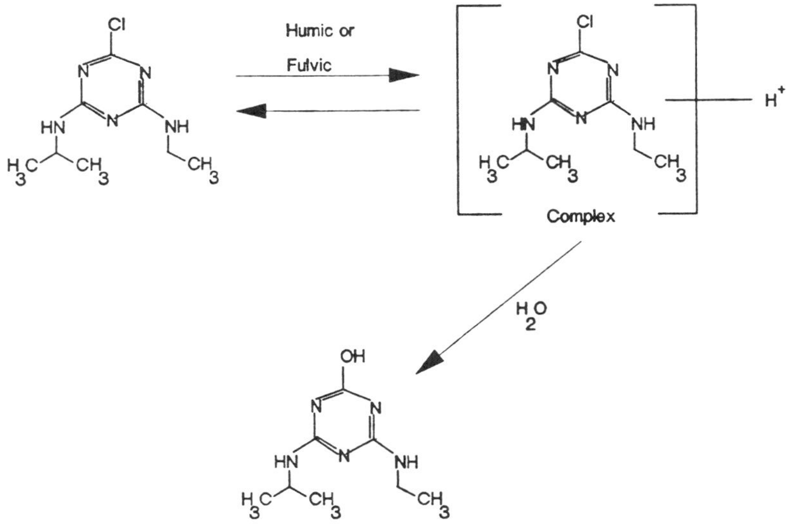


Fig. 5–2. Humic and fulvic acid-mediated hydrolysis of atrazine to give hydroxyatrazine.

Temperatures of natural waters range from a low of 0 °C in lakes, ponds, and rivers in the temperate zone in the winter to as high as 45 °C in shallow ponds and stagnant waters in the summer. For reactions with activation energies as low as 10 kcal/mol, this temperature range corresponds to 14-fold difference in the half-life of the pesticide. For a reaction with an activation energy around 30 kcal/mol, this corresponds to a factor of about 2500-fold difference in the half-life.

### 5–1.1.3 Dissolved Organics and Suspended Particulates

Concentrations of humic acids in natural waters range from nearly nil in oligotrophic lakes to as high as 30 mg dissolved organic C (DOC) per liter, for example in swamp lands or in southeastern coastal rivers (Perdue & Wolfe, 1983). In general, most rivers and lakes have concentrations ranging from 2 to 15 mg DOC/L.

Both dissolved and suspended organic material can affect abiotic transformations (Khan, 1978). For example, dissolved fulvic acids have been shown to accelerate the hydrolysis of atrazine in water (Fig. 5–2). On the other hand, the alkaline hydrolysis of the hydrophobic n-octyl ester of 2,4-D (Fig. 5–3) was shown to be retarded by humic materials (e.g., Perdue & Wolfe, 1982). This was believed to be a heterogeneous phenomenon in which the ester partitioned to the suspended humic material and was thus protected from alkaline hydrolysis (Fig. 5–3).

### 5–1.1.4 Metal Ions

The metals most often proposed to catalyze the hydrolysis of pesticides in natural waters are $Cu^{2+}$, $Fe^{3+}$, $Mn^{2+}$, $Mg^{2+}$, $Ca^{2+}$ (Blanchet, 1982).

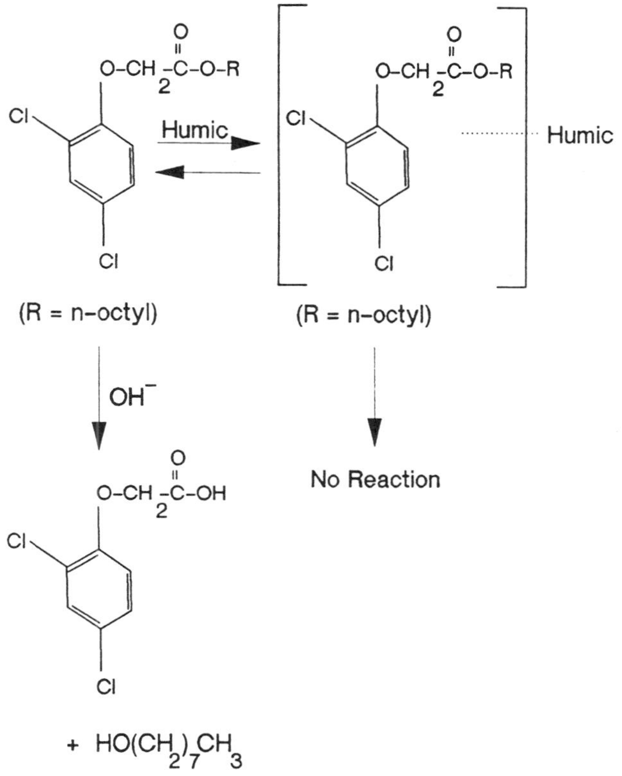


Fig. 5–3. The effect of humic acid on the alkaline hydrolysis of 2,4-D esters (Perdue & Wolfe, 1982).

Laboratory studies on the interaction of selected organophosphate esters with $Cu^{2+}$ and $Mn^{2+}$ (Blanchet & St-George, 1982) have demonstrated metal-catalyzed hydrolysis and have provided rate constants for this pathway (Fig. 5–4). Similar studies, with $Mg^{2+}$ and $Ca^{2+}$ did not demonstrate any measurable catalysis.

### 5–1.1.5 Redox State of the Water Column

The redox state in the water column of large lakes and reservoirs is likely to vary with depth and may be defined in terms of the limnion and the epilimnion (Hutchinson, 1957). In the limnion, the water is nearly anoxic and reducing conditions exist. In the epilimnion, $O_2$ is present and aerobic conditions can exist. In small ponds and rivers, mixing is fast enough that oxidative conditions may exist throughout the volume of the water body.

Although many pesticides were shown to undergo reduction in the anaerobic environments found in natural waters, cases in which a clear distinction between biological and chemical reduction pathways were demonstrated are few (Bromilow et al., 1986; Smelt et al., 1983; Macalady et al., 1986).

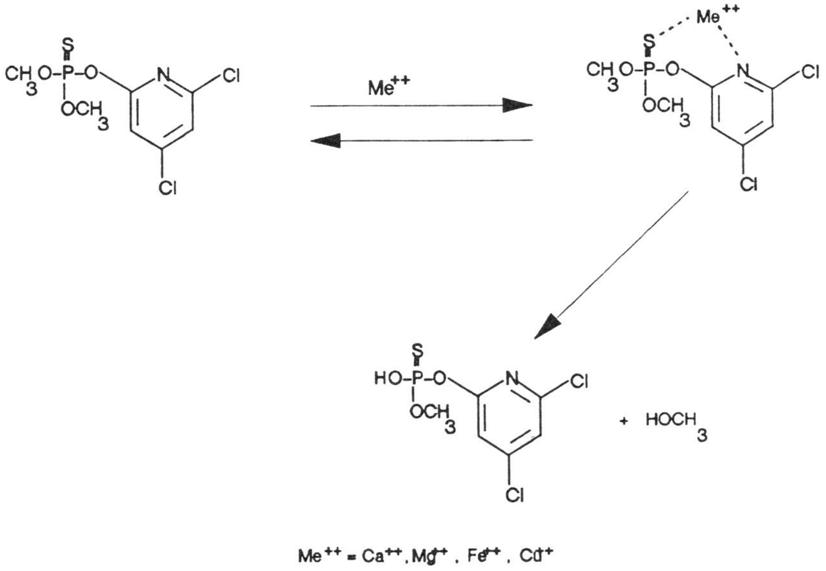


Fig. 5–4. Mechanism for the metal-catalyzed hydrolysis of chlorpyrifos (Blanchet & St.-George, 1980).

## 5–1.2 Abiotic Reactions Occurring in Natural Waters

Among the numerous abiotic transformations that pesticides may undergo in natural waters, hydrolysis and redox reactions are the most prevalent.

### 5–1.2.1 Hydrolysis

In general, the term *hydrolysis* refers to the cleavage of a bond of the pesticide and the formation of a new bond with the O atom of water. Hydrolysis, however, is a loosely used term and often erroneously includes reactions in which water serves only as a solvent. In numerous types of reactions (e.g., elimination reactions, decarboxylation, or isomerization) water is not incorporated into the transformation product. Therefore, these reactions are not hydrolysis reactions even if they did occur in water.

It is convenient to classify hydrolysis reactions into three categories: acid-mediated, base-mediated, and neutral (or pH/independent) reactions. In the case of acid-mediated hydrolysis, an acid, usually a proton ($H^+$), catalyzes the bond breaking-bond making process. Because the proton is not consumed in the reaction, the process is referred to as *acid-catalyzed* hydrolysis. The rate of the reaction is dependent on the proton concentration and, therefore, increases as the pH decreases. In the case of base-mediated hydrolysis, hydroxyl ($OH^-$) behaves as a nucleophile and is consumed in the reaction. This pathway is often referred to as *alkaline hydrolysis*. The rate of the reaction is dependent on the hydroxyl concentration and increases with increasing pH. In the third type of hydrolysis, the rate of reaction is independent of the acid-base concentration (i.e., it is pH independent). This process is often referred to as *neutral hydrolysis.* Neutral and alkaline hydrolysis are the most common reactions over the pH's common to the environment. The relation between the first-order hydrolysis rate constants and pH is often presented as a pH-rate profile.

The hydrolysis of pesticides has been studied extensively and many of the details of the mechanisms and kinetics of such reactions have been elucidated in well-defined systems such as buffered, distilled water under laboratory conditions (e.g., Wolfe et al., 1977). To what extent the results of these laboratory studies can be extrapolated to the water column of aquatic ecosystems is, in many cases, not certain.

Given sufficiently drastic reaction conditions (i.e., extremely low or high pH and high temperatures), almost all pesticides will undergo hydrolysis, unless another reaction preempts the transformation of the pesticide. The actual contribution of hydrolysis to the overall degradation of a pesticide will depend on the rates of other competing processes such as photolysis, bioloysis, and redox reactions.

Functional groups that commonly occur in pesticides and are susceptible to hydrolysis include:

1. Carboxylic acid esters
2. Organophosphate and organophosphorothioate esters
3. Amides
4. Anilides
5. Carbamates
6. Organohalides
7. Triazines
8. Oximes
9. Nitriles

Carboxylic acid esters break down through hydrolysis into a carboxylic acid and an alcohol (e.g., Zepp et al., 1975; Wolfe et al., 1977). The wide use of pesticides such as 2,4-D and 2,4,5-T and pyrethoid esters makes the hydrolysis of carboxylic acid esters an important reaction in the environment.

Organophosphate and organophosphorothioate esters undergo both alkaline and neutral hydrolysis (Smith et al., 1978). Alkaline hydrolysis results in the cleavage of a P-O bond, producing a phosphoric or thiophosphoric acid and an alcohol moiety. The neutral hydrolysis is via the cleavage of a C-O bond. The products are again a phosphoric or thiophosphoric acid and an alcohol (Greenhalgh et al., 1980; Macalady & Wolfe, 1983).

Alkaline hydrolysis of amides and anilides was extensively studied (e.g., Bergon & Calmon). These compounds undergo hydrolysis through cleavage of the carbonyl C-N bond. The products of the reaction are a carboxylic acid and the corresponding amine.

Carbamates are susceptible to alkaline hydrolysis and considerably less so to acid-catalyzed hydrolysis (Banks & Tyrell, 1984). The carbamates may be hydrolyzed through the breakdown of both the carbonyl C-O and the carbonyl C-N bonds, yielding an alcohol, $CO_2$, and an amine.

Organohalides may lose their halogen substituents by a myriad of pathways, some of which are hydrolytic, either pH dependent or pH independent (Jeffers et al., 1989). Hydrolysis through nucleophilic substitution and addition reactions (Cooper, 1987) are examples of pathways through which hydrolysis may occur.

The hydrolysis of triazine-based pesticides has received considerable attention and mechanisms for the reaction were proposed (Grayson, 1980). Pesticides containing oxime and nitrile functional groups have recieved less attention.

Although most studies of the hydrolysis of pesticides focused on specific acid-base mediated reactions, other pathways may also be important in natural waters. These pathways include general acid-base catalysis (by both inorganic and organic species), metal catalysis, and surface-mediated heterogeneous catalysis. Heterogeneous catalysis will be discussed in section 5-3.2.

General acid-base catalyzed hydrolysis occurs when an acid or base, other than a proton or a hydroxide ion, mediates the reaction. For example, bicarbonate or dihydrogenphosphate can act either as Brønsted acids or as bases and thus facilitate hydrolysis reactions. The rate of the hydrolysis reaction is dependent not only on the concentration of the acido-basic species, but also on their strength as acids or bases.

Laboratory studies and theoretical considerations suggest that, over the concentration ranges of general acids and bases common to natural water systems, hydrolysis mediated by the general acido-basic species is far less important than specific acid-base hydrolysis. Under the most favorable conditions, the maximum contribution of the general acid-base catalyzed hydrolysis is likely to be $<10\%$ of the total breakdown by hydrolysis (Perdue & Wolfe, 1983).

The hydrolysis of pesticides in the presence of humic and fulvic acids was studied under laboratory conditions (Kahn, 1978). General acid-base catalysis by these organic acids was demonstrated and kinetic expressions derived. Extrapolation of the data to environmental concentrations of the humic materials suggests, however, that this will not be an important process in most natural waters.

Several studies described metal catalyzed hydrolysis of pesticides. Blanchet and St-George (1982) discussed the hydrolysis of organophosphorothioate esters by $Cu^{2+}$ (Fig. 5-4). Under laboratory conditions, the rate of hydrolysis of these pesticides was considerably faster in $Cu^{2+}$ solutions than in distilled water at the same pH. Extrapolation of the data to metal concentrations common to natural waters indicates that metal-catalyzed hydrolysis is not likely to contribute significantly to the hydrolytic breakdown of most pesticides.

### 5-1.2.2 Redox Reactions

Redox reactions of pesticides are well documented in the literature and are known to occur under a variety of laboratory conditions. It was suggested that abiotic oxidation of pesticides occurs in surface waters, and abiotic reduction is believed to occur in bottom sediments and anaerobic waters (Macalady et al., 1986).

Many oxidations that occur in surface waters are believed to be light mediated. Both direct oxidative photolysis and indirect light induced oxidation through photolytic processes that generate reactive species take place. Photoactivated species that may participate in redox reactions include singlet molecular $O_2$, hydroxyl-free radicals, hydrated electrons, super oxide radical anion, and hydrogen peroxide. Photolytic reactions of pesticides are discussed in section 5–4.1.

Various pathways for the nonphotolytic oxidation of pesticides were also described. They include direct oxidation by ozone (Spencer et al., 1980b) as well as autoxidation enhanced by metals (Stone & Morgan, 1987) and peroxides (Mill et al., 1980). Biologically produced species can either oxidize or catalyze the oxidation of pesticides. Monooxygenases, dioxygenases, peroxidases, and lactones were among the species implicated as the oxidizing agents. (Burns & Edwards, 1980).

Many pesticides may be reduced under environmental conditions through reactions such as (Macalady et al., 1986):

1. Dehalogenation to alkanes.
2. Nitroreduction to the corresponding amine.
3. Azo reduction to an hydrazo or an amino group.
4. Quinone reduction to semiquinones or hydroquinones.
5. Sulfone reduction to the sulfoxide or sulfide.

The reducing agents postulated to bring about these reductions include inorganic species, with Fe and Mn species being the most important ones. Bioorganics such as flavins, porphyrins, and extracellular enzymes common to aquatic environments are also responsible for reduction of pesticides (Wade & Castro, 1972).

### 5–1.3 Kinetic Models Used to Describe Transformations of Pollutants in Natural Waters

The rate of specific acid-base and neutral hydrolysis in natural waters can be described by Eq. [3] and [4]:

$$-d[\mathrm{C}]/dt = k_{\mathrm{H}}[\mathrm{C}][\mathrm{H}^{+}] + k_{\mathrm{H_2O}}[\mathrm{C}] + k_{\mathrm{OH}}[\mathrm{C}][\mathrm{OH}^{-}] \qquad [3]$$

where the definition of the symbols is as follows:

$C$ = Concentration of the hydrolyzable compound,
$t$ = Time,
$H^{+}$ = Hydrogen ion concentration,
$OH^{-}$ = Hydroxide ion concentration,
$k_H$ = Acid hydrolysis rate constant,
$k_{H_2O}$ = Neutral hydrolysis rate constant, and
$k_{OH}$ = Alkaline hydrolysis rate constant.

$$k_{OBS} = k_H[H^+] + k_{H_2O} + k_{OH}[OH^-], \qquad [4]$$

where $k_{OBS}$ is the observed first-order disappearance rate constant. From Eq. [3], it is apparent that the rate of hydrolysis may be strongly pH dependent. Equations [3] and [4] allow a given pesticide to exhibit maxima in its rate of hydrolysis at both low and high pH values. This was actually observed in the case of many pesticides (e.g., Worthing, 1983).

## 5-2 SEDIMENTS

### 5-2.1 Relevant Properties of Sediment

#### 5-2.1.1 pH

The pH of sediments in ponds and rivers is generally within one pH unit of that of the overlying water. pH values measured in sediments are a composite of the pH of the interstitial water and the pH in the vicinity of the charged surfaces. Most charged solids in the environment possess a negative charge and, hence, the pH near their surface is lower than in the bulk aqueous phase (see section 5-3.1.2). The biological activity in the sediments (i.e., production of carbonate) will also affect the pH. To the extent that the biological activity in the sediments differs from the biological activity in the overlying water, so will the pH. Finally, the electric field near charged surfaces may interfere with the reading of the glass electrode that is used for most pH measurements. This interference may contribute to the observed difference in pH between the water body and the sediment at its bottom.

#### 5-2.1.2 Redox

Two classifications are widely used to define the redox state of natural waters and sediments. The most common classification defines the redox state according to the Eh value as measured with a platinum electrode or by an equivalent procedure. The following is an operational definition of the relative redox states of aqueous systems according to their measured Eh:

800 to 400 mV = Strongly oxidizing,
400 to 200 mV = Moderately oxidizing,
200 to $-50$ mV = Moderately reducing,
$-50$ to $-200$ mV = Reducing, and
$-200$ to $-400$ mV = Strongly reducing.

It should be emphasized that the above-defined boundaries between the various redox states are arbitrary and, in many cases, other boundaries between the redox states are set.

The Eh of aqueous systems is bound on the upper side by the oxidation of water ($O_2$ formation) and on the lower side by the reduction of water ($H_2$ formation). At pH 7 and at unit partial pressures of $O_2$ and $H_2$, the limiting Eh values are about 800 and $-400$ mV, respectively.

Table 5-2. Order of utilization of principle electron acceptors in soils and sediments, equilibrium potentials of the half-reactions at pH 7, and measured potentials of these reactions in soils (Bohn et al., 1985).

| Reaction | Eh at pH 7, V | Measured redox Potential in soils, V |
|---|---|---|
| $O_2$ Disappearance | | |
| $1/2\ O_2 + 2e^- + 2H^+ = H_2O$ | 0.82 | 0.6–0.4 |
| $NO_3$ Disappearance | | |
| $NO_3^- + 2e^- + 2H^+ = NO_2^- + H_2O$ | 0.54 | 0.5–0.2 |
| $Mn^{2+}$ Formation | | |
| $MnO_2 + 2e^- + 4H^+ = Mn^{2+} + 2H_2O$ | 0.40 | 0.4–0.2 |
| $Fe^{3+}$ Formation | | |
| $FeOOH + e^- + 3H^+ = Fe^{2+} + 2H_2O$ | 0.17 | 0.3–0.1 |
| $HS^-$ Formation | | |
| $SO_4^- + 6e^- + 9H^+ = HS^- + 4H_2O$ | −0.16 | 0– −0.15 |
| $H_2$ Formation | | |
| $H^+ + e^- = 1/2H_2$ | −0.41 | −0.15– −0.22 |
| $CH_4$ Formation (example of fermentation) | | |
| $(CH_2O)_n = n/2CO_2 + n/2CH_4$ | -- | −0.15– −0.22 |

A second classification of the redox states of aqueous systems is based on the redox activity of the microflora in the system (e.g., Rogers, 1986). The redox state is defined by the electron acceptor predominantly used by the microorganisms in the system. The redox half-reactions that define the redox states according to this classification are presented in Table 5-2.

Most sediments appear to be strongly electron buffered (poised). Titration with chemical oxidants such as hydrogen peroxide shows that the sediments have a high capacity to supply electrons. Although molecular $O_2$ can oxidize sediments, thus depleting the sediments' reducing capacity, the reaction occurs slowly (Macalady et al., 1986). The low rate of oxidation of the sediments may result either from the intrinsically slow rate of reaction of $O_2$ with the reducing moieties in the sediment or from the slow diffusion of $O_2$ into the sediment.

Measurements of Eh in sediments of ponds and rivers show that the Eh stabilizes at a depth of approximately 1 cm and stays constant to a depth of at least 5 to 6 cm (Wolfe et al., 1986). The half-reactions believed to be the most important in determining the redox state of sediments are presented in Table 5-2.

### 5-2.2 Abiotic Reactions Occurring in Sediments

Pesticides in sediments are exposed to an environment characterized by a close proximity to solid surfaces under permanently saturated conditions. As in the case of the water body itself, hydrolysis and redox reactions probably dominate the abiotic transformations of pesticides in sediments.

#### 5-2.2.1 Hydrolysis

Although pesticide hydrolysis has been observed in sediment samples, microcosms, and sediments in situ (Zepp et al. 1975), no experimental distinction was made in most cases between biologically mediated and abiotic hydrolysis. The assignment of a hydrolysis reaction as biotic or abiotic is often carried out as follows. The disappearance rate constant in distilled water buffered to the appropriate pH is compared with the disappearance rate constant in the system under study. If both disappearance rate constants are similar, then it is assumed that abiotic hydrolysis was dominant. If the rate constant is larger in the investigated system, then biolysis (enzyme-mediated hydrolysis) was invoked as the dominant transformation process. If the disappearance rate constant in the studied system is lower than in the buffered distilled water, then sorption to the solid particles is invoked to account for the observation. The above analysis neglects, however, surface-enhanced hydrolysis in sediments.

The role of sediments in the hydrolysis of pesticides is not well defined. Recent studies (Macalady & Wolfe, 1985) suggest that the manner in which sediments affect hydrolysis reactions is dependent on the nature of the hydrolyzed pesticide and on the type of hydrolysis. For example, neutral, or pH independent, hydrolysis of organophosphorothioate pesticides was not drastically affected by sorption. The hydrolytic disappearance rate constant of these pesticides in the aqueous phase was, within experimental error, the same as in the sorbed state (Macalady & Wolfe, 1985). A lack of heterogeneous catalysis was also reported for some pesticides that underwent acid and alkaline promoted hydrolysis reactions (Macalady & Wolfe, 1983). Surface-catalyzed hydrolysis in saturated systems was reported, on the other hand, for some pesticides (e.g., ciodrin; Konrad & Chester, 1969).

The products of abiotic hydrolysis in sediment systems were found to be, in many cases, the same as the products of hydrolysis in clear water (Macalady & Wolfe, 1985). Thus, while the solid phase may alter the kinetics of hydrolysis, it does not appear to alter, as a rule, the nature of the overall, net reaction.

In sediment systems, pesticides may adsorb on the solid phase. The fraction of the pesticide sorbed at equilibrium will be determined by the tendency of the pesticide molecule to accumulate at the surface of the solid phase and by the sediment to water ratio. Often, the sorption can be described to a good approximation as being at its equilibrium level, since both the adsorption and desorption processes are usually faster than diffusive or degradative processes. The ratio between the amount of pesticide adsorbed per unit weight of solid and the concentration of the pesticide in the liquid phase at equilibrium is defined as the partition coefficient ($K_p$). At sufficiently low concentrations (or in a sufficiently narrow concentration range), $K_p$ is well approximated by a constant. Furthermore, it was demonstrated (Karickhoff et al., 1979; Karickhoff & Morris, 1985) that, in many cases, nonionic pesticides are sorbed predominantly on the organic fraction of the sediment. The adsorption coefficient, $K_p$ can, in these cases, be normalized to the organic matter content of the sediment, giving an adsorption coefficient per unit weight of organic matter ($K_{oc}$). The $K_{oc}$ of many nonionic pesticides was shown not to vary significantly between sediments throughout a wide range of sediments (Karickhoff et al., 1979; Karickhoff, 1980).

Thus, the tendency of a pesticide to hydrolyze can be influenced in two major ways by sediments. First, the environment the pesticide experiences in the interstitial water (e.g., the activity of reactants such as water molecules, protons, and hydroxide ions) is affected by the proximity to solid surfaces and by the porous nature of the medium. A significant part of the interstitial water is close enough to the surface to be influenced by it. Secondly, the direct effect by the solid particles on the susceptibility of adsorbed pesticides to hydrolysis (e.g., through participation of acid groups belonging to the solid organic matter in heterogeneous hydrolysis; Armstrong & Konrad, 1974) may be significant.

### 5-2.2.2 Redox Reactions

A large variety of redox reactions may occur in sediments. Heterogeneous catalysis, and other surface effects as well as the restricted diffusion of $O_2$ through the porous medium, will all affect the rates of the redox transformations that may occur in the sediments. Redox reactions in sediments have been reported recently, for example, for methyl parathion and aldicarb (Wolfe et al., 1986; Macalady et al., 1986) (Fig. 5-5). Although many redox

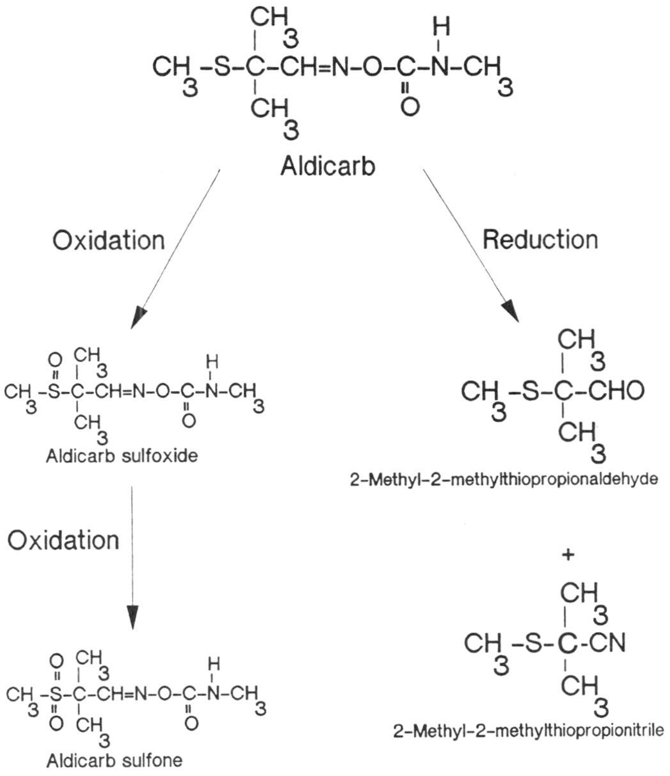


Fig. 5-5. Pathways and products for the redox transformations of aldicarb.

transformations of pesticides have been reported, in most cases there was no effort made to distinguish between abiotic and biotic pathways.

## 5-3 SOILS

### 5-3.1 Composition and Properties of the Medium and Their Relationship to Abiotic Transformation

The composition of the solid phase exerts a strong influence on the types and rates of the abiotic transformations that can occur in the soil. This is especially true for heterogeneous catalysis, since for a surface reaction to occur, catalytically active adsorption sites must be available. Most pesticides display a strong affinity to the soil organic matter. Yet, the high clay content in many soils and the large surface area and high charge of clay minerals make the contribution of the mineral clay fraction to surface-enhanced abiotic transformations in soils as important as the contribution of the organic matter if not more so. Abiotic transformations in the liquid phase can also be affected by the composition of the solid phase because components of the solid phase may control the chemical make-up of the soil solution. For example, carbonates, can act as a buffer and compounds containing redox active metals such as Fe or Mn, can poise the redox potential of the liquid phase.

#### 5-3.1.1 Mineral Fraction

Information on clay minerals prevalent in soils has been presented in detail previously (e.g., Dixon & Weed, 1977). Consequently, the properties of individual clays will not be further discussed here and only properties of soil minerals as a whole that are relevant to their role in abiotic transformations of pesticides will be reviewed. Clay minerals dominate the clay size fraction of mineral soils and at times the coarser fractions as well. In some soils, more than half of the solid mass is composed of clay minerals. Clay minerals are characterized by a relatively high surface area and charge density,and interact readily with any molecule that possesses a permanent or induced electric pole. Because of competition with water for adsorption sites, the potentially strong interaction does not always result in extensive adsorption. The sources and locations of charges in the clay mineral will affect the strength of the electric field at the surface and, hence, the strength of potential electrostatic interactions between the mineral and charged, polar, or polarizable adsorbates.

Many soils in arid areas (or in more humid areas if formed from the appropriate parent material) contain significant quantities of the relatively soluble calcium carbonate ($CaCO_3$) or calcium sulfate ($CaSO_4$). Some agricultural soils may contain more than 50% $CaCO_3$. The importance of $CaCO_3$ is mainly in its buffering of the pH of the soil solution to values

>7.5. Of course, this pH will strongly affect reactions such as hydrolysis in the liquid phase. The surface of the mineral itself may be a site of specific surface reactions (Mingelgrin & Yaron, 1974).

Iron and other metal oxides are found in soils in various crystalline and amorphous forms (Schwertmann & Taylor, 1977). The surface of iron oxides in soils is often hydroxylated, either structurally or through hydration of the Fe atoms. The hydroxylated system may adsorb further layers of water. The presence of the hydroxyls and highly polarized hydration water molecules on the oxide surfaces makes these surfaces effective heterogeneous catalysts for abiotic hydrolysis. Depending on the soil pH, anions, cations, and various nonionic species can interact with these oxides. Transition metal ions tend to form complexes with many ligands. Consequently, specific adsorption and surface-enhanced transformations of many pesticides may occur on the surface of oxides of transition metals such as Fe (Schwertmann et al., 1986). The ability of Fe and some other metal ions to undergo redox reactions further increases the role of the metal oxides in abiotic transformations of pesticides in soils. Aside from the direct interactions that may occur between metal oxides and pesticides, metal oxides may affect both the redox potential (especially under sufficiently reducing conditions) and the pH in the soil solution. Under tropical conditions, soils with significant quantities of metal oxides are prevalent. In oxisols, 20 to 80% of the solid phase was reported to be Fe as $Fe_2O_3$ (Soil Survey Staff, 1975).

It is hard to estimate the contribution of amorphous metal oxides to surface-enhanced transformations. Gorbunov et al. (1961) demonstrated that an amorphous iron oxide adsorbed 109 times more phosphate than crystallized oxides. Not only metal oxides, but also other minerals, such as aluminosilicates (e.g., allophane) and carbonates, may appear in soils in amorphous forms. The amorphous materials often coat crystals present in the soil. Therefore, aside from their direct contribution to the abiotic transformations in soils, the amorphous materials may affect these transformations by altering the surface properties of crystalline substances. The effect of the amorphous mineral fraction on abiotic interactions can be considerably larger than suggested by its content in the soil.

### 5-3.1.2 Organic Fraction

Most pesticides tend to adsorb on the organic matter in soils. Except for soils that are poor in organic matter, a rather good correlation is often observed between the organic matter content and the retention of many pesticides (e.g., Hamaker & Thompson, 1972). Extensive adsorption does not necessarily imply enhancement of abiotic transformations, but the organic fraction of the soil has no doubt a major effect on the behavior of pesticides in soils. Khan (1978) cited examples in which the organic fraction of the soil enhanced abiotic transformations of pesticides.

Humic substances are predominantly amorphous polymers that can exist either as three-dimensional particles or as a coating on mineral particles. The cation exchange capacity (CEC) of the soil organic matter is relatively high at the pH range prevalent in soils but is pH dependent. Thus, the soil organic matter possesses both hydrophilic and hydrophobic sites. The hydrophobic sites of the organic fraction often dominate the retention of nonionic pesticides from the soil solution. Competition with water reduces the role of the hydrophilic sites in the retention of pesticides. Because of the large variety of potential interaction sites at the surface of the organic matter, numerous types of surface interactions with a wide range of adsorbates may occur at that surface.

Plant and animal remains at various stages of decomposition are present in the solid phase. The relatively low surface area of these remains make them less important in affecting surface transformations in soils. Microbial surfaces are also present in the soil both as living and as dead biomass. Aside from their obvious role in biodegradation, microorganisms have surfaces for retention of molecules that may either enhance or hinder their abiotic transformations. Also, active biomass has a substantial effect on the chemical-composition of the liquid and gaseous phases of the soil (e.g., on the $CO_2$ and $O_2$ concentrations in these phases). Both the redox potential and the pH in the soil solution are strongly influenced by the microbial activity in the soil.

#### 5-3.1.3 Interactions between Components of the Solid Phase

The interaction between various components of the solid phase of the soil strongly affects the surface-enhanced transformations of pesticides. The surface of the soil solid phase is inhomogeneous. It is characterized by the multicomponent association between humic substances, clays, metal oxides, $CaCO_3$, and other minerals. In some cases, up to 90% of the soil organic matter was found to be associated with the mineral fraction of the soil (Greenland, 1965). On the other hand, there is evidence that much of the surface of clay minerals in soils, specifically in the interlayer spaces of smectites, is not covered by organic matter (Ahlrichs, 1972). Metal oxides are also likely to coat the external surfaces of clay minerals and intercalation of oligomeric hydroxyaluminum species with clays was reported (e.g., Ahlrichs, 1972). Cationic aluminum hydroxy-oxides, the charge of which is pH dependent, may coat clay surfaces thus reducing the contribution to CEC from the clay minerals. This phenomenon is more important in acid soils. As the pH increases, the hydroxyaluminum polymers lose their positive charge and their effect on the CEC is reduced.

Although coated or ill-defined particles may contribute most of the surface area in many soils, much of the more detailed work on surface-related abiotic transformations was conducted with pure soil constituents. In many cases, interactions between different solid components poison the surface for abiotic transformations. For example, coating clay surfaces with organic matter, and even more so with mineral oxides, binds or replaces exchangeable cations that are the sites of many surface-enhanced transformations (e.g., Mingelgrin et al., 1977). Coating may also block access to active sites that are not coated themselves. The presence in soil of numerous types of adsorption sites increases the likelihood of a competing adsorption at sites other than the reactive ones. A surface reaction that is enhanced by an isolated soil component may not occur in the soil to a significant extent, therefore, even if an exposed surface of that component is present.

### 5–3.1.4 Properties of the Soil Relevant to Abiotic Transformations

**5–3.1.4.1 The Exchange Complex.** Counterions must be located on or near the surface of the charged components of the soil. These ions create a strong electric field about them that causes the polarization and, in some cases, the dissociation of molecules adsorbed at the counterions. Water is the dominant molecule adsorbed on the counterions in soils. The polarization and dissociation of water molecules in the hydration shell of exchangeable cations is responsible for the enhancement of many hydrolytic and nonhydrolytic transformations of pesticides.

Much of the CEC of mineral soils arises from isomorphic substitution in clay minerals and is pH independent. The broken edges of clay crystals and the amorphous oxide coatings display, however, pH-dependent cation and anion exchange capacities. The organic matter in the soil possesses a high pH-dependent CEC. Most of the CEC of soils sufficiently rich in organic matter may be contributed by the organic fraction. A largely pH dependent anion exchange capacity is also displayed by the organic matter and by metal (especially iron) oxides in the soil.

Many abiotic transformations are strongly affected by the nature of exchangeable cations. The soil environment will determine which cations will dominate the exchange complex. In arid areas, Ca, Mg, and Na (and to a lesser extent K) are dominant. In more acid soils, such as those common in humid areas, Ca, Al, Mg, K and at sufficiently low pH even H become the more important exchangeable cations. Less frequently, $NH_4^+$ may also be important.

Continuous equilibration takes place between the bulk solution and the solid-liquid interface. Changes in the liquid phase (e.g., by irrigation with saline water or effluents) will, therefore, bring about changes in the composition of the exchange complex. Similarly, the composition of the exchange complex can affect the properties of the liquid phase. For example, as the exchangeable sodium percentage increases, the pH of the soil solution can also increase because of Donnan hydrolysis (e.g., Kamil & Shainberg, 1968).

Heavy metal cations often form slightly soluble hydroxides or salts. Negatively charged surfaces may serve as an efficient sink for such cations that might have otherwise precipitated in various insoluble forms. Heavy metal ions are well known as catalysts for many chemical reactions. Adsorbed ions are more likely than precipitated species to enhance or participate in abiotic transformations due to the higher accessibility of the adsorbed ions to potential reactants. Land disposal of wastes (e.g., through sludge application) and fertilization may introduce to the soil environment heavy metal ions in a free or in a complexed form. The high selectivity of the exchange complex for cations of heavy metals may bring about the cations' accumulation at the negatively charged surfaces and the consequent retention of at least some of their catalytic capacity.

**5-3.1.4.2 Liquid and Gaseous Phases.** A large fraction of the soil volume consists of pores that can contain both liquid and gas. The relative volume occupied by the liquid and by the gaseous phases is constantly changing as the moisture content of the soil changes. Although relatively little abiotic breakdown of pesticides takes place in the soil's gaseous phase, that phase can affect the kinetics of abiotic reactions that occur in the other phases. The $O_2$ and $CO_2$ concentrations in the soil air, for example, affect the redox potential and the pH in the soil solution.

The liquid phase of the soil is an aqueous suspension rich in mineral and organic solutes. Numerous abiotic reactions can occur in the liquid phase (the soil solution). The kinetics of some of these reactions are controlled by general properties of the soil solution (e.g., pH or redox potential), while other reactions are catalyzed by suspended or dissolved species such as transition metal ions. Thus, the less-abundant constituents of the soil solution, may have a stronger influence on the abiotic transformations of some pesticides than the major constituents.

There is a strong interdependence between the different phases as well as between biotic and abiotic activity in the soil. Some treatments, such as certain agricultural practices, may increase the bioactivity of the soil. The increased bioactivity can in turn reduce the redox potential of the soil by fast utilization of the $O_2$ in the system. The interdependence between the soil solution and the exchange complex was already mentioned above. Fluctuations in the composition of the liquid phase will bring about fluctuations in the surface properties of the solid phase through precipitation-dissolution and adsorption processes. Thus, as the concentration of the soil solution increases (through drying of the soil, for example), the fraction of the exchange complex that is occupied by cations with a lower charge increases at the expense of cations with a higher charge.

Despite their abundance in the soil solid phase, the concentration of polyvalent cations such as Fe and Al in the soil solution is relatively low. This concentration increases, however, with soil acidity. The redox potential may also affect the solubility of species of oxidizable elements. Thus, the pH and redox potential of the soil solution, aside from directly affecting the kinetics of reactions such as hydrolysis and oxidation or reduction transformations of pesticides, may control the concentration of species that can catalyze or participate in abiotic reactions.

The chemical composition of the soil solution may vary strongly with depth. Evaporation may result in high concentrations of nonvolatile solutes at the surface, whereas the soil solution at the lower horizons can be poor in externally added solutes because of volatilization, precipitation, adsorption, or degradation of the solutes at shallower depths. The pH, redox potential, and organic matter content will change with the depth. Thus, a pesticide may undergo different abiotic transformations at different depths of the same soil.

Organic substances such as fulvic or humic acids can be either dissolved or suspended in the soil solution. The fulvic and humic acids in the liquid phase have a lower molecular weight than those found in the solid phase.

Their molecular weight may be as low as 500 (Thurman, 1985). Soluble humic substances can interact with slightly soluble pesticides, thereby increasing the concentration of the pesticides in the soil solution. Aquatic humic substances increased, for example, the content of DDT in solution two to four times (e.g., Carter & Suffet, 1982). Similarly, Schnitzer and Khan (1972) demonstrated that the solubility of phthalates increased through complexation with aqueous fulvic acids. Geschwend and Wu (1985) discussed the interaction of hydrophobic organic substances with soluble macromolecules (or non-settling particles).

Interaction with soluble humic substances can enhance abiotic transformations of pesticides. The tendency of humic acids to complex catalytically active metal cations, suggests that abiotic transformations similar to those occurring at the surface of solids can be catalyzed by metal-pesticide-soluble humic associations. The chemical hydrolysis of atrazine in aqueous solution was strongly enhanced by the presence of fulvic acid (e.g., Khan, 1980).

The soil solution can also contain nonhumic organic substances originating from both indigenous biological activity and the application to soil of manure, effluent, sludge, fertilizers, or other agrochemicals. Among the more prominent organic solutes found in the soil solution are monosaccharides and polysaccharides and simple organic acids (e.g., Kaurichev et al., 1963, Takijima et al., 1961). Organic acids can complex transition metal cations, keeping them in solution and possibly increasing these cations' catalytic contribution to the transformation of dissolved pesticides. Conversely, interactions between organic solutes and transition metal ions may modify the effect of the organic species on the transformations of pesticides. Thus, although the association of atrazine with protonated fulvic acids in the soil solution enhances the pesticide's hydrolysis, the presence of Cu(II) retards the catalytic effect of the fulvic acids (Haniff et al., 1985).

**5-3.1.4.3. Adsorbed Water and the Diffuse Electric Double Layer Region.** Water molecules adsorb strongly on clay minerals, on many sites in the humic complex, and on other surfaces in the soil. As a result, water molecules may both compete for adsorption sites with other adsorbates and serve as adsorption sites (e.g., through water bridges). Adsorbed water, and in particular hydration water of adsorbed ions, participates in numerous surface-enhanced hydrolysis reactions. Aside from acting as a reactant in surface reactions, the strongly polarized hydration water catalyzes various transformations of adsorbed pesticides. The interaction of water with clays affects the kinetics of abiotic transformations of pesticides in yet another way: The moisture content of the soil affects the extent of swelling or disperion of expandable clays. This swelling, in turn, will influence the accessibility of reactive sites at the interlayer surfaces to pesticides.

The properties of the liquid phase in the vicinity of the surface of a charged solid are different from the properties of the bulk liquid phase, because of the electric field emanating from the solid. The volume of the liquid phase in which the strength of the electric field is not negligible is defined as the diffuse electric double-layer region. In that volume, the concentration of both anions and cations is significantly different from their concentration in the bulk solution. Near negatively charged surfaces, the cations' concentration steeply increases as the distance to the surface decreases. The concentration of the anions behaves in the opposite way. The thickness of the double layer is an inverse function of the valence of the exchangeable counterions and of the ionic concentration in the soil solution.

Under the conditions prevalent in soils, the double layer at field capacity or at higher moisture contents may, in many cases, be considerably thicker than 1.0 nm (van Olphen, 1966). The kinetics ofthe abiotic transformations that occur in the double-layer region are often different from the kinetics of the transformations that occur in either the bulk liquid phase or at specific adsorption sites on the surface. The diffuse electric double-layer region is distinct not only in its ionic composition, but also in the reactivity of water molecules found in that region. The electric field in the double-layer region, arising from either the negative charge of the solid or from the positive charge of the exchangeable cations, increases the polarization and the dissociability of water molecules in that region (Mortland, 1970).

Accordingly, the proton concentration in the diffuse electric double-layer region near negatively charged surfaces is higher and the hydroxyl ion concentration is lower than in the bulk solution. This difference in concentrations results in a difference in the rate of pH controlled reactions, such as hydrolysis, between the bulk solution and the double-layer region. Just as the electric field at the surface enhances the polarization and dissociability of water, it can enhance the polarization and dissociability of many solutes, such as carboxylic acids or phenols. Such a modification in the properties of the solutes must affect the solutes' tendency to undergo both biotic and abiotic transformations.

**5–3.1.4.4 pH and Eh.** Many abiotic transformations in the soil require the presence of either a specific solute or a specific adsorption site. Some parameters, however, can control the lability of pesticides to abiotic breakdown independently of the species that determine the level of these parameters. Probably the two most important such parameters are the pH and Eh.

The pH of the soil solution has a critical effect on the rate of many abiotic transformations. Chapman and Cole (1982) measured the pH dependence in the pH range of 4.5 to 8.0 of the half-life of 24 insecticides in sterile solutions. Most pesticides had a much shorter half-life at the higher pH values. The half-life of five pesticides was hardly affected by the pH, and in one case, the half-life was shortest at the lower pH range. Thus, the effect of the pH on the rate of many pesticide transformations, as strong as it is, is hard to predict a priori.

As mentioned above, the proton and hydroxyl concentrations in the soil solution are strong functions of the distance from charged surfaces (e.g., Bailey et al., 1968). The efficiency of the soil solution in enhancing pH-catalyzed transformations may therefore, depend on the moisture content of the soil. The dependency of the rates of the pH-catalyzed transformations on the moisture content should increase, as the moisture content decreases.

In dry soils, where a liquid phase is absent, surface acidity is the parameter analogous to the pH in solution as far as abiotic transformations are concerned. Surface acidity (the capacity of the surface to act as a Brønsted or a Lewis acid) may also be an important factor in determining the kinetics of surface-enhanced reactions in the presence of a liquid phase. Some of the surface acidity results from the increased ease of dissociation of adsorbed water molecules. Because the dissociation of water contributes as many hydroxyls as protons, those reactions in which sorbed water molecules are involved can be both acid and base catalyzed. The designation of surface acidity as the parameter controlling these reactions is not strictly accurate (e.g., Mingelgrin et al., 1977). The capacity of the adsorption site to polarize the adsorbed water molecule is, in some cases, more appropriate a parameter than the site's acidity.

The $CO_3^{2-}$-$HCO_3^-$-$CO_2$ (g) system buffers, as a rule, the pH of calcareous soils at pH above 7.5. In soils where $CaCO_3$ is not present, the exchange complex often buffers the soil's pH. Protons, being cations themselves, can exchange with other exchangeable cations. Negatively charged surfaces can act, therefore, as partially dissociated weak acids. Exchangeable polyvalent cations such as Al, tend to keep the pH of the soil solution low by undergoing hydrolysis. In acid soils, values below pH 4 may be reached. After draining, water-logged, ferrous sulfide containing soils in the tropics displayed a pH of 2. This low value resulted apparently from the oxidation of both $S^{2-}$ and Fe(II). Sodium-rich soils may reach pH values above 10. The pH of a soil may vary substantially with depth, and horizontal variations between different microenvironments may also exist.

The term *Eh* is defined as the redox potential relative to the standard hydrogen half-cell. In the following, the terms *redox potential* and Eh will be used interchangeably. Under most conditions, the redox potential of the soil is controlled by the concentration of $O_2$ in the gas and liquid phases. This potential is a function of the ease of gas exchange with the atmosphere and the level of respiration in the soil. The level of respiration, in turn, is dictated by such factors as fresh organic matter content, temperature, and moisture content.

Addition of fresh organic matter to the soil (e.g., manure) can, therefore, cause a reduction in the Eh of the soil. Treatments that change the metabolic activity in the soil can affect not only the biotic degradation of pesticides, but also their tendency to become oxidized or reduced abiotically. The upper layer of the soil displays faster rates of both drying and gas exchange than lower layers. Accordingly, the prevalance of reduced conditions and with it the probability that pesticides undergo reduction will increase with depth.

The redox potential determines the ratio between the activities of the oxidized and the reduced members of a redox couple at equilibrium. It says nothing about the rate of the redox reaction. High activation energies can inhibit some redox reactions, even if the redox potential of the system indicated that these reactions should occur. In many cases, a catalyst is required for the redox reaction to occur. Probably most redox processes in the soil are microbially catalyzed.

Typical Eh values in well-aerated soil systems are between 0.8 and 0.4 V. In less well-aerated soils, the rate of entry of $O_2$ from the outer atmosphere is insufficient to compensate for the uptake of $O_2$ by aerobic respiration (and to a lesser extent by chemical oxidation) that prevails under well-aerated conditions and the Eh, consequently, is reduced. Soils having Eh values between 0.4 and 0.1 V are considered to be moderately reduced. Those soils having an Eh value down to $-0.1$ V are termed reduced and those with yet lower Eh values are considered highly reduced (Patrick & Mahapatra, 1968). Highly reduced conditions, with typical Eh values between $-0.1$ and $-0.3$ V, occur in flooded soils such as rice paddies and in some histosols. In the case of histosols, however, oxidation is often limited by low temperature and pH rather than by an insufficient $O_2$ concentration. In sufficiently reduced soils, Mn, and at lower Eh values Fe, tend to get reduced and thus, poise the redox potential.

The reduced forms of polyvalent cations are usually much more soluble than their oxidized forms. As a result, whenever the aeration is poor, appreciable concentrations of reduced cationic species, such as $Fe^{2+}$, can be present in the soil solution. This reduction and consequent dissolution can have a strong influence on the abiotic transformations of pesticides in the liquid phase, for example, through the capacity of many species of heavy metals to catalyze, or participate in, abiotic transformation. Under sufficiently reducing conditions, sulfate may be reduced, especially in acidic soils, producing such S species as sulfides that can be involved in the degradation of certain pesticides (e.g., Wahid & Sethunathan, 1979).

The values of the Eh and the pH parameters in soils are strongly dependent on each other. The evolution of $CO_2$, the most common end product of the reduction of $O_2$, has considerable influence on the soil's pH. When a soil system that was previously under reduced conditions becomes oxidized (e.g., by drainage), its pH may decrease drastically due to the oxidation of iron to Fe(III) and the subsequent hydrolysis of the iron or the oxidation of sulfite to sulfate, which is accompanied by the release of protons. Lowering the Eh of the soil (e.g., by submerging) often will result in a rise in the pH, because many reduction reactions (such as the reduction of sulfate to sulfite) involve the uptake of protons or the release of hydroxyls.

When the reaction of a couple that controls the redox potential of a given soil system involves protons or hydroxyls, a change in the pH of the system will directly cause a change in its Eh. The pH may affect the rate and direction of a redox reaction by determining the concentrations of members of the redox couple in the soil solution. Acidification of the soil, for example, is likely to strongly increase the solubility of trivalent iron and of other oxidized transition metal species, but will have a smaller effect on the solubility of the reduced species of these metals. The redox reactivity of a pesticide in soil, can, therefore, depend on the pH. Manipulating the pH of the soil (e.g., by liming) can affect redox reactions of pesticides, just as it affects other pH-dependent reactions such as hydrolysis.

## 5-3.2 Types and Kinetics of Abiotic Transformations

### 5-3.2.1 Abiotic Transformations in the Liquid Phase

A large variety of chemical transformations of pesticides may occur in the soil solution. The rate of some of these transformations is controlled by general environmental parameters such as pH and temperature, whereas other transformations may be catalyzed by one of the many species that exist in the soil solution, often at trace levels. Chemical transformations of pesticides in soil solution resemble the transformation pesticides undergo in natural water bodies (section 5-1). In the following discussion, emphasis will, therefore, be placed on aspects more characteristic to the soil solution.

Perhaps the most common reaction pesticides undergo in the liquid phase is hydrolysis. Pesticide hydrolysis produces, as a rule, acids or bases. This implies that the products of hydrolysis are often more polar than the parent material and hence more soluble in water. The interaction with the solid phase and the mobility of these products may thus be very different from that of the original pesticide. Although ester hydrolysis is probably the most important hydrolysis reaction of pesticides, hydrolysis of other groups may also be of great importance. One example of the many nonester hydrolysis reactions is that of dalapon. In basic media, dalpon-sodium interacts with water to give pyruvic acid, NaCl, and hydrochloric acid. The rate of this reaction at 50 °C is rather rapid. Temperatures above 40 °C are not uncommon during the summer in the top soils of warmer regions. Hydrolysis reactions in water were discussed in sections 5-1.2 and 5-1.3.

Microorganisms play a dominant role in many of the oxidation reactions that pesticides undergo in the soil. Yet, in the case of some pesticides, nonbiological oxidation dominates the pesticide's degradation. Pesticides that contain oxidizable metal cations, for example, can be oxidized in the soil abiotically. Kaufman et al. (1968) suggested that the chemical oxidation of the herbicide amitrole by $OH\cdot$ or by another free radical is probably the major pathway for the pesticide's disappearance in soils. The final product of this chemical oxidation is $CO_2$.

The soil solution contains numerous species, even if in trace quantities, that can produce free radicals through chemical or photochemical processes. Riboflavin (Plimmer et al., 1967) and ascorbate in the presence of Cu (Castelfranco et al., 1963) are examples of such species. Accordingly, a relative abundance in soils of free radicals was claimed (Steelink & Tollin, 1967). It is expected that transformations of pesticides in which free radicals participate will occur in the soil solution.

Free radicals were indeed implicated in many transformations of pesticides in the soil (e.g., Khan, 1980; Kaufman et al., 1968). The reduction of DDT under anaerobic conditions (Glass, 1972) can proceed through the transfer of an electron from a ferrous ion to a DDT molecule. The DDT molecule then releases a chloride ion and becomes a free radical. The free-radical strips a proton from an organic donor to become DDD and the organic donor becomes a free radical itself. The high reactivity of free radicals results in a considerable increase in the rate of a reaction in which free radicals participate as compared with the rate of the same transformation through pathways in which free radicals do not participate.

Another abiotic oxidation that occurs in soils is that of organomercury pesticides with the general formula RHgX, where R is a hydrocarbon radical and X is an acid group. These pesticides undergo reaction in the soil:

$$RHgX + H_2O + ox \rightarrow ROH + HgXOH \quad [5]$$

where ox is an oxidizing agent. The mercury-containing product continues to degrade in the soil with the ultimate formation of HgS (Melnikov, 1971).

The main pathway of trifluralin degradation in many soils is, probably, through a nonbiological reductive process (Probst & Tepe, 1969), that involves the conversion of the nitro groups into amino groups. It is not clear, however, to what extent that reductive process occurs in the liquid phase. Under sufficiently reducing conditions, pesticides in the soil solution can participate in redox reactions with Fe(II) ions. The insecticides mirex and toxaphene, for example, are reduced by $Fe^{2+}$ through partial dechlorination (Holmstead, 1976; Khalifa et al., 1976).

Many transformations, other than hydrolysis and redox, can occur in the soil solution. Glyphosate undergoes nitrosation to form N-nitrosoglyphosate in the presence of nitrite (Young & Kahn, 1978). The nitrite ion is frequently found in the soil solution, usually at concentrations below 3 mg $L^{-1}$. Plimmer et al. (1970) suggested that nitrite also participates in the degradation of propanil in soil. Chlorotriazine herbicides readily undergo a displacement of the chlorine atom by various nucleophiles (Knuesli et al., 1969). The replacement of Cl by OH, for example, is both acid and base catalyzed (Burkhard & Guth, 1981). Alkyl mercaptans and alcohols can enhance the displacement of the chlorine atom in chlorotriazines. Some compounds, such as *N*-hydroxy-benzoxazinones, that are found in plants and, therefore, may be released to the soil solution at low concentrations, are also capable of enhancing this displacement. At any rate, hydroxyatrazine was reported as the major product of atrazine degradation in sterilized soils (Knuesli et al., 1969).

Organophosphate esters in solution undergo many heavy-metals catalyzed transformations including hydrolysis (Mortland & Raman, 1967; Bowman & Sans, 1980). The transition complexes in these transformations (e.g., Fig. 5–6) are examples of short-lived complexes between a pesticide and a heavy metal ion. Stable complexes between such ions and pesticides are also formed in the soil and can affect the mobility, stability, and bioactivity of the pesticide.

The tendency of a reaction to occur, as expressed by the law of mass action, should be distinguished from the rate at which this reaction will actually proceed. In most reactions, there is an energy barrier that has to be overcome for the reaction to take place. The higher the barrier, the slower the reaction. In dynamic systems such as soils, which are never at equilibri-

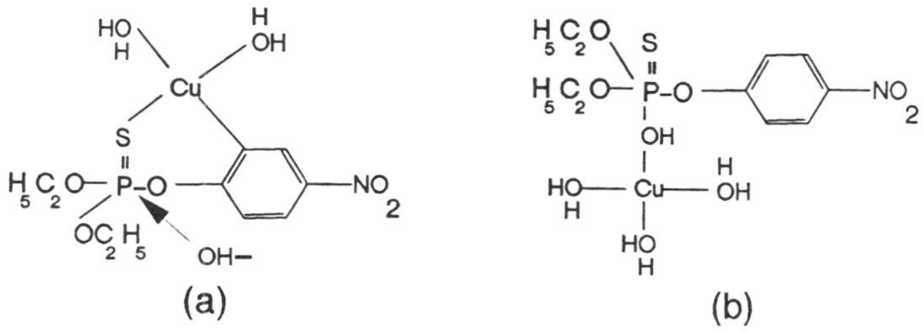

Fig. 5–6. Possible transition complexes for the Cu-catalyzed hydrolysis for parathion in aqueous solution [complex (a) from Mortland & Raman, 1967].

um, the rate of the reaction may be, in many cases, more important than the equilibrium constant in determining which species will be present at significant concentrations.

Redox reactions that are not biologically catalyzed often proceed in the soil solution at a slow rate. Thus, although oxidation of organic substances by $SO_4^{2-}$ is thermodynamically possible in many sulfate-containing soils and especially under acid conditions, this oxidation usually proceeds slowly without microbial mediation. Some redox reactions of pesticides that would proceed at a slow rate in the absence of a suitable catalyst, may, on the other hand be enhanced by one of the many potential catalysts and inductors that are among the numerous species continuously formed in the soil environment, or released into it (e.g., the above discussed oxidation of amitrole by free radicals).

The rates of chemical reactions in the liquid phase may be strongly affected by the moisture content of the soil. The moisture content influences the total solute concentration in the soil solution, thus affecting the relationship between concentrations and activities. A solute whose concentration is below saturation can be concentrated by partial drying and diluted by wetting. The concentration of a solute that is near saturation may be kept relatively constant by precipitation and dissolution (if a solid component that can release that solute into the soil solution is present). Adsorption-desorption also can dampen the fluctuations in concentration upon partial wetting and drying processes. Changes in the moisture content of the soil can, therefore, modify the activities of different reactants and products of a reaction in different ways. This, in turn, will affect both the direction and the rate of that reaction. As mentioned above, the moisture content can have a strong effect on the redox potential of the soil by regulating the rates of gas exchange and respiration. Sediments, which are permanently saturated, can be viewed as a limiting case of soils as far as the effect of the moisture content on abiotic transformations is concerned.

### 5–3.2.2 Abiotic Transformations at the Solid-Liquid Interface

The environment in the proximity of a charged solid surface is different from that in the bulk solution. As discussed above, both the spatial distribution of ions and the charge distribution within polarizable species (and hence their dipole moments and dissociation constants) are strongly influenced by the electric field emanating from charged surfaces. Consequently, the tendency of pesticides to undergo abiotic transformations is often much higher near the surface than in the bulk solution. Molecules in direct contact with the surface can undergo many transformations catalyzed by the adsorption sites.

Heterogeneous catalysis that requires adsorption on specific sites on the surface, is similar in many respects to enzymatic catalysis. Just as in the case of enzymatic reactions, small changes in the nature of the surface or in the environment around it may strongly affect the efficiency of the surface as a catalyst even if the active site itself remains unaltered.

Although many abiotic transformations progress faster at the interface than in the bulk solution, other reactions are inhibited by proximity to the surface. A higher persistence of pesticides in the adsorbed state than in the free state was reported by many authors (e.g., Hurle & Walker, 1980; Macalady & Wolfe, 1985).

In some cases, the surface is modified during the reaction with a pesticide. The reaction product can, for example, remain bound to the surface. Thus, the term *catalysis* does not strictly apply to all cases of acceleration of a reaction by adsorption. Enhancement of abiotic reactions by adsorption will be referred to here as *catalysis,* however, as is the practice of many authors.

Among the numerous pesticides that are reported to break down by surface catalysis are trifluralin (Probst & Tepe, 1969), diazinon (Konrad et al., 1967), ciodrin (Konrad & Chesters, 1969), parathion (e.g., Mingelgrin et al., 1977), azido-triazines (Barnsley & Gabbott, 1966), and others (e.g., Morrill et al., 1982). Hance (1970) concluded that adsorption catalyzes the hydrolysis of chlorotriazines in soils and Spencer et al. (1980a) reported the conversion of paration to paraoxon on soil dusts in the presence of atmospheric oxidants.

**5-3.2.2.1 Clays.** Catalytic reactions on clay surfaces were reported by many authors and several reviews on that subject were published (e.g., Mortland, 1970; Theng, 1974) (Table 5-3). Mortland and Raman (1967) showed that Cu(II), which catalyzes the hydrolysis of organophosphate esters in solution, retained its catalytic capacity as an exchangeable cation on montmorillonite. Magnesium-montmorillonite exhibited a much weaker catalytic effect than the Cu-clay. Copper-beidellite and nontronite exhibited a lower catalytic capacity than Cu-montmorillonite. These results support the suggestion that, as in the case of enzymes, the catalytic activity of a surface is often strongly influenced by small changes in the properties of the surface, even when the active site (the exchangeable Cu[II] ion) remains the same.

Rosenfield and Van Valkenburg (1965) suggested that the degradation of ronnel adsorbed on bentonite is catalyzed by Al. The mechanism of surface-catalyzed transformations of organophosphate ester pesticides (such as ronnel) and the dependence of these transformations on the exchangeable cations was studied in detail (e.g., Mingelgrin et al., 1977; Mingelgrin & Saltzman, 1979).

Table 5–3. Some reactions catalyzed by clay surfaces.

| Substrate | Reaction | Type of clay | Remarks | References |
|---|---|---|---|---|
| Ethyl acetate | Hydrolysis | Acid clays | -- | McAuliffe & Coleman, 1955 |
| Sucrose | Inversion | Acid clays | -- | McAuliffe & Coleman, 1955 |
| Alcohols, alkene | Ester formation | Al-montmorillonite | -- | Adams et al., 1983 |
| Organophosphate esters | Hydrolysis | Cu- and Mg-montmorillonite, Cu-Bidellite, Cu-Nontronite | Cu-Mont. better catalyst than other Cu-clays or Mg-Mont. | Mortland & Raman, 1967 |
| Organophosphate esters | Hydrolysis and rearrangement | Na-, Ca- and Al-kaolinites and bentonites | Room and other temperatures. No liquid phase | Mingelgrin et al., 1975, 1977, 1978; Saltzman et al., 1974, 1976; Mingelgrin & Saltzman, 1979 |
| Phosmet | Hydrolysis | Homoionic montmorillonite | In suspension | Sanchez Camazano & Sanchez Martin, 1983b |
| Ronnel | Hydrolysis and rearrangement | Acidified bentonite; dominant exchangeable cations: H, Ca, Mg, Al, Fe(III) | Al catalyzed reaction suggested | Rosenfield & Van Valkenburg, 1965 |
| *s*-Triazines | Hydrolysis | Montmorillonitic clay | Cl-Analog degrades faster than methoxy or methoxy-thio compounds | Brown & White, 1969 |
| Atrazine | Hydroxylation | H-Montmorillonite | -- | Russel et al., 1968a |
| Atrazine | Hydrolysis | Al- and H-montmorillonite; montmorillonitic soil clay | Ca- and Cu clays much weaker catalysts | Skipper et al., 1978 |
| 1-(4-methoxyphenyl)-2,3-epoxypropane | Hydrolysis | Homoionic montmorillonites; Na-Kaolinite | -- | El-Amamy & Mill, 1984 |
| DDT | Transformation to DDE | Homoionic clays | Na-Bentonite is a better catalyst than H-bentonite | Lopez-Gonzales & Valenzuela-Calahorro, 1970 |
| Urea | Ammonification | Cu-Montmorillonite | Air-dryed clay at 20 °C | Boyd & Mortland, 1985 |
| Aromatic amines | Redox | Clays (review) | In presence of metal ions that are good redox agents | Theng, 1974 |
| 3,3′,5,5′ Tetra methyl benzidine | Redox | Hectorite | $O_2$ oxidizing agent | McBride, 1985 |
| Pyridine derivatives, olefines, dienes | Oligomerization | Clays (review) | -- | Theng, 1974 |
| Glycine | Oligomeryzation | Na-Kaolinite, Na-Bentonite | -- | Lahav et al., 1978 |
| Styrene | Polymerization | Palygorskyte, kaolinite, and montmorillonite | Weakest catalyst is montmorillonite | Solomon & Rosser, 1965 |
| Fenarimol | Dialdehyde formation | Homoionic montmorillonite | -- | Fusi et al., 1983 |

Brown and White (1969) discussed the effect of the substituents of *s*-triazines on the transformation of these pesticides at montmorillonitic clay surfaces. The methoxy and methoxy-thio compounds hydrolyzed less rapidly than their Cl analogs. Although the protonation and adsorption of the triazines were directly related to their basicity, the rate of degradation of these pesticides was inversely related to the basicity. The above authors also observed that after a certain part of the adsorbed pesticide decomposed, no further degradation took place. This behavior, typical of many surface-enhanced transformations will be discussed below.

Russell et al. (1968a) reported the catalytic hydroxylation of atrazine on H-montmorillonite. This reaction involved the substitution of the chlorine atom by a hydroxyl, and the degradation product remained apparently adsorbed on the clay as the keto form of the protonated hydroxy analog. Skipper et al. (1978) showed that while Al- and H- montmorillonite and a montmorillonitic soil clay promoted the hydrolysis of atrazine, the Ca- and Cu-clays enhanced that transformation to a far lesser extent.

There are many examples of nonbiological redox reactions of pesticides occurring in soils (Table 5–4). Redox transformations of aromatic amines were observed on clay surfaces in the presence of those metal ions that are good reducing or oxidizing agents (Theng, 1974). In some cases, $O_2$ rather than metal ions, oxidizes organic molecules adsorbed on clay surfaces (e.g., 3,3′,5,5′-tetramethyl benzidine on hectorite, McBride, 1985).

Clays can catalyze (and in other cases inhibit) oligomerization reactions (e.g., Lahav et al., 1978; Theng, 1974). Carbonium ion formation by surface protons was implicated in the polymerization of styrene, and accordingly, H- and Al-containing montmorillonites displayed a catalytic polymerization (Solomon & Rosser, 1965).

Any organic adsorbate, especially if it contains an unsaturated C, can interact with acidic surface sites to form a carbonium ion. The unstable carbonium ion can act then as an intermediate in numerous transformations. The fungicide fenarimol, for example, partially transformed at 100 °C when adsorbed on montmorillonite into a carbonium ion, probably by interaction with the protons of the solvation water of the exchangeable cations (Fusi et al., 1983).

The catalysis of various types of rearrangement reactions of pesticides by clays has also been reported. Mingelgrin et al. (1978) and Mingelgrin and Saltzman (1979) described the rearrangement of parathion to *O,S*-diethyl *O*-*p*-nitrophenyl phosphate at room temperature when adsorbed on bentonite or kaolinite in the absence of a liquid phase. Parathion adsorbed on kaolinite, degraded in two parallel pathways. While some of the pesticide hydrolyzed, some of it underwent rearrangement followed by hydrolysis. The catalytically active sites were probably the hydration shells of the exchangeable cations. The rate of the rearrangement reaction increased with the polarity of the water of hydration. Lopez-Gonzales and Valenzuela-Calahorro (1970) reported an interesting clay-catalyzed dehydrochlorination reaction. These authors stated that DDT is transformed into DDE at clay mineral surfaces and that Na-bentonite catalyzes the transformation more efficiently than the corresponding H-clay.

Table 5-4. Some nonbiological redox reactions of pesticides occurring in soils.

| Pesticide | Reaction | Remarks | Reference |
|---|---|---|---|
| Arsenious acid | Oxidation | Slow, in acid media | Worthing, 1983 |
| Amitrole | Oxidation to $CO_2$ | Free radicals participation | Kaufman et al., 1968 |
| Organomercury pesicides | Oxidation $RHgX \rightarrow HgXOH$ | R = Hydrocarbon radical X = acid group | Melnikov, 1971 |
| Trifluralin | Reduction | Nitro groups → → Amino groups | Probst & Tepe, 1969 |
| Mirex; toxaphene | Reduction | By Fe(II), partial dechlorination | Holmstead, 1976; Khalifa et al., 1976 |
| DDT | Reduction to DDD (Cl removal) | By $Fe^{2+}$ | Glass, 1972 |
| Phorate | *s*-Oxidation | On soil organic matter (review) | Khan, 1978 |
| Phenoloxidase substrates | Oxidative coupling | Catalyst: Manganese-citrate complex | Loll & Bollag, 1985 |

**5-3.2.2.2 Organic Matter.** Organic matter often dominates the adsorption of pesticides in soils (e.g., Hamaker, 1975; Burns, 1975) and Stevenson (1982) suggested that the soil organic matter is capable of promoting abiotic transformations of many pesticides. Stevenson based this suggestion on two facts: (i) the organic fraction contains many reactive groups that are known to enhance chemical changes in several families of organic substances and (ii) humic substances possess a strong reducing capacity (Stevenson, 1985). The presence of relatively stable-free radicals in the fulvic and humic acid fractions of the soil organic matter further supports the assumption that the organic matter enhances abiotic transformations of many pesticides (Khan, 1978; Stevenson, 1982).

The role of humic substances in the abiotic transformations of pesticides was studied in less detail than the role of clay surfaces. Khan (1978) cited some abiotic transformations of pesticides that were enhanced by the soil organic matter or substances that are likely components of the soil organic matter. These transformations include the hydrolysis of organophosphorus esters, the dehydrochlorination of chlorinated hydrocarbons such as DDT and lindane, the degradation of 3-aminotriazole, the *S*-oxidation of phorate and the conversion of aldrin to dieldrin.

Armstrong and Konrad (1974), as well as other investigators, suggested that hydrolysis of chloro-*s*-triazines (the substitution of the chlorine by a hydroxyl) in soils occurs through the interaction of the pesticides with carboxyl groups of the soil organic matter. Stevenson (1972, 1982) suggested that after the removal of the side chain of phenoxyalkanoates, possibly by enzymatic action, the phenolic metabolites transform into humic-like substances by interaction with the amino groups of the soil organic matter. The above author also suggested that amines produced in the biological degradation of phenylcarbamates and other related pesticides may undergo an analogous transformation through interaction with carbonyl groups in the soil organic matter. The last two reactions are examples of abiotic transformations of intermediates produced through biotic degradation of pesticides.

**5-3.2.2.3 Other Solid Components.** Adsorption by substances other than clays and organic matter can also enhance transformations of pesticides. Metal oxides, for example, may enhance redox and other reactions of species adsorbed on their surfaces (e.g., Schwertmann et al., 1986). Nash et al. (1973) reported that MgO enhanced the degradation of DDT even at −5 °C in dry soils, thus demonstrating that the degradation is an abiotic, surface-catalyzed reaction. Aluminum oxide catalyzed the hydrolysis and rearrangement of parathion (Mingelgrin & Saltzman, 1979). $Na_2CO_3$, $CaSO_4$, and to a lesser extent $CaCO_3$ also enhanced the hydrolysis of parathion (Mingelgrin et al., 1977).

**5-3.2.2.4 Nonspecifically Enhanced Surface Transformations.** Heterogeneously catalyzed transformations, require specific sites at the surface with which the reacting pesticides must be in contact for the transformation to occur. Charged surfaces can, however, enhance the transformations of pesticides that are found in the vicinity of the surface but are not adsorbed at specific sites. The strong effect the electric field originating from the charged surface has on the properties of that part of the liquid phase that is under the field's influence was already pointed out above. Probably the most outstanding phenomenon in the interfacial region near the surface of charged solids is the strong dependence of the concentration of charged solutes on the distance from the surface. The concentration in the interfacial region of charged inductors, catalysts, reactants and products can, therefore, be different from their concentration in the bulk solution. Skujins (1967) has pointed out the importance of the excess of protons in the diffuse double-layer region near negatively charged surfaces for enzymatic reactions near such surfaces. Another factor that can affect pesticide transformations in the interfacial region near charged particles is the aforementioned influence of the electric field on the polarization and dissociation of both solutes and solvents.

Polar species will orient near the charged surface according to the direction of the electric field. The well-defined orientation of polar molecules relative to each other, as opposed to the nearly random orientation in the bulk solution, can, in some cases, hinder and, in other cases, enhance their abiotic transformation.

Adsorption measurements determine the accumulation of a solute in the interfacial region in excess of the amount required to maintain in the interface the same concentration as in the bulk liquid phase. The measurements do not distinguish between specific adsorption and nonspecific excess concentration in the interfacial region (e.g., of counter ions in the diffuse double layer). Hence, some of the transformations that were observed in the presence of a liquid phase and shown to be adsorption catalyzed (e.g., Konrad & Chesters, 1969) may have been enhanced by the special conditions prevalent in the interfacial volume and not by direct interaction with specific sites at the solid surface. It is possible that some transformations that were attributed to surface acidity (the ability of the surface to act as either a Brønsted or a Lewis acid) could have actually been enhanced by the increase in proton concentration as the surface of a negatively charged solid is approached.

### 5-3.2.3 Kinetics of Abiotic Transformations at the Solid-Liquid Interface

Environmental factors, such as temperature and moisture content, influence the kinetics of surface reactions differently from the way they influence the kinetics of homogeneous reactions. Attempts to assay the importance of abiotic transformations of pesticides in the soil by extrapolating rates of reaction at elevated temperatures to ambient temperatures (Hance, 1967) may have, therefore, underestimated the contribution of the abiotic reactions. A continuous increase in the rate of degradation of amitrole with temperature was observed by Ecregovich and Frear (1964) between 8 and 100 °C. This temperature dependence indicates the presence of an abiotic transformation. Yet, taking into consideration the complexity of the temperature dependence of the kinetics of surface-catalyzed reactions, the absence of an increase in the reaction rate with temperature up to 100 °C would not have proven that the reaction is biologically controlled. Because reactions that are enhanced by extracellular enzymes are heat labile as are microbial transformations, one cannot easily distinguish between microbial transformations and extracellular enzymatic transformations by their temperature dependence.

**5-3.2.3.1 Temperature.** The adsorbed reactant and the site at the surface to which it is adsorbed form the intermediate complex through which the surface-catalyzed reaction must pass. Either the formation or the breakdown of this complex is the rate-limiting step of the surface-catalyzed reaction. It is evident that the adsorption of a pesticide in a suitable mode is a prerequisite for the occurrence of heterogeneous catalysis. The temperature dependence of surface-catalyzed transformations is influenced, therefore, by the effect of the temperature on the adsorption. Frequently, the adsorption is exothermic and the adsorption coefficient is, thus, negatively correlated with the temperature (e.g., Bansal, 1983), whereas the rate of the surface catalyzed degradative step is positively correlated with the temperature.

A sufficiently large adsorbate (as most pesticides are) can interact with various sites on the surface simultaneously. Nondegradative interactions may strongly affect the rate of the surface-catalyzed reaction, for example, by modifying the orientation of the adsorbate species relative to the surface or by affecting the electron charge distribution in that part of the molecule in which the surface-catalyzed transformation occurs. The various simultaneous interactions between the surface and the pesticide can each depend on the temperature in a different way.

The net temperature dependence of the kinetics of heterogeneously catalyzed transformations is, thus, hard to predict. It is, accordingly, difficult to extract reliable thermodynamic parameters from the temperature dependence of the rate of many such transformations. Although the rate of hydrolysis of the organophosphate ester pirimiphos ethyl on Na-kaolinite was practically temperature independent between 23 and 47 °C, the rate of that reaction on Na-bentonite increased strongly with the temperature (Mingelgrin et al., 1975).

**5-3.2.3.2 Moisture Content.** The moisture content can have a drastic effect on the rate of surface-catalyzed transformations in soil. It may also determine the relative importance of biotic and abiotic transformations. In sufficiently dry soils, microbial activity is inhibited, and the relative contribution of abiotic transformations to the dissipation of many pesticides is likely to increase.

When a free-liquid phase is absent, the moisture content will determine which sites on the surface will be exposed and which will be hydrated. As discussed below, many catalytically efficient interactions between pesticides and surfaces in soils are through the exchangeable cations. The hydration level of the exchangeable cation may determine whether an adsorbate will interact with the cation directly or through a water bridge. The mode of interaction with the cation will, in turn, determine the tendency of the adsorbed pesticide to undergo any transformation enhanced by the cation or by its water of hydration.

Some types of ester hydrolysis were shown to be enhanced by the hydration water of exchangeable cations, and their rates were controlled, accordingly, by the cation's hydration status. Oven-dried Na-kaolinite, which is almost anhydrous, catalyzed parthion hydrolysis much less efficiently than the partially hydrated Na-clay (Saltzman et al., 1976). The less water molecules in the hydration shell of the exchangeable cation, the more polarized these molecules become and the higher their ability to act as a Brønsted acids (e.g., Sposito, 1984). Thus, the capacity of hydrated cations to catalyze some transformations may diminish when the moisture content increases above a certain level. A sharp decrease in the rate of hydrolysis of parathion on some homoionic kaolinites (Saltzman et al., 1976) and of 1-(4-metoxyphenyl)-2,3-epoxypropane on some homoionic montmorillonites and on Na-kaolinite (El-Amamy & Mill, 1984) was observed when a certain critical level of moisture content was reached.

When the surface is in contact with a free-liquid phase, its catalytic properties may differ strongly from what they are in the absence of such a phase. A liquid phase is present not only in water-logged soils. Even at moisture contents below saturation, a three-dimensional liquid phase may be present, for example in the capillary pores of the soil.

Because of competition for adsorption sites with water molecules, the pesticide may, in the presence of a liquid phase, not adsorb at those sites with which it interacts the strongest in the absence of a liquid phase. From the above discussion, it is evident that the dissociability of hydration water molecules will be lower in the presence of a free-aqueous phase than in dry soils, where the surface (and in particular the exchangeable cations) is only partially hydrated. Accordingly, the surface hydrolysis of organophosphate esters on homoionic kaolinites (e.g., Saltzman et al., 1976) that seemed to require hydration water to take place, proceeded at moisture contents above that required to saturate the surface at a considerably slower rate than on the partially hydrated clays. Under saturated conditions, expandable clays may disperse or at least swell. Bulky solute molecules may then interact with interlayer surface sites that, under dry conditions, were inaccessible to these molecules.

The moisture content (and specifically the presence or absence of a liquid phase) may determine whether acidic groups in the solid fraction of the soil will preferentially interact in their dissociated or in their associated state with adsorbate molecules. Konrad and Chesters (1969) studied the surface-catalyzed degradation of ciodrin in water-saturated soils. They observed considerably higher degradation rates in neutral and slightly acid soils than in strongly acid soil. These authors suggested that a major reason for the difference in rates may have been the much larger fraction of the acid groups in the organic matter that was dissociated in the less acidic soils.

The rate and mechanism of transport of the reactants and products of a reaction may strongly affect the reaction's kinetics. When a liquid phase is present, the mobility of the reactants and the products of surface reactions is often considerably higher than in the absence of a liquid phase. In the latter case, the slow surface diffusion might dominate the transport. An important possible exception is the transport of volatile species. In the absence of a liquid phase, those reactant molecules that are adsorbed at the catalytically active sites may transform promptly. Additional reactant molecules are often transformed, however, much more slowly because of the low rate at which reactant molecules reach the reactive surface sites and product molecules leave these sites. Saltzman et al. (1974) observed that, in the absence of a liquid phase, the kinetics of parathion degradation on Ca-kaolinite could be described in terms of two first-order stages. The rate of degradation during the first stage was much higher than during the second one. The pesticide molecules initially adsorbed at the catalytically active sites degraded fast. The degradation rate at the second stage was apparently limited by the slow exchange between the products at the active sites and reactants adsorbed on other sites (or at the same site in a different mode of adsorption). In the presence of a liquid phase, continuous, relatively fast equilibration between the bulk solution and the surface may keep the rate of reaction high even after the total number of molecules transformed surpassed the number of reactant molecules initially adsorbed at the reactive sites.

The capacity of the liquid phase to enable fast transport of species participating in a surface reaction may in some cases diminish due to the fixation (binding) of the reaction products to the reactive site. Such a fixation was observed in the degradation of parathion on homoionic kaolinites (e.g., Saltzman et al., 1976). Sanchez Camazano and Sanchez Martin (1983b) showed that the hydrolysis of phosmet is strongly enhanced by suspensions of homoionic montmorillonites. Two stages were again observed in the kinetics of the surface degradation and the rate of hydrolysis in the first stage was significantly higher than in the second. A sufficiently strong interaction between a product of the reaction and the catalytically active site on the surface could account for the two-stage kinetics observed. Soluble Cu and to a lesser extent, Ca also catalyzed the hydrolysis of phosmet. Only a single, continuous stage was observed, however, in the kinetics of these homogeneously catalyzed transformations.

The drier the soil, the more its volume is occupied by the gas phase. The relative volumes of the liquid and gas phases affect the distribution of volatile pesticides between those phases. The distribution of a pesticide between the liquid and gas phases and the ratio between the solid surface area in contact with the gas phase and the surface area in contact with the liquid phase will influence both the nature and the rates of the pesticide transformations.

Many of the abiotic transformations that pesticides may undergo in soils are acido-basic catalyzed. In the absence of a liquid phase, the capacity of the soil to enhance acido-basic reactions is determined by the number, dissociability, and extent of dissociation of the acido-basic groups exposed at the surfaces of the soil solid phase. Surface acidity is the parameter that is most frequently used to define the efficiency of a surface as an acido-basic catalyst. When a liquid phase is present, its pH rather than the surface acidity may become the most important parameter defining the tendency of acido-basic catalysis to occur in the soil. Because of the equilibration between the solid surfaces and the liquid phase, the extent of dissociation of the acido-basic groups on the surfaces varies with the pH of the liquid phase. Thus, the pH of the liquid phase controls both the kinetics of acido-basic catalyzed reactions in the soil solution and the catalytic efficiency of the soil for many acido-basic surface reactions. The influence of the pH in a soil suspension on the rate of surface-catalyzed hydrolysis is suggested by the study of Konrad and Chesters (1969) on the degradation of ciodrin in soils.

5–3.2.3.3 **Properties of the Surface.** Adsorbed pesticides can be transformed either by a component of the solid (e.g., oxidation by Fe-containing adsorbers) or by another adsorbed species (e.g., hydrolysis by water of hydration or catalytic transformations by adsorbed enzymes). In either case, the intrinsic properties of the surface, such as its surface area or CEC will affect both the adsorption and the ensuing transformation of the pesticide. Thus, for the many transformations that are enhanced by adsorption on the exchangeable cations or on their hydration water, the CEC will dictate the capacity of the surface for the catalysis.

The strength of the electric field exerted by charged solids affects the catalytic efficiency of the solid itself as well as the catalytic activity in the diffuse electric double-layer region near the surface of the charged solid. Mortland and Raman (1967) suggested that the efficiency of exchangeable Cu(II) as a catalyst in the surface-enhanced hydrolysis of organophosphate pesticides is controlled by the strength of the electric field exerted by the charged solid. This field will polarize the cation and its hydration shell and thereby affect the catalytic efficiency of the cation. In the case of the Cu(II)-catalyzed hydrolysis of organophosphates, an increase in the strength of the electric field apparently reduced the catalytic efficiency. The catalytic activity of a hydrated cation near a charged surface may be affected by specific interactions with various groups at the surface and not only by the general process of polarization brought about by the electric field. For example, a cation may interact with the O atoms in the O surfaces of layer silicates, directly or through a water bridge. The properties of the interacting cation and its hydration shell will be determined, in part, by the location of the isomorphic substitution. In the case of octahedral substitution in 2:1 clay minerals the charge distribution throughout the O plane is more uniform than in the case of tetrahedral substitution. For the same extent of isomorphic substitution, the average field strength at the surface of 2:1 clay minerals is higher in the case of tetrahedral substitution. The charge distribution throughout the O plane will, no doubt, affect the properties of the water molecules or the cations interacting with the O atoms (e.g., Farmer, 1978).

The above-defined surface acidity was suggested as the factor responsible for the surface-catalyzed degradation of many pesticides (e.g., Brown & White, 1969). Intrinsic properties of the solid (such as its surface charge density and the presence of exposed structural acidic groups at the surface) as well as the nature of the exchangeable cation, the moisture content and the temperature, may all strongly influence the surface acidity.

Mortland (1970) suggested that the dissimilarity between the capacity of H-montmorillonite and allophane to catalyze the hydroxylation of atrazine resulted from the difference in surface acidity between the two minerals. Fripiat (1968) discussed the role of the high electric field at clay mineral surfaces and the related, enhanced acidity of water molecules located in the vicinity of these surfaces in promoting the surface catalysis of reactions that involve protons. Both Brønsted acidity and Lewis (electron-accepting) acidity are involved in the adsorption and surface transformations of organic substances (e.g., Theng, 1974). Solomon (1968) suggested that Al species at broken crystal edges, as well as certain transition metals located in the silicate layers of clay minerals in their oxidized state, can act as electron acceptors in surface-catalyzed polymerization of organic monomers. Aluminum is not likely, however, to act as an oxidizing agent and is more apt to act as a Lewis acid.

Ponec et al. (1974) discussed the role of surface acidity in some adsorption catalyzed reactions. Two pathways were proposed for reactions that involve acido-basic surface catalysts: the formation of a carbonium ion intermediate and the formation of a covalent alkoxy-metal bond with the surface. The second pathway was suggested, for example, for dehydration reactions of alcohols. The surface acidity of the soil organic matter was implicated in many pesticide transformations, such as the hydrolysis of propazine in slurries (Nearpass, 1972) and the hydroxylation of chloro-*s*-triazines (e.g., Khan, 1978). In the case of some surface transformations, the acidity of the surface retards the reaction. The aforementioned transformation of DDT to DDE, which was catalyzed more efficiently by Na- than by H-bentonite, is one such transformation. It is not surprising that a dehydrochlorination reaction, such as the conversion of DDT to DDE, is slowed by an acid medium.

**5-3.2.3.4 Exchangeable Cations.** The high, localized charge and the relatively accessible position of the exchangeable cation, result in a strong interaction between numerous pesticides, especially polar and anionic ones and the exchangeable cations or their water of hydration (e.g., Praffit & Mortland, 1968; Bowman & Sans, 1977; Ristori et al., 1981; Sanchez Camazano & Sanchez Martin, 1983a). Under saturated conditions, competition with the polar water molecules may considerably reduce the extent of adsorption on the exchangeable cations, both directly or through a water bridge. Yet, the pesticide molecules that do reach the vicinity of the cation will undergo a strong electrostatic interaction which may, in turn, perturb the adsorbate's electron charge distribution and thus weaken some bonds in the adsorbate, making it more likely to break down.

Because of their high, localized charge, the exchangeable cations strongly polarize their hydration water. The dissociability of the water molecules in the first hydration shell of the exchangeable cations is, thus, higher than that of water molecules adsorbed at practically any other site on the charged surfaces in soils (e.g., Sposito, 1984; Theng, 1974). Accordingly, the hydration shells of the exchangeable cations acting as Brønsted acids, may be the sites of many surface-enhanced transformations (e.g., various types of hydrolysis). The strong effect of the composition of the exchange complex on the kinetics of the surface hydrolysis of organophosphate or thiophosphate esters has been reported (e.g., Mingelgrin et al., 1977; Sanchez Camazano & Sanchez Martin, 1983b). Some transition metals, such as Fe, may catalyze redox transformations as exchangeable cations, just as they were reported to do when embedded in the silicate layers of clay minerals (e.g., Solomon, 1968).

The radius of the exchangeable cation and the arrangement of ligand water molecules about it may have a dominant effect on the catalytic efficiency of the cation. When the radius of the cation is sufficiently small, the surface may sterically hinder the adsorbed pesticide molecule from assuming the configuration that is most favorable for the occurrence of the transformation. Theng (1974) pointed out the importance of steric factors in charge transfer interactions on clay surfaces. Aluminum (III) with one hydration shell has a smaller radius than the similarly hydrated $Ca^{2+}$. This difference in size was suggested as the cause for the much lower efficiency of Al-kaolinite as compared to that of Ca-kaolinite in catalyzing the hydrolysis of parathion in the absence of a liquid phase (Mingelgrin et al., 1977). A weaker steric hindrance should be exerted by the surface on methyl parathion than on parathion, because the ethyl groups of parathion are replaced in its analog by the less bulky methyl groups. Accordingly, the difference between the rates of hydrolysis of methyl parathion on Al- and on Ca-kaolinite was smaller than the difference between the rates of hydrolysis of parathion on those homoionic clays (Saltzman et al., 1974).

Agricultural practices such as fertilization or liming, may influence the composition of the exchange complex. The importance of exchangeable cations in many surface-catalyzed transformations implies that such practices can also affect the nature and kinetics of the abiotic transformations that occur in the soil.

**5-3.2.3.5 Analogy between Surface Enhanced and Enzymatic Reactions.** The aforementioned analogy between heterogeneous catalysis and enzymatic reactions is demonstrated by the effect that small changes in the properties of the surface (other than the nature of the catalytically active site itself) have on the efficiency of the catalysis. Water molecules located in the first hydration shell of the exchangeable cations are, most likely, the sites responsible for catalyzing the hydrolysis of organophosphate pesticides at the surface of charged solids (Mingelgrin et al., 1977; Mingelgrin & Saltzman, 1979). Yet, the rates of hydrolysis at the surface of different minerals were dissimilar even when the exchangeable cations and their hydration status were the same. The rates of hydrolysis of parathion on the surface of two types of kaolinite were rather different, and the rate of hydrolysis on the surface of a montmorillonite was considerably lower than on either kaolinite.

The accessibility of the catalytically active surface sites may strongly affect the efficiency of the catalysis. Groups at the surface, other than those responsible for the catalysis, if bulky enough, may sterically hinder the adsorbate molecule from reaching the active site or from assuming the orientation required for the catalysis to occur. Restrictions on the access and orientation of adsorbate molecules in the interlayer spaces of montmorillonite may strongly modify the catalytic activity of this clay.

Various interactions between the adsorbate molecule and the surface may occur simultaneously. Even if only one such interaction is directly responsible for the catalysis, the other interactions may still affect the reaction kinetics. Thus, the nitrophenyl moiety of parathion interacts with the hydroxyl-dominated surface of the octahedral layer of kaolinite (e.g., Sahay & Low, 1974) differently than with the O-dominated surface of the tetrahedral layer of that clay. In montmorillonite, only the latter type of surface is exposed, except at broken edges. The nature of the interaction between the surface and the nitrophenyl moiety affects the electron charge distribution about the phosphate ester bond and hence the lability of this bond to hydrolysis. The aforementioned difference between the kinetics of parathion hydrolysis on kaolinite and montmorillonite is in agreement with the above discussion.

Correspondingly, the rate of surface hydrolysis of phosphate esters was influenced strongly by the structure of the hydrolyzed molecule as a whole and not only by the nature of the hydrolyzed bond. The difference in the reaction kinetics between the various phosphate esters may arise not only from the intrinsic difference in the strength of the hydrolyzed bond between these esters, but also from the considerable effect the overall structure of the molecule has on the interaction between the ester bond and the surface. The effect of the overall structure of the phosphate ester molecule on the hydrolysis of the ester bond is illustrated by the difference between the kinetics of hydrolysis of various phosphate ester pesticides on clays (e.g., Mingelgrin et al., 1975; Saltzman et al., 1976).

**5-3.2.3.6 Fixation.** In some cases, one or more of the products of the surface-catalyzed transformation remain chemically sorbed (or otherwise strongly attached) to the reaction site. Such a product is defined as fixed, or bound, to the surface. Fixation of the phosphate moiety produced in the degradation of parathion on kaolinite was reported, for example, by Saltzman et al. (1974). In some cases, the fixed product may retain some toxic activity. Stevenson (1982) suggested that pesticide-derived residues form stable chemical linkages with the soil organic matter and that these linkages greatly increased the persistence of the residues. As described above for the degradation of phenoxyalkanoates, actual incorporation of products of pesticide transformations in the newly formed fulvic and humic acids takes place during the humification process (Stevenson, 1982). At least some of the bound residues of pesticides reported to exist in soils (e.g., Khan, 1982) may be the result of this incorporation phenomenon.

Bartha (1971) reported the fixation of chloroanilines, the products of the partial degradation of phenylamide herbicides, to the soil organic matter, for example. Bound residues are more often the degradation products of pesticides than the pesticides themselves. This preferred binding of the degradation products may be due to the fact that the partial degradation of many pesticides by either microbial or abiotic pathways leads to the formation of species having a higher chemical reactivity than the parent material (Stevenson, 1982). These chemically active species may strongly interact with different sites on the organic or mineral surfaces in soils.

**5-3.2.3.7 Rate Equations.** Rate expressions that have been used to describe pesticide transformations in soil systems are given in Table 5-5. The kinetics of surface-catalyzed reactions are more difficult to describe with the aid of rate equations than the kinetics of homogeneous reactions. Limitations on the free access of pesticide molecules to some catalytically active sites at the surface and the slow exchange of reactants and products at the active sites when a liquid phase is not present are among the factors that contribute to this difficulty. Ponec et al. (1974) discussed a formal description of the kinetics of surface reactions at solid-gas interfaces. A similar treatment of surface reactions in soils will only rarely be useful because of the ideal conditions assumed by those authors (i.e., a mass flow rapid enough to avoid interference with the reaction kinetics).

In some cases, a rate equation that approximates the kinetics of a given transformation in the soil can be formulated. For example, the rate of a surface-catalyzed reaction of a nonvolatile pesticide which takes place in the absence of a liquid phase, will be directly proportional to the surface concentration of the pesticide adsorbed on catalytically active sites. The reactions follows, in other words, first-order kinetics with respect to that part of the adsorbate that is adsorbed at the catalytically active sites. If a significant part of the adsorbed pesticide is held through nonreactive adsorption processes, a simple first-order expression in terms of the total surface concentration of the pesticide will not describe the reaction kinetics adequately.

Table 5-5. Summary of some rate equations.

| Type of Eq. | Differential form ($-dC/dT=$) | Integrated form | Half-life | Linear form in $T$ |
|---|---|---|---|---|
| Zero order | $k$ | $C=C(0)-kT$ (at $C>0$) | $0.5C(0)/k$ | $C=C(0)-kT$ |
| First order | $kC$ | $C=C(0)e^{-kT}$ | $-LN(0.5)/k$ | $LN(C)=LN[C(0)]-kT$ |
| Simple second order | $kC^2$ | $C=C(0)/[1+C(0)kT]$ | $1/[kC(0)]$ | $1/C=1/C(0)+kT$ |
| Hyperbolic | $k_1C/(k_2+C)$ | $e^C C^{k_2}=e^{[C(0)-k_1T]}\times C(0)^{k_2}$ | $[0.5C(0)-k_2\times LN(0.5)]/k_1$ | $A(C)=A[C(0)]-k_1T$ where $A=C+k_2LN(C)$ |

† $C$, concentration, $T$, time, and $k_i$, constants.

The hydrolysis rate of parathion on kaolinite in the absence of a liquid phase (Saltzman et al., 1974) followed to a good approximation first-order kinetics, in terms of the total concentration of the adsorbed parathion at sufficiently low initial surface concentrations. After most of the adsorbed pesticide degraded, however, a discontinuity in the rate was observed and the remaining pesticide molecules (those adsorbed in noncatalytically active modes) degraded at a low rate. The fraction of the pesticide adsorbed by noncatalytically active modes of adsorption was apparently small enough to retain the validity of the approximation of first-order kinetics in the total surface concentration for the hydrolysis of most of the pesticide (which was adsorbed in the catalytically active mode).

A liquid phase is often absent in the soil and many nonvolatile pesticides adsorb through many competing processes. A two- or multi-stage kinetic pattern, similar to that observed for parathion on kaolinite may, therefore, describe many abiotic transformations of pesticides in unsaturated soils.

In most cases, an empirical approach was adapted to define rate equations for the transformation of pesticides in soils. The expressions used included first-order (e.g., Konrad & Chesters, 1969) higher-order, and hyperbolic (e.g., Hamaker, 1972) relationships. An empirical approach is suitable for fitting experimental data but it is less appropriate for extrapolation purposes. Graham-Bryce (1981) pointed out the risk involved in applying empirically derived first-order rate equations to a pesticide at concentrations that are far from the concentration range at which the empirical fit was performed. The half-life values calculated from such equations is similarly valid only at pesticide concentrations sufficiently close to the concentration range at which kinetic measurements were made. A concentration independent half-life is not likely to exist unless the reaction kinetics is truly first-order.

Konrad et al. (1969) demonstrated that first-order kinetics described well their data on the adsorption catalyzed degradation of ciodrin in soils. Hamaker (1972) suggested that surface-catalyzed reactions may be well described in some cases by a hyperbolic rate model

$$-cd/dt = k_1c/(k_2 + c), \qquad [6]$$

where $k_1$ is the maximum rate achieved at sufficiently high concentrations. The constant $k_2$ could be assigned a physical significance by analogy to the Michaelis-Menten rate equation to which the hyperbolic model bears a formal resemblance. The Michaelis-Menten formulation that describes enzymatic reactions is based, however, on many assumptions that do not hold for many surface-enhanced transformations in soils and especially in the absence of a liquid phase. Farmer and Aochi (1987) did point out that only few surface-catalyzed reactions were shown to follow hyperbolic kinetics.

**5-3.2.3.8 Mechanisms Proposed for Selected Surface-Catalyzed Transformations of Pesticides.** Armstrong and co-workers (Armstrong & Konrad, 1974; Armstrong & Chesters, 1968) proposed a mechanism for the hydrolysis of chloro-*s*- triazines on the surface of the soil organic matter. A schematic representation of this mechanism is given in Fig. 5-7. It was postulated that the catalytic interaction was between a protonated carboxyl group at the surface and a ring N atom of the triazine. This postulation is in agreement with the suggestion (Russell et al., 1968a, b) that when 3-aminotriazole or *s*-triazines are adsorbed on montmorillonite they protonate at the ring N atoms rather than at N atoms in the side chains.

The carbon atom of the chloro-*s*-triazine to which the Cl atom is bound, is surrounded by electronegative N and Cl atoms and is, thus, exposed to

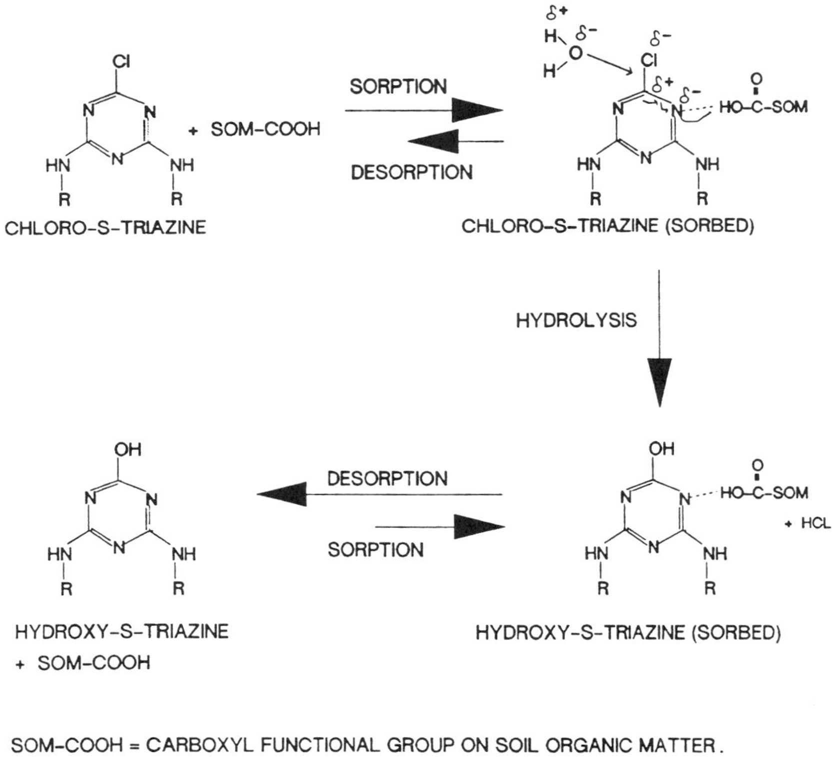


Fig. 5-7. A proposed mechanism for the surface-catalyzed hydrolysis of chloro-*s*-triazines in the soil (Armstrong & Konrad, 1974).

attack by nucleophilic agents such as hydroxyls. Further withdrawal of electrons from the C atom results from H-bonding between the ring N atom next to it and the surface carboxyl group (Fig. 5–7). The withdrawal of electrons increases, in turn, the attraction between the O atom of the attacking water molecule and the electron-deficient C atom, thus enhancing the hydrolytic replacement of the Cl by a hydroxyl.

Konrad et al. (1967) discussed the surface-enhanced hydrolysis of organophosphate ester pesticides in soils. Their results were summarized by Armstrong and Konrad (1974). Diazinon underwent direct hydrolysis of the phosphate ester bond. Malathion hydrolyzed first at the phosphate ester bond and then proceeded to hydrolyze at the carboxyl ester bond. Ciodrin, which also contains both phosphate ester and carboxyl ester bonds, hydrolyzed-first at the carboxyl ester bond and only then at the phosphate ester bond. Detailed mechanisms for the surface-enhanced transformations were not offered by the authors, but a possible role of soil bound metal ions in the catalysis was suggested.

Mingelgrin and co-workers (Mingelgrin et al., 1977; Mingelgrin & Saltzman, 1979) proposed a mechanism for the surface-enhanced hydrolysis of organophosphate esters. The proposed mechanism involves the interaction of the phosphate moiety of the ester with the strongly polarized water molecules in the first hydration shell of exchangeable cations. In the case of phosphorothioate esters, such an interaction catalyzes both rearrangement and hydrolysis of the pesticides (Fig. 5–8). Parathion degradation on ben-

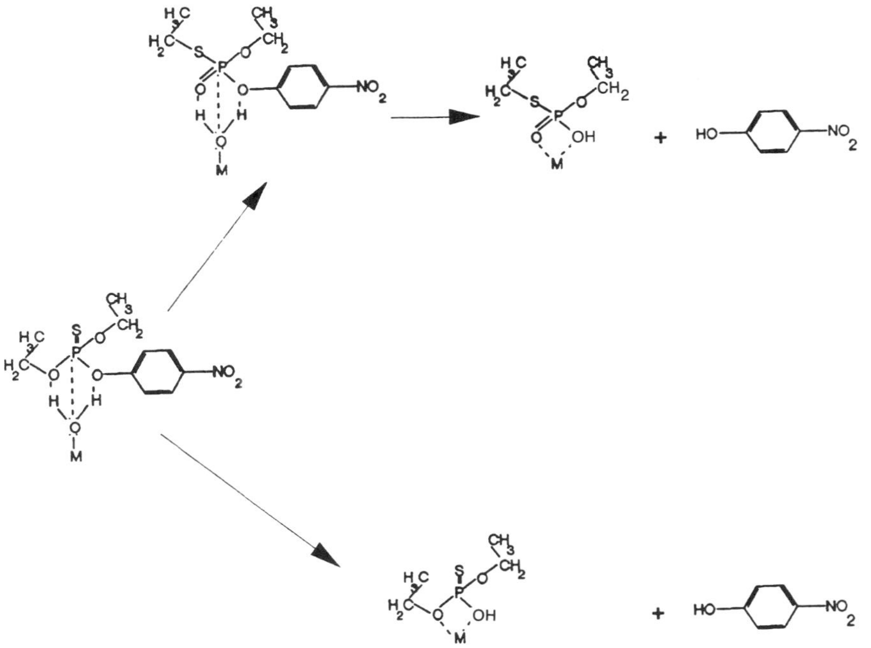

Fig. 5–8. A proposed model for the surface-catalyzed degradation of phosphorothioate esters exemplified by parathion. M represents an exchangeable cation. (From Mingelgrin & Saltzman, 1979.)

tonite in the absence of a liquid phase, for example, proceeded predominantly through a rearrangement transformation, which was followed at elevated temperatures by hydrolysis. The model described in Fig. 5–8 accounts for the observed fixation of the phosphate (or thiophosphate) hydrolysis product.

### 5–3.2.4 Transformations Catalyzed by Enzymes Extracellularly

Extracellular enzymes, as well as enzymes released by plasmolysis, are products of biological activity, but once released into the soil environment their action is governed by the same principles that govern the action of abiotic catalysts. Microorganisms, plant roots, and the soil's fauna all release enzymes into the soil. The proteinaceous nature of enzymes suggests that at the pH range prevalent in soils, most of the enzyme content of the soil is adsorbed on clay and organic surfaces (e.g., Skujins, 1967). The adsorbing organic surfaces may belong not only to humic particles but also to plant parts, such as roots, and to dead or alive microorganisms. Yet, activity of enzymes in the solution was suggested in the past. Suflita and Bollag (1980), for example, described an oxidative coupling activity in soil extracts which displayed the highest rate at pH values around 7.0. They assigned this activity to the presence of phenoloxidases.

The above authors (Suflita & Bollag, 1980) suggested the possible role of extracellular enzymes in the incorporation of xenobiotics in newly formed humic matter through enzyme-catalyzed oxidative coupling. The tendency of enzymes and of many pesticides to exist in soil in the adsorbed state suggests the possibility of enzyme-catalyzed formation of bound residues. For example, Liu et al. (1985) reported that the binding of amino acid esters to phenolic humus constituents was catalyzed by an extracellular enzyme.

There are reports of enzymes extracted from soils (e.g., Ladd, 1985; Burns & Edwards, 1980). Many other studies, on the other hand, such as that of Suflita and Bollag (1980), did not actually demonstrate the presence of an enzyme, but rather described enzyme-like activity and behavior in the soil (Skujins, 1967). The extraction of enzymes from soils is often difficult, probably because of the aforementioned tendency of enzymes to adsorb on the solid surfaces in the soil. The resulting reliance on circumstantial evidence to prove enzymatic activity might have caused an erroneous assignment to enzymes of observed catalytic activity. Thus, Loll and Bollag (1985) suggested that the oxidative coupling activity observed by Suflita and Bollag (1980) in soil extracts may be assigned to a Mn-citrate complex rather than to an enzyme.

Getzin and Rosefield (1968) demonstrated that malathion and dichlorvos decomposed faster in soils that were sterilized by irradiation than in soils that were sterilized by autoclaving. Burns and Gibson (1980) also reported that malathion, which degraded at a high rate in a nonsterile soil, still displayed a high rate of degradation after that soil was sterilized by irradiation. The rate of degradation of the pesticide declined sharply after the soil was autoclaved. Those authors concluded, on the basis of the heat lability of the transformation, that malathion degraded in soil by an extracellular enzyme and suggested that the active enzyme was adsorbed on the organic matter of the soil.

Getzin and Rosefield (1971) extracted from a soil, both before and after sterilization by irradiation, and partially purified a substance that possessed enzyme-like properties and that catalyzed the degradation of malathion. Adsorbed enzymes do often withstand nonthermal sterilization procedures such as irradiation. Adsorption may also decrease in some cases the heat lability of enzymes, although severe procedures such as drastic autoclaving are believed to eliminate all enzymatic activity (Graham-Bryce, 1981). Some nonenzyme catalysts may, on the other hand, be heat labile. It is difficult to conclusively determine from the sensitivity of a catalytic process in the soil to heat whether such a process is enzymatic or nonenzymatic.

Skujins (1967) listed more than 30 types of enzymes that were reported to be active in the soil. Some of these enzymes (e.g., phosphatases) participate in transformations of important groups of pesticides. Other studies (e.g., Brown, 1981) suggested additional enzymes to be active in soils. Despite the numerous reports of extracellular enzymatic reactions in soils, the above discussion demonstrates the difficulty in estimating the relative contribution of the extracellular enzymatic reactions in the soil to the dissipation of pesticides.

Enzymes are frequently more stable in the soil than in vitro. Their role in pesticide transformations in soils may be, therefore, more important than implied by the usual lability of enzymes to microbial and other degradation processes. The added stability of the enzymes in the soil results, most likely from their strong association with surfaces of the soil's solid phase. This association may, however, affect not only the stability, but also the activity of the enzymes. While, in many cases, adsorption reduces the enzymatic activity, in other cases adsorption may enhance this activity, or at least leave it unchanged. The activity of urease, for example, was hardly affected by adsorption on bentonite (Durand, 1965). Hoffmann (1959) demonstrated that urease released from lysed cells adsorbed on the clay fraction of the soil and remained active. Boyd and Mortland (1985) demonstrated that urease did not lose its activity upon adsorption on a smectite-alkyl ammonium complex and suggested that the same may be true for natural clay-organic matter complexes. Kroll and Kramer (1955) showed that the addition of montmorillonite to soil did not affect the activity of phosphatase.

Not only enzymes, but also many of the pesticides that serve as their substrates, adsorb strongly on the solid constituents of the soil. The adsorption of the substrates can have a considerable influence on the catalytic efficiency of extracellular enzymes. Enzymatic degradation of adsorbed substrates was reported to proceed in some cases at a higher rate, but more frequently at a lower rate, than that of the enzymatic degradation of the nonadsorbed substrates (e.g., Skujins, 1967; Estermann et al., 1959; Estermann & McLaren, 1959).

The enzymatic activity near the surfaces of charged solids is affected by the above discussed, special conditions that exist in the diffuse electric double-layer region. The pH at negatively charged surfaces, for example, is considerably lower than the pH of the bulk soil solution. The activity of enzymes is, as a rule, sensitive to the pH and hence the distance from a charged surface may strongly affect the enzymatic activity. The soil pH at which an enzymatic reaction displays a maximum rate may not be the same as the optimum pH for that reaction in a soil-free aqueous solution. One reason for this discrepancy is that the pH values measured in soils are mean values for the total soil system and thus are higher than the pH values near negatively charged surfaces (e.g., Skujins, 1967). Ions other than $H^+$ and $OH^-$, such as ions of the transition elements, may play an important role in enzymatic reactions (e.g., Juma & Tabatabai, 1977; Press et al., 1985). The change in the concentration of these ions with distance from charged surfaces further complicates the effect of proximity to a charged surface on the activity of enzymes.

## 5-4 PHOTOLYSIS

Sunlight is one of the primary forces affecting the loss of pesticides from exposed environmental compartments. Photolysis is the process in which ultraviolet (UV) or visible light causes transformation of compounds. Sunlight and environmental reactants can chemically transform pesticides to substances with different (generally reduced) toxicity and begin the process towards mineralization. Photolysis rates and products vary considerably depending on the compartment exposed to sunlight. In some systems, photolysis may be the primary degradative pathway of a pesticide, but in another it may not be significant. Assessing how photochemical processes affect the fate of a particular pesticide requires not only an understanding of the basic photochemistry of the pesticide but also an understanding of the chemistry of the environmental compartment in which it resides. Photolysis is but one of the myriad of processes that can affect pesticide stability, and the significance of this process should be determined only in relation to other degradative processes.

### 5-4.1 Photolysis in Water

Of all the environmental compartments in which photolysis can occur, water has received the most attention (Miller & Zepp, 1983; Miller & Crosby, 1983a; Zepp, 1980, 1982). A useful bibliography on natural water photochemistry has been compiled by Zafiriou (1984). Leifer (1988) presented a thorough review and discussion of the calculations used to predict photolysis rates. The interest in aqueous photolysis is because of the environmental and health-related importance of water, as well as to water being a condensed, homogeneous system that behaves in a generally predictable fashion, except when suspended substances are present.

Photolysis of pesticides in water occurs by two processes: direct and indirect photolysis. Direct photolysis requires absorption of sunlight by the pesticide, which is then transformed into products. Indirect photolysis begins by adsorption of light by some substance other than the pesticide. That absorption initiates a series of reactions that finally results in the transformation of the pesticide. Although the products of the two types of reactions are often the same, the kinetics are generally different, and of the two, only direct photolysis has been successfully modeled in environmental waters.

### 5-4.1.1 Direct Photolysis

Direct photolysis requires absorption of sunlight by the pesticide. The photolysis rate is directly dependent on the overlap of the absorption spectrum of the pesticide with the spectral distribution of sunlight striking the earth. Radiation at wavelengths below 290 nm is efficiently absorbed by ozone in the atmosphere, and hence, pesticides not absorbing radiation above 290 nm will not undergo direct photolysis. The intensity of solar irradiance ($E_o$) is dependent on the thickness of the atmosphere through which the light must travel and varies with the angle of incident radiation and the surface of the earth (Peterson, 1976). The irradiance changes with latitude, season, and time-of-day. Of the three attenuating processes that determine the optical thickness of the atmosphere, ozone absorption, and molecular scattering are strongly wavelength dependent. Particulate diffusion is the third process. Consequently, a substantial difference exists between the attenuation of intensity at different wavelengths. The thickness of the atmosphere through which sunlight must penetrate has a greater effect on the intensity of radiation at the shortest sunlight wavelengths (290–320 nm range) than on solar radiation at longer wavelengths.

Figure 5–9 presents the spectral distribution of solar irradiance ($E_o$) at four different dates of the year (40° N) and demonstrates how sunlight intensity varies seasonally, especially at the shorter wavelengths. These sunlight spectra were calculated using the GC SOLAR program (available from the USEPA's Laboratory, Athens, GA). Similarly, sunlight intensities may be estimated for any date at any time globally (Zepp & Cline, 1977).

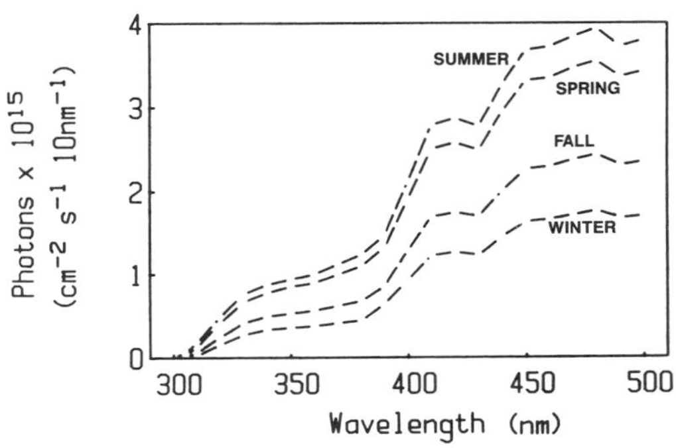


Fig. 5–9. Solar irradiance reaching the earth at four times of the year.

Modeling the direct photolysis rate of a pesticide requires measurement of the extinction coefficients ($\epsilon_\lambda$) of the pesticide at the wavelengths occurring in the solar radiation reaching the earth's surface. Figure 5-10 presents the absorption spectra of four pesticides. For modeling photolysis, extinction coefficients are determined by computations exemplified by the model calculation of Zepp and Cline (1977). Trifluralin and methyl parathion are examples of pesticides that absorb sunlight strongly. The pesticide 3,4-dichloroaniline absorbs at shorter wavelengths, and methoxychlor absorbs sunlight only weakly. Although exceptions exist, most pesticides do not photolyze strongly, because if they did, their lifetimes would most likely be insufficient for efficient application. For example, naturally occurring pyrethrin insecticides are rarely used in outdoor applications, in part due to their rapid sunlight degradation (Ruzo, 1982). Similarly, the insecticidally potent nitromethylene compounds are not sufficiently stable in sunlight for general outdoor applications (Klehr et al., 1983).

For weak light absorption by a chemical, the specific light absorption rate ($k_{\alpha,\lambda}$) is:

$$k_{\alpha,\lambda} = 2.303\ E_{o,\lambda}\epsilon_\lambda j^{-1}. \qquad [7]$$

The term $j$ is 1 Einstein/mol and renders the units consistent with $E_o$.

In addition to the absorption spectra and sunlight intensity the quantum yield, ($\phi_r$) must be determined in each chemical matrix of interest. The quantum yield is a unitless number that gives the ratio of the number of molecules transformed divided by the number of photons absorbed. For most environmental applications in condensed media, quantum yields are assumed to be invariant with wavelength, although this assumption should be experimentally verified. Quantum yields have classically been determined using monochromatic light in optically opaque solutions. Solubility restrictions of many xenobiotics in water preclude use of concentrated solutions that com-

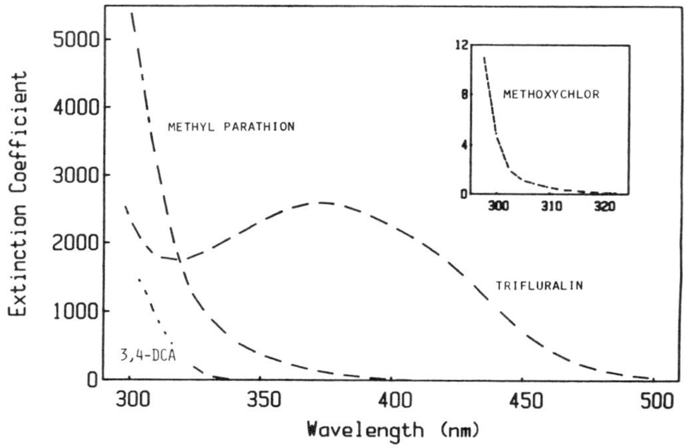


Fig. 5-10. Absorption spectra of four pesticides.

pletely absorb the incident radiation. Instead, dilute solutions are used for measuing quantum yields in which the light absorption of the solution is <5% through the entire pathlength (Zepp, 1978). Determination of quantum yields requires that the light intensity be accurately determined. This is accomplished with the aid of actinometers (Miller & Zepp, 1983; Dulin & Mill, 1982). Actinometers are compounds that are photochemically transformed with accurately known quantum yields over the wavelengths of interest.

Selection of the light source for establishing the quantum yield requires consideration of several variables. Of primary importance is the spectral distribution of the absorbed light. Generally, monochromatic light is preferred, and the mercury lines, 313 and 366 nm, have been used extensively with the appropriate filters to screen unwanted wavelengths (Herbert & Miller, 1990; Miller et al., 1989). Another option is xenon arc sources that offer a continuous selection of wavelengths when passed through large monochrometers. Although more expensive, these have the advantage of determining quantum yields at a variety of wavelengths. Alternatively, less expensive broad band light sources can be used to estimate quantum yields. An interlaboratory comparison of various methods for determining quantum yields demonstrated that, in many cases, broad-band light sources will enable sufficiently accurate measurements (Turner, 1984). The spectral distribution must be carefully measured, however, to ensure proper consideration of each wavelength absorbed by the compound being examined. Draper (1985) has also described a procedure for determination of an average quantum yield between 310 and 410 nm using broad band light with the aid of actinometers.

Quantum yields vary depending on the solvent system. For instance, the quantum yield for photolysis of 3,4-dichloroaniline is 0.052 in water; 0.0046 in methanol; and <0.0005 in acetonitrile (Miller et al., 1979). Thus, the quantum yield must be determined in the solvent applicable to the environmental compartment of interest, most often water. Solubility limitations for many xenobiotics require use of dilute solutions, and homogeneity of the solutions must be established.

Finally, the light transmission characteristics of the aqueous system must be determined to factor in the degree to which components of that system other than the pesticide attenuate the radiation in the water column (Zepp & Cline, 1977).

With the above data available, the direct photolysis rate can be determined. Summing over all the wavelengths absorbed by the pesticide gives the total rate constant of light absorption:

$$k_{\mathrm{a}} = 2.303 j^{-1} \Sigma \epsilon_{\lambda} E_{\mathrm{o},\lambda}. \qquad [8]$$

Direct photolysis rates are simply the rate of light absorption multiplied by the quantum yield (Zepp & Cline, 1977). The following equation will give useful estimates for photolysis rates on clear days in near-surface, pure water.

$$-d[\mathrm{P}]/dt = \phi_r k_{\mathrm{a}}\ [\mathrm{P}]. \qquad [9]$$

This calculation has been validated for several compounds and provides a reliable gauge of the direct photolysis contribution to the overall loss of a compound from a water system (Zepp & Cline, 1977). Many factors can affect (usually reduce) the observed photolysis rate. Clouds and other shading factors will reduce the radiation available to the compound, and various absorbing or scattering species such as suspended sediments (Miller & Zepp, 1979) may affect the radiation reaching the pesticide molecule in a fashion difficult to predict. With these limitations in mind, the above calculations agree with rates observed in several field tests (Wolff & Crossland, 1985; Crossland & Wolff, 1985).

### 5-4.1.2 Indirect Photolysis

Indirect photolysis occurs when a species other than the pesticide absorbs sunlight and initiates a series of reactions that result in the transformation of the pesticide. Naturally occurring organic and inorganic species absorb the radiation required for the subsequent reactions. Although the exact identities of these photosensitizers are generally unknown, humic substances, clay minerals, and transition metals are believed to be involved in indirect photolysis.

Oxidation is the predominant photolytic process, and several oxidants that cause the loss of pesticides have been identified. These oxidants are activated primarily by two processes; energy transfer to $O_2$-producing singlet oxygen (Zepp et al., 1981b) and radical production that, in turn results in the formation of radical as well as nonradical active oxidants.

**5-4.1.2.1 Singlet Oxygen.** Singlet oxygen ($^1O_2$) is formed when an excited sensitizer ($^3S^*$) transfers its energy to ground state triplet oxygen ($^3O_2$),

$$S \xrightarrow{hv} {}^1S^* \rightarrow {}^3S^* \quad [10]$$

$$^3S^* + O_2 \rightarrow S + {}^1O_2. \quad [11]$$

The sensitizers in water are generally believed to be related to dissolved humic substances, although any substance that absorbs sunlight (hv) and can undergo intersystem crossing to a triplet state is potentially capable of producing singlet oxygen. Some ground-state reactants, including the hydrogen perioxide-hypochlorite system, also can produce singlet oxygen, although their involvement in environmental reactions is not considered to be substantial.

Singlet oxygen is a selective oxidant; it reacts at appreciable rates only with electron-rich groups, such as tri- and tetra-alkyl substituted olefins, sulfides, furans, and some polynuclear aromatics. Singlet oxygen has been detected in natural waters (Zepp et al., 1977; Wolff et al., 1981), and has been shown to oxidize various substrates including thioether pesticides such as disulfoton (Zepp et al., 1981a), resmethrin (Ruzo, 1982), and the pesticide transformation product, 2-dimethylamino-5,6-dimethylpyrimidin-4-ol (Dixon & Wells, 1983). In contrast, singlet oxygen does not oxidize thiophosphates to phosphates at appreciable rates.

**5-4.1.2.2 Other Oxidants.** Several other apparently photochemically generated oxidants have been detected in natural waters exposed to sunlight, including hydrogen peroxide, superoxide, alkyl peroxy radicals, and hydroxyl radical. Hydrogen peroxide is produced in natural waters containing dissolved humic substances when exposed to sunlight (Draper & Crosby, 1983; Cooper & Zika 1983). Mill et al. (1980) have shown that photooxidation of cumene and pyridine in dilute solution in natural waters gives products characteristic of reactions with alkylperoxy and hydroxy radicals. They suggest that the concentration of alkylperoxy radicals is large enough to make oxidation by these radicals an important process in natural waters for some classes of reactive chemicals. The hydroxyl radical is an extremely reactive oxidant, reacting with a variety of organic molecules at near diffusion-controlled rates. Among the processes for generating hydroxyl radicals are the nonphotolytic reduction of hydrogen peroxide and the photolysis of peroxides. Draper and Crosby (1981) have demonstrated that irradiation of dilute hydrogen peroxide in the presence of various nonsunlight absorbing herbicides results in enhanced oxidation of those substances.

## 5-4.2 Photolysis on Soils

The photolysis of pesticides on soil surfaces is poorly understood, and only a few studies are available that provide a basis for estimating photolysis rates on environmental surfaces. Unlike solutions, which are homogeneous and generally well mixed, soils are highly heterogeneous and unmixed and illumination produces a light field difficult to accurately define. The same models used in aqueous systems are thus not applicable to soils.

Klehr et al. (1983) have suggested an experimental procedure for examining the photolysis of pesticides on soil surfaces using a filtered xenon arc source to eliminate radiation below 290 nm. The chamber used to contain the soil layers was water cooled and dark controls were treated identically, except that they were covered with aluminum foil. Soils were irradiated as 0.5-mm thick layers, and a stream of air was passed through the chamber to collect volatile products in a solution of ethylene glycol. For this study, the pesticide examined was the defoliant thiadiazuron. It was applied using a thin layer spotter that minimized penetration into the soil. Using this procedure, they were able to distinguish between the photochemical and non-photochemical transformation of the compound. The observed photochemical half-life of 0.5 h demonstrated that this compound was rapidly transformed.

Burkhard and Guth (1979) have also suggested methods for assessing photolysis rates on soil surfaces using artificial light sources. They examined the photolysis rates of three organophosphorus pesticides in an air-cooled chamber, also using filtered xenon arc sources. For this study, the pesticides were applied in a solvent uniformly to the soil. Because this technique allows reproducible exposures under controlled conditions, it has been used for evaluating pesticide photolysis on soils. Unfortunately, comparisons of photolysis rates observed for each the above methods to that in sunlight were not made, and environmental interpretation of these data is difficult.

Few systematic studies have examined the effect of various soil properties on photolysis processes, including organic content, particle size, and soil mineralogy. This is somewhat surprising, because soil surfaces receive a large fraction of surface-applied pesticides. Takahashi et al. (1985) examined the photolysis of the pyrethroid fenpropathrin on three soils using 0.5-mm thick layers of soil. The photolysis rate of fenpropathrin was highest on a clay soil containing 15.3% organic matter. The photolysis rates were slower on a sandy loam soil containing 1.9% organic matter and on a sandy clay loam soil containing 2.5% organic matter. Half of the originally applied insecticide on the three soils was gone in 1, 4, and 5 d, respectively. In a separate sunlight study on another pyrethroid, fenvalerate, using the same soils, the times for half of the compound to be lost were 2, 6, and 18 d for light-colored clay, sandy loam, and sandy clay loam soils, respectively (Mikami et al., 1980). In both studies, the most rapid photolysis occurred on soils with the highest organic and clay content.

Smith et al. (1978), however, found no correlation between photolysis rates of methidathion and organic matter content using six soils with organic matter content between 0.1 to 20%. A study by Mikami et al. (1985) also indicated that half of the organophosphate fenitrothion was photolyzed in approximately 1 d on both a silty loam containing 19% organic matter and a silty loam containing 2.5% organic matter. In each of these studies, dark reactions accounted for $<10\%$ of the loss of pesticide.

Modeling direct photolysis rates on soil surfaces requires an estimation of both the depth of light penetration in the soil and also the distribution of the pesticide in the soil. Simple absorption measurements using a xenon arc through a thin soil layer suggest that the light is more than 90% attenuated in the top 0.2 mm (Herbert & Miller, 1990; Miller et al., 1989). Thus, direct photolysis would be significant only near the exposed soil surface. Although these measurements were not sufficiently sensitive to accurately determine small differences, they suggest that organic matter content did not affect light penetration dramatically. The above results were consistent with those obtained when pesticides were irradiated in suspensions of sediment in water (Miller & Zepp, 1979). The primary effect was light attenuation.

Because of the light-attenuating effect, the rates of direct photolysis of pesticides on soil surfaces are substantially slower, relative to the rate of photolysis in distilled water. Nilles and Zabik (1974) found that more than 80% of fluchloralin degraded in 60 h of irradiation in water, while more than 80% of Basalin, a formulation of fluchloralin, remained after 80 h of irradiation on a 0.5-mm deep sandy loam soil. Similarly, they observed more than 80% loss of bentazone when irradiated in distilled water for 24 h while more than 50% of Basagran, a formulation of bentazone, remained after 120 h of irradiation on the soil surface (Nilles & Zabik, 1975).

A reduction in photolysis rate was observed when the herbicide flumetralin was irradiated on soil compared to the rate on leaf surfaces, on silica gel, on glass plates, and in solution. The photolysis rate was substantially slower on soils and leaf surfaces than in the other systems (Miller & Zepp, 1983). These differences were likely because of differences in light attenuation.

Soil attenuates light strongly, as do leaf surfaces once the pesticide penetrates the waxy layer. Glass and pure water do not absorb light and silica gel primarily scatters and does not absorb the light. This simple comparison demonstrates the problems with using glass or silica gel as models for soils.

Competitive light attenuation by soils is probably responsible for the relatively low rate of xenobiotic photolysis on soils. It is unclear, however, whether this light reduction is because of bulk attenuation or to an inner filter phenomenon described by Yokley et al. (1986) for the photolysis of polynuclear aromatic hydrocarbons on coal stack ashes. The importance of the two effects is particularly great for modeling photolytic processes. For attenuation by bulk processes, the soil is assumed to be homogeneous, with light attenuation described in a Beer-Lambert fashion with depth. The inner filter phenomenon, on the other hand, occurs when a xenobiotic becomes entrapped inside a soil particle and is protected from photolysis in the interior of the particle. Although proving such a phenomenon is difficult, Yokley et al. (1986), using diffuse reflectance spectroscopy and pore-size determinations, demonstrated that polynuclear aromatic hydrocarbons could fit within the pores of the ash, and that, accordingly, the darker ashes generally provided the most efficient inhibition of photolysis. Modeling of photolysis in this case will be more complicated, because the photolysis rate will depend not only on the depth of light penetration, but also on the type of sorption sites that exist within the particle.

Evidence for this inner filter effect also was provided by Miller and Zepp (1979) by demonstrating that the photolysis rate of DDE on suspended sediments is dependent on the time allowed for sorption on the sediments. The DDE equilibrated with suspensions of two different sediments for 4 d underwent photolysis more rapidly than DDE equilibrated with the sediments for 60 d. In both cases, photolysis did not follow simple first-order kinetics, and the rate constants decreased throughout the irradiation period. For the longer equilibration period, the decrease in rate was more rapid and approached zero when about 50% of the DDE had been degraded. For the shorter equilibration period, as much as 90% of the DDE was photodegraded. It was suggested that approximately 50% of the DDE migrated to a microenvironment where the chemical may have not been exposed to radiation. These findings are consistent with those of Karickhoff and Morris (1985), who found that sorption could be described by a two compartment model. Xenobiotics could either exit in a labile fraction or in a nonlabile fraction, corresponding to a surface sorption or an interior sorption. For hydrophobic compounds, those authors found that the labile fraction was generally less than half of the total amount adsorbed.

Hautala (1978) suggested an additional reason for the slower direct photolysis rates of pesticides on soil surfaces. The photo-excited compound may be quenched by intermolecular energy transfer to humics or other substances. The fluorescence of carbaryl, for example, was not observed when the pesticide was sorbed on thin layers of soil. The relative significance of this quenching process, however, is unknown.

Until a better understanding of the basic factors affecting photolysis in real soils is obtained, it is doubtful that modeling efforts will provide reliable predictions. Thibodeaux and Lipsky (1985) have reported a fate and transport model for 2,3,7,8-TCDD in fly ash on soil and urban surfaces. The mechanisms incorporated into this study included dust fallout, volatilization, resuspension by wind, run-off by water, evaporation into air, photochemical reaction, and bioturbation. The model did not consider processes that would move the 2,3,7,8-TCDD downward or upward in the soil. In this study, photolysis was the most important mechanism for dissipation of 2,3,7,8-TCDD from the soils. Inputs for photolysis were based on the work of Crosby and Wong (1977) in which they demonstrated a 15% loss of 2,3,7,8-TCDD on a Sacramento loam soil (very-fine, montmorillonitic, thermic Vertic Haplaquolls) exposed to sunlight for 6 h. Thibodeaux and Lipsky extrapolated a first-order photolysis rate and obtained a rate constant of 0.64/d. This assumed not only first-order kinetics, but also used a photolysis rate from experiments in which 2,3,7,8-TCDD was irradiated in the presence of a good hydrogen donor (Agent Orange formulation). Hydrogen-donating solvents are known to substantially increase the photolysis rate over that observed in unamended soils (Dulin et al., 1986). In addition, photolysis rates on even very thin soils do not follow first-order kinetics and rates are generally observed to become slower as the photolysis proceeds. The same rate effect was observed by Miller and Zepp (1979) in experiments with DDE in suspended sediment.

The results of the Thibodeaux and Lipsky model indicated that the second most important dissipation pathway was volatilization. This process was predicted to be only half as fast as photolysis even at 40 °C, which was the highest temperature considered.

Temperature of the soil not only affects volatilization rates of 2,3,7,8-TCDD, but also photolysis rates (Liberti et al., 1978). Temperatures at the soil surface exposed to sunlight generally are higher than in the bulk soils. Temperature of soil surface will also have greater diurnal variations than the bulk soil. For example, surface soil temperatures measured in Athens, GA rose from an overnight low of 24 °C to an afternoon high of 40 °C. At a depth of 15 cm, the maximum temperature was only 32 °C.

#### 5–4.2.1 Singlet Oxygen Production

Because soil contains humic substances that are known to photosensitize the production of singlet oxygen in natural waters, similar photochemistry was expected to occur on irradiated soil surfaces. Using singlet oxygen traps, including tetramethylethylene and 1,2-dimethylcyclohexene, Gohre and Miller (1983) and Gohre et al. (1986) demonstrated the production of singlet oxygen on surfaces irradiated by both sunlight and laboratory lamps. Neither compound absorbs sunlight directly. The expected products, 2,3-dimethyl-2-buten-1-ol and 2-methyl-1-methylenecyclohexan-2-o1, respectively, were observed. The rate of loss of both compounds in sunlight was rapid, with half-lives of <5 h for each compound on several soils.

Initially, the assumption was made that only humic substances, and perhaps inorganic transition metals, contributed to the production of singlet oxygen, while silica gel and alumina were used as inactive controls (Gohre & Miller, 1983). Irradiation of the same singlet oxygen traps on various grades of silica gel and alumina, however, also demonstrated production of singlet oxygen at significant rates. The photochemical production of singlet oxygen was conclusively demonstrated on these nontransition metal oxides (Gohre & Miller, 1985). The authors suggested that the inorganic components of the soil can also be active in producing singlet oxygen, and perhaps other oxidizing species. Although clay minerals were not examined, the possibility exists that they also may produce singlet oxygen. This work demonstrated that both the inorganic and organic components of the soil contribute to the photosensitizing capacity of the soil.

As mentioned above, singlet oxygen reacts rapidly with sulfide-containing pesticides. Many common surface-applied insecticides contain sulfide moieties, including fenthion, butocarboxim, and disulfoton. Thus, when disulfoton was placed on soil surfaces and exposed to September sunlight, half of the original pesticide was converted to the corresponding sulfoxide within 2 d (Gohre & Miller, 1986).

### 5–4.2.2 Other Oxidants

Little work is published on the production of oxidants other than singlet oxygen on irradiated soil surfaces. Spencer et al. (1980a, b) found that parathion was efficiently converted to paraoxon on irradiated soil surfaces in the presence of ozone. Gohre et al. (1986) found that olefins were converted to compounds other than those expected from singlet oxygen reaction. Epoxides were particularly evident on low organic soils, whereas products expected from free radical reactions were not observed.

### 5–4.2.3 Applicability of Experimental Simulation and In Situ Studies to Understanding Photolysis in Soils

Diurnal and seasonal variability in sunlight can be duplicated only with elaborate equipment that can simultaneously alter the wavelength distribution and light intensity. Even if a particular constant-intensity lamp is sunlight calibrated for one compound, there is no assurance that the same calibration will be adequate for a different compound. Particularly on soils, indirect processes may be more important than direct processes, because of the competitive absorption of light by the various components of the soils. Obtaining a representative light source is difficult. Thus, sunlight should be used whenever possible for obtaining quantitative information relevant to photolytic processes in the environment.

Even in settings where clouds cause variability in sunlight intensity, sunlight experiments are most useful for predicting rates. Simultaneous irradiation of sunlight actinometers can provide an estimation of sunlight intensity even under cloudy conditions.

The use of silica gel as a model for soil has been frequently proposed. The advantages of silica gel are that it is uniform, easy to obtain, can provide reproducible results and rarely causes analytical interferences. The disadvantage of silica gel is that it bears little or no resemblance to real soils. Although silica can be an important constituent of the inorganic fraction of many soils, other constituents, including clay minerals and transition metals can affect the photochemistry. Silica is not an inert surface and, as discused previously, can catalyze photochemical events that may not be duplicated on soils. The greatest difference between soil and silica gel surfaces, however, is the presence of an organic fraction and transition metal oxides in soil that act as a light screen and effectively limit direct photolysis. The light that is absorbed by humic substances can either initiate photolytic reactions or be dissipated as heat. Indeed, because silica bears little resemblance to most soils and does not contain an organic component, it would be surprising if photolysis rates were not substantially faster on silica than on soils.

Glass surfaces also have been used as a surface model. Glass has several of the characteristics of silica gel, particularly in its homogeneous nature and the ease with which reproducibility can be maintained. Any similarity to soil surfaces with respect to photolysis rates is coincidental, however, because glass surfaces have none of the complex array of characteristics of soils. Unless some specific reason exists for the use of silica and glass, such as generation of photoproducts for identification, the use of either should be avoided if environmental modeling is being attempted.

Modeling soil photolysis is likely to be successful only after the basic chemistry and transport processes are more completely understood. Standard methods for examining photolysis on soils surfaces have not been developed (Lemaire et al., 1982). Yet, when performing photolysis experiments, six factors should be considered.

1. Compound photochemistry—The fate of an electronically excited molecule depends on the chemical environment in which it resides. Quenching processes, spin conversions, and available reactants can all affect photolysis rates and reaction products. An understanding of the photochemistry of each compound will greatly aid in interpreting the results of soil photolysis experiments.

2. Soil type—The soil type chosen should be representative of the soils that will be receiving the pesticide. The soil should be characterized as to organic content, mineral constituents, pH, and particle-size distribution. At present, the extent to which these factors affect photolysis is unknown.

3. Pesticide application method—Pesticides reach the surface of soils in several ways: application to fields, vapor deposition, run-off from leaves, and mass transport from bulk soil. Although the effect of various application methods on photolysis rates is also unknown, modeling surface photolysis rates will almost certainly depend on the method of application. If the pesticide is surface applied with a carrier solvent, the quantity and type of solvent used are likely to affect penetration into the soil. If the pesticide is incorporated into the soil the photolysis rate will be a function of depth of incorporation.

4. Water content—Water content of the soil affects the degree of sorption on soils, because water can displace xenobiotics from sorption sites. In addition, it can allow greater movement of the chemical into the light field and allow photolysis to proceed. Some chemicals have significantly different photolysis rates and products, depending on the chemical environment. An example is 2,3,7,8-TCDD which undergoes more rapid photolysis in organic solvent than in water (Dulin et al., 1986). On the other hand, 3,4-dichloroaniline undergoes substantially more rapid photolysis in water than in organic solvents (Miller et al., 1979).

5. Volatilization and advection—Affecting the photolysis will be processes that move the contaminant to and from the soil surface. Volatilization and advection are probably the two most important processes. Volatilization can be significant for many pesticides. Furthermore, transport to the surface through volatilization can be the rate-limiting step in photolysis under certain circumstances.

6. Pre-exposure of the surface—Also unknown, but probably important is the effect of photolytic and other weathering processes of the soil surface. Although cultivated lands are often turned over, other soils are not often mixed as a rule. The photic zone of the soil, therefore, can be exposed to sunlight for widely different durations. The photooxidative environment may change the chemistry of the soil surface, particularly in regard to the organics. The extent of preexposure of the soil surfaces to sunlight before the pesticide was applied may be important for estimating photolysis rates in the environment.

### 5-4.3 Pesticide Photoproducts

Phototransformation products of pesticides are often similar to those derived from metabolic and from other abiotic processes. Because $O_2$ availability is usually high wherever sunlight strikes the earth, photooxidation is of primary importance. Both direct and indirect photooxidations are common, and the distinction between the two is often difficult.

A toxicologically significant oxidation process is the conversion of thiophosphate to phosphate, which generally results in compounds having enhanced toxicity (Eto, 1974). Parathion, for example, is photooxidized to paraoxon directly (Grunwell & Erickson, 1973), as well as indirectly on soil dusts in the presence of ozone and UV light (Spencer et al., 1980a; Eq. [12]). Oxidation of sulfides to sulfoxides can also result in more toxic products. Disulfoton reacts with singlet oxygen to give the corresponding sulfoxide in high yields (Zepp et al., 1981a; Eq. [13]).

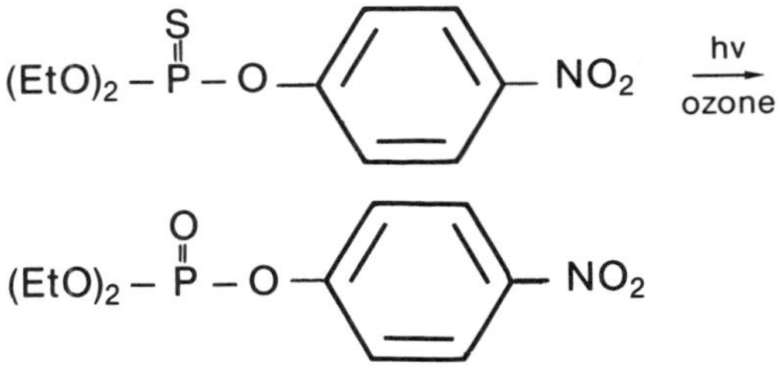


[12]

$$(H_3C-O)_2-\overset{\overset{\displaystyle S}{\|}}{P}-S-CH_2-CH_2-S-CH_2-CH_3 \xrightarrow{O_2} (H_3C-O)_2-\overset{\overset{\displaystyle S}{\|}}{P}-S-CH_2-CH_2-\overset{\overset{\displaystyle O}{\|}}{S}-CH_2-CH_3 \quad [13]$$

This conversion may be relevant to the large number of pesticides that contain sulfide linkages and are, therefore, susceptible to singlet oxygen oxidation. For surface-applied pesticides, this conversion is probably common, because of the singlet oxygen generating capacity of natural waters (Zepp et al., 1981a) and of soil surfaces (Gohre & Miller, 1983). Thiocarbonyl pesticides are also susceptible to singlet oxygen reactions (Ando, 1981). For example, the fungicides thiophanate and thiram are photooxidized on soil surfaces with a half-life <1 d (Gohre, 1985). Other oxidative pathways include ring oxidations, alkyl side chain oxidation, oxidation of carbons adjacent to N and amino to nitro oxidations (Miller & Crosby, 1983a).

Products characteristic of hydrolysis are commonly observed for pyrethroids (Ruzo, 1982), ureas (Crosby & Tang, 1969) and organophosphates such as prothiophos (Takase et al., 1982) and fenitrothion (Ohkawa et al., 1974) exposed to light. At least two hydrolytic processes appear to be uniquely photochemical. Nitrofen and related compounds undergo photonucleophilic substitution to the corresponding phenols (Eq. [14]).

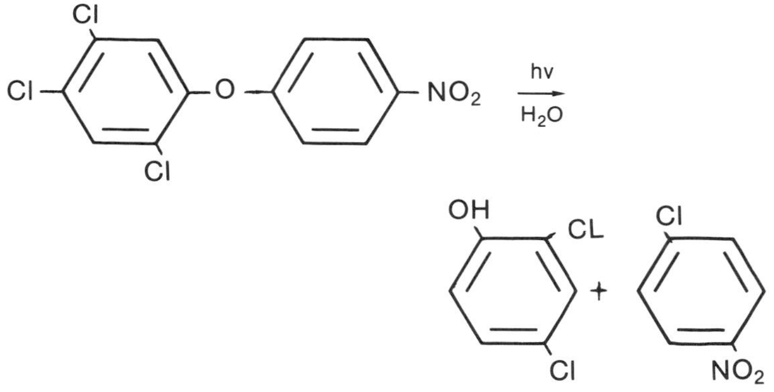


[14]

Likewise, 3,4-dichloroaniline undergoes photohydrolysis as the primary transformation pathway in water to give 2-chloro-5-animophenol (Miller et al., 1979; Eq. [15]).

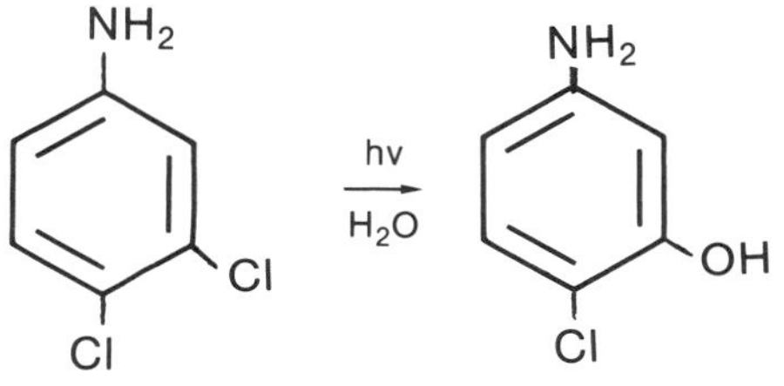


[15]

Although photoreduction can occur in $O_2$-saturated systems, thermal and most metabolic reductions require anaerobic conditions. Aromatic nitrocompounds are photoreduced to the corresponding amines as well as various other products. Hydrogen substitution for chlorine is observed for direct, but not for indirect, processes. For example, octachlorodibenzodioxin is photoreduced to 2,3,7,8-TCDD, albeit at low yields (0.05%) on soil surfaces (Miller et al., 1988). This photoreduction is not observed in solution.

Other uniquely photochemical environmental pathways have been observed, including isomerization or extrusions. These reactions are compound specific, however, and few generalizations can be made, except that the molecules must be electronically excited, either by direct light absorption or by intermolecular energy transfer from some other electronically excited chromophore. Examples include intramolecular chlorine transfer in DDE (Zepp et al., 1977) and extrusion of $CO_2$ from pyrethroids (Ruzo, 1982).

Product formation is wavelength dependent, thus the spectral distribution of the light source used should simulate the solar spectrum. For example, although recent photolysis studies using 254 nm radiation on aldicarb (Freeman & McCarthy, 1984) and methomyl (Freeman & Nipp, 1984) may provide information on the short wavelength photochemistry of these compounds, they give no information on the environmental fate of either compound. Because neither compound absorbs sunlight, direct photolysis will not occur. By analogy with other thioether compounds, both compounds will likely react with singlet oxygen to produce sulfoxides and thus be susceptible to indirect photolysis. Even for compounds that absorb sunlight, photolysis using 254 nm light may produce products not observed in sunlight. For example, a primary photoproduct of Sustar, a plant growth regulator, using 254 nm light is a Fries rearrangement product that was not observed in sunlight (Miller & Crosby, 1978).

## 5-5 SUMMARY: PREVALENCE AND SIGNIFICANCE OF ABIOTIC TRANSFORMATIONS

Both abiotic and biotic transformations of pesticides occur simultaneously in the environment. Although the dominance of abiotic transformations was demonstrated in some cases, it was often assumed that biotic transformations control the degradation of pesticides in natural systems. It is actually difficult to offer a general estimate of the relative importance of abiotic transformations under conditions that enable both biotic and abiotic transformations to occur. Conditions under which abiotic transformations dominate are described below.

Pesticide residues may be leached below the root zone and, at times, reach the groundwater. The concentration of nutrients, the organic matter content, and the rate of gas exchange usually decrease with depth. As a result, the microbial population and, with it, the probability of occurrence of most microbial transformations decline with depth as well. Thus, abiotic transformations are likely to control the degradation of many pesticides at sufficient depths.

An ever-increasing number of cases of pollution resulting from the leakage of toxic organic substances, including pesticides, from concentrated waste-disposal sites is being detected. The high concentration of toxicants in these wastes often inhibits microbial degradation. Physicochemical processes such as abiotic transformations may then be responsible for the disappearance of the toxic organic substances that could endanger groundwater or surface water bodies in the vicinity of the source of the leak.

Below some critical moisture content, microbial activity is strongly reduced. Under dry conditions, abiotic transformations and physical-phenomena such as volatilization may determine, therefore, the rate of disappearance of pesticides. In arid zones, where a long dry season exists, damage to crops by herbicide residues that remained in the soil from the previous growing season can occur. Abiotic transformations may reduce or even eliminate such damage.

It is apparent that abiotic transformations can play an important role in the detoxification of contaminated soil and water systems, especially when biotic conversions are not likely to be significant. Detoxification of contaminated containers, water bodies, or soils may be achieved by manipulating the contaminated system to enhance the abiotic transformation of the contaminant. Wastes containing high concentrations of organophosphate pesticides, for example, may be cleaned up by adding a $Cu^{2+}$ salt at a low concentration and by adjusting the pH.

Abiotic transformations may not always reduce the biocidal activity in contaminated systems. Products of abiotic transformations of pesticides may still be toxic. Joiner et al. (1973) demonstrated that some photoalteration products of parathion are more toxic to mammals than the parent material.

In recent years, pesticide application through the irrigation systems has become a common practice. Because the residence time of the pesticides in the irrigation system is usually no more than a few hours, the establishment of the microbial population required for a specific biotic degradation is not likely to occur. If the appropriate conditions exist, it is possible, however, that certain abiotic transformations will occur. The properties of the irrigation water, such as its pH, may strongly affect the stability of the applied pesticide. The method of irrigation may also influence the fate of the pesticide. Flooding or ditch irrigation, for example, make photodegradation possible.

Mixed application of pesticides and other chemicals, such as fertilizers, through the irrigation system is also commonly practiced. One important consideration in the use of this procedure is the compatibility between the components of the mixture. Among the reactions that pesticides may undergo in the irrigation water during mixed application are acid or base-hydrolysis and transformations catalyzed by transition metal cations.

Compatibility is also an important factor in the design of pesticide formulations. Formulation chemists are well aware, for example, that pesticide degradation may be catalyzed by some clay carriers. Formulants may, on the other hand, increase the persistence of pesticides, for example, by protective associations between the carrier and the pesticide. Such associations may hinder the adsorption of the pesticide, once it reaches the soil, on sites that catalyze its degradation. Formulation may affect the behavior of the pesticide both in the soil and aqueous systems. The Cu-catalyzed hydrolysis of some pesticides in solution, for example, was inhibited by their formulants (Chapman & Harris, 1984).

Considerable research effort is devoted to the development of predictive models for the fate of pesticides in the environment. Such models, if successful, will help make the application of pesticides more efficient and less environmentally hazardous. A recognition of the important role abiotic transformations play in the degradation of pesticides in soils and water and a better understanding of the transformations are essential for the successful development of predictive models.

## REFERENCES

Adams, J.M., T.V. Clapp, and D.E. Clement. 1983. Catalysis by montmorillonites. Clay Miner. 18:411–421.

Ahlrichs, J.L. 1972. The soil environment. p. 3–43. *In* C.A.I. Goring and J.W. Hamaker (ed.) Organic chemicals in the soil environment. Vol. 1. Marcel Dekker, New York.

Ando, W. 1981. Photooxidation of organosulfur compounds. Sulfur Rep. 1:147–207.

Armstrong, D.E., and G. Chesters. 1968. Adsorption catalyzed chemical hydrolysis of atrazine. Environ. Sci. Technol. 2:683–689.

Armstrong, D.E., and J.G. Konrad. 1974. Nonbiological degradation of pesticides. p. 123–130. *In* W.D. Guenzi (ed.) Pesticides in soil and water. SSSA, Madison, WI.

Bailey, G.W., J.L. White, and T. Rothberg. 1968. Adsorption of organic herbicides by montmorillonite. Role of pH and chemical character of adsorbate. Soil Sci. Soc. Am. Proc. 32:222–234.

Banks, S., and J.R. Tyrell. 1984. Kinetics and mechanism of alkaline and acidic hydrolysis of aldicarb. J. Agric. Food Chem. 32:1223–1232.

Bansal, O.P. 1983. Adsorption of oxamyl and dimecron in montmorillonite suspensions. Soil Sci. Soc. Am. J. 47:877–883.

Barnsley, G.E., and P.A. Gabbott. 1966. A new herbicide 2-azido-4-ethylamino-6-*t*-butylamino-1,3,5-triazine. p. 372–376. *In* Proc. 8th Br. Weed Control Conf.

Bartha, R. 1971. Fate of herbicide-derived chloroanilines in soil. J. Agric. Food Chem. 19:385–387.

Barug, D., and J.W. Vonk. 1980. Studies on the degradation of bis(tributyltin) oxide in soil. Pestic. Sci. 11:77–82.

Bender, M.L., and M.S. Silver. 1963. The hydrolysis of substituted 2-phenyl-1,3-dioxane. J. Am. Chem. Soc. 85:3006–3010.

Blanchet, P.F., and A. St-George. 1982. Kinetics of chemical degradation of organophosphorus pesticides. Hydrolysis of chlorpyrifos and chlorpyrifos-methyl in the presence of copper (II). Pestic. Sci. 13:85–91.

Bohn, H.L., B.L. McNeal, and G.A. O'Connor. 1985. Soil chemistry. John Wiley and Sons, New York.

Bowman, B.T., and W.W. Sans. 1977. Adsorption of parathion, fenitrothion, methyl-parathion, aminoparathion and paraoxon by $Na^+$, $Ca^{2+}$ and $Fe^{3+}$ montmorillonite suspensions. Soil Sci. Soc. Am. J. 41:514–519.

Bowman, B.T., and W.W. Sans. 1980. Stability of parathion and DDT in dilute iron solutions. J. Environ. Sci. Health B15:233–246.

Boyd, S.A., and M.M. Mortland. 1985. Urease activity on a clay-organic complex. Soil Sci. Soc. Am. J. 49:619–622.

Bromilov, R.H., G.G. Briggs, M.R. Williams, J.H. Smelt, L.G.M. Th. Tuinstra, and W.A. Traag. 1986. The role of ferrous ions in the rapid degradation of oxamyl, methomyl and aldicarb in anaerobic soils. Pestic. Sci. 17:535–547.

Brown, C.B., and J.L. White. 1969. Reactions of 1,2-*s*-triazines with soil clays. Soil Sci. Soc. Am. Proc. 33:863–867.

Brown, K.A. 1981. Biochemical activities in peat sterilized by gamma-irradiation. Soil Biol. Biochem. 13:469–474.

Burkhard, N., and J.A. Guth. 1979. Photolysis of organophosphorus insecticides on soil surfaces. Pestic. Sci. 10:313–319.

Burkhard, N., and J.A. Guth. 1981. Chemical hydrolysis of 2-chloro-4,6-bis(alkylamino)-1,3,5-triazine herbicides and their breakdown in soil under the influence of adsorption. Pestic. Sci. 12:45–52.

Burns, L.A., and D.M. Cline. 1985. Exposure analysis modeling system: Reference manual for EXAMS II. EPA-600/3-85/038, p. 1–83. USEPA, Athens, GA.

Burns, R.G. 1975. Factors affecting pesticide loss from soil. p. 103–141. *In* E.A. Paul and A.D. McLearen (ed.) Soil biochemistry. Vol. 4. Marcel Dekker, New York.

Burns, R.G., and J.A. Edwards. 1980. Pesticide breakdown by soil enzymes. Pestic. Sci. 11:506–512.

Burns, R.G., and W.P. Gibson. 1976. The disappearance of 2,4-D, diallate and malathion from soil and soil components. p. 5–13 *In* A. Banin and U. Kaffear (ed.) Agrichemicals and soils. Pergamon, Oxford.

Carter, W.C., and I.H. Suffet. 1982. Binding of DDT to dissolved humic materials. Environ. Sci. Technol. 16:735–740.

Castelfranco, P., A. Oppenheim, and S. Yamaguchi. 1963. Riboflavin mediated photodecomposition of amitrole in relation to chlorosis. Weeds 11:111–117.

Chapman, R.A., and M. Cole. 1982. Observations on the influence of water and soil pH on the persistence of insecticides. J. Environ. Sci. Health B17:487–504.

Chapman, R.A., and C. Harris. 1984. The chemical stability of formulations of some hydrolyzable insecticides in aqueous mixtures with hydrolysis catalysts. J. Environ. Sci. Health B:19:397–407.

Cooper, C.W, and R.G. Zika. 1983. Photochemical formation of hydrogen peroxide in surface and ground water exposed to sunlight. Science 220:711–717.

Cooper, W.J. 1987. Abiotic transformations of halogenated organics. I. Elimination reaction of 1,1,2,2-tetrachloroethane and formation of 1,1,2-trichloroethene. Environ. Sci. Technol. 21:1112–1114.

Crosby, D.G. 1970. The nonbiological degradation of pesticides in soils. p. 86–94. *In* Pesticides in soil: Ecology, degradation, and movement. Proc. Int. Symp. Pesticides in Soil. Michigan State Univ., East Lansing.

Crosby, D.G., and A.S. Wong. 1977. Environmental degradation of 2,3,7,8-tetrachlorodibenzo-*p*-dioxin (TCDD). Science 195:1337–1338.

Crosby, D.G., and C.S. Tang. 1969. Photodecomposition of 3-(*p*-chlorophenyl)-1,1-dimethylurea (Monuron). J. Agric. Food Chem. 17:1041–1044.

Crossland, N.O., and C.J.M. Wolff. 1985. Fate and biological effects of pentachlorophenol in outdoor ponds. Environ. Toxicol. Chem. 4:73–86.

Deuel, L.E., Jr., K.W. Brown, J.D. Price, and F.T. Turner. 1985. Dissipation of carbaryl and the 1-naphthol metabolite in flooded rice fields. J. Environ. Qual. 14:349–354.

Dixon, J.B., and S.B. Weed (ed.). 1977. Minerals in soil environments. SSSA, Madison, WI.

Dixon, S.R., and C.H.J. Wells. 1983. Dye-sensitized photo-oxidation of 2-dimethylamino-5,6-dimetriylpyrimidin-4-ol in aqueous solution. Pestic. Sci. 14:444–448.

Draper, W.M. 1985. Determination of wavelength-averaged, near UV quantum yields for environmental chemicals. Chemosphere 14:1195–1203.

Draper, W.M., and D.G. Crosby. 1981. Hydrogen peroxide and hydroxyl radical: Intermediates in photolysis reactions in water. J. Agric. Food. Chem. 29:699–702.

Draper, W., and D.G. Crobsy. 1983. The photochemical generation of hydrogen peroxide in natural waters. Arch. Environ. Contam. Toxicol. 12:121–126.

Dulin, D., H. Drossman, and T. Mill. 1986. Products and quantum yields for photolysis of chloroaromatics in water. Environ. Sci. Technol. 20:72–77.

Dulin, D. and T. Mill. 1982. Development and evaluation of sunlight actinometers. Environ. Sci. Technol. 16:815–820.

Durand, G. 1965. Enzymatic splitting of urea in the presence of bentonite. Ann. Inst. Pasteur 109 (Suppl. 3):121–132.

El-Amamy M.M., and T. Mill. 1984. Hydrolysis kinetics of organic chemicals on montmorillonite and kaolinite surfaces as related to moisture content. Clays Clay Miner. 32:67–73.

Emerson, S. 1976. Early diagenesis in anaerobic lake sediments: Chemical equilibriums in interestitial waters. Geochem. Cosmochim. Acta 40:925–934.

Eto, M. 1974. Organophosphorus pesticides: Organic and biological chemistry. CRC Press, Cleveland.

Ercegovich, C.D., and D.E.H. Frear. 1964. The fate of 3-amino-1,2,4 triazole in soils. J. Agric. Food Chem. 12:26–31.

Estermann, E.F., and A.D. McLaren. 1959. Simulation of bacteria proteolysis by adsorbents. J. Soil Sci. 10:64–78.

Estermann, E.F., G.H. Peterson, and A.D. McLaren. 1959. Digestion of clay-protein, lignin-protein, and silica-protein complexes by enzymes and bacteria. Soil Sci. Soc. Am. Proc. 23:31–36.

Farmer, V.C. 1978. Water on particle surfaces. *In* D.J. Greenland and M.H.B. Hayes (ed.) The chemistry of soil constituents. John Wiley and Sons, Chichester, England.

Farmer, W.J., and Y. Aochi. 1987. Chemical conversion of pesticides in the soil water environment p. 69–74. *In* J.W. Biggar and J.M. Siber (ed.) Fate of pesticides in the environment. California Agric. Exp. Stn. Spec. Publ. 3320.

Freeman, P.K., and K.D. McCarthy. 1984. Photochemistry of oxime carbamates. I. Phototransformations of aldicarb. J. Agric. Food Chem. 32:873–877.

Freeman, P.K., and E.M.N. Nipp. 1984. Photochemistry of oxime carbamates. 2. Phototransformations of methomyl. J. Agric. Food Chem. 32:877–881.

Fripiat, J.J. 1968. Surface fields and transformation of adsorbed molecules in soil colloids. Trans. 9th Int. Congr. Soil Sci. 1:679–689.

Fusi, P., G.G. Ristori, S. Cecconi, and M. Franci. 1983. Adsorption and degradation of fenarimol on montmorillonite. Clays Clay Miner. 31:312–314.

Geschwend, P.M., and S.-C. Wu. 1985. On the constancy of sediment-water partition coefficients of hydrophobic organic pollutants. Environ. Sci. Technol. 19:90–96.

Getzin, L.W., and I. Rosefield. 1968. Organophosphorus insecticide degradation by heat-labile substances in soil. J. Agric. Food Chem. 16:598–601.

Getzin, L.W., and I. Rosefield. 1971. Partial purification and properties of a soil enzyme that degrades the insecticide malathion. Biochim. Acta 235:442–453.

Glass, B.L. 1972. Relation between the degradation of DDT and the iron redox system in soils. J. Agric. Food Chem. 20:324–327.

Gohre, K. 1985. Singlet oxygen production on metal oxide and soil surfaces: Relation to sulfur containing pesticide photodegradation. Ph.D. thesis. Univ. of Nevada, Reno (Diss. Abstr. 46:07-B).

Gohre, K., and G.C. Miller. 1983. Singlet oxygen generation on soil surfaces. J. Agric. Food Chem. 31:1104–1108.

Gohre, K., and G.C. Miller. 1985. Photochemical generation of singlet oxygen on nontransition-metal oxide surfaces. J. Chem. Soc. Faraday Trans. I. 81:793–800.

Gohre, K., R. Scholl, and G.C. Miller. 1986. Singlet oxygen reactions on irradiated soil surfaces. Environ. Sci. Technol. 20:934–938.

Gohre, K., and G.C. Miller. 1986. Photooxidation of thioether pesticides on soil surfaces. J. Agric. Food Chem. 34:709–713.

Gorbunov, N.I., G.S. Dzyadevich, and B.M. Tunik. 1961. Methods of determining non-silicate amorphous and crystalline sesquioxides. Soviet Soil Sci. 11:1251–1259.

Graham-Bryce, I.J. 1981. The behaviour of pesticides in soil. p. 621–671. *In* D.J. Greenland and M.H.B. Hayes (ed.) The chemistry of soil processes. John Wiley and Sons, Chichester, England.

Grayson, B.T. 1980. Hydrolysis of cyanazine and related diaminochloro-1,3,5-triazines. Part II: Hydrolysis in sulphuric acid solutions. Pestic. Sci. 11:493–505.

Greenhalgh, R., K.L. Dhawan, and P. Weinberger. 1980. Hydrolysis of fenitrothion in model and natural aquatic systems. J. Agric. Food Chem. 28:102–105.

Greenland, D.J. 1965. Interaction between clays and organic compounds in soils. Part 1. Mechanisms of interaction between clays and defined organic compounds. Soils Fert. 28:415–420.

Grunwell, J.R., and R.H. Erickson. 1973. Photolysis of parathion [*O,O*-diethyl-*O*-(4-nitrophenyl)thiophosphate]. New products. J. Agric. Food Chem. 21:929–931.

Hamaker, J.W. 1972. Decomposition: Quantitative aspects. p. 253–341. *In* C.A.I. Goring and J.W. Hamaker (ed.) Organic chemicals in the soil environment. Vol. 1. Marcel Dekker, New York.

Hamaker, J.W. 1975. The interpretation of soil leaching experiments. p. 115–133. *In* V.H. Freed (ed.) Environmental dynamics of pesticides. Plenum Press, New York.

Hamaker, J.W., and J.M. Thompson. 1972. Adsorption. p. 51–145. *In* C.A.I. Goring and J.W. Hamaker (ed.) Organic chemicals in the soil environment. Vol. 1. Marcel Dekker, New York.

Hance, R.J. 1967. Decomposition of herbicides in the soil by non-biological chemical processes. J. Sci. Food Agric. 18:544.

Hance, R.J. 1970. Influence of adsorption on the decomposition of pesticides. p. 92–104. *In* Sorption and transport processes in soils. SCI Monogr. 37. Soc. Chem. Ind., London.

Haniff, M.I., R.H. Zienius, C.H. Langford, and D.S. Gamble. 1985. The solution phase complexing of atrazine by fulvic acid: Equilibria at 25 °C. J. Environ. Sci. Health B20:215–262.

Hautala, R. 1978. Surfactant effect on pesticide photochemistry in water and soil. EPA-600/3-78-060. USEPA, Athens, GA.

Herbert, V.R., and R.G. Miller. 1990. Depth dependence of direct and indirect photolysis on soil surfaces. J. Agric. Food Chem. (In press.)

Hoffmann, G. 1959. Verteilung und Herkunft einiger Enzyme im Boden. Z. Pflanzenernaehr. Dung. Bodenk. 85:97–104.

Holmstead, R.L. 1976. Studies on the degradation of mirex with an iron(II) prophyrin model system. J. Agric. Food Chem. 24:621–624.

Hurle, K., and A. Walker. 1980. Persistence and its prediction. p. 83–122. *In* R.J. Hance (ed.) Interactions between herbicides and the soil. Academic Press, London.

Hutchinson, G.E. 1957. The thermal properties of lakes. A treatise on limnology. p. 426–540. *In* Vol. 1. John Wiley and Sons, New York.

Jeffers, P.M., L.M. Ward, L.M. Woybowltch, and N.L. Wolfe. 1989. Homogeneous hydrolysis rate constants for selected chlorinated methanes, ethanes, ethenes and propanes. Environ. Sci. Technol. 23:965–969.

Joiner, R.L., H.W. Chambers, and K.P. Baetcke. 1973. Comparative inhibition of boll weevil, golden shiner and white rat cholinesterases by selected photoalteration products of parathion. Pestic. Biochem. Physiol. 2:371–376.

Juma, N.J., and K.A. Tabatabai. 1977. Effect of trace elements on phosphatase activity in soils. J. Soil Sci. Soc. Am. 41:343–346.

Kamil, J., and I. Shainberg. 1968. Hydrolysis of sodium montmorillonite in sodium chloride solutions. Soil Sci. 106:193–199.

Karickhoff, S.W. 1980. Sorption kinetics of hydrophobic pollutants in natural sediments. p. 193–205. *In* R.A. Baker (ed.) Contaminants and sediments. Vol. 2. Ann Arbor Sci., Ann Arbor, MI.

Karickhoff, S.W., D.S. Brown, and T.A. Scott. 1979. Sorption of hydrophobic pollutants on natural sediments. Water Res. 13:241.

Karickhoff, S.W., and K.R. Morris. 1985. Sorption dynamics of hydrophobic pollutants in sediment suspensions. Environ. Toxicol. Chem. 4:469–479.

Kaufman, D.D., J.R. Plimmer, P.C. Kearney, J. Blake, and F.S. Guardia. 1968. Chemical versus microbial decomposition of amitrole in soil. Weed Sci. 16:266–272.

Kaurichev, I.S., T.N. Ivanova, and E.M. Nozdrunova. 1963. Content of low-molecular organic acids in the composition of water-soluble organic matter of soils. Pochvovedenie 3:27–35.

Kearney, P.C., D.D. Kaufman (ed.) 1969. Degradation of herbicides. Marcel Dekker, New York.

Khalifa, S., R.L. Holmstead, and J.E. Casida. 1976. Toxaphene degradation by iron(II) protoporphyrin system. J. Agric. Food Chem. 24:277–282.

Khan, S.U. 1978. Kinetics of hydrolysis of atrazine in aqueous fulvic acid solution. Pestic. Sci. 9:39–43.

Khan, S.U. 1978. The interaction of organic matter with pesticides. p. 137–173. *In* M. Schnitzer and S.U. Khan (ed.) Soil organic matter. Developments in soil science 8. Elsevier Sci. Publ. Co., Amsterdam.

Khan, S.U. 1980. Pesticides in the soil environment. Elsevier Sci. Publ. Co., Amsterdam.

Khan, S.U. 1982. Bound pesticide residues in soil and plants. Residue Rev. 84:1–25.

Klehr, M., J. Iwan, and J. Riemann. 1983. An experimental approach to the photolysis of pesticides absorbed on soil: thiodiazuron. Pestic. Sci. 14:359–366.

Knuesli, E., D. Berrer, G. Dupuis, and H. Esser. 1969. *s*-Triazines. p. 51–79. *In* P.C. Kearney and D.D. Kaufman (ed.) Degradation of herbicides. Marcel Dekker, New York.

Konrad, J.G., D.E. Armstrong, and G. Chesters. 1967. Soil degradation of diazinon, a phosphorothioate insecticide. Agron. J. 59:591–594.

Konrad, J.G., and G. Chesters. 1969. Degradation in soils of ciodrin, an organophosphate insecticide. J. Agric. Food Chem. 17:226–230.

Kroll, L., and M. Kramer. 1955. Der Einfluss der Tonmineralein auf die Enzym-Aktivitat der Bodenphosphatase. Naturwissenschaftern. 42:157–158.

Ladd, J.N. 1985. Soil enzymes. p. 175–223. *In* D. Vaughan and R.E. Malcolm (ed.) Soil organic matter and biological activity. Nijhoff/Junk Publ., Dordrecht.

Lahav, N., D. White, and S. Chang. 1978. Peptide formation in the prebiotic ear: Thermal condensation of glycine in fluctuating clay environments. Science 201:67–69.

Leifer, A. 1988. The kinetics of environmental aquatic photochemistry. Am. Chem. Soc., Washington, DC.

Lemaire, J., I. Campbell, H. Hulpke, J.A. Guth, W. Merz, J. Philip, and C. von Waldow. 1982. An assessment of test methods for photodegradation of chemicals in the environment. Chemosphere 11:119–164.

Liberti, A., D. Brocco, I. Alligrini, A. Cecinato, and M. Possanzini. 1978. Solar and UV photo decomposition of 2,3,7,8-tetrachlorodibenzo-*p*-dioxin in the environment. Sci. Total Environ. 10:97–104.

Liu, S.-Y., A.J. Freyer, R.D. Minard, and J.-M. Bollag. 1985. Enzyme-catalyzed complex-formation of amino acid esters and phenolic humus constituents. Soil Sci. Soc. Am. J. 49:337–342.

Loll, M.J., and J.—M. Bollag. 1985. Characterization of a citrate-buffer soil extract with oxidative coupling activity. Soil Biol. Biochem. 17:115–117.

Lopez-Gonazles, J.De D., and C. Valenzuela-Calahorro. 1970. Associated decomposition of DDT to DDE in the diffusion of DDT on homoionic clays. J. Agric. Food Chem. 18:520–523.

Macalady, D.L., and N.L. Wolfe. 1983. New perspectives on the hydrolytic degradation of the organophosphorothioate insecticide chlorpyrifos. J. Agric. Food Chem. 31:1139–1147.

Macalady, D.L., and N.L. Wolfe. 1985. Effects of sediment sorption and abiotic hydrolyses. I. Organophosphorothioate esters. J. Agric. Food Chem. 33:167–173.

Macalady, D.L., P.G. Tratnyek, and T.J. Grundy. 1986. Abiotic reduction reactions of anthropogenic organic chemicals in anaerobic systems: A critical review. J. Contam. Hydrol. 1:1–28.

McBride, M.B. 1985. Surface reactions of 3,3′,6,6′-tetramethyl benzidine on hectorite. Clays Clay Miner. 33:510–516.

Melnikov, N.N. 1971. Chemistry of pesticides. Springer-Verlag, New York.

Mikami, N., N. Takahashi, K. Haytashi, and J. Miyamoto. 1980. Photodegradation of fenvalerate (sumicidin) in water and on soil surface. J. Pestic. Sci. 5:225–236.

Mikami, N., K. Imanishi, H. Yamada, and J. Miyamoto. 1985. Photodegradation of fenitrothion in water and on soil surface, and its hydrolysis in water. J. Pestic. Sci. 10:263–272.

Mill, T., D.G. Hendry, and H. Richardson. 1980. Free radical oxidants in natural waters. Science 207:886–887.

Miller, G.C., and D.G. Crosby. 1978. Photodecomposition of sustar in water. J. Agric. Food Chem. 26:1316–1321.

Miller, G.C., and D.G. Crosby. 1983a. Pesticide photoproducts: Generation and significance. J. Toxicol. Clin. Chem. 19:707–727.

Miller, G.C., and D.G. Crosby. 1983b. Photooxidation of 4-chloroaniline and *N*-(4-chlorophenyl)benzenesulfonamide to nitroso and nitro products. Chemosphere 12:1217–1227.

Miller, G.C., M.J. Miille, D.G. Crosby, S. Sontum, and R.G. Zepp. 1979. Photosolvolysis of 3,4-dichloroaniline in water. Evidence for an arylcation intermediate. Tetrahedron 35:1797–1800.

Miller, G.C., and R.G. Zepp. 1979. Photoactivity of aquatic pollutants sorbed on suspended sediments. Environ. Sci. Technol. 13:860–863.

Miller, G.C., and R.G. Zepp. 1983. Extrapolating photolysis rates from the laboratory to the environment. Residue Rev. 85:89–110.

Miller, R.G., V.R. Herbert, M.J. Millie, R. Mitzel, and R.G. Zepp. 1989. Photolysis of octachlorodibenzyl-*p*-dioxin on soils: Production of 2,3,7,8-TCDD. Chemosphere 18:1265–1274.

Mingelgrin, U., Z. Gerstl, and B. Yaron. 1975. Pirimiphos ethyl-clay surface interactions. Soil Sci. Soc. Am. Proc. 39:834–837.

Mingelgrin, U., and S. Saltzman. 1979. Surface reactions of parathion on clays. Clays Clay Miner. 27:72–78.

Mingelgrin, U., S. Saltzman, and B. Yaron. 1977. A possible model for the surface-induced hydrolysis of organophosphorus pesticides on kaolinite clays. Soil Sci. Soc. Am. J. 41:519–523.

Mingelgrin, U., S. Yariv, and S. Saltzman. 1978. Differential infrared spectroscopy in the study of parathion-bentonite complexes. Soil Sci. Soc. Am. J. 42:664–665.

Mingelgrin, U., and B. Yaron. 1974. The effect of calcium salts on the degradation of parathion in sand and soil. Soil Sci. Soc. Am. Proc. 38:914–917.

Moore, J.W., and R.G. Pearson. 1981. Kinetics and mechanism. 3rd ed., John Wiley and Sons, New York.

Morrill, L.G., B.C. Mahilum, and S.H. Mohiuddin. 1982. Organic compounds in soils. Ann Arbor Sci., Ann Arbor, MI.

Mortland, M.M. 1970. Clay-organic complexes and interactions. Adv. Agron. 22:75–117.

Mortland, M.M., and K.V. Raman. 1967. Catalytic hydrolysis of some organic phosphate pesticides by copper(II). J. Agric. Food Chem. 15:163–167.

Nash, R.G., W.G. Harris, and C.C. Lewis. 1973. Soil pH and metallic amendment effects of DDT conversions to DDE. J. Environ. Qual. 2:390–394.

Nearpass, D.C. 1972. Hydrolysis of propazine by the surface acidity of organic matter. Soil Sci. Soc. Am. Proc. 36:606–610.

Nilles, G.P., and M.J. Zabik. 1974. Photochemistry of bioactive compounds. Multiphase photodegradation of basalin. J. Agric. Food Chem. 22:684–688.

Nilles, G.P., and M.J. Zabik. 1975. Photochemistry of bioactive compounds. Multiphase photodegradation and mass spectral analysis of basagran. J. Agric. Food Chem. 23:410–415.

Ohkawa, H., N. Mikami, and J. Miyamoto. 1974. Photodecomposition of sumithion (*O,O*-dimethyl *O*-(3-methyl-4-nitrophenyl) phosphorothioate). Agric. Biol. Chem. 38:2247–2255.

Patrick, W.H., Jr., and I.C. Mahapatra. 1968. Transformation and availability to rice of nitrogen and phosphorus in waterlogged soils. Adv. Agron. 20:323–359.

Perdue, E.M., and N.L. Wolfe. 1982. Modification of pollutant hydrolysis kinetics in the presence of humic substances. Environ. Sci. Technol. 16:847–852.

Perdue, E.M., and N.L. Wolfe. 1983. Prediction of buffer catalysis in field and laboratory studies. Environ. Sci. Technol. 17:635–642.

Peterson, J.T. 1976. Calculated actinic fluxes (290–700 nm) for air pollution photochemistry applications. EPA-600/4-76-025. USEPA, Athens, GA.

Plimmer, J.R., P.C. Kearney, D.D. Kaufman, and F.S. Guardia. 1967. Amitrole decomposition by free radical-generating systems and by soils. J. Agric. Food Chem. 15:996–999.

Plimmer, J.R., P.C. Kearney, H. Chisaka, J.B. Yount, and U.I. Klingebiel. 1970. 1,3-bis (3,4-dichlorophenyl)triazine from propanil in soils. J. Agric. Food Chem. 18:859–861.

Ponec, V., Z. Knor, and S. Cerny. 1974. Adsorption on solids. Butterworths, London.

Praffit, R.L., and M.M. Mortland. 1968. Ketone adsorption on montmorillonite. Soil Sci. Soc. Am. Proc. 32:355–363.

Press, M.C., J. Henderson, and J.A. Lee. 1985. Arylsulphatase activity in peat in relation to acidic deposition. Soil Biol. Biochem. 17:99–103.

Probst, G.W., and J.B. Tepe. 1969. Trifluralin and related compounds. p. 255–282. *In* P.C. Kearney and D.D. Kaufman (ed.) Degradation of herbicides. Marcel Dekker, New York.

Ristori, G.G., P. Fusi, and M. Franci. 1981. Montmorillonite-asulam interactions. II. Catalytic decomposition of asulam adsorbed on Mg, Ba, Ca, Li, Na, K and Cs-clay. Clay Miner. 16:125–137.

Rogers, J.E. 1986. Anaerobic transformation processes: A review of the microbial literature. EPA-600/S3/042. November USEPA, Athens, GA.

Rosenfield, C., and W. Van Valkenberg. 1965. Decomposition of (*O,O*-dimethyl-O-2,4,5-trichlorophenyl) phosphorothioate (ronnel) adsorbed on bentonite and other clays. J. Agric. Food Chem. 13:68–72.

Russell, J.D., M. Curz, and J.L. White. 1968a. Mode of chemical degradation of *s*-triazines by montmorillonite. Science 160:1340–1342.

Russell, J.D., M. Curz, and J.L. White. 1968b. The adsorption of 3-aminotriazole by montmorillonite. J. Agric. Food Chem. 16:21–24.

Ruzo, L.A. 1982. Photochemical reactions of the synthetic pyrethroids. Progress Pestic. Biochem. 2:1–33.

Sahay, B.K., and M.J.D. Low. 1974. Interactions between surface hydroxyl groups and adsorbed molecules. V. Fluorobenzene adsorption on Germania. J. Colloid. Interface Sci. 48:20–31.

Saltzman, S., B. Yaron, and U. Mingelgrin. 1974. The surface catalyzed hydrolysis of parathion on kaolinite. Soil Sci. Soc. Am. Proc. 38:231–234.

Saltzman, S., U. Mingelgrin, and B. Yaron. 1976. Role of water in the hydrolysis of parathion and methylparathion on kaolinite. J. Agric. Food Chem. 24:739–743.

Sanchez Camazano, M., and M.J. Sanchez Martin. 1983a. Factors influencing interactions of organophosphorous pesticides with montmorillonite. Geoderma 29:107–118.

Sanchez Camazano, M., and M.J. Sanchez Martin. 1983b. Montmorillonite catalyzed hydrolysis of phosmet. Soil Sci. 136:89–93.

Schnitzer, M., and S.U. Khan. 1972. Humic substances in the environment. Marcel Dekker, New York.

Schwertmann, U., and R.M. Taylor. 1977. Iron oxides. p. 145–180. *In* J.B. Dixon and S.B. Weed (ed.) Minerals in soil environments. SSSA, Madison, WI.

Schwertmann, U., H. Kodoma, and W.R. Fischer. 1986. Mutual interactions between organics and iron oxides. p. 223–250. *In* P.M. Huang and M. Schnitzer (ed.) Interactions of soil minerals with natural organics and microbes. SSSA Spec. Publ. 17. SSSA, Madison, WI.

Skipper, H.D., V.V. Volk, M.M. Mortland, and K.V. Raman. 1978. Hydrolysis of atrazine on soil colloids. Weed Sci. 26:46–51.

Skujins, J.J. 1967. Enzymes in soil. p. 371–416. *In* A.D. McLaren and G.H. Peterson (ed.) Soil biochemistry. Marcel Dekker, New York.

Smelt, J.H., A. Dekker, M. Leistra, and N.W.H. Houx. 1983. Conversion of four carbamoyloximes in soil samples from above and below soil water table. Pestic. Sci. 14:173–181.

Smith, C.A., Y. Iwata, and F.A. Gunther. 1978. Conversion and disappearance of methidathion on thin layers of dry soil. J. Agric. Food Chem. 26:959–962.

Smith, J.H., W.R. Mabey, N. Bohonos, B.R. Holt, S.S. Lee, T.W. Chou, D.C. Bomberger, and T. Mill. 1978. Environmental pathways of selected chemicals in freshwater systems. Part II. Laboratory studies. EPA-600/7-78-074. USEPA, Athens, GA.

Soil Survey Staff. 1975. Soil taxonomy: A basic system of soil classification for making and interpreting soil surveys. USDA-SCS Agric. Handb. 436. U.S. Gov. Print. Office, Washington, DC.

Solomon, D.H. 1968. Clay minerals as electron acceptors and/or electron donors in organic reactions. Clays Clay Min. 16:31–39.

Solomon, D.H., and M.J. Rosser. 1965. Reactions catalyzed by minerals. I. Polymerization of styrene. J. Appl. Polymer Sci. 9:1261–1271.

Spencer, W.F., J.D. Adams, T.D. Shoup, and R.C. Spear. 1980a. Conversion of parathion to paraoxon on soil dusts and clay minerals as affected by ozone and UV light. J. Agric. Food Chem. 28:367–371.

Spencer, W.F., T.D. Shoup, and R.C. Spear. 1980b. Conversion of parathion to paraoxon on soil dusts as related to atmospheric oxidants at three California locations. J. Agric. Food Chem. 28:1295–1300.

Sposito, G. 1984. The surface chemistry of soils. Oxford Univ. Press, New York.

Steelink, C., and G. Tollin. 1967. Free radicals in soil. p. 147–172. *In* A.D. McLaren and G.H. Peterson (ed.) Soil biochemistry. Marcel Dekker, New York.

Stevenson, F.J. 1972. Role and function of humus in soil with emphasis on adsorption of herbicides and chelation of micronutrients. BioScience 22:643–650.

Stevenson, F.J. 1982. Humus chemistry. Wiley Interscience, New York.

Stevenson, F.J. 1985. Geochemistry of soil humic substances. p. 13–52. *In* R.G. Aiken et al. (ed.) Humic substances in soil, sediment and water. John Wiley and Sons, New York.

Stone, T.A., and J.J. Morgan. 1987. Reductive dissolution of metal oxides. p. 221–254. *In* W. Stumm (ed.) Aquatic surface chemistry: Chemical processes at the particle-water interface. John Wiley and Sons, New York.

Suflita, J.M., and J.-M. Bollag. 1980. Oxidative coupling activity in soil extracts. Soil Biol. Biochem. 12:177–183.

Takahashi, N., N. Mikami, H. Yamada, and J. Miyamoto. 1985. Photodegradation of the pyrethroid insecticide fenpropathrin in water, on soil and on plant foliage. Pestic. Sci. 16:119–131.

Takase I., H. Ohama, and I. Uelyama. 1982. Photodecomposition of prothiofos (*O*-2,4-dichlorophenyl *O*-ethyl *S*-propyl phosphorodithioate). Pestic. Chem. 7:463–471.

Takijima, Y., H. Sakuma, and M. Chiba. 1961. Metabolism of organic acids in soils of paddy fields and their inhibitory effects on ice growth. 6. The accumulation of organic acids in soil presence of sucrose and its relationship to growth inhibition of rice seedlings. Soil Sci. Plant Nutr. (Jpn) 7:167–168.

Theng, B.K.G. 1974. The chemistry of clay-organic reactions. Adam Hilger, London.

Thibodeaux, L.J., and D. Lipsky. 1985. A fate and transport model for 2,3,7,8-tetrachlorodibenzo-*p*-dioxin in fly ash on soil and urban surfaces. Hazard. Waste Hazard Materials 2:225–235.

Thurman, E.M. 1985. Humic substances in ground water. p. 87–104. *In* G.R. Aiken et al. (ed.) Humic substances in soil, sediment, and water. John Wiley and Sons, New York.

Turner, L. (ed.). 1984. The phototransformation of chemicals in water. Results of a ring-test. European Chem. Ind. Ecol. and Toxicol. Ctr., Brussels, Belgium.

Turekian, K.K. 1969. The oceans, streams, and atmosphere. p. 297-323. *In* K.H. Wedepohl (ed.) Handbook of geochemistry. Vol. 1. Springer-Verlag, New York.

Van Olphen, H. 1966. An introduction to clay colloid chemistry. Interscience Publ., New York.

Wade, R.S., and C.E. Castro. 1972. Oxidation of iron(II) prophyrins by alkyl halides. J. Am. Chem. Soc. 95:226-230.

Wahid, P.A., and N. Sethunathan. 1979. Involvement of hydrogen sulfide in the degradation of parathion in flooded acid sulphate soil. Nature (London) 282:401-402.

Wolfe, N.L., B.E. Kitchens, D.L. Macalady, and T.J. Grundl. 1986. Physical and chemical factors that influence the anaerobic degradation of methyl parathion in sediment systems. Environ. Sci. Technol. 5:1019-1026.

Wolfe, N.L., D.F. Paris, W.C. Steen, and G.L. Baughman. 1980. Correlation of microbial degradation rates with chemical structure. J. Environ. Sci. 14:1143-1144.

Wolfe, N.L., R.G. Zepp, J.A. Gordon, G.L. Baughman, and D.M. Cline. 1977. Kinetics of chemical degradation of malathion in water. Environ. Sci. Technol. 11:88-93.

Wolff, C.J.M., M.T.H. Halmans, and H.B. van der Heijde. 1981. The formation of singlet oxygen in surface waters. Chemosphere 10:59-62.

Wolff, C.J.M., and N.O. Crossland. 1985. Fate and effects of 3,4-dichloaniline in the laboratory and in outdoor ponds: I. Fate. Environ. Toxicol. Chem. 4:481.

Worthing, C.R. 1983. The pesticide manual. 7th ed. The British Crop Protection Counc. Lathenham, Suffolk, Great Britain.

Yokley, R.A., A.A. Garrison, E.L. Wehry, and G. Mamantov. 1986. Photochemical transformation of pyrene and benzo[a]pyrene vapor deposited on eight coal stack ashes. Environ. Sci. Technol. 20:86-90.

Young, J.C., and S.U. Khan. 1978. Kinetics of nitrosation of the herbicide glyphosate. J. Environ. Sci. Health B13:59-72.

Zafiriou, O.C. 1984. A bibliography of references in natural water photochemistry. Tech. Mem. WHOI-2-84. Woods Hole Oceanographic Inst. Woods Hole, MA.

Zepp, R.G. 1978. Quantum yields for reaction of pollutants in dilute aqueous solution. Environ. Sci. Technol. 12:327-339.

Zepp, R.G. 1980. Assessing the photochemistry of organic pollutants in aquatic environments. p. 69-110. *In* R. Haque (ed.) Dynamics, exposure and hazard assessment of toxic chemicals. Ann Arbor Science, Ann Arbor, MI.

Zepp, R.G. 1982. Experimental approaches to environmental photochemistry. p. 19-41. *In* O. Hutzinger (ed.) Handbook of environmental chemistry. Springer-Verlag, Berlin.

Zepp, R.G., G.L. Baughman, and P.F. Schlotzhauer. 1981a. Comparison of photochemical behavior of various humic substances in water: I. Sunlight induced reactions of aquatic pollutants photosensitized by humic substances. Chemosphere 10:109-117.

Zepp, R.G., G.L. Baughman, and P.F. Schlotzhauer. 1981b. Comparison of photochemical behavior of various humic substances in water: II. Photosensitized oxygenation. Chemosphere 10:119-126.

Zepp, R.G., and D.M. Cline. 1977. Rates of direct photolysis in the aquatic environment. Environ. Sci. Technol. 11:359-366.

Zepp, R.G., N.L. Wolfe, L.V. Azarraga, R.H. Cox, and C.W. Pape. 1977. Photochemical transformation of the DDT and methoxychlor degradation products, DDE and DMDE, by sunlight. Arch. Environ. Contam. Toxicol. 6:305-314.

Zepp, R.G., N.L. Wolfe, G.L. Baughman, and R.C. Hollis. 1977. Singlet oxygen in natural water. Nature (London) 267:421-423.

Zepp, R.G., N.L. Wolfe, J.A. Gordon, and G.L. Baughman. 1975. Dynamics of 2,4-D esters in surface waters. Hydrolysis, photolysis, and vaporization. Environ. Sci. Technol. 9:1144-1150.

# Chapter 6

# Biological Transformation Processes of Pesticides

**J.-M. BOLLAG AND S.-Y. LIU,** *Pennsylvania State University, University Park, Pennsylvania*

Once introduced into the environment, pesticides and other anthropogenic pollutants are subjected to biological and nonbiological transformation processes. Among biological processes, microbial metabolism is the primary force in pesticide transformation or degradation. Indeed, it has been established that, in many cases, microbes are more important in the degradation of a pesticide than are physical or chemical mechanisms (Munnecke et al., 1982). The importance of microorganisms is not surprising since the diversity and uniqueness of their metabolic activities enable them to thrive in ecological niches that are otherwise uninhabitable. Microorganisms are key agents in the degradation of a vast array of organic pesticide molecules in terrestrial and aquatic ecosystems because of such processes as aerobic, anaerobic, and chemolithotrophic metabolism, fermentation, and metabolism via extracellular enzymes. Some pesticides, however, are resistant to microbial degradation and persist longer in the environment. Others are only transformed to intermediates, occasionally with increased toxicity. Therefore, to determine the potential recalcitrance or toxicity of pesticides or their metabolites, it is essential to obtain basic knowledge concerning the action of microorganisms and the fate of individual pesticides. Only then will it be possible to work towards a resolution of the problem of pesticide pollution.

In this review, we discuss the metabolism of pesticides by microorganisms. We also refer to the potential applications of new technological advances in pollution management, that is, the use of enrichment procedures to isolate pesticide-degrading microorganisms or their enzymes for the decontamination of soil or water and the utilization of genetic engineering for the creation of microbial strains with novel catabolic capabilities.

## 6-1 MODES OF MICROBIAL PESTICIDE TRANSFORMATION

The degradation of pesticides through microbial metabolic processes is considered to be the primary mechanism of biological transformation. Since microorganisms can proliferate in virtually any imaginable environment

 *Pesticides in the Soil Environment*—SSSA Book Series, no. 2.

because of their remarkable powers of mutation and adaptation, there appears to be great potential for acquiring degradative capabilities when exposed to xenobiotics. Some microbial transformation mechanisms are unique and cannot be found in other organisms. The ability of certain microorganisms to ferment carbohydrates, to grow under anaerobic conditions, and to produce enzymes that are active extracellularly exemplifies the enormous diversity of microbial metabolism.

In nature and axenic cultures, microbial degradation of pesticides can be due to direct metabolism (a primary metabolic reaction) or to an indirect effect of microbes on the chemical or physical environment, resulting in a secondary transformation reaction.

Basically, five processes are involved in the microbial transformation of pesticides:

1. *Biodegradation,* in which the pesticide can serve as a substrate for growth.
2. *Cometabolism,* in which the pesticide is transformed by metabolic reactions but does not serve as an energy source for the microorganism.
3. *Polymerization* or *Conjugation,* in which pesticide molecules are linked together with other pesticides, or with naturally occurring compounds.
4. *Accumulation,* in which the pesticide is incorporated into the microorganism.
5. *Secondary effects of microbial activity,* in which the pesticide is transformed because of changes in the pH, redox conditions, reactive products, etc., in terrestrial or aquatic environments brought about by microorganisms.

The microbial transformation of a pesticide may involve more than one type of mechanism, and under different conditions, various products can be derived from the same initial compound depending on the environmental parameters. The transformation processes can be mediated by one organism or can result from the combined effects of several organisms.

### 6–1.1 Biodegradation (Mineralization)

The most interesting and environmentally valuable aspect of pesticide transformation by microbes is the complete biodegradation of a certain xenobiotic molecule. If a pesticide can be used in such a way by one or more interacting microorganisms, it will be metabolized into $CO_2$ and other inorganic components, and the microorganisms can obtain their requirements for growth and energy from the pesticidal molecules. From an environmental point-of-view, the complete metabolism of an organic pesticide is desired if one is interested in avoiding the generation of potentially hazardous pesticidal intermediates.

There are many organic pesticides whose complete decomposition could be demonstrated under certain conditions by adequately labeling them with

$^{14}C$-carbon. The presence of biodegradable pesticides in a natural ecosystem can cause the proliferation of the active microbial flora and concurrently increase the rate of decomposition of the applied pesticide.

### 6–1.2 Cometabolic Transformation

The degradation of synthetic compounds in the environment largely involves cometabolism; it is the prevalent form of microbial metabolism. In cometabolism, microorganisms, while growing at the expense of a growth substrate, are able to transform a compound without deriving any nutrient or energy for growth from the process. Thus, cometabolism is a fortuitous metabolism, in which enzymes involved in catalyzing the initial reaction are often lacking substrate specificity.

Many microorganisms are capable of causing cometabolic transformations. Examples of pesticide cometabolism are increasing with the continuous introduction of newly synthesized chemicals. The reader is referred to the reviews of Horvath (1971), Alexander (1980), and Bollag (1982).

Cometabolism generally does not result in an extensive degradation of a certain substrate, but it is possible that different microorganisms can transform a molecule by sequential cometabolic attacks or that cometabolic products of one organism can be used by another as a growth substrate. For instance, the herbicide 2,4,5-T and 2,3,6-trichlorobenzoate are converted by a cometabolic oxidation to 3,5-dichlorocatechol by a *Brevibacterium* sp. (Horvath, 1970a, 1971), whereas an *Achromobacter* sp. was capable of further transforming 3,5-dichlorocatechol to 3,5-dichloro-2-hydroxymuconic semialdehyde (Horvath, 1970b). It was found that cometabolism could account for complete mineralization of 3,4-dichloroaniline when an unchlorinated analogue substrate was supplied to a microbial culture (You & Bartha, 1982). A cometabolic transformation in a mixed culture may form intermediate products that can be mineralized by other organisms.

Cometabolic transformation may lead to the accumulation of intermediate products with decreased or increased toxicity, and thus cause some adverse environmental impacts and, in some cases, inhibit the microbial growth as well as their metabolism. Tranter and Cain (1967) have proposed that the failure of many halogenated aromatic compounds to support bacterial growth could be because of an accumulation of metabolic inhibitors. Gibson et al. (1968) described the conversion of 4-chlorotoluene to 3-chloro-6-methylcatechol that could not be further metabolized. They suggested that the catehol inhibited dioxygenase enzymes that are essential in the degradative pathways of aromatic molecules.

The nature of metabolism (i.e., mineralization or cometabolism) can be influenced largely by environmental factors. Therefore, the microorganisms can metabolize a pesticide differently according to environmental conditions. Wang et al. (1984) found that isopropyl *N*-phenylcarbamate (IPC) is mineralized at a low concentration, but at a higher concentration it is only converted to organic products by cometabolism.

### 6-1.3 Conjugation and Polymerization

In many cases, pesticides are transformed not by biodegradation, but by microbially mediated polymerization or conjugation.

Polymerization can be a process of oxidative coupling reactions by which a pesticide or its intermediate combines with itself, with other xenobiotic residues, or with naturally occurring compounds to form a larger molecular polymer. In conjugation reactions, pesticides or their intermediates are linked together with other endogenous substrate(s) resulting in the formation of methylated, acetylated, or alkylated compounds, glycosides, or amino acid conjugates.

Considerable evidence shows that microbial polymerization plays a significant role in the incorporation of xenobiotics into soil organic matter (Bollag & Loll, 1983). This biochemical process not only affects the efficacy and biodegradability of the applied chemical, but also raises concern about the environmental impact of the bound residues after repeated application.

Various examples of polymers and conjugate formation are described under the section of synthetic reaction.

### 6-1.4 Microbial Accumulation of Pesticides

Cellular accumulation of pesticides by target and nontarget microorganisms represents another type of microbial interference with xenobiotics. The primary microbial uptake of pesticides is attributed to a passive physical process of absorption rather than active metabolism. Many studies indicate that dead autoclaved cells accumulate similar amounts or even more pesticides as compared to live cells and that actual metabolic factors are not always involved in the accumulation process. In this sense, the active absorption refers to the mechanism for living cells, and passive accumulation (nonactive transport), when referring exclusively to dead cells, appears to be incorrect. Johnson and Kennedy (1973) found that the accumulation of DDT and methoxychlor by autoclaved cells of *Aerobacter aerogenes* was twice as much as that by living cells. The absorption capacity for lindane and dieldrin in a yeast, *Saccharomyces cerevisiae,* increased after boiling the organism (Voerman & Tammes, 1969). When methoxychlor was added to autoclaved cultures of bacteria, fungi, and algae, these cells sorbed as much as or slightly more pesticide than the viable cells, indicating that the sorption of methoxychlor was not the result of a metabolically active process (Paris & Lewis, 1976). Kikuchi et al. (1984) reported that bioaccumulation of both fenitrothion and DDT in dead cells of three species of algae was significantly higher than in living cells.

The rate of accumulation of pesticides differs in various organisms and depends on the type and concentration of the pesticide in the surrounding medium. For instance, *A. aerogenes* and *Bacillus subtilis* require only 30 s to accumulate 80 to 90% of the DDT residues accumulated in 24 h, whereas, *Agrobacterium tumefaciens* accumulates 100% of the DDT and 90% of the dieldrin after 4 h of incubation (Chacko & Lockwood, 1967). *Flavobacterium harrisonii, Bacillus subtilis,* and *Chlorella pyrenoidosa* accumulate methoxychlor rapidly, reaching equilibrium within 30 min, but *Aspergillus* sp. accumulate the pesticide slowly and require 16 h to reach the equilibrium (Paris & Lewis, 1976). Other organochlorine insecticides are accumulated less readily than DDT. Accumulation of $\gamma$-BHC by *C. pyrenoidosa* was slower than that of DDT (Sodergren, 1971). An even slower accumulation of dieldrin by *C. pyrenoidosa* was reported (Sodergren, 1968).

Studies on retention or excretion of the accumulated pesticides have provided some interesting information. Burchfield and Storrs (1957) reported that spores of *Neurospora sitophila* rapidly absorb *s*-triazine. They suggested that the pesticide probably combined irreversibly with protoplasmic constituents because the amount recoverable on immediate extraction was only 20 to 25%. Lindane and dieldrin, absorbed by the yeast *Saccharomyces cerevisiae,* could be removed by washing the cells with fresh water (Voerman & Tammes, 1969). The algae *Cladophora* accumulated DDT and retained it without any apparent excretion (Ware et al., 1968). Dekoning and Mortimer (1971) also found that the flagellated protozoan *Euglena gracilis* accumulated DDT without eliminating it. It was suggested that such species can be used as indicators of DDT contamination in water. Kruglov and Paromenskaya (1970) reported that both simazine resistant and sensitive algae culture adsorbed and bound the herbicide to cellular protein and also metabolized it. Transference of algae, which have accumulated a significant amount of fenitrothion, to a fenitrothion-free medium rapidly decreased the concentrations of fenitrothion and its products in the algae. In contrast, DDT showed higher accumulation ratios and longer retention times under similar conditions. Kikuchi et al. (1984) suggested that these differences are because of the low lipid affinity and the high water solubility of the fenitrothion.

Many species of axenic algae are able to accumulate a high degree of DDT and also degrade it to some extent to DDE (Bowes, 1972; Neudorf & Khan, 1975). *Anabaena flos-aquae* accumulated and metabolized fenitrothion most actively to its oxon and demethyl analog and its phenols, whereas *Chlorella vulgaris* accumulated and decomposed the chemical to demethylfenitrothion (Kikuchi et al., 1984).

Most studies on microbial accumulation of chlorinated hydrocarbons have been discussed in a review by Lal and Saxena (1982). Mycelia of actinomycetes and fungi were able to accumulate dieldrin, DDT and pentachloronitrobenzene from soil to levels above ambient concentration (Ko & Lockwood, 1968). Absorption and concentration of aldrin was determined for floc-forming bacteria isolated from Lake Erie, and it has been suggested that the adsorption capacity of flocculant bacteria might even be evaluated for removal of pesticides in aqueous environments (Leshniowsky et al., 1970).

The organophosphate pesticide fensulfothion was metabolized by the bacterium *Klebsiella pneumoniae,* and the transformation product fensulfothion sulfide was rapidly bound by both living and heat-killed cells (Timms & MacRae, 1982, 1983). Three species of algae—*C. vulgaris, Nitzschia closterium,* and *Anabaena flos-aquae*—rapidly absorbed fenitrothion from the medium, and the maximum bioaccumulation ratios in terms of dry weight were 40, 105, and 53, respectively (Kikuchi et al., 1984).

Besides playing a role in the removal of toxic chemicals from the surrounding medium, microbial accumulation of pesticides can also be considered as a process in pesticide translocation. Many microorganisms are important food sources for a broad spectrum of filter-feeding organisms. The presence of pesticide-containing microorganisms in the aquatic environment pollutes the food chain for fish and higher vertebrates, and the latter can be transported to a new environment. Dead microbial cells containing pesticides eventually reach the soil or aquatic environment. Later, sediment-bound pesticides can be released by biological or nonbiological processes, thus posing a delayed pollution problem.

### 6–1.5 Nonenzymatic Transformation Caused by Microbial Activity

Considerable evidence exists that microbial activities contribute to the formation of reactive products and also to the alteration of environmental parameters such as pH, redox potential, or other factors that are conducive to the secondary or nonenzymatic transformation of pesticidal molecules. Incorporation of pesticide molecules or their intermediates into soil humus often takes place by interaction between enzymatic and nonenzymatic processes. Hsu and Bartha (1974) found that microbial activity is responsible for the release of aniline moieties, and their binding to humus is a spontaneous and relatively rapid chemical reaction that occurs in sterile as well as in natural soil. Laccase of *Rhizoctonia praticola* catalyzed the formation of quionoid dimers from 2,4-dichlorophenol. Subsequent formation of hybrid trimers between the quinonoid dimer and 2,4-dichloroaniline was proved to be nonenzymatic (Minard et al., 1981; Liu et al., 1981).

The pH of a milieu can be the dominating factor that causes the transformation of a pesticide. Drastic pH changes are often associated with metabolic activities of microorganisms. This observation was relevant in aqueous media as well as field conditions. Alteration can occur towards the alkaline region, particularly under anaerobic conditions, or by the degradation of proteins; or toward acidic pH values, usually during the metabolism of carbohydrates, oxidation of organic N to nitrite or nitrate, sulfide to elemental S or sulate-ferrous to ferric ion, or other metal oxidations.

Plimmer et al. (1970), who treated soil with propanil, determined the formation of a triazene. They proposed that soil nitrite reacts with the herbicide intermediate 3,4-dichloroaniline to form an intermediate diazonium cation, which subsequently couples with free 3,4-dichloroaniline resulting in the formation of a corresponding triazine. A bacterium *Paracoccus* sp. converts nitrate to nitrite under anaerobic conditions and simultaneously lowers the pH of the medium. This leads to the formation of a triazene, but no triazine was formed at pH 7.0 or above (Minard et al., 1977).

Through microbial activities a reducing environment can be created, particularly in a flooded soil. Many pesticides are degraded by reductive reactions that proceed nonenzymatically under anaerobic conditions. DDT, methoxychlor, and heptachlor readily break down in an anaerobic flooded ecosystem (Sethunathan, 1973). Chemical transformation of toxaphene was attributed indirectly to microbial activities which lower the redox potential in the soil to a range of 0 to −100 mV (Parr & Smith, 1976).

Certain microbial products act as photosensitizers by absorbing energy from light and passing it to a pesticidal molecule that otherwise could not be activated. A number of natural products are known to be efficient photosensitizers, and several interactions with pesticides were determined: riboflavin in the oxidation of 2,4-dichlorophenol (Plimmer & Klingebiel, 1971) and chloroaniline (Rosen et al., 1970); flavoprotein in sulfoxide formation from carboxin (Lyr et al., 1974); mexacarbate transformation (Matsumura & Esaac, 1979); rotenone in the alteration of chlorinated cyclodienes; and chlorophylls in the photodegradation of various types of insecticide chemicals (Ivie & Casida, 1971).

## 6-2 MAJOR BIOCHEMICAL REACTIONS OF PESTICIDE METABOLISM

The principal processes involved in microbial pesticide metabolism are oxidation, reduction, hydrolytic, and synthetic reactions.

### 6-2.1 Oxidative Reactions

Oxidation of pesticides that occur frequently in microorganisms is one of the most important and basic metabolic reactions. Enzymes involved in pesticide oxidation reactions belong to the various groups of known oxidative enzymes such as peroxidases, laccases, and mixed function oxidases. The major oxidation reactions are presented in Table 6-1.

Table 6-1. Oxidation reactions in microbial pesticide metabolism.†

| |
|---|
| Hydroxylation |
| $RCH \rightarrow RCOH$ |
| $ArH \rightarrow ArOH$ |
| *N*-Dealkylation |
| $RNCH_2CH_3 \rightarrow RNH + CH_3CHO$ |
| $ArNRR' \rightarrow ArNH_2$ |
| β-Oxidation |
| $ArO(CH_2)_nCH_2CH_2COOH \rightarrow ArO(CH_2)_nCOOH$ |
| Decarboxylation |
| $RCOOH \rightarrow RH + CO_2$ |
| $ArCOOH \rightarrow ArH + CO_2$ |
| $Ar_2CH_2COOH \rightarrow Ar_2CH_2 + CO_2$ |
| Ether cleavage |
| $ROCH_2R' \rightarrow ROH + R'CHO$ |
| $ArOCH_2R' \rightarrow ArOH + R'CHO$ |
| Epoxidation |
| $RCH{=}CHR' \rightarrow RCH{-}CHR'$ (with O bridging the two carbons, epoxide) |
| Oxidative coupling |
| $ArOH \rightarrow (Ar)_2(OH)_2$ |
| Sulfoxidation |
| $RSR' \rightarrow RS(O)R'$ or $RS(O_2)R'$ |

† R = organic moiety, AR = aromatic moiety.

### 6–2.1.1 Hydroxylation

Microbial hydroxylation is often the first step in pesticide degradation. The addition of a hydroxyl group into the pesticide molecule makes the compound more polar and hence more soluble in water, and makes the substance biologically more reactive. Enzymes catalyzing this reaction are hydroxylases, monooxygenases, or mixed function oxidases.

Hydroxylation can occur at the aromatic ring, at aliphatic groups, as well as at aralkyl side chains. In all cases, it was found that molecular oxygen and NADPH are necessary in order for the mixed function oxygenases to be able to attack the compound.

The occurrence of both *N*-alkyl- and aromatic ring-hydroxylation of carbaryl was shown with the fungus *Gliocladium roseum* (Liu & Bollag, 1971). Metabolism of metolachlor by a strain of actinomycetes gave rise to several products that were hydroxylated at *N*-alkyl as well as aralkyl side chains (Fig. 6–1; Krause et al., 1985).

Desulfuration of the *s*-triazine herbicides prometryne or ametryne by bacterial cultures leads to the formation of the corresponding hydroxy-derivative (hydroxyametryne or hydroxyprometryne) (Cook & Hütter, 1982).

### 6–2.1.2 *N*-Dealkylation

*N*-Dealkylation was a significant detoxification reaction for phenylureas, acylanilides, carbamates, *s*-triazines, dinitroanilines, and bipyridyl pesticides. Alkyl substitutions on an aromatic molecule are often the first place where microorganisms initiate the catabolic transformation of a xenobiotic molecule. The enzyme involved is a mixed function oxidase requiring a reduced nicotinamide nucleotide as a H donor.

Giardina et al. (1980, 1982) found that the primary route of atrazine metabolism by a culture of *Nocardia* appeared to be *N*-dealkylation (Fig. 6–2). Zeyer and Kearney (1983) demonstrated dealkylation of trifluralin, an dinitroaniline herbicide, by a *Candida* sp. Only a trace ($<1\%$) of *N*-dealkylated trifluralin was accumulated in the culture, indicating further degradation of the dealkylated product, but no ring cleavage of the aromatic ring was observed.

### 6–2.1.3 $\beta$-Oxidation

Many aromatic pesticides, especially phenoxyalkanoate herbicides, have fatty acid side chains that can be metabolized by $\beta$-oxidation. This reaction proceeds by the stepwise cleavage of two-carbon fragments from a fatty acid; the shortened fatty acid can be further degradaded in the same manner until the chain length is four or two carbons. Substitution on the aromatic ring of an $\omega$-phenoxyalkanoic acid or especially on the side chain has a strong influence on the ease of $\beta$-oxidation.

For the normal functioning of $\beta$-oxidation, two protons are required on both the $\alpha$- and $\beta$-carbons. The reactions is prevented if the $\alpha$- and $\beta$-carbon atoms have a substitution, but substituents on other carbons of the aliphatic chain may merely inhibit $\beta$-oxidation (Dias & Alexander, 1971; Hammond & Alexander, 1972).

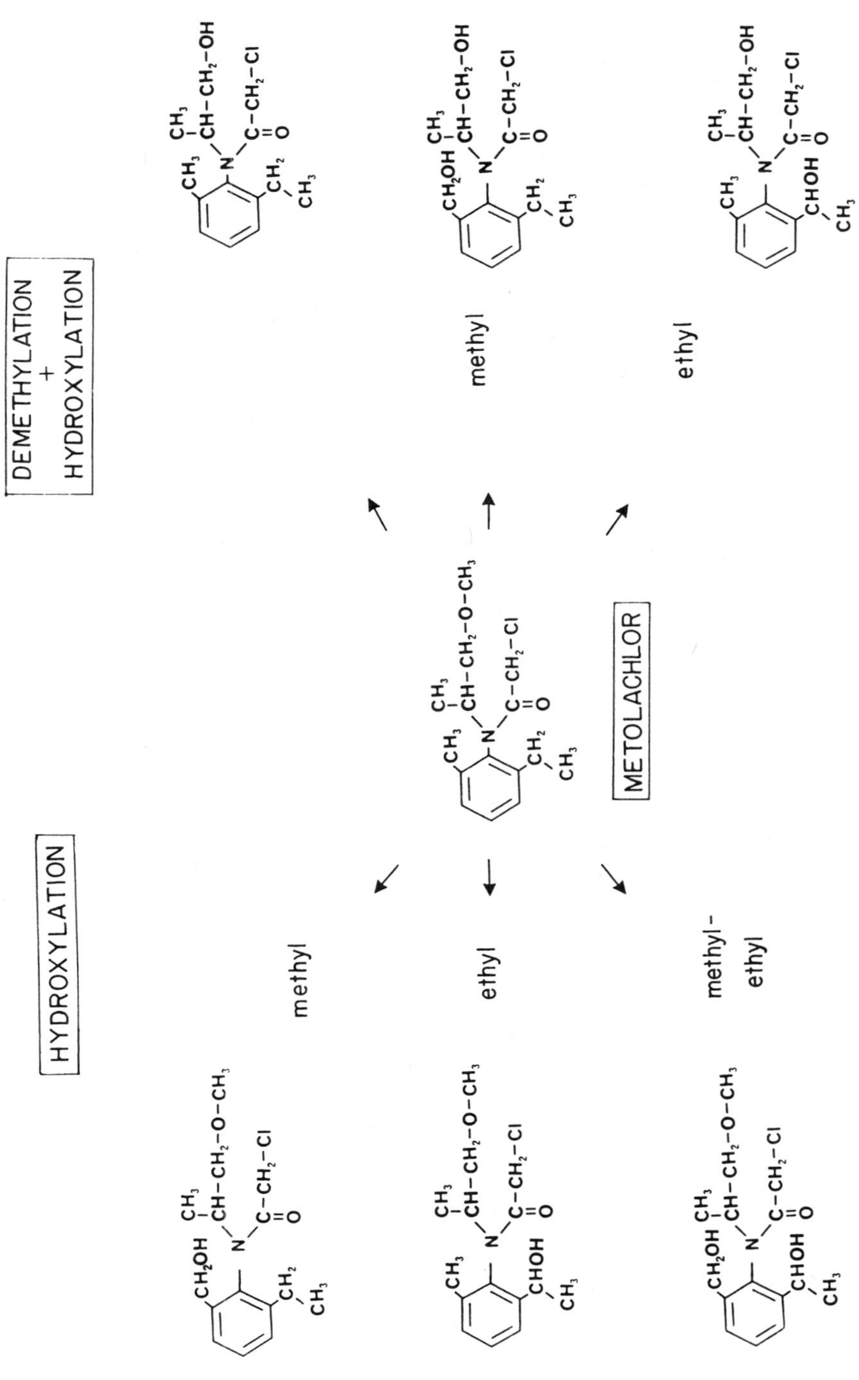


Fig. 6-1. Hydroxylation of metolachlor by a soil actinomycete.

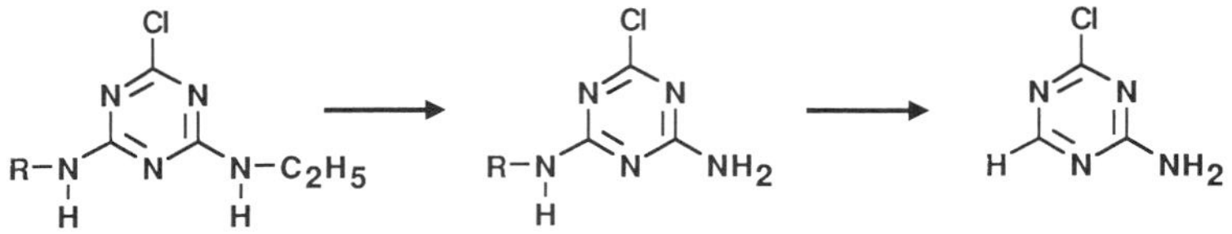


Fig. 6–2. N-Dealkylation of atrazine by a *Nocardia* sp.

Since β-oxidation was found and studied mostly with ω-phenoxyalkanoic acids, and the reaction is the same in bacteria (Taylor & Wain, 1962), actinomycetes (Webley et al., 1958), fungi (Byrde & Woodcock, 1957), and even plants, the general pathway described by Loos (1975) is depicted (Fig. 6–3). The acids with an odd number of carbon atoms in the side chain are transformed to the half-ester of 2,4-dichlorophenol with carbonic acid that degrades to 2,4-dichlorophenol and $CO_2$. Oxidation of 2,4-dichlorophenoxyalkanoic acids with an even number of carbon atoms in the side chain results in the formation of 2,4-dichlorophenoxyacetate.

### 6–2.1.4 Decarboxylation

The replacement of a carboxyl group by H is another common enzymatic reaction achieved by microbial activity. In the case of aliphatic carboxylic acids, the carboxyl group may be more or less readily degraded depending on the influence of the configuration and substituents of the molecule. Decarboxylation is a widespread, microbially catalyzed reaction for naturally occurring compounds, and various examples have been detected in pesticide metabolism. Several indications of decarboxylation caused by the soil microflora were found with benzoic acid herbicides (Frear, 1975), bipyridyls (Cripps & Roberts, 1978), and chlorinated hydrocarbon acaricides.

Miyazaki et al. (1969) reported an example of decarboxylation. In the degradation of 4,4'-dichlorobenzilic acid, which is a hydrolysis product of the acaricides chlorobenzilate and chloropropylate, a decarboxylation reaction was demonstrated using the yeast *Rhodotorula gracilis* (Fig. 6–4). During

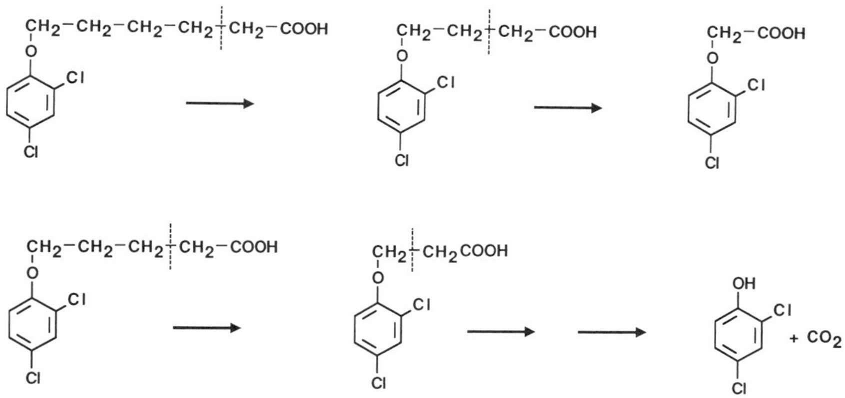


Fig. 6–3. β-Oxidation of 2,4-dichlorophenoxyalkanoic acid.

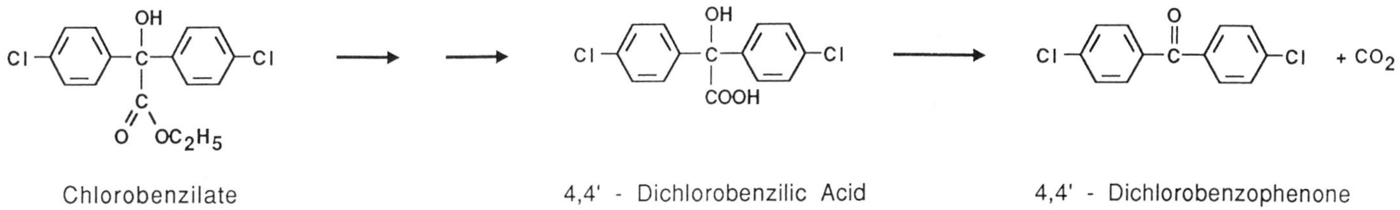


Fig. 6-4. Decarboxylation of 4,4′-dichlorobenzilic acid by the yeast *Rhodotorula gracilis*.

this metabolic process, the acid underwent a combined decarboxylation-dehydrogenation reaction, resulting in the formation of 4,4'-dichlorobenzophenone and $CO_2$. When citric acid is used as the main C source, this decarboxylation reaction is stimulated, whereas α-ketoglutarate has an inhibitive effect.

### 6–2.1.5 Cleavage of Ether Linkage

Cleavage of ether linkages in pesticides may significantly decrease their toxicity towards the target organisms. Such cleavage, the separation of a hydrocarbon from an oxygen atom that functions to link it with the other moiety of a molecule, is thus important in studying the microbial transformation of pesticides. However, the specific mechanism of this reaction is still unclear. Studies suggest that the cleavage is catalyzed by mixed function oxidases in the presence of reduced pyridine nucleotides and molecular oxygen. *O*-dealkylation also represents an ether cleavage reaction in which only an alkyl group is removed as opposed to larger molecular groups.

Many pesticides, such as benzoic acid herbicides, organophosphates, carbamates, methoxy-*s*-triazines, phenylureas, and phenoxyalkanoates, have an ether linkage or an alkoxy group. The cleavage of the ether linkage was investigated particularly with the phenoxyalkanoic herbicides (Loos, 1975) that produce the corresponding phenols as intermediates. This transformation could be shown with cell-free extracts from an *Arthrobacter* sp. (Loos et al., 1967; Tiedje & Alexander, 1969) and a *Pseudomonas* sp. (Gamar & Gaunt, 1971).

When the *Pseudomonas* sp. was cultured in a basal salt medium with MCPA as the sole source of carbon, a cell-free extract could be prepared that catalyzed the cleavage of the ether linkage of MCPA between the aliphatic side chain and the ether-oxygen atom (Gamar & Gaunt, 1971). The MCPA could only be oxidized by the enzyme preparation in the presence of a reduced nicotinamide nucleotide (NADH or NADPH) to 2-methyl-4-chlorophenol and glyoxylic acid (Fig. 6–5).

### 6–2.1.6 Epoxidation

Epoxidation, the insertion of an oxygen atom into a carbon-carbon double bond, frequently results in the formation of products with greater

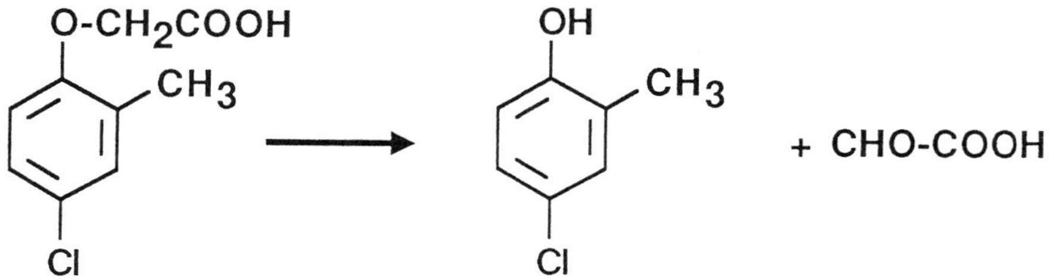


Fig. 6–5. Ether cleavage of MCPA by a cell extract from *Pseudomonas* sp.

environmental toxicity. For instance, various microorganisms can catalyze the reaction of the chlorinated cyclodiene insecticides aldrin, isodrin, and heptachlor to their toxic epoxide derivatives. Korte et al. (1962) were the first to demonstrate such an epoxidation; *Aspergillus niger, A. flavus, Penicillium chrysogenum,* and *P. notatum* coverted aldrin to the epoxide dieldrin.

Heptachlor has also been found to be epoxidized (Miles et al., 1969). Heptachlor epoxide concentrations of up to 6% were produced by most of the bacteria, actinomycetes, and fungi tested. An additional pathway for epoxidation was observed following the bacterial dechlorination of heptachlor to chlordene. When incubated in an aqueous medium, specific bacteria were able to form chlordene epoxide from chlordene. If heptachlor was first hydrolyzed chemically to 1-hydroxychlordene, it could also be converted by many fungi (e.g., *Rhizopus, Trichoderma, Penicillium,* and *Fusarium*) to 1-hydroxy-2,3-epoxychlordene.

### 6–2.1.7 Oxidative Coupling

Oxidative coupling or condensation reactions are catalyzed by phenol oxidases such as laccases or peroxidases. In the oxidative coupling of phenol, for example, aryloxy or phenolate radicals are formed by removal of an electron and a proton from the hydroxyl group. The resulting phenolate radicals then couple with phenolic or other compounds to yield dimerized or polymerized products (Brown, 1967). This process, which is important during humus formation, links together reactive compounds through enzymatic activity, resulting in a complex mixture of polymerized molecules (Martin & Haider, 1971; Flaig et al., 1975). In particular, phenolic degradation products of lignins produced by fungi and other microorganisms react among themselves, with amino acids, peptides, and with pesticidal compounds possessing similar functional groups (Stevenson, 1972; Sjoblad & Bollag, 1981).

The herbicide 2,4-D, for example, is degraded to 2,4-dichlorophenol, which can be oxidative coupled by phenol oxidases. Minard et al. (1981) showed that a fungal laccase isolated from the soil fungus *Rhizoctonia praticola* polymerized 2,4-dichlorophenol to dimeric (Fig. 6–6) and higher oligoremic products. In this investigation, it could also be established that the herbicidal phenol intermediate formed cross-coupling products with phenolic humus constituents such as orcinol, syringic acid, vanillic acid, and vanillin (see section 6–2.4 Synthetic Reactions).

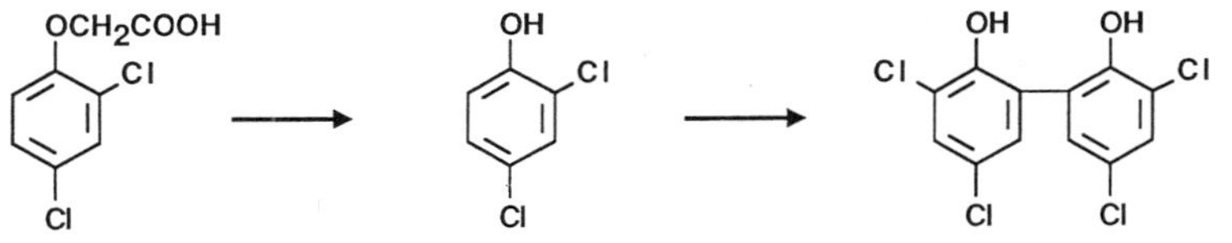


Fig. 6–6. Oxidative coupling of 2,4-dichlorophenol by a laccase from the fungus *Rhizoctonia praticola.*

### 6-2.1.8 Aromatic Ring Cleavage

Most organic pesticides have one or more aromatic rings, and consequently, their catabolism is only possible if ring cleavage takes place. While numerous microorganisms, especially bacteria, were isolated that are capable of cleaving a benzene ring, it was soon realized that aromatic pesticides with multiple and diverse substitutions are often quite resistant to microbial attack, depending on the particular molecular configuration. The type of linkage, the specific substituent(s), their position, and their number determine the susceptibility to fission of an aromatic ring. Although certain generalizations concerning probable decomposition can be made—such as stronger resistance to degradation by increasing halogen substitutions, or slower degradation of *meta*-substituted molecules as compared to *ortho*- or *para*-substituted analogs—each pesticide has to be investigated independently to comprehend and evaluate the probability for microbial cleavage of its aromatic ring. Usually the substituents have to be modified or removed and hydroxyl groups inserted in appropriate positions before oxygenase enzymes can cause ring cleavage.

Dihydroxylation is usually essential for enzymatic cleavage of the benzene ring under aerobic conditions. The hydroxyl groups must be placed either *ortho* or *para* to each other, an arrangement that is probably required in order to facilitate the shifts of electrons involved in the ring-fission reaction. Dioxygenases are the enzymes responsible for ring cleavage, and they can cause *ortho* (intra-diol) or *meta* (extra-diol) fission of a catechol forming *cis, cis*-muconic acid or 2-hydroxymuconic semialdehyde, respectively.

Several comprehensive reviews have been published that discuss pathways or ring degradation for both monocyclic and polycyclic aromatic compounds, the specific enzymes involved, and the regulation of enzyme synthesis (Dagley, 1971; Chapman, 1972; Stanier & Ornston, 1973; Gibson, 1984).

Only a few pesticides, the chlorinated phenoxyalkanoate herbicides (Loos, 1975), or pyrazon (Sauber et al., 1977; Eberspächer & Lingens, 1978), for example, were investigated intensively in relation to ring cleavage by isolation of the responsible enzymes, but several important groups of aromatic pesticides still pose many questions. One does not yet know much about the degradation pathway of halogenated anilines, which are products of many pesticides, the decomposition of the phenyl rings from DDT and related compounds, or the fission of the heterocyclic *s*-triazines to name a few.

As an example of the cleavage of substituted phenol, the report from Engelhardt et al. (1977) was selected which describes the fission of 4-(methylmercapto)-phenol, a hydrolysis product from various organophosphorus and methylcarbamate insecticides (Fig. 6-7). A *Nocardia* sp., isolated from

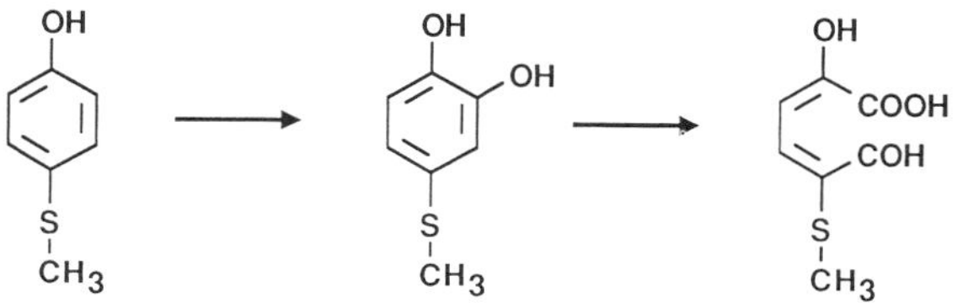


Fig. 6-7. Aromatic ring cleavage of 4-(methylmercapto)-phenol by a *Nocardia* sp.

soil, oxidized the substituted phenol through a hydroxylation reaction to 4-(methylmercapto)-catechol and subsequently cleaved the benzene ring between C atoms 2 and 3 to yield 2-hydroxy-5-methylmercaptomuconic semialdehyde. Fission of the ring occurred in this case by *meta* cleavage and was caused by a 2,3-dioxygenase. It appears that the enzymes responsible for *ortho* or *meta* cleavage can be induced with this *Nocardia* strain depending on the chemical nature of the substrates employed.

### 6–2.1.9 Heterocyclic Ring Cleavage

Like compounds with aromatic carbocyclic molecules, those with heterocyclic rings are subject to metabolism by microorganisms. These heterocyclic compounds, however, have attracted less attention because of the difficulty in tracing their intermediate metabolites. In pesticides with such heterocyclic rings, the path followed by the degradation process is complicated by the hetero atoms, usually N, O, and S, which contribute individual characteristics to the decomposition reactions. These compounds may contain one or more rings (mostly aromatic), and the rings generally have five or six members. Although the most common heterocyclic ring pesticides contain one six-membered ring, as do the pyridines, the triazines, and the pyrimidines, pesticides with other heterocyclic ring forms are also produced. Although ring cleavage among these pesticides is not uncommon, only rarely have the decomposition routes successfully been traced. For example, all attempts of establishing clear studies of the cleavage of pyridines by cell-free extracts have been fruitless.

Wright and Cain (1972) proposed a decomposition pathway of the bipyridyl herbicide paraquat. By photolytic degradation, paraquat is transformed to *N*-methylisonicotinate and methylamine. A gram-positive bacterium (Orpin et al., 1972) was isolated that is capable of using *N*-methylisonicotinate as the primary C source. Whole cells and cell-free extracts of this bacterium degraded the substrate by hydroxylation and N-demethylation, followed by further hydroxylation and subsequent cleavage of the pyridine ring and resulting in the formation of maleamic acid (Fig. 6–8).

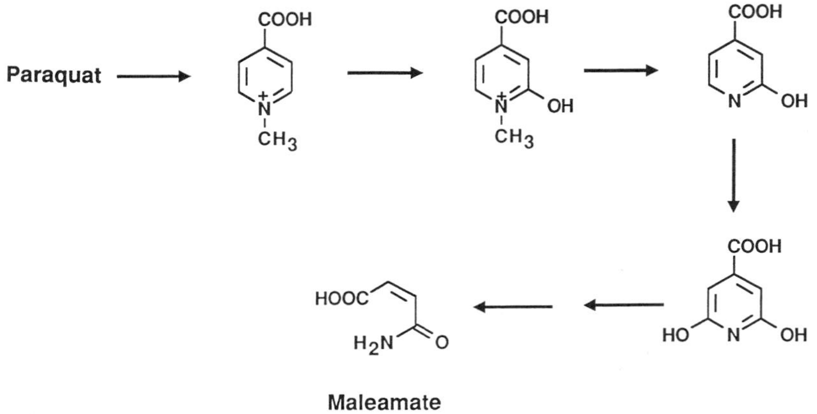


Fig. 6–8. Degradation of paraquat and cleavage of pyridine ring by a bacterium.

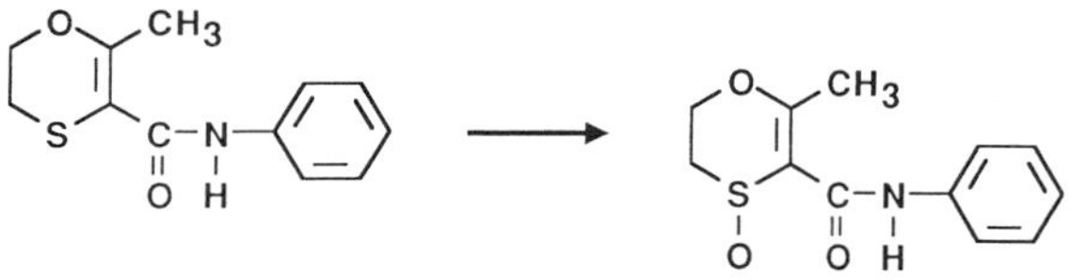


Fig. 6–9. Sulfoxidation of carboxin by the fungus *Ustilago maydis*.

### 6–2.1.10 Sulfoxidation

The sulfoxidation reaction implies the enzymatic conversion of a divalent compound to a sulfoxide, and occasionally to a sulfone:

$$S \rightarrow SO \rightarrow SO_2.$$

The oxidation of organic sulfides (thioethers) and sulfites to the corresponding sulfoxides and sulfates can be catalyzed by minerals (Crosby, 1976). Often it is difficult to distinguish between a biological and a chemical reaction. The observation of a sulfoxidation reaction in soil cannot be attributed to microbial activity without establishing that the catalyst is indeed of biological origin.

For example, the systemic fungicide carboxin has been observed in soil to undergo sulfoxidation to the nonfugitoxic S-oxide (carboxin sulfoxide), but since the reaction also occurred in sterile soil it is apparently not because of microbial activity (Chin et al., 1970) in this case. However, Lyr et al. (1974) established that the fungus *Ustilago maydis* could form the sulfoxide from carboxin, and even mitochondria isolated from this fungus could oxidize carboxin (Fig. 6–9). Usually the formation of carboxin sulfoxide occurs slowly in soil, while the reaction proceeds much faster inside organisms.

## 6–2.2 Reduction Reactions

The major reduction reactions are presented in Table 6–2. The reduction of the nitro group to amine involves the intermediate formation of a nitroso and a hydroxyamino group. This type of reduction reaction has been found during the microbial metabolism of various pesticides. Organophosphorus pesticides like parathion, paraoxon, EPN, or fenitrothion are often reduced to amino compounds, which are mostly inactive as pesticides (Miyamoto et al., 1966; Matsumura & Benezet, 1978). Sulfoxide can be reduced to sulfide. Timms and MacRae (1982) found that washed cell suspensions of *K. pneumoniae* reduced the organophosphorus pesticide fensulfothion to fensulfothion sulfide. Other reductive reactions frequently found are saturation of double bonds, reduction of aldehydes to alcohols, reduction of ketones to secondary alcohols, and reduction of certain metals.

Many pesticides possess halogen atoms that are the actual cause of their activity on target organisms; therefore, the removal of the halogen is of major importance, and this reaction often has a reductive character. Halogen substituents on an aromatic ring that withdraw electrons from carbons render a nucleus less susceptible to microbial attack. Usually, the greater the num-

Table 6-2. Reduction reactions in microbial pesticide metabolism.†

| |
|---|
| Reduction of nitro group |
| $RNO_2 \rightarrow ROH$ |
| $RNO_2 \rightarrow RNH_2$ |
| Reduction of double bond or triple bond |
| $Ar_2C{=}CH_2 \rightarrow Ar_2CHCH_3$ |
| $RC{\equiv}OH \rightarrow RCH{=}CH_2$ |
| Sulfoxide reduction |
| $RS(O)R' \rightarrow RSR'$ |
| Reductive dehalogenation |
| $Ar_2CHCCl_3 \rightarrow Ar_2CHCHCl_2$ |

† R = organic moiety, Ar = aromatic moiety.

ber of chlorine substitutions on an aromatic ring, the greater is a compound's resistance to microbial attack. However, some compounds with only one chlorine, for instance 2-chloroaniline, are not readily metabolized (Rochkind-Dubinsky et al., 1987). Mikesell and Boyd (1985) also found that a highly chlorinated compound such as pentachlorophenol is rapidly dehalogenated to persistent dichlorophenols. Thus, it is not only the number and position of substituents but also the type of the molecule itself that affects biodegradability.

Pesticides like DDT, heptachlor, lindane, or methoxychlor can be reductively dechlorinated by various microorganisms. Wedemeyer (1966) studied enzymatic dehalogenation with a cell-free extract from *Aerobacter aerogenes* that anaerobically catalyzed the reduction of DDT to TDE (DDD). They suggest that reduced Fe (II) cytochorome oxidase was responsible for the reductive dechlorination. Incubation of methoxychlor with *K. pneumoniae* (*A. aerogenes*) resulted in the reductive dechlorination of the substrate and gave rise to 19% methoxydichloro compound (Fig. 6–10) (Baarschers et al., 1982). The initial step of lindane ($\gamma$-BHC) degradation by cell-free extracts of *Clostridium rectum* was also shown to be a reductive dechlorination (Kurihara et al., 1981). Tiedje et al. (1987) reviewed studies on reductive dehalogenation of chlorinated aromatic hydrocarbons by anaerobic microorganisms.

### 6–2.3 Hydrolytic Reactions

Hydrolytic reactions are common in the microbial metabolism of pesticides. Many pesticides can be degraded by a hydrolytic reaction by the simultaneous addition of water. Enzymes involved in the hydrolytic reaction include esterase, acrylamidase, phosphatase, hydrolase, and lyase. Often it is difficult to determine the original catalyst of the reaction, since specific environmental circumstances or secondary effects of microbial metabolism can create physical conditions conducive to hydrolysis, which makes it hard for the detection of actual enzymatic hydrolytic activity. Pesticides that have ether, ester, or amide linkages can easily undergo hydrolysis with the resulting products usually losing their pesticidal effect (Table 6–3). In the case of hydrolytic dehalogenation, a halogen is exchanged with a hydroxyl group generated from the water, thus the process can also be categorized as a hydrolytic reaction (Table 6–3).

Fig. 6–10. Reductive dechlorination of methoxychlor by *Klebsiella pneumoniae.*

Many microorganisms, particularly the fungi, excrete hydrolytic enzymes outside the cells. Therefore, it is not surprising to find exoenzymes that have hydrolytic properties in soil. An esterase capable of degrading malathion to its monoacid was extracted with 0.2 *M* NaOH from soil (Getzin & Rosefield, 1971). Another enzyme capable of hydrolyzing the organophosphorus insecticide crotoxyphos at the P-O-C linkage was also isolated from soil (Getzin & Satyanarayana, 1979).

Munnecke et al. (1982) have reviewed in detail enzymatic hydrolysis of six classes of pesticides. They claim that the hydrolysis of organophosphates, phenylcarbamates, and dithioate resulted in a great decrease in the toxicity of the pesticide molecules, while only small differences in toxicity between the parent compound and their metabolites was observed with phenoxyacetate, acylanilides, and phenylurea compounds.

The herbicide metamitron is degraded by *Arthrobacter* sp. by hydrolytic cleavage of the amide bond in the triazinone ring (Engelhardt et al., 1982). Hydrolysis reactions are initial stages of the organophosphate insecticide Phosalone transformation by *Acinetobacter calcoaceticus* (Golovleva et al., 1983).

### 6–2.4 Synthetic Reactions

In microbial mediated synthetic reactions, pesticides or their intermediates are linked together among themselves or with other compounds. Binding takes place at functional groups that may be present either in original molecules, in its intermediate, or at a substituent added during metabolic or chemical transformation. The results of such synthetic reactions are the formation of larger products. Synthetic reactions can be divided into: (i) conjugation reactions, which involve the union of two substances, and (ii) con-

Table 6–3. Hydrolysis reactions in microbial pesticide metabolism.†

| |
|---|
| Ether hydrolysis |
| $ROR' + H_2O \rightarrow ROH + R'OH$ |
| Ester hydrolysis |
| $RC(O)OR' + H_2O \rightarrow RC(O)OH + R'OH$ |
| Phosphor-ester hydrolysis |
| $(RO)_2P(O)OR' + H_2O \rightarrow (RO)_2P(O)OH + R'OH$ |
| Amide hydrolysis |
| $RC(O)NR'R'' + H_2O \rightarrow RC(O)OH + HNRR''$ |
| Hydrolytic dehalogenation |
| $RCl + H_2O \rightarrow ROH + HCl$ |

† R = organic moiety, Ar = aromatic moiety.

densation reactions, which yield oligomeric or polymeric products. Conjugation reactions, such as methylation and acetylation, commonly occur during microbial metabolism of xenobiotics, while conjugations like glycoside formation, reactions with amino acids, or S that have been frequently observed in plants and animals, have rarely been found in the metabolism of microorganisms.

Cserjesi and Johnson (1972) found methylation of pentachlorophenol. They isolated a methylated product, pentachloroanisole, from the culture medium of *Trichoderma viride* containing the phenolic compound.

Acylation resulting in formylated or acetylated products is another conjugation mechanism that was mostly found in the microbial metabolism of anilines originating from various pesticides. Tweedy et al. (1970) demonstrated that *p*-bromoaniline was acetylated by the culture of *Talaromyces wortmanii* as well as *Fusarium oxysporium* to form an *N*-acetyl product. Kaufman et al. (1973) detected a small amount of 4-chloroacetanilide in the culture medium of *F. oxysporum* amended with 4-chloroaniline, and the corresponding acetylated anilines from aniline; 2-, 3-, and 4-chloroaniline were major products in the growth medium of a *Paracoccus* sp. (Bollag & Russel, 1976). Microbial formylation of 4-chloroaniline and the respective acetylation reaction was determined in the growth medium of a *Streptomyces* sp. (Russel & Bollag, 1977).

It is known that soil microorganisms play an important role in the binding of pesticidal residues into soil organic matter. In soil phenolic and quinonoid compounds originating either from lignin or from other organic residues, as well as those synthesized by microorganisms, are oxidatively coupled either alone or with other substances possessing a specific functional group to form polymers that constitute the base of humus material (Flaig et al., 1975; Haider et al., 1975; Sjoblad & Bollag, 1981). This process can occur through spontaneous chemical reactions, autooxidation, or oxidation by microbial enzymes, the latter being considered the primary mechanism in synthetic reactions. Since many xenobiotics or pesticides, when degraded, yield phenol- or aniline-like chemicals analogous to naturally occurring compounds, it is not surprising that many xenobiotics can be incorporated by cross-coupling into soil humus.

Microbial phenoloxidases and peroxidases are the major biological agents that catalyze the transformation of anilinic or phenolic compounds to polymerized products. For example, phenols, are transformed to phenoxy-radicals by removal of a H ion and an electron from the hydroxyl group. The resulting phenoxy-radicals are reactive and may be converted to stable products either by self-coupling or by cross-coupling with other molecules. Similarly, arylamino radicals that are formed during the oxidation of anilines are reactive and can couple with themselves or with phenolic and quinoidal compounds. Enzymatic polymerization of several phenolic and aromatic amine pesticide derivatives and their copolymerization with naturally occurring phenolic humus monomers have been intensively studied by our research group (Bollag, 1983). In our approach, we used fungal laccases to synthesize lower molecular weight polymers and have gained some insight into the

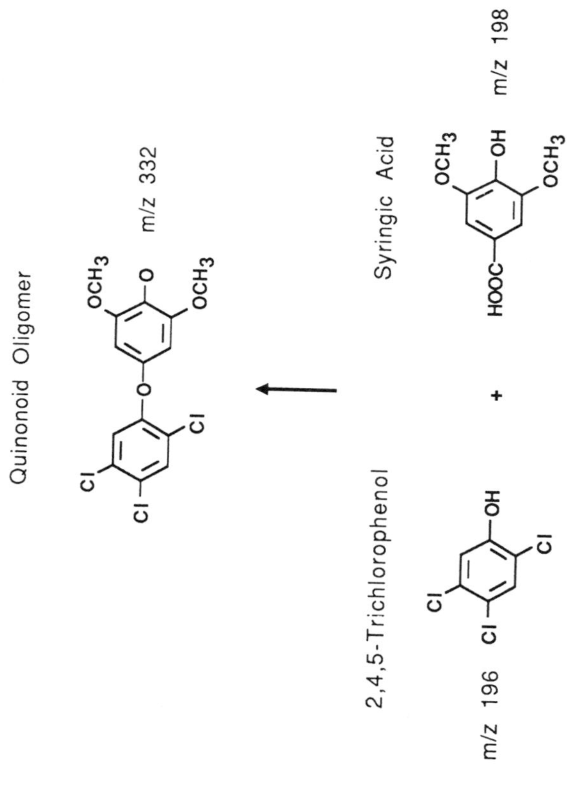
Quinonoid Oligomer
m/z 332
Syringic Acid
m/z 198
2,4,5-Trichlorophenol
m/z 196

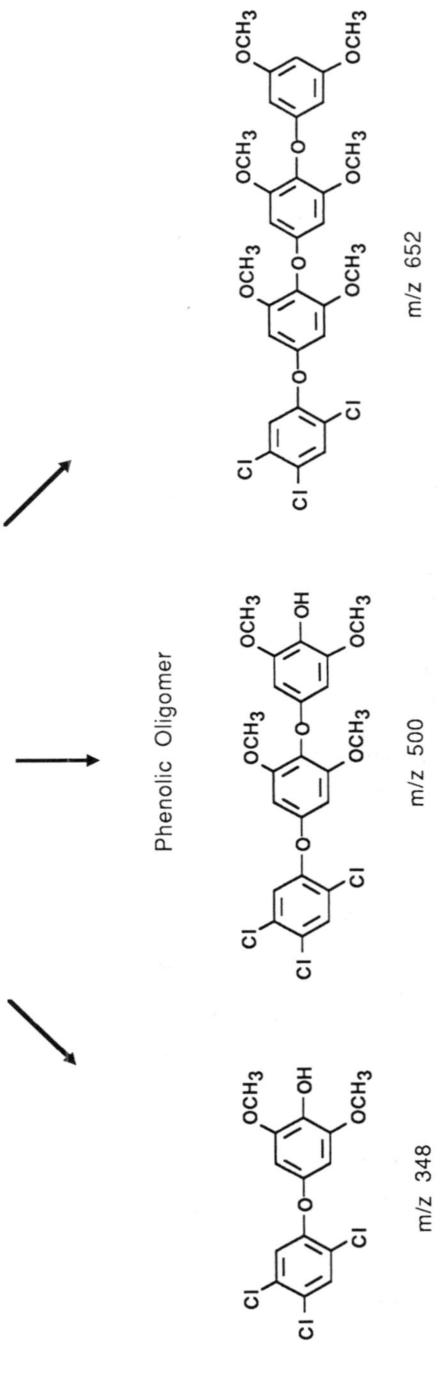


Fig. 6-11. Cross-coupling of 2,4,5-trichlorophenol and syringic acid.

nature and extent of polymerization. This information can be of value in elaborating highly complex polymers such as humus-bound pesticide residues that occur in the natural environment.

2,4-Dichlorophenol, an intermediate of 2,4-D, was oxidatively coupled by a lacase of *Rhizoctonia praticola* to yield dimeric to pentameric products, two of the dimeric products were dichlorophenoxy-benzoquinones, molecules in which chlorobenzoquinones are bound to chlorobenzene by ether linkages. Under acidic conditions, phenoxyquinones may probably cyclize to form dioxin (Bollag et al., 1980; Minard et al., 1981). In the presence of the laccase, 2,4-dichlorophenol reacted with various humus constituents, such as syringic acid, vanillic acid, orcinol, or vanillin to form various hybrid or cross-coupling products (Bollag et al., 1980).

When incubated together with a laccase, higher chlorinated phenols (2,3,6- and 2,4,5-trichlorophenol; 2,3,5,6-tetrachlorophenol; and pentachlorophenol) cross-coupled with syringic acid, and two types of hybrid oligomers were determined: (i) quinonoid oligomers (dimers to trimers) in which the chlorophenols were bound through either linkage to an orthoquinone derived from syringic acid, and (ii) phenolic oligomers (dimers to pentamers) in which the chlorophenols were bound through either ether linkages to decarboxylated products of syringic acid. All hybrid products contained only one halogenated phenol molecule, and no dehalogenation took place. A model is presented in Fig. 6–11 in which the structure configuration of phenolic oligomers from the reaction of 2,4,5-trichlorophenol and syringic acid is proposed. An orthoquinone product derived from syringic acid is shown to couple with 2,4,5-trichlorophenol yielding a 3-methoxy-t-(2′, 4′, 5′-trichlorophenoxy)-benzoquinone (1.2) (Bollag & Liu, 1985).

Enzymatic copolymerization of pesticide-derived aromatic amines with phenolic humus monomers such as syringic, vanillic, protocatechuic, and ferulic acids also resulted in the formation of several hybrid dimers, trimers, and tetramers (Bollag et al., 1983). A quinone product originating from syringic acid was bound via an imine linkage to 2,6-diethylaniline or 3,4-dichlorophenol yielding *N*-(2,6-diethylphenyl)-2,6-dimethoxy-*p*-benzoquinone, and *N*-(2,4-dichlorophenyl)-2,6-dimethoxy-*p*-benzoquinone, respectively. Cross-coupling between 2,6-diethylaniline and protocatechuic acid yielded a hybrid tetramer that consisted of three 2,6-diethylaniline monomers and a protocatechuic acid derivative. Structure determination of the tetramer indicated that one 2,6-diethylaniline molecule was bound to a protocatechuic derivative through an imine linkage, while the other two aniline molecules were connected to aromatic C through a C-N single bond. Figure 6–12 depicts the products resulting from the combined incubation of 2,6-diethylaniline with syringic and protocatechuic acids. Liu et al. (1981) also observed enzyme-catalyzed hybridization between halogenated anilines and 2,4-dichlorophenol.

Although a laccase of *R. praticola* was active in cross-coupling of anilines and phenols, it was not able to oxidize aniline along as a substrate, with the exception of *p*-methoxyaniline (Sjoblad & Bollag, 1977). Hoff et al. (1985), however, found that the laccase of *Trametes versicolor* catalyzed the

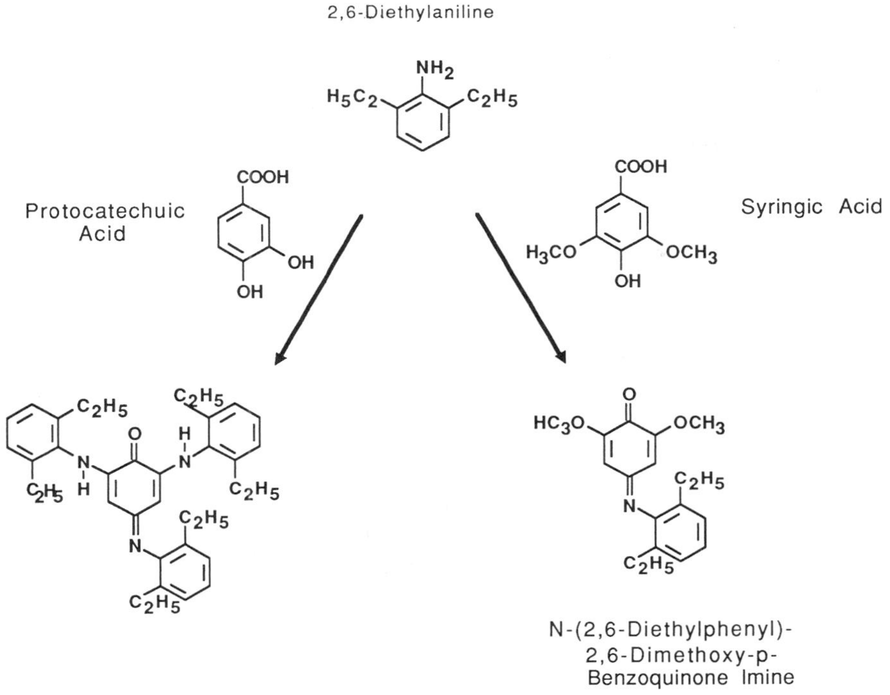


Fig. 6–12. Cross-coupling of 2,6-diethylaniline with syringic acid or with protocatechuic acid.

oxidative coupling reaction of a variety of halogen-, alkyl-, and alkoxy-substituted anilines, causing the formation of oligomeric products (dimers to pentamers). Other condensation products of anilines that appear to be mediated by microbial activity include azobenzenes (Bartha et al., 1968), azoxybenzenes (Kaufman et al., 1972), anilino-azobenzenes (Linke, 1970), diphenylamines (Briggs & Ogilvie, 1971), and phenoxazinones (Briggs & Walker, 1973; Abrosi et al., 1977).

## 6–3 ENZYMES INVOLVED IN PESTICIDE METABOLISM

In the scientific literature, many microbial pesticide transformation reactions have been described using the expression metabolism, but only in a relatively small number of investigations was an active enzyme preparation demonstrated. Few in-depth biochemical studies have been performed with pesticide-transforming enzymes as compared to the large number of investigations of the many enzymes involved in general metabolic activities. Most enzyme preparations described in pesticide metabolism are enzymes previously described as active in the transformation of naturally occurring compounds that possess a broad substrate specificity. Similar observations

Table 6-4. Isolated enzymes from microorganisms involved in pesticide metabolism.

| Substrate | Enzyme | Isolated from: | Reference |
|---|---|---|---|
| Carboxin | Aryl acylamidase | *Nocardia* sp. | Bachofer & Lingens, 1983 |
| Carbofuran | Hydrolase | *Achromobacter* sp. | Derbyshire et al., 1987 |
| Chloridazon | Rhodanese | *Acinetobacter calcoaceticus, Pseudomonas aeruginosa* | Layh et al., 1982 |
| Chloropicrin | Cytochrome P-450 cam | *Pseudomonas putida* | Castro et al., 1985 |
| Chloropropham (CIPC) | Hydrolase (Esterase) | *Pseudomonas striata* | Kearney, 1965 |
| Dalapon | Dehalogenase | *Arthrobacter* sp. | Kearney et al., 1964 |
| Dalapon | Dehalogenase | *Rhizobium* | Berry et al., 1979 |
| Diazinone | Hydrolase | *Pseudomonas* sp. | Barik & Munnecke, 1982 |
| DDT | Dehalogenase-reduced cytochrome oxidase | *Aerobacter aerogenes* | Wedemeyer, 1967 |
| Fenthion | 1,2-Dioxygenase | *Nocardia* sp. | Rast et al., 1979 |
| Karsil | Aryl acylamidase | *Penicillium* sp. | Sharabi & Bordeleau, 1969 |
| Karsil | Aryl acylamidase | *Bacillus sphaericus* | Engelhardt et al., 1973 |
| Lindane | Dehalogenase | *Clostridium rectum* | Ohisa et al., 1980 |
| Linuron | Hydrolase aryl acylamidase | *Bacillus sphaericus* ATCC 12123 | Engelhardt et al., 1973<br>Wallnöffer & Bader, 1970 |
| Malathion | Hydrolase | *Trichoderma viride Pseudomonas* sp. | Matsumura & Boush, 1966 |
| Metobromuron | Hydrolase aryl acylamidase | *Bacillus sphaericus* ATCC 12123 | Engelhardt et al., 1973<br>Wallnöffer & Bader, 1970 |
| Parathion | Phosphohydrolase | *Escherichia coli* | Zech & Wigand, 1975 |
| Parathion | Parathion hydrolase | *Pseudomonas alcaligenes* | Munnecke, 1980 |
| Parathion | Hydrolase | *Pseudomonas* sp. ATCC 29343 | Brown, 1980 |
| Parathion (Paraoxon) | Hydrolase, Phosphotriesterase | *Flavobacterium* sp. ATCC 27551 | Brown, 1980 |
| Propanil | Acylamidase | *Fusarium solani* | Lanzilotta & Pramer, 1970 |
| Propanil | Acylamidase | *Fusarium oxysporium* | Blake & Kaufman, 1975 |
| Propanil | Aryl acylamidase | *Bacillus sphaericus* | Engelhardt et al., 1973 |
| Propham | Aryl acylamidase | *Bacillus sphaericus* | Engelhardt et al., 1973 |
| Propham (IPC) | Hydrolase (Esterase) | *Pseudomonas striata* | Kearney, 1965 |
| Pyrazon | Pyrazon dioxygenase | Bacterial strain E | Sauber et al., 1977 |

were made with drug metabolism in higher organisms where it was determined that most anthropogenic compounds are metabolized by nonspecific enzymes.

However, since pesticides possess chemical structures or configurations that mark them as unnatural or xenobiotic, it is not surprising that the formation of microbial enzymes was provoked through internal DNA sequence changes (mutations) and recombinations (external DNA additions, gene transfer between organisms). As Kearney and Kellogg (1985) outlined in a review on the adaptation of microorganisms to pesticides, some pesticide-degrading enzymes evolved through the participation of extrachromosomal elements. Dehalogenases that catalyze initial reactions in halogen-substituted alkanoic acid herbicides (e.g., Dalapon and TCA) and oxygenases that attack phenoxyalkanoates (e.g., 2,4-D, 2,4,5-T, and MCPA) are known to be encoded on plasmids (see section 6–4.2 Genetic manipulation—Construction of Microorganisms with Novel Catabolic Activities). Kawasaki et al. (1984) have isolated and characterized a conjugative plasmid pU01 from a *Moraxella* sp. that encoded for two haloacetic dehalogenases. A strain of *Alcaligenes* was shown to contain a 58-MDa conjugal plasmid, pJPI, encoding the enzyme required to degrade part or all of both 2,4-D and MCPA (Fisher et al., 1978; Don & Pemberton, 1981).

Several enzymes isolated from microorganisms that are responsible for the transformation of specific pesticides are listed in Table 6–4. Despite the great interest in the use of enzymes for the detoxication of pesticides, the number of the isolated enzymes is rather limited.

## 6–4 BIOTECHNOLOGICAL ASPECTS OF PESTICIDE METABOLISM

### 6–4.1 Use of Microorganisms and Enzymes for Treatment of Pesticidal Soil Contamination

Pollution arises from a wide variety of people's activities, with agricultural practices being an important contributor. Through run-off, drift, or soil erosion, pollutants such as pesticides may move from their point of application to the surrounding environment. Larger amounts of these chemicals may be released by accidental spillage, improper clean-up of storage containers, leaks at pesticide dump sites, or as waste discharged from production and formulation facilities. A greater awareness of the dangers of pesticide pollution has brought about the need to develop appropriate technologies to control the problem.

The realization that the mineralization of pesticides in natural ecosystems is largely because of biodegradative processes mediated by microorganisms (Alexander, 1980; Bollag, 1982) has suggested the possibility of using microorganisms or their enzymes to degrade the pollutants. This strategy for pollution control may prove to be a feasible alternative to, or complement of, existing physicochemical methods (Munnecke, 1980; Kobayashi & Rittman, 1982; Finn, 1983).

The idea of using biological processes to degrade waste products is not new. For a long time, people have exploited microorganisms to detoxify human sewage. Microbial systems have been used in the treatment of manufacturing wastes to remove parathion (Coley & Stutz, 1966), phenols, and trifluralin (Howe, 1969).

That a microbial inoculant could enhance the rate of degradation of a pesticide in soil was perhaps first suggested by Audus in 1951 for the removal of 2,4-D from soil. His approach has since proven successful for a number of other researchers. Kearney et al. (1969) were able to demonstrate the disappearance of DDT from soil after heavy inoculation with a DDT-degrading *Aerobacter aerogenes.* In 1970, Clark and Wright observed the detoxification of isopropyl-*N*-phenyl-carbamate propham (IPC) in soil inoculated with IPC-utilizing *Arthrobacter* and *Achromobacter* spp.

Inoculation of greenhouse soil with a propham-degrading culture of coryneform bacteria reduced or eliminated the toxicity of propham, chlorpropham, and Swep (McClure, 1972). Sethunathan (1973) and Rajaram and Sethunathan (1975) were able to accelerate the hydrolysis of parathion in flooded soils by inoculating with a parathion-hydrolyzing bacterial culture. Parathion was also detoxified by a combined culture of *Pseudomonas stutzeri* and *P. aeruginosa.* When added to parathion-contaminated soil, this two-component culture mineralized 85% of the parathion in 4 d (Daughton & Hsieh, 1977). In a separate experiment, these same authors observed the degradation of nearly 100%, up to 5 g $L^{-1}$, of parathion in silt loam treated with a parathion-acclimated bacterial culture.

More recent experiments have proven equally successful. Inoculation of contaminated soil with a pentachlorophenol (PCP) degrading *Arthrobacter* led to a 10-fold increase in the rate of PCP degradation, reducing its half-life from 2 wk to $<1$ d. A second dose of PCP resulted in a further enhancement in the rate of degradation (Edgehill & Finn, 1983). The PCP concentration in contaminated sewage was also successfully reduced by the introduction of the *Arthrobacter.* Chatterjee et al. (1982) demonstrated the microbial detoxification of 2,4,5-T. In their experiment, a strain of *P. cepacia* removed as much as 95% of 2,4,5-T within 1 wk from soil treated with the chemical at 1 g $kg^{-1}$ of soil. Repeated application of the culture resulted in 10% removal of 2,4,5-T at concentrations of as much as 20 g $kg^{-1}$ of contaminated soil (Kilbane et al., 1983).

An alternative method of microbial control of pollution is suggested by recent investigations into the degradation of 2,4-D. Kunc et al. (1984) and Kunc and Rybarova (1984) have reported the stimulation of 2,4-D mineralization in soil enriched with saccharides and other organic substrates. Thus, pesticides may be biologically detoxified in two ways: by inoculating contaminated soil with pesticide-degrading cultures or by enhancing the growth rate of pesticide degradation by microorganisms already present at the site.

Nevertheless, there are certain limitations to the use of acclimated microbial cultures to degrade toxic chemicals. Biological methods for waste treatment involve whole cells, which are subjected to chemical shock, extremes of pH and temperature, and metabolic inhibition. In some cases, these factors may destroy or harm the microorganisms and prevent their degradation of a pesticide. For instance, Barles et al. (1979) found that the ability of an acclimated culture to degrade parathion deteriorated at application rates $>1250$ kg $ha^{-1}$. They suggested that the accumulation of *p*-nitrophenol and diethyl thiophosphate adversely affected the culture. Therefore, for successful pesticide detoxification, not only must the proper microbial population be selected but also suitable environmental conditions be maintained. Goldstein et al. (1985) speculated the reasons why microorganisms cannot metabolize pollutants in the environment: (i) the concentration of a xenobiotic may be too low to support growth; (ii) the organisms may be susceptible to toxins or predators; (ii) microorganisms may use other organic chemicals in preference to the pollutants; or (iv) they may be unable to move through the soil to the polluted site.

Under in situ conditions, it is not always possible to generate ideal conditions for microbial growth. For this reason, seeding of contaminated soil or container with enzymes obtained from metabolically active microorganisms may prove to be a more successful means of pesticide detoxication. Munnecke (1976) and Munnecke et al. (1982) have suggested the potential use of an enzyme for the detoxification of organophosphate pesticides, particularly parathion, in industrial waste-waters, pesticide containers, and contaminated soil following accidental spills. The addition of parathion hydrolase to a 200-L metal drum resulted in the hydrolysis of 76.5 g, or 90% of the residual parathion present in the drums as a 48% emulsifiable concentrate (Munnecke, 1980). The hydrolysis mediated by such enzymes is not restricted to organophosphate pesticides. Other pesticides that are reported to be hydrolyzed by a cell-free microbial enzyme include phenylcarbamates (Kearney, 1965); phenoxyacetates (Loos et al., 1967); acylanilides (Lanzilotta & Pramer, 1970); and phenylureas (Engelhardt et al., 1971, 1973). According to Munnecke et al. (1982), once the initial enzymatic reaction has occurred, many pesticides lose their biospecificity and toxicity. For instance, the hydrolysis products of parathion and paraoxon are 60 to 200 times less toxic than the parent pesticide. The toxicity of the phenylcarbamate carbaryl was decreased 920-fold as a result of enzymatic hydrolysis. Detoxification by the enzymatic hydrolysis of phenoxyacetate, anilide, and phenylurea compounds is much less significant. The hydrolysis product of 2,4-D was found to be only seven times less toxic than the parent compounds; however, the biospecificity was completely removed.

Honeycutt et al. (1984) reported that the enzyme parathion hydrolase is active in degrading high concentrations of diazinon in greenhouse soil. The half-life of diazinon in soil treated at a concentration of 5 g $L^{-1}$ was 5.6 h. Furthermore, more than 98% of 10 g $L^{-1}$ diazinon in soil can be removed within 24 h by a parathion hydrolase (EC 3.1.3) isolated from *Pseudomonas* sp. (Barik & Munnecke, 1982).

Furthermore, there are certain advantages to using enzymes for pollution control rather than whole microbial cells. Enzymes can often withstand environmental extremes that would be lethal to whole cells. Such extremes of pH, temperature, or of salt or solvent concentrations are often present in pesticide production wastewaters. For instance, parathion hydrolase can tolerate temperatures of up to 50 °C, salt concentrations of up to 10%, and solvent concentrations of 10 g $L^{-1}$, conditions that would prevent the growth of the *Pseudomonas* sp. that produces the hydrolase. Hydrolases may be particularly attractive for use as enzyme inoculants for other reasons: they are noted for their stability, lack of cofactor requirements, and ability to act on more than one substrate.

In some cases, enzymes may also be immobilized to maintain their stabilities and to enhance the efficiency of substrate transformation. Munnecke (1979) reported using immobilized enzymes in a continuous flow column for detoxifying effluents from pesticide production facilities. Parathion levels were reduced from 10 to 0.5 mg $L^{-1}$ in an approximately 1-min passage through the column. The enzyme remained active on the column for more than 70 d of continuous use.

Shuttleworth and Bollag (1986) immobilized a laccase on Celite and examined its ability to transform and remove halogenated and alkylated phenols. They showed that the enzymatic transformation was dependent on the substituent group and position of substitution. The overall ability of the immobilized enzyme was consistent with the ability of the free enzyme to transform the same compounds.

Co-immobilization of two or more enzymes for more complete conversion of pesticides has also been investigated. However, until operating conditions suitable for all enzymes used can be established, the success of this technique is not assured.

At the present time, if the metabolism of a pesticide requires a cofactor, the use of whole cells is preferred over the immobilized enzyme methods because of difficulties in regenerating or producing the cofactors cheaply and efficiently. Successful application of immobilized microcells in the transformation of certain pesticides has been demonstrated by several researchers. As early as 1959, Mills designed a unit comprised of a trickling filter and an activated sludge system for the detoxification of wastewater from a 2,4-D production facility. More recent attempts have also been successful. Pore and Sorenson (1981) found that algae immobilized on an agar column removed kepone from the aqueous eluate. Etzel and Kirsch (1975) used a continuous-flow fiber-wall biological reactor to treat wastewater containing pentachlorophenol. Elution through an activated carbon column with fixed-film bacteria resulted in the removal and biodegradation of chlorinated organic compounds (Bouwer & McCarty, 1982). Salkinoja-Salonen et al. (1983) reported that a specific biofilm reactor containing a mixed microbial consortium could be adapted for the mineralization of organochlorine compounds and chlorophenols.

It should be noted that there are difficulties associated with mirobial or enzymatic detoxification of pesticides. As mentioned previously, in situ environmental conditions may not be suitable for microbial growth or the enzyme reaction. Also, products of the activity of microorganisms or their enzymes may sometimes be more toxic than the parent compound. Obviously, it is essential to have a basic understanding of microbial metabolism of pesticides before pollution control via biodegradation can be accomplished. Nevertheless, advances such as the development of enrichment techniques for the isolation of microorganisms with pesticide-degrading capabilities has made the future use of microorganisms or their enzymes for the detoxification of pesticides a distinct possibility.

### 6-4.2 Genetic Manipulation—Construction of Microorganisms with Novel Catabolic Activities

It seems reasonable to assume that xenobiotics with structures resembling natural compounds are more likely to be efficiently degraded, whereas synthetic substances that differ considerably from those found in nature persist for a longer time. Mechanistically, microorganisms might exploit an indigenous pathway to metabolize xenobiotics of the former class, but would need to evolve novel degradative functions for compounds of the latter class. Under most conditions, the acquisition of such capabilities via the processive accumulation of single mutations would not be adequate. Rather, such a process can be accelerated enormously by an additional mechanism of genetic diversification—gene transfer between organisms.

Microorganisms capable of degrading 2,4-D are abundant in both soil and water; accordingly, the herbicide persists for only a few days or weeks. On the other hand, 2,4,5,-T is resistant to microbial degradation and often lingers for 24 wk or more. This phenomenon is illustrative of the classifications noted above and has generated considerable interest in genetic manipulations leading to the creation of microorganisms capable of degrading recalcitrant molecules.

Traditionally, genetic manipulation involved random mutation and selection, whereby microbial populations are subjected to selective pressure under which only the desired variant is able to survive and proliferate. However, this approach relies on unique mutational events that modify preexisting genes by genetic rearrangement, which alter regulatory genes controlling a degradative pathway, or which may activate silent genes that already exist in the microorganism. When novel pathways are required, a far more promising approach has been to add distinct functions by introducing a cluster of degradative genes on a single genetic unit, such as a plasmid or bateriophage.

The two most promising methods for the construction of hybrid metabolic pathways are:

1. In vivo genetic engineering—the use of naturally cloned catabolic DNA located on degradative plasmids. These naturally occurring modes of genetic exchange, namely conjugation, transduction, and transformation have been used successfully to construct a variety of new strains of bacteria (Pemberton, 1981).

2. In vitro genetic engineering—the cloning of catabolic genes onto plasmid or bacteriophage by recombinant DNA techniques. The procedure consists of several distinct biochemical and biological manipulations: breaking and joining DNA molecules derived from different sources; identifying a suitable gene carrier that can replicate both itself and foreign DNA segments linked to it; introducing composite DNA molecules into a functional bacterial cell by transformation; and selecting among the transformants a clone that has acquired the composite molecules (Cohen, 1980).

During the past 10 yr, numerous genes involved in the degradation of organic compounds have been shown to be plasmid-encoded. Moreover, it has been increasingly evident that special characteristics exhibited by bacterial strains that are important in medicine, agriculture, and the environment are often plasmid-determined. The finding that plasmids carry genes for catabolic enzymes helps to explain microorganisms' remarkable capacity for enzymatic adaptation to xenobiotic substances. Plasmids have enabled a greater degree of natural exchange of genetic information between different strains, species, or genera of microorganisms, and this genetic exchange promotes the evolution of new degradative capacities, especially during selective challenges in the chemostat (Williams, 1981; Kellogg et al., 1981).

Involvement of plasmids in the degradation of many naturally occurring substances has been reported as in the cases of octane (Chakrabarty, 1976); camphor (Reinwald et al., 1973); naphthalene (Dunn & Gunsalus, 1973); and toluene and xylenes (Williams & Worsey, 1976; Kunz & Chapman, 1981). Plasmids encoding degradative genes for more complex compounds such as lignin have also been observed (Salkinoja-Salonen et al., 1979).

Pemberton and Fisher (1977) and Fisher et al. (1978) first reported the existence of plasmids that code for pesticide degradation. These were degradative plasmids for the pesticides 2,4-D and MCPA, isolated from a strain of *Alcaligenes paradoxus*. The authors emphasized that the presence of pesticide-degrading plasmids are important in the degration of 2,4-D in soil. Six other plasmids encoding degradative genes for 2,4-D and MCPA were later isolated from strains of *A. paradoxus* and *A. entrophus* (Don & Pemberton, 1981). Pierce et al. (1981, 1982) also isolated *Pseudomonas* spp. possessing plasmids capable of degrading 2,4-D.

It has been shown that chloridazon (pyrazon or 5-amino-4-chloro-2-phenol-3-pyridazine), the active component of the herbicide Pyramin and an analogue, antipyrin (2,3-dimethyl 1-phenyl-pyrazolone), were degraded by strains of soil bacteria containing at least two and up to six plasmids, ranging in molecular weight from 8 to 300 MDa (Kreis et al., 1981).

Single plasmids ranging in molecular weight from 109 to 190 MDa were isolated from four *Pseudomonad* spp. and two *Alcaligenes* spp. (Slater & Bull, 1982). Curing the four *Pseudomonad* spp. with ethidium bromide resulted in concomitant loss of the plasmids and dehalogenase activities. This demonstrated that dehalogenase genes responsible for degrading halogenated alkanoic acids, a few of which are widely used herbicides such as dalapon, were encoded on the plasmids. Since these strains can express up to four different dehalogenases and no dehalogenase activity was present after curing of plasmids, it indicates that *Pseudomonas* plasmids encode more than one type of dehalogenase (Hardman & Slater, 1981). This result represents an example of isoenzyme gene manipulation on plasmids. Beeching et al. (1983) have isolated a plasmid carrying a dehalogenase gene from *P. putida* capable of degrading chloropropionate.

Using a similar approach, Serdar et al. (1982) demonstrated that parathion hydrolysis by *P. diminuta* was plasmid-encoded. Expression of enzymatic activity was lost after treatment of cells with mitomycin C. Hydrolase-negative derivatives were missing a plasmid present in the parental strain.

Polychloro-aromatic insecticides and their derivatives such as DDT and kelthane were degraded by strains of *P. aeruginosa* carrying biodegradative plasmids that appear to encode metapyrocatechase and salicylate hydroxylase activity (Golovleva et al., 1982).

Pertsova et al. (1984) studied the degradation of 3-chlorobenzoate, a structural analog of chlorinated aromatic pesticides, in soil columns that were inoculated with *Pseudomonad* spp. carrying biodegradative plasmids. They suggested the likelihood of genetic transfer of biodegradation capabilities between the indigenous microbiota and the plasmid-carrying strains of *P. aeruginosa* and *P. putida.*

Although the discovery of degradative plasmids for the majority of pesticides has yet to come, many of the more exotic catabolic pathways in the saprophytic microbial population may eventually be demonstrated to be plasmid coded (Williams & Worsey, 1976). The existence of pesticide degradative plasmids raises many interesting possibilities. Most important, an understanding of the evolution and spread of pesticide degradative plasmids may provide the basis for producing soil microbial populations that carry plasmids conferring abilities to degrade recalcitrant pesticide molecules such as 2,4,5-T and its potent contaminant 2,3,7,8-tetrachloro-dibenzo-*para*-dioxin (Pemberton, 1979; Kilbane et al., 1982).

It is not uncommon for degradative pathways to be encoded partly by a plasmid and partly by chromosomal genes in a given strain. The ability of *Alcaligenes paradoxus* to degrade the herbicide 2,4-D is encoded by the conjugal plasmid pJP1 (Pemberton & Fisher, 1977; Fisher et al., 1978), thereby enabling the conversion of 2,4-D to 2,4-dichlorophenol. Subsequent catabolism of the 2,4-dichlorophenol is encoded on chromosomal genes (Pemberton, 1981). Similarly, the conversion of octane to octanaldehyde is under the control of OCT plasmid genes, but the further oxidation reactions are specified by the *P. putida* or *P. aeruginosa* chromosome (Fennewald et al., 1979).

An important aspect in the genetic manipulation of degradative pathways as pointed out by Franklin et al. (1981) is to increase the yields of enzymes that carry out particular transformations. They further indicated that the cloning of determinants for individual enzymes will be of great value for the construction of new catabolic pathways and for "recruiting" enzymes of one pathway into another to expand their substrate specificities.

It is even possible to construct a strain that features a combination of co-metablic and catabolic genes necessary for a particular pesticide degradative pathway. The co-metabolic genes code for the initial enzyme that transforms the compound to an intermediate that the catabolic enzymes are able to degrade. Introduction of degradative plasmids into bacteria lacking particular catabolic genes would be a way of conferring novel catabolic ability to strains that populate a given soil.

Reineke and Knackmuss (1979, 1980) demonstrated the transfer of the TOL plasmid from *P. putida* mt-2 (WR101), to *Pseudomonas* sp. B13 (WR1), which is capable of degrading halocatechol. This enables new strains to be isolated that are able to use 4-chloro- and 3,5-dichlorobenzate as their sole source of C and energy. The combination of methylene-using strains providing high co-metabolic activities for halosubstituted analogues by their initial enzymes, together with halocatechol-degrading enzymes of *Pseudomonas* sp. B13, constitutes a general concept for the construction of new strains that could degrade halosubstituted benzenes, phenols, anilines, and naphthalenes. Similarly, Chatterjee and Chakrabarty (1981) obtained a new plasmid, apparently coding for 4-chlorobenzoate degradation, by recombination between the TOL plasmids and a 3-chlorobenzoate degradative plasmid (pAC25) during selection in a chemostat.

Kellogg et al. (1981) described the potential use of degradative plasmids to construct bacterial strains for the clean-up of industrial pollution. They inoculated into a chemostat microorganisms from various waste-dump sites, along with microorganisms harboring a variety of plasmids such as CAM, TOL, SAL, pAC21, and pAC25 to provide appropriate genes to establish degradative pathways. The mixed microbial culture was adapted to grow in the presence of increasing concentration of 2,4,5-T for 32 to 40 wk. Continuous subculturing of microbial population in flasks with higher concentrations of 2,4,5-T shortened the growth lag and increased the ability of microorganisms to degrade higher concentrations of the xenobiotic. This technique of plasmid-assisted molecular breeding has led to the development of a bacterial strain, *P. cepacia,* capable of using the recalcitrant molecule 2,4,5-T as a sole source of C at a relatively high concentration (Kilbane et al., 1982). Furthermore, the plasmid encoding 2,4-D degradative capability described previously was transferred to this 2,4,5-T-degrading strain that subsequently possessed the ability to metabolize both substrates (Ghosal et al., 1985).

Schwien and Schmidt (1982) demonstrated improved degradation of monochlorophenols by a constructed strain. *Alcaligenes* sp. strain A7, which can use high concentrations of phenol, acquired halocatechol-degrading capacities from *Pseudomonas* sp. strain B13. The transconjugant *Alcaligenes* sp. strain A7-2 was able to use all three isomeric chlorophenols; this property was not possessed by the donor or the recipient. Quensen and Matsumura (1984) have established that degradative capabilities of a gram-positive bacterium, *Bacillus megaterium,* can be transferred to *B. subtilis* by the plasmic transfer technique.

Plasmids often carry transposons and insertion sequences (IS), which are both discrete pieces of DNA capable of translocating freely between plasmids and chromosomes in many species of bacteria. In recent years, transposon mutagenesis has been developed for the analysis and manipulation of gene structure and function (Eaton & Timmis, 1984). Transposition mutagenesis, the generation of mutations caused by the insertion of transposons into functional genes, is particularly valuable for the isolation of mutants of biodegradative pathways (Franklin et al., 1981). Transposable elements, which encode selective functions and insert at high frequency into target DNA sequences by transposition from a donor replicon to a recipient replicon in the same cell, are directly or indirectly responsible for some chlorinated hydrocarbon degradative pathways (Slater & Bull, 1982).

There are only a few studies that concern themselves with in vitro gene cloning of pesticidal degradative plasmids. Cloning into *Escherichia coli* of the haloacetate dehalogenase genes from a *Moraxella* plasmid pU01 was first carried out by Kawaski et al. (1984), and two hybrid strains of E. coli that contained the haloacetate dehalogenase H-1 gene were found to be viable. Serdar and Gibson (1985) reported the cloning of parathion hydrolase genes and their expression in *E. coli* and *Pseudomonas* strains. Using stepwise cloning and conjugation techniques, they were able to construct the strain *P. diminuta* MG (pCMS 55), which had approximately four and one-half times the parathion hydrolases activity of the wild-type strain, *P. diminuta* MG (pCHSI). Similar experiments were carried out with the *Achromobacter,* which produces carbofuran hydrolase (Kearney et al., 1986).

The use of plasmids to create pollutant degrading microorganisms has many advantages. Degradative pathways are usually complex and require several enzymes along with cofactors and regulatory mechanisms. Plasmid-associated degradation combined various traits in one. Many plasmids are transmissible—they can be transmitted between strains or even between species simply by culturing donor and recipient cells together. Moreover, plasmids may be used to genetically alter microorganisms already present in a polluted environment so that they gain the ability to better degrade a pollutant without losing their ability to survive under field conditions.

Several problems exist with plasmid engineering. Little is known about the stability of recombinant DNA molecules in bacteria. The new host cells may recognize the introduced DNA as foreign and destroy or modify it in such a way that it becomes nonfunctional. Also, many plasmids are unstable and maybe lost without constant selective pressure in the presence of a xenobiotic. Certain plasmids are incompatible and cannot coexist in the same cell. In addition, the lack of expression of plasmid genes in the new host presents a problem about which little is known. Only by field trials can the degradative capability of a novel strain as well as the fate and impact of genetically altered bacterial populations in the environment be fully assessed.

Despite the considerable effort which has been directed toward basic research in genetic engineering and several constructed microbial strains believed to be of value for degrading chemical waste or pesticide residues, the potential application of genetically manipulated microorganisms for pollution control in natural environments is still somewhat speculative. For one thing, efforts to use genetic engineering for degrading xenobiotics have been minimal, and to date, genetically engineered strains have not been actually used for pollution control.

Among the first highly publicized uses of genetic engineering was that of the group working with Chakrabarty who engineered two strains of *Pseudomonas* by combining the qualities of four different strains into a single bacterial strain that acquired the ability to degrade four chemicals found in oil spills (Friello et al., 1976). Although this strain led the U.S. Supreme Court to permit the patenting of an experimentally constructed strain of *Pseudomonas* to degrade oil, it has not been applied under field conditions and is not commercially available. Restricting factors include the problems of liability in the event of health and environmental damage, and also the economic uncertainty. Thus, to which extent the genes expressed in these genetically engineered organisms could function in nature ecosystems is still not known.

Despite these concerns, scientists believe that genetic manipulation can be used to improve the degradative abilities of microbes. The potential exists that these new biotechnologies will alter or replace conventional physical-chemical pollutant disposal processes. Careful evaluation, however, of the relevant biological systems and the acquisition of basic knowledge on the microbial metabolism, genetics and natural ecology is essential before genetic engineering can be implemented to construct novel strains for pollution control.

## 6–5 CONCLUSION

To achieve peak efficacy, an applied pesticide must retain its biospecificity for a certain time. To minimize possible environmental hazards, however, it is desirable that pesticides are degraded to inorganic products after adequate pest control. Currently, considerable emphasis and efforts have been placed on microbially mediated pollution control, since high expenditure of pollutant disposal by means of physicochemical methods makes the use of alternative processes appear more attractive.

With the development of in vivo and in vitro gene manipulation techniques, the ability to construct a novel strain with desired traits for xenobiotic degradation has become a reality. This new technology not only facilitates fundamental and applied research on the biochemistry of microbial degradation, but also provided a new dimension in pollution control.

Release of a multitude of xenobiotic chemicals into the environment appears to have produced a response from microbial communities in the evolution of degradative functions. Thus, the conventional approach (selection and isolation of degradative mutants) is still playing an important role in the manipulation of degradative genes for obtaining microorganisms able to carry out novel reactions.

In laboratory as well as in natural environments, the behavior of mixed microbial populations have also attracted considerable attention. Often, success in developing a biodegradative system for certain pesticides depends on synergistic microbial communities that function cooperatively. Furthermore, microbial communities are capable of substantial exchange of genetic information between strains, species, or genera. Therefore, the interaction among microorganisms may be more important for the evolution of a novel catabolic pathway than a series of evolutionary steps occurring within one organism.

If microorganisms able to degrade pesticides or other toxic chemicals are found, then it may be possible to mass inoculate them into the soil with the proper maintenance of environmental conditions most conducive to their metabolism. This is obviously only possible if these organisms do not cause a threat to human health or cause serious disruption in soil ecology.

While we are concerned with the potential ecological threat of xenobiotics persisting in the environment, we must also be aware of a newly observed and continuing phenomenon, the so-called problem soil, in which enhanced degradation of some pesticides was reported to be caused by microbial activity. It is necessary to evaluate this problem of accelerated and undesirable loss of pesticide efficacy.

Microorganisms may transform pesticides by other than degradative pathways. A number of xenobiotic molecules or their degradative intermediates can be transformed by microorganisms or their enzymes to methyl, acyl, nitro, and other derivatives. In addition, oxidative coupling and cross-coupling of many phenolic and aromatic amine pesticide intermediates among themselves or with naturally occurring phenolic humus constituents results in the formation of polymeric products. Indeed, synthetic reactions further complicate the fate of pesticides in the environment, since pesticides may lose their biospecificity once being incorporated into soil organic matter. Employing copolymerization reactions as a process for containing or removing hazardous chemicals from aquatic or terrestrial environments might be a useful method of pollution control. However, the possibility of the subsequent release of bound xenobiotics by microbial activities may result in a delayed pollution threat.

## REFERENCES


Alexander, M. 1980. Biodegradation of chemicals of environmental concern. Science 211:132–138.

Alexander, M. 1985. Biodegradation of organic chemicals. Environ. Sci. Technol. 19:106–111.

Ambrosi, D., P.C. Kearney, and J.A. Macchia. 1977. Persistence and metabolism of phosalone in soil. J. Agric. Food Chem. 25:342–347.

Audus, L.J. 1951. The biological detoxication of hormone herbicides in soil. Plant Soil 3:170–192.

Baarschers, W.H., A.I. Bharath, J. Elvish, and M. Davies. 1982. The biodegradation of methoxychlor by *Klebsiella pneumonia*. Can. J. Microbiol. 28:176–179.

Bachofer, R., and F. Lingens. 1983. Degradation of carboxanilide fungicide by a *Nocardia* species. Hoppe-Seyler's Z. Physiol. Chem. 364:21–29.

Barik, S., and D.M. Munnecke. 1982. Enzymatic hydrolysis of concentrated diazinon in soil. Bull. Environ. Contam. Toxicol. 29:235–239.

Barles, R.W., C.T. Daughton, and D.P.H. Hsieh. 1979. Accelerated parathion degradation in soil inoculated with acclimated bacteria under field conditions. Arch. Environ. Contam. Toxicol. 8:647–660.

Bartha, R. 1969. Pesticide interaction creates hybrid residue. Science 166:1299–1300.

Bartha, R., H.A.B. Linke, and D. Pramer. 1968. Pesticide transformations: Production of chlorozaobenzenes from chloro-anilines. Science 161:582–583.

Beeching, J.R., A.J. Weightman, and J.H. Slater. 1983. The formation of an R-prime carrying the fraction I dehydrogenase gene for *Pseudomonas putida* PP3 using the IncP plasmic R64.44 J. Gen. Microbiol. 129:2071–2078.

Benezet, H.J., and F. Matsumura. 1974. Factors influencing the metabolism of mexacarbate by microorganisms. J. Agric. Food Chem. 22:427–430.

Berry, E.K.M., N. Allison, A.J. Skinner, and R.A. Cooper. 1979. Degradation of the selective herbicide 2,2-dichloropropionate (Dalapon) by a soil bacterium. J. Gen. Microbiol. 110:39–45.

Berry, D.F., and S.A. Boyd. 1985. Decontamination of soil through enhanced formation of bound residues. Environ. Sci. Technol. 19:1132–1133.

Blake, J., and D.D. Kaufman. 1975. Characterization of acylanilide-hydrolyzing enzyme(s) from *Fusarium oxysporum* Schlecht. Pestic. Biochem. Physiol. 5:305–313.

Bollag, J.-M. 1982. Microbial metabolism of pesticides. p. 126–168. *In* J.C. Rosazza (ed.) Microbial transformations of bioactive compounds. Vol. 2. CRC Press, Boca Raton, FL.

Bollag, J.-M. 1983. Cross-coupling of humus constituents and xenobiotic substances. p. 127–141. *In* F.R. Christman and E.T. Gjessing (ed.) Aquatic and terrestrial humic materials. Ann Arbor Sci. Publ., Ann Arbor, MI.

Bollag, J.-M., and M. Alexander. 1971. Bacterial dehalogenation of chlorinated aliphatic acids. Soil Biol. Biochem. 3:91–96.

Bollag, J.-M., C.S. Helling, and M. Alexander. 1968. 2,4-D metabolism: Enzymatic hydroxylation of chlorinated phenols. J. Agric. Food Chem. 16:826–828.

Bollag, J.-M., and S.-Y. Liu. 1985. Copolymerization of halogenated phenols and syringic acid. Pest. Biochem. Physiol. 23:261–272.

Bollag, J.-M., S.-Y. Liu, and R.D. Minard. 1980. Cross-coupling of phenolic humus constituents and 2,4-dichlorophenol. Soil Sci. Soc. Am. J. 44:52–56.

Bollag, J.-M., and M.J. Loll. 1983. Incorporation of xenobiotics into soil humus. Experientia 39:1221–1231.

Bollag, J.-M., R.D. Minard, and S.-Y. Liu. 1983. Cross-linkage between anilines and phenolic humus constituents. Environ. Sci. Technol. 17:72–80.

Bollag, J.-M., and S. Russel. 1976. Aerobic versus anaerobic metabolism of halogenated anilines by a *Paracoccus* sp. Microb. Ecol. 3:65–73.

Bouwer, E.J., and P.L. McCarty. 1982. Removal of trace chlorinated organic compounds by activated carbon and fixed-film bacteria. Environ. Sci. Technol. 16:836–843.

Bowes, G.W. 1972. Uptake and metabolism of 2,2-bis(*p*-chlorophenyl)-1,1,1-trichloroethane (DDT) by marine phytoplankton and its effect on growth and chloroplast electron transport. Plant Physiol. 49:172–176.

Briggs, G.G., and S.Y. Ogilvie. 1971. Metabolism of 3-chloro-4-methoxyaniline and some *N*-acyl derivatives in soil. Pest. Sci. 2:165–168.

Briggs, G.G., and N. Walker. 1973. Microbial metabolism of 4-chloroaniline. Soil Biol. Biochem. 5:695–697.

Brown, B.R. 1967. Biochemical aspects of oxidative coupling of phenols. p. 167. *In* W.I. Taylor and A.R. Battersby (ed.) Oxidative coupling of phenols. Marcel Dekker, New York.

Brown, K.A. 1980. Phosphotriesterases of *Flavobacterium* sp. Soil Biol. Biochem. 12:105–112.

Burchfield, H.P., and E.E. Storrs. 1957. Effect of chlorine substitution and isomerization on the interaction of *s*-triazine derivatives with conidia of *Neurospora sitophila.* Contrib. Boyce Thompson Inst. 18:429–462.

Byrd, R.J.W., and D. Woodcock. 1957. Fungal detoxication. 2. The metabolism of some phenoxy-*n*-alkylcarboxylic acids by *Aspergillus niger.* Biochem. J. 65:682–686.

Castro, C.E., R.S. Wade, and N.O. Belser. 1985. Biodehalogenation: Reactions of cytochrome P-450 with polyhalomethanes. Biochemistry 24:204–210.

Chacko, C.I., and J.L. Lockwood. 1967. Accumulation of DDT and dieldrin by microorganisms. Can. J. Microbiol. 13:1123–1126.

Chakrabarty, A.M. 1976. Plasmids in pseudomonas. Ann. Rev. Genet. 10:7–30.

Chapman, P.J. 1972. An outline of reaction sequences used for the bacterial degradation of phenolic compounds. p. 17–55. *In* Degradation of synthetic organic molecules in the biosphere. Natl. Acad. Sci., Washington, DC.

Chatterjee, D.K., and A.M. Chakrabarty. 1981. Plasmids in the biodegradation of PCBs and chlorobenzoates. p. 213–219. *In* T. Leisinger et al. (ed.) Microbial degradation of xenobiotics and recalcitrant compounds. Academic Press, London.

Chatterjee, D.K., J.J. Kilbane, and A.M. Chakrabarty. 1982. Biodegradation of 2,4,5-trichlorophenoxyacetic acid in soil by a pure culture of *Pseudomonas cepacia.* Appl. Environ. Microbiol. 44:514–516.

Chibata, I., and T. Tosa. 1979. Transformation of organic compounds by immobilized microbial cells. Adv. Appl. Microbiol. 22:1–27.

Chin, W.T., G.M. Stone, and A.E. Smith. 1970. Degradation of carboxin (Vitavax) in water and soil. J. Agric. Food Chem. 18:731–732.

Clark, C.G., and S.J.L. Wright. 1970. Detoxication of isopropyl *N*-phenylcarbamate (IPC) and isopropyl *N*-3-chlorophenylcarbamate (CIPC) in soil, and isolation of IPC-metabolizing bacteria. Soil Biol. Biochem. 2:19–26.

Cohen, S.N. 1980. The manipulation of genes. p. 321–330. *In* Molecules to living cells. Scientific American, W.H. Freeman & Co., San Francisco.

Coley, G., and C.N. Stutz. 1966. Parathion waste treatment and other organics. J. Water Pollut. Control Fed. 38:1345–1349.

Cook, A.M., and R. Hütter. 1982. Ametryne and prometryne as sulfur sources for bacteria. Appl. Environ. Microbiol. 43:781–786.

Cripps, R.E., and T.R. Roberts. 1978. Microbial degradation of herbicides. p. 669. *In* I.R. Hill and S.J.L. Wright (ed.) Pesticide microbiology. Academic Press, London.

Crosby, D.G. 1976. Nonbiological degradation of herbicides in the soil. p. 65. *In* L.J. Audus (ed.) Herbicides, physiology, biochemistry, ecology. Vol. 2. 2nd ed. Academic Press, London.

Cserjsei, A.J., and E.L. Johnson. 1972. Methylation of pentachlorophenol by *Trichoderma virgatum*. Can. J. Microbiol. 18:45–49.

Dagley, S. 1971. Catabolism of aromatic compounds by microorganisms. Adv. Microb. Physiol. 6:1–46.

Daughton, C.G., and D.P.H. Hsieh. 1977. Accelerated parathion degradation in soil by inoculation with parathion-utilizing bacteria. Bull. Environ. Contam. Toxicol. 18:48–56.

DeKoning, H.W., and D.C. Mortimer. 1971. DDT uptake and growth of *Euglena gracilis*. Bull. Environ. Contam. Toxicol. 6:244–248.

Derbyshire, M.K., J.S. Karns, P.C. Kearney, and J.O. Nelson. 1987. Purification and characterization of an *N*-methylcarbamate pesticide hydrolyzing enzyme. J. Agric. Food Chem. 35:871–877.

Dias, F.F., and M. Alexander. 1971. Effect of chemical structure on the biodegradability of aliphatic acids and alcohols. Appl. Microbiol. 22:114–118.

Don, R.H., and J.M. Pemberton. 1981. Properties of 6 pesticide degradation plasmids isolated from *Alcaligenes paradoxus* and *Alcaligenes eutrophus*. J. Bacteriol. 145:681–686.

Dunn, N.W., and I.C. Gunsalus. 1973. Transmissible plasmid coding early enzymes of naphthalene oxidation in *Pseudomonas putida*. J. Bacteriol. 114:974–979.

Eaton, R.W., and K.N. Timmis. 1984. Genetics of xenobiotic degradation. p. 694–703. *In* M.J. Klug and C.A. Reddy (ed.) Current perspectives in microbial ecology. Proc. of the 3rd Int. Symp. on Microbial Ecol. Michigan State Univ., East Lansing. 7–12 Aug. 1983. Am. Soc. Microbiol., Washington, DC.

Eberspächer, J., and F. Lingens. 1978. Reinigung und Eigenschaften von zwei Chloridazondihydrodiol-Dehydrogenasen aus Chloridazon-Chlor abbauenden Bakterien. Hoppe-Seyler's Z. Physiol. Chem. 359:1323–1334.

Edgehill, R.U., and R.K. Finn. 1983. Microbial treatment of soil to remove pentachlorophenol. Appl. Environ. Microbiol. 45:1122–1125.

Engelhardt, G., H.G. Rast, and P.R. Wallnöfer. 1977. Bacterial metabolism of substituted phenols. Oxidation of 4-(methylmercapto)- and 4-(methylsulfinyl)-phenol by *Norcardia* spec. DSM 43251. Arch. Microbiol. 114:25–33.

Engelhardt, G., P.R. Wallnöfer, and R. Plapp. 1971. Degradation of linuron and some other herbicides and fungicides by a linuron-inducible enzyme obtained from *Bacillus sphaericus*. Appl. Microbiol. 22:284–288.

Engelhardt, G., P.R. Wallnöfer, and R. Plapp. 1973. Purification and properties of an aryl acylamidase of *Bacillus sphaericus,* catalyzing the hydrolysis of various phenylamide herbicides and fungicides. Appl. Microbiol. 26:709–710.

Engelhardt, G., W. Ziegler, P.R. Wallnöfer, H.J. Jarczyk, and L. Oehlmann. 1982. Degradation of the triazinone herbicide metamitron by *Arthrobacter* sp. DSM 20389. J. Agric. Food Chem. 30:278–282.

Etzel, J.E., and E.J. Kirsch. 1975. Biological treatment of contrived and industrial wastewater containing pentachlorophenol. Dev. Ind. Microbiol. 16:287–295.

Fennewald, M., S. Benson, M. Oppici, and J. Shapiro. 1979. Insertion element analysis and mapping at the *Pseudomonas* plasmid *alk* regulon. J. Bacteriol. 139:940–952.

Finn, R.K. 1983. Use of specialized microbial strains in the treatment of industrial waste and in soil decontamination. Experientia 39:1231–1236.

Fisher, P.R., J. Appleton, and J.M. Pemberton. 1978. Isolation and characterization of the pesticide-degrading plasmid pJP1 from *Alcaligenes paradoxus*. J. Bacteriol. 135:798–804.

Flaig, W., H. Beutelspacher, and E. Rietz. 1975. Chemical composition and physical properties of humic substances p. 1–211. *In* J.E. Gieseking (ed.) Soil components. Vol. 1. Organic compounds. Springer-Verlag New York, New York.

Fox, J.L. 1983. Soil microbes pose problems for pesticides. Science 221:1029–1031.

Franklin, F.C.H., M. Bagdasarian, M.M. Bagdasarian, and K.N. Timmis. 1981. Molecular and functional analysis of the TOL plasmid pWWO from *Pseudomonas putida* and cloning of genes for the entire regulated aromatic ring *meta* cleavage pathway. Proc. Natl. Acad. Sci. USA. 78:7458–7462.

Frear, D.S. 1975. The benzoic acid herbicides. p. 541. *In* P.C. Kearney and D.D. Kaufman (ed.) Herbicides: Chemistry degradation and mode of action. Marcel Dekker, New York.

Friello, D.A., J.R. Mylroie, and A.M. Chakrabarty. 1976. Use of genetically engineered multiplasmid microorganisms for rapid degradation of fuel hydrocarbons. *In* Biodeterioration of materials. Vol. 3. Proc. of the 3rd Int. Biodegradation Symp., Applied Sci. Publ., London.

Gamar, Y., and J.K. Gaunt. 1971. Bacterial metabolism of 4-chloro-2-methylphenoxyacetate, formation of glyoxylate by side-chain cleavage. Biochem. J. 122:527–531.

Getzin, L.W., and I. Rosefield. 1971. Partial purification and properties of a soil enzyme that degrades the insecticide malathion. Biochim. Biophys. Acta 235:442–453.

Getzin, L.W., and T. Satyanarayana. 1979. Isolation of an enzyme from soil that degrades the organophosphorus insecticide crotoxyphos. Arch. Environ. Contam. Toxicol. 8:661–672.

Ghosal, D., I.-S. You, D.K. Chatterjee, and A.M. Chakrabarty. 1985. Microbial degradation of halogenated compounds. Science 228:135–142.

Giardina, M.C., M.T. Giardi, and G. Filacchioni. 1980. 4-Amino-2-chloro-1,3,5-triazine: A new metabolite of atrazine by a soil bacterium. Agric. Biol. Chem. 44:2067–2072.

Giardina, M.C., M.T. Giardi, and G. Filacchioni. 1982. Atrazine metabolism by Nocardia: Elucidation of initial pathway and synthesis of potential metabolites. Agric. Biol. Chem. 46:1439–1445.

Gibson, D.T., and V. Subramarian. 1984. Microbial degradation of aromatic compounds. p. 181–252. *In* D.T. Gibson (ed.) Microbial degradation of organic compounds. Marcel Dekker, New York.

Gibson, D.T., J.R. Koch, C.L. Schud, and R.E. Kallio. 1968. Oxidative degradation of aromatic hydrocarbons by microorganisms. II. Metabolism of halogenated aromatic hydrocarbons. Biochemistry 7:3795–3802.

Goldstein, R.M., L.M. Mallory, and M. Alexander. 1985. Reasons for possible failure of inoculation to enhance biodegradation. Appl. Environ. Microbiol. 50:977–983.

Golovleva, L.A., R.N. Pertsova, A.M. Boronin, V.G. Baskunov, and A.V. Polyvakov. 1982. Degradation of polychloro aromatic insecticides and their derivatives by *Pseudomonas aeruginosa* strains containing biodegradation plasmids. Mikrobiologiya 51:973–978.

Golovleva, L.A., B.P. Baskunov, Z.I. Finkel'stein, and M. Yu. Nefedova. 1983. Microbial degradation of the organophsophorous insecticide fosalone. Biol. Bull. Acad. Sci. USSR 10:44–51.

Haider, K., J.P. Martin, and Z. Filip. 1975. Humus biochemistry. Vol. 4. p. 195–244. *In* E.A. Paul and A.D. McLaren (ed.) Soil biochemistry. Marcel Dekker, New York.

Hammond, M.W., and M. Alexander. 1972. Effect of chemical structure on microbial degradation of methyl-substituted aliphatic acids. Environ. Sci. Technol. 6:732–735.

Hardman, D.J., and J.H. Slater. 1981. Dehydrogenase in soil bacteria. J. Gen. Microbiol. 123:117–128.

Heritage, A.D., and I.C. MacRae. 1977. Degradation of lindane by cell-free preparations of *Clostridium sphenoides*. Appl. Environ. Microbiol. 34:222–224.

Hoff, T., S.-Y. Liu, and J.-M. Bollag. 1985. Transformation of halogen-, alkyl- and alkoxy-substituted anilines by a laccase of *Trametes versicolor*. Appl. Environ. Microbiol. 49:1040–1045.

Honeycutt, R., L. Ballantine, H. LeBaron, D. Paulson, V. Seim, C. Ganz, and G. Milad. 1984. Degradation of high concentrations of a phosphorothioic ester by hydrolase. p. 343–352. *In* R.F. Krueger and J.N. Seiber (ed.) Treatment and disposal of pesticide wastes. Am. Chem. Soc., Washington, DC.

Horvath, R.S. 1970a. Microbial cometabolism of 2,4,5-trichlorophenoxyacetic acid. Bull. Environ. Contam. Toxicol. 5:537–541.

Horvath, R.S. 1970b. Co-metabolism of methyl- and chloro-substituted catechols by an *Achromobacter* sp. possessing a new *meta*-cleaving oxygenase. Biochem. J. 119:871–876.

Horvath, R.S. 1971. Cometabolism of the herbicide 2,3,6-trichlorobenzoate. J. Agric. Food Chem. 19:291–293.

Howe, R.H.L. 1969. Toxic wastes degradation and disposal. Process Biochem. 4:25–28.

Hsu, T.-S., and R. Bartha. 1974. Interaction of pesticide-derived chloroaniline residues with soil organic matter. Soil Sci. 116:444–452.

Ivie, G.W., and J.E. Casida. 1971. Sensitized photodecomposition and photosensitizer activity of pesticide chemicals exposed to sunlight on silica gel chromatoplates. J. Agric. Food Chem. 19:405–409.

Johnson, B.T., and J.O. Kennedy. 1973. Biomagnification of *p,p'*-DDT and methoxychlor by bacteria. Appl. Microbiol. 26:66–71.

Johnson, L.M., and H.W. Talbot. 1983. Detoxification of pesticides by microbial enzymes. Experientia 39:1236–1246.

Kaufman, D.D., Y. Katan, D.F. Edwards, and E.G. Jordan. 1982. Microbial adaptation and metabolism of pesticides. p. 437–451. *In* J.L. Hilton (ed.) Agricultural chemicals of the future. Rowman and Allenheld, Totowa, NJ.

Kaufman, D.D., J.R. Plimmer, J. Iwan, and U.I. Klingebiel. 1972. 3,3', 4,4'-Tetrachloroazoxybenzene from 3,4-dichloroaniline in microbial cultures. J. Agric. Food Chem. 20:916–919.

Kaufman, D.D., J.R. Plimmer, and U.I. Klingebiel. 1973. Microbial oxidation of 4-chloroaniline. J. Agric. Food Chem. 21:127–132.

Kawasaki, H., H. Yahara, and K. Tonomura. 1984. Cloning and expression in *Escherichia coli* of the haloacetate dehalogenase genes from *Moraxella* plasmid pU01. Agric. Biol. Chem. 48:2627–2632.

Kearney, P.C. 1965. Purification and properties of an enzyme responsible for hydrolyzing phenylcarbamates. J. Agric. Food Chem. 13:561–564.

Kearney, P.C., J.S. Karns, and W.W. Mulbry. 1986. Biochemical and genetic aspects of pesticide microbial metabolism. Am. Chem. Soc. 192nd ACS Natl. Meet., Anaheim, CA. 7–12 Sept. Abstr. (AEQI). Am. Chem. Soc., Washington, DC.

Kearney, P.C., D.D. Kaufman, and M.L. Beall. 1964. Enzymatic dehalogenation of 2,2-dichloropropionate. Biochem. Biophys. Res. Commun. 14:29–33.

Kearney, P.C., and S.T. Kellogg. 1985. Microbial adaptation to pesticides. Pure Appl. Chem. 57:389–403.

Kearney, P.C., E.A. Woolson, J.R. Plimmer, and A.R. Isensee. 1969. Decontamination of pesticides in soil. Residue Rev. 29:137–149.

Kellogg, S.T., D.K. Chatterjee, and A.M. Chakrabarty. 1981. Plasmid-assisted molecular breeding: New technique for enhanced biodegradation of persistent toxic chemicals. Science 212:1133–1135.

Kikuchi, R., T. Yasutaniya, Y. Takimoto, H. Yamada, and J. Miyamoto. 1984. Accumulation and metabolism of fenitrothion in three species of algae. J. Pestic. Sci. 9:331–337.

Kilbane, J.J., D.K. Chatterjee, J.S. Karns, S.T. Kellogg, and A.M. Chakrabarty. 1982. Biodegradation of 2,4,5-trichlorophenoxyacetic acid by pure culture of *Pseudomonas cepacia.* Appl. Environ. Microbiol. 44:72–78.

Kilbane, J.J., D.K. Chatterjee, and A.M. Chakrabarty. 1983. Detoxification of 2,4,5-trichlorophenoxyacetic acid from contaminated soil by *Pseudomonas cepacia.* Appl. Environ. Microbiol. 45:1697–1700.

Ko, W.H., and J.L. Lockwood. 1968. Accumulation and concentration of chlorinated hydrocarbon pesticides by microorganisms in soil. Can. J. Microbiol. 14:1075–1078.

Kobayashi, M., and B.E. Rittman. 1982. Microbial removal of hazardous organic compounds. Environ. Sci. Technol. 16:107A–183A.

Korte, F., G. Ludwig, and J. Vogel. 1962. Umwandlung von Aldrin-[$^{14}C$] und Dieldrin-[$^{14}C$] durch Mikroorganismen, Leberhomogenate und Moskito-Larven. Liebigs Ann. Chem. 656:135–140.

Krause, A., W.G. Hancock, R.D. Minard, A.J. Freyer, R.C. Honeycutt, H.M. LeBaron, D.L. Paulson, S.-Y. Liu, and J.-M. Bollag. 1985. Microbial transformation of the herbicide metolachlor by a soil actinomycete. J. Agric. Food Chem. 33:584–589.

Kreis, M., J. Eberspächer, and F. Lingens. 1981. Detection and characterization of plasmids in chloridazon and antipyrin degrading bacteria. Zbl. Bakt. Hyg. I. Abstr. Orig. C:2 45–60.

Kruglov, Yu. V., and L.N. Paromenskaya. 1970. Detoxication of simazine by microscopic algae. Microbiology (Mikrobiologiya) 39:139–142.

Kunc, F., and J. Rybarova. 1984. Mineralization of 2,4-dichlorophenoxyacetic acid in soil previously enriched with organic substrates. Folia Microbiol. 29:156–161.

Kunc, F., J. Rybarova, and J. Lasik. 1984. Mineralization of 2,4-dichlorophenoxyacetic acid in soil simultaneously enriched with saccharides. Folia Microbiol. 29:148–155.

Kunz, D.A., and P.J. Chapman. 1981. Isolation and characterization of spontaneously occurring TOL plasmid mutants of *Pseudomonas* HS1. J. Bacteriol. 146:952–964.

Kurihara, N., N. Ohisa, M. Nakajima, T. Kakutani, and M. Senda. 1981. Relationship between microbial degradability and polarographic half-wave potential of polychlorocyclohexenes and BHC isomers. Agric. Biol. Chem. 45:1229–1235.

Lal, R., and D.M. Saxena. 1982. Accumulation, metabolism, and effects of organochlorine insecticides on microorganisms. Microbiol. Rev. 46:95–127.

Lanzilotta, R.P., and D. Pramer. 1970. Herbicide transformation. II. Studies with an acylamidase of *Fusarium solani*. Appl. Microbiol. 19:307–313.

Layh, G., J. Eberspächer, and F. Lingens. 1982. Rodanese in chloridazon-degrading bacteria. FEMS Microbiol. Lett. 15:23–26.

Leshniowsky, W.O., P.R. Dugan, R.M. Pfister, J.I. Frea, and C.I. Randles. 1970. Aldrin: Removal from lake water by flocculent bacteria. Science 169:993–995.

Linke, H.A.B. 1970. 3,3′,4′-Trichloro-4-(3,4-dichloroanilino)-azobenzol, ein Abbauprodukt des Herbizides Propanil im Boden. Naturwissenschaften 57:307–308.

Liu, S.-Y., and J.-M. Bollag. 1971. Metabolism of carbaryl by a soil fungus. J. Agric. Food Chem. 19:487–490.

Liu, S.-Y., R.D. Minard, and J.-M. Bollag. 1981. Coupling reaction of 2,4-dichlorophenol with various anilines. J. Agric. Food Chem. 29:253–257.

Loos, M.A. 1975. Phenoxyalkanoic acids. p. 1–128. *In* P.C. Kearney and D.D. Kaufman (ed.) Herbicides: Chemistry, degradation and mode of action. Vol. 1. Marcel Dekker, New York.

Loos, M.A., J.-M. Bollag, and M. Alexander. 1967. Phenoxyacetate herbicide detoxication by bacterial enzymes. J. Agric. Food Chem. 15:858–860.

Lyr, H., G. Ritter, and L. Banasiak. 1974. Detoxication of carboxin. Z. Allg. Mikrobiol. 14:313–320.

Martin, J.P., and K. Haider. 1971. Microbial activity in relation to soil humus formation. Soil Sci. 111:54–63.

Mathur, S.P., and H.V. Morley. 1975. Incorporation of methoxychlor $^{14}C$ in model humic acids prepared from hydroquinone. Bull. Environ. Contam. Toxicol. 20:268–273.

Matsumura, F., and H.J. Benezet. 1978. Microbial degradation of insecticides. p. 648–667. *In* I.R. Hill and S.J.L. Wright (ed.) Pesticide microbiology. Academic Press, London.

Matsumura, F., and G.M. Boush. 1966. Malathion degradation by *Trichoderma viride* and a *Pseudomonas* species. Science 153:1278–1280.

Matsumura, F., and E.G. Esaac. 1979. Degradation of pesticides by algae and aquatic microorganisms. p. 371–387. *In* Pesticide and xenobiotic metabolism in aquatic organisms. ACS Symp. Ser. 99. Am. Chem. Soc., Washington, DC.

McClure, G.W. 1972. Degradation of phenylcarbamates in soil by mixed suspension of IPC-adapted microorganisms. J. Environ. Qual. 1:177–180.

Mikesell, M.D., and S.A. Boyd. 1985. Reductive dechlorination of the pesticides 2,4-D, 2,4,5-T and pentachlorophenol in anaerobic sludges. J. Environ. Qual. 14:337–340.

Miles, J.R.W., C.M. Tu, and C.R. Harris. 1969. Metabolism of heptachlor and its degradation products by soil microorganisms. J. Econ. Entomol. 62:1334.

Mills, R.E. 1959. Development of design criteria for biological treatment of an industrial effluent containing 2,4-D wastewater. p. 340–358. *In* Eng. Ext. Seminar Ser. No. 104, Proc. 14th Indiana Waste Conf., 5–7 May. Purdue Univ., Lafayette, IN.

Minard, R.D., S.-Y. Liu, and J.-M. Bollag. 1981. Oligomers and quinones from 2,4-dichlorophenol. J. Agric. Food Chem. 29:250–253.

Minard, R.D., S. Russel, and J.-M. Bollag. 1977. Chemical transformation of 4-chloroaniline to a triazene in a bacterial culture medium. J. Agric. Food Chem. 25:841–844.

Miyamoto, J., K. Kitagawa, and Y. Sato. 1966. Metabolism of organophosphorus insecticides by *Bacillus subtilis* with special emphasis on sumithion. Jpn. J. Exp. Med. 36:211–225.

Miyazaki, S., G.M. Bousch, and F. Matsumura. 1969. Metabolism of $^{14}C$-chlorobenzilate and $^{14}C$-chloropropylate by *Rhodotorula gracilis*. Appl. Microbiol. 18:972–976.

Munnecke, D.M. 1976. Enzymatic hydrolysis of organophosphate insecticides, a possible pesticide disposal method. Appl. Environ. Microbiol. 32:7–13.

Munnecke, D.M. 1979. Hydrolysis of organophosphate insecticides by an immobilized-enzyme system. Biotechnol. Bioeng. 21:2247–2261.

Munnecke, D.M. 1980. Enzymatic detoxification of waste organophosphate pesticides. J. Agric. Food. Chem. 28:105–111.

Munnecke, D.M., L.M. Johnson, H.W. Talbot, and S. Barik. 1982. Microbial metabolism and enzymology of selected pesticides. p. 1–32. *In* A.M. Chakrabarty (ed.) Biodegradation and detoxification of environmental pollutants. CRC Press, Boca Raton, FL.

Neudorf, S., and M.A.Q. Khan. 1975. Pick-up and metabolism of DDT, dieldrin and photodieldrin by a fresh water alga (*Ankistrodesmus amalloides*) and a microcrustacean (*Daphnia pulex*). Bull. Environ. Contam. Toxicol. 13:443–450.

Ohisa, N., M. Yamaguchi, and N. Kurihara. 1980. Lindane degradation by cell-free extracts of *Clostridium rectum*. Arch. Microbiol. 125:221–225.

Orpin, C.G., M. Knight, and W.C. Evans. 1972. The bacterial oxidation of *N*-methylisonicotinate, a photolytic product of paraquat. Biochem. J. 127:833–844.

Paris, D.F., and D.L. Lewis. 1976. Accumulation of methoxychlor by microorganisms isolated from aqueous systems. Bull. Environ. Contam. Toxicol. 15:24–32.

Parr, J.F., and S. Smith. 1976. Degradation of toxaphene in selected anaerobic soil environments. Soil Sci. 121:52–57.

Pemberton, J.M. 1979. A biological answer to environmental pollution by phenoxyherbicides. AMBIO 8:202–205.

Pemberton, J.M. 1981. Genetic engineering and biological detoxification of environmental pollutants. Residue Rev. 78:1–11.

Pemberton, J.M., and P.R. Fisher. 1977. 2,4-D plasmids and persistence. Nature (London) 268:732–733.

Pertsova, R.N., F. Kunc, and L.A. Golovleva. 1984. Degradation of 3-chlorobenzoate in soil by *Pseudomonas* carrying biodegradative plasmids. Folia Microbiol. 29:242–247.

Pfaender, F.K., and M. Alexander. 1972. Extensive microbial degradation of DDT in vitro and DDT metabolism by natural communities. J. Agric. Food Chem. 20:842–846.

Pierce, G.E., T.J. Facklam, and J.M. Rice. 1981. Isolation and characterization of plasmids from environmental strains of bacteria capable of degrading the herbicide 2,4-D. Dev. Ind. Microbiol. 22:401–408.

Pierce, G.E., J.B. Robinson, T.J. Facklam, and J.M. Rice. 1982. Physiological and genetic comparison of environmental strains of *Pseudomonas* capable of degrading the herbicide 2,4-D. Dev. Ind. Microbiol. 23:407–417.

Plimmer, J.R., P.C. Kearney, H. Chisaka, J.B. Yount, and U.I. Klingebiel. 1970. 1,3-Bis(3,4-dichlorophenyl) triazine from propanil in soils. J. Agric. Food Chem. 18:859–861.

Plimmer, J.R., and U.I. Klingebiel. 1971. Riboflavin photosensitized oxidation of 2,4-dichlorophenol: An assessment of possible chlorinated dioxin formation. Science 174:407–408.

Pore, R.S., and W.G. Sorenson. 1981. Kepone removal from aqueous solution by immobilized algae. J. Environ. Sci. Health A16:51–63.

Quensen, J.F., and F. Matsumura. 1984. Transfer of degradative capabilities *Bacillus megaterium* to *Bacillus subtilis* by plasmid transfer techniques. p. 327–341. *In* R.T. Krueger and J.N. Seiber (ed.) Treatment and disposal of pesticide wastes. ACS Symp. Ser. 259. Am. Chem. Soc., Washington, DC.

Rajaram, K.P., and N. Sethunathan. 1975. Effect of organic sources on the degradation of parathion in flooded soil. Soil Sci. 119:296–300.

Rast, H.G., G. Engelhardt, P.R. Wallnöfer, L. Oehlmann, and K. Wagner. 1979. Bacterial metabolism of substituted phenols. Oxidation of 3-methyl-4-(methylthio)phenol by *Nocardia* sp. DSM 43251. J. Agric. Food Chem. 27:699–702.

Reineke, W., and H.-J. Knackmuss. 1979. Construction of haloaromatics utilizing bacteria. Nature (London) 277:385–386.

Reineke, W., and H.-J. Knackmuss. 1980. Hybrid pathway for chlorobenzoate metabolism in *Pseudomonas* sp. B13 derivatives. J. Bacteriol. 142:467–473.

Rheinwald, J.G., A.M. Chakrabarty, and I.C. Gunsalus. 1973. A transmissible plasmid controlling camphor oxidation in *Pseudomonas putida*. Proc. Natl. Acad. Sci. USA 70:885–889.

Rochkind-Dubinsky, M.L., G.S. Sayler, and J.W. Blackburn. 1987. Microbiological decomposition of chlorinated aromatic compounds. Marcel Dekker, New York.

Rosen, J.D., M. Siewierski, and G. Winnett. 1970. FMN-Sensitized photolyses of chloroanilines. J. Agric. Food Chem. 18:494–496.

Russel, S., and J.-M. Bollag. 1977. Formylation and acetylation of 4-chloroaniline by a *Streptomyces* sp. Acta Microbiol. Pol. 26:59–64.

Salkinoja-Salonen, M.S., R. Hakulinen, R. Valo, and J. Apajalahti. 1983. Biodegradation of recalcitrant organochlorine compounds in fixed film reactors. Water Sci. Technol. 15:309–319.

Salkinoja-Salonen, M.S., E. Vaisanen, and A. Paterson. 1979. Involvement of plasmids in the bacterial degradation of lignin-derived compounds. p. 301–314. *In* K.N. Timmis and A. Pühler (ed.) Plasmids of medical, environmental and commercial importance. Elsevier, Amsterdam.

Sauber, K., C. Fröhner, G. Rosenberg, J. Eberspächer, and F. Lingens. 1977. Purification and properties of pyrazon dioxygenase from pyrazon-degrading bacteria. Eur. J. Biochem. 74:89–97.

Schwien, U., and E. Schmidt. 1982. Improved degradation of monochlorophenols by a constructed strain. Appl. Environ. Microbiol. 44:33–39.

Serdar, C.M., D.T. Gibson, D.M. Munnecke, and J.H. Lancaster. 1982. Plasmid involvement in parathion hydrolysis by *Pseudomonas diminuta.* Appl. Environ. Microbiol. 44:246–249.

Serdar, C.M., and D.T. Gibson. 1985. Cloning and expression of the parathion hydrolase gene from plasmid pCMS1. p. 199 (K169). *In* Abstr. Am. Soc. Microbiol., Washington, DC.

Sethunathan, N. 1973. Degradation of parathion in flooded acid soils. J. Agric. Food Chem. 21:602–604.

Sharabi, N.E., and L.M. Bordeleau. 1969. Biochemical decomposition of the herbicide *N*-(3,4-dichlorophenyl)-2-methylpentanamide and related compounds. Appl. Microbiol. 18:369–375.

Shuttleworth, K.L., and J.-M. Bollag. 1986. Soluble and immobilized laccase as catalysts for the transformation of substituted phenols. Enzyme Microbiol. 44:246–249.

Sjoblad, R.D., and J.-M. Bollag. 1981. Oxidative coupling of aromatic compounds by enzymes from soil microorganisms. p. 113–152. *In* E.A. Paul and J.N. Ladd (ed.) Soil biochemistry. Vol. 5. Marcel Dekker, New York.

Slater, J.H., and A.T. Bull. 1982. Environmental microbiology: Biodegradation. Philos. Trans. R. Soc. London, B 297:575–597.

Sodergren, A. 1968. Uptake and accumulation of $^{14}$C-DDT by *Chlorella* sp. (Chlorophyceae). Oikos 19:126–138.

Sodergren, A. 1971. Accumulation and distribution of chlorinated hydrocarbons in cultures of *Chlorella pyrenvidosa* (Chlorophyceae). Oikos 22:215–220.

Stainer, R.Y., and L.N. Ornston. 1973. The $\beta$-ketoadipate pathway. Adv. Microb. Physiol. 9:89–151.

Stevenson, F.J. 1972. Organic matter reactions involving herbicides in soil. J. Environ. Qual. 1:333–343.

Taylor, H.F., and R.L. Wain. 1962. Side-chain degradation of certain $\omega$-phenoxyalkane carboxylic acids by *Nocardia coeliaca* and other microorganisms isolated from soil. Proc. R. Soc. London, B 156:172–186.

Tiedje, J.M., and M. Alexander. 1969. Enzymatic cleavage of the ether bond of 2,4-dichlorophenoxyacetate. J. Agric. Food Chem. 17:1080.

Tiedje, J.M., S.A. Boyd, and B.Z. Fathepure. 1987. Anaerobic degradation of chlorinated aromatic hydrocarbons. Dev. Ind. Microbiol. 27:117–127.

Timms, P., and I.C. MacRae. 1982. Conversion of fensulfothion by *Klebsiella pneumoniae* to fensulfothion sulfide and its accumulation. Aust. J. Biol. Sci. 35:661–668.

Timms, P., and I.C. MacRae. 1983. Reduction of fensulfothion and accumulation of the products fensulfothion sulfide by selected microbes. Bull. Environ. Contam. Toxicol. 31:112–115.

Tranter, E.K., and R.B. Cain. 1967. The degradation of fluoro aromatic compounds to fluorocitrate and fluoroacetate by bacteria. Biochem. J. 103:22p--23p.

Tweedy, B.G., C. Loeppky, and J.A. Ross. 1970. Metobromuron: Acetylation of the aniline moiety as a detoxification mechanism. Science 168:482–483.

Van Alfen, N.K., and T. Kosuge. 1976. Metabolism of the fungicide 2,6-dichloro-4-nitroaniline in soil. J. Agric. Food Chem. 25:584–588.

Voerman, S., and R.M.C. Tammes. 1969. Absorption and desorption of lindane and dieldrin by yeast. Bull. Environ. Contam. Toxicol. 4:271–277.

Wallnöfer, P.R., and J. Bader. 1970. Degradation of urea herbicides by cell-free extracts of *Bacillus sphaericus.* Appl. Microbiol. 19:714–717.

Wang, Y.-S., R.V. Subba-Rao, and M. Alexander. 1984. Effect of substrate concentration and organic and inorganic compounds on the occurrence and rate of mineralization and cometabolism. Appl. Environ. Microbiol. 47:1195–1200.

Ware, G.W., M.K. Dee, and W.P. Cahill. 1968. Water florae as indicators of irrigation water contamination by DDT. Bull. Environ. Contam. Toxicol. 3:333–338.

Webley, D.M., R.B. Duff, and V.C. Farmer. 1958. The influence of chemical structure on $\beta$-oxidation by soil Nocardias. J. Gen. Microbiol. 18:733–746.

Wedemeyer, G. 1966. Dechlorination of DDT by *Aerobacter aerogenes.* Science 152:647.

Wedemeyer, G. 1967. Dechlorination of 1,1,1-trichloro-2,2-bis (*p*-chlorophenyl) ethane by *Aerobacter aerogenes.* Appl. Microbiol. 15:569–574.

Williams, P.A. 1981. Genetics of biodegradation. p. 97–107. *In* T. Leisinger et al. (ed.) Microbial degradation of xenobiotics and recalcitrant compounds. Academic Press, London.

Williams, P.A., and M.J. Worsey. 1976. Uniquity of plasmids in coding for toluene and xylene metabolism in soil bacteria: Evidence for existence of new TOL plasmids. J. Bacteriol 125:818–828.

Worsey, M.J., and P.A. Williams. 1975. Metabolism of toluene and xylene by *Pseudomonas putida( arvilla)* mt-2: Evidence for a new function of the TOL plasmid. J. Bacteriol. 124:7–13.

Wright, K.A., and R.B. Cain. 1972. Microbial metabolism of pyridinium compounds, metabolism of 4-carboxy-1-methylpyridinium chloride, a photolytic product of paraquat. Biochem. J. 128:543–559.

Wright, K.A., and R.B. Cain. 1972. Microbial metabolism of pyridinium compounds. Radioisotope studies of the metabolic fate of 4-carboxy-1-mehylpyridinium chloride. Biochem. J. 128:561–568.

You, I.S., and R. Bartha. 1982. Stimulation of 3,4-dichloroaniline mineralization by aniline. Appl. Environ. Microbiol. 44:678–681.

Zech, R., and K.D. Wigand. 1975. Organophsophate-detoxication enzymes in *E. coli.* Gel filtration and isoelectric focusing of DFPase, paraoxonase and unspecific phosphohydrolases. Experientia 31:157–158.

Zeyer, J., and P.C. Kearney. 1983. Microbial dealkylation of trifluralin in pure culture. Pestic. Biochem. Physiol. 20:10–18.

# Chapter 7

# Volatilization and Vapor Transport Processes[1]

**A. W. TAYLOR,** *University of Maryland, College Park, Maryland*

**W. F. SPENCER,** *USDA-ARS, Riverside, California*

Volatilization is the dominant process that controls the dispersal of many pesticides into the general environment as well as the length of their effective lifetime in the target area. The factors that influence volatilization are, therefore, of major interest for understanding the role it plays in controlling the efficiency and environmental distribution and impact of pesticides.

In this chapter, we discuss the physics and chemistry of the volatilization process. The discussion is confined to processes acting at or close to the surfaces of the soil or plant leaves and in the atmosphere a few meters above them. The long-distance transport of pesticides in the atmosphere, including their chemical and photochemical stability in air and their rates of redeposition, touches upon more general aspects of environmental chemistry that will not be discussed here in detail.

Immediately after application, pesticides begin to disappear from the target area either by chemical degradation with the formation of new compounds or by physical removal by the action of air or water. Chemical degradation reflects the stability of the compound in the chemical or biochemical environment existing on the soil or plant surface. Degradation reactions can be predicted from data obtained under controlled conditions in the laboratory or greenhouse because they are sensitive to environmental factors such as temperature, light intensity, or humidity. When such data are used to predict the rate of disappearance of the residues in the field, it is often found that physical losses must also be taken into account. The rate of physical loss is often faster than chemical degradation.

Rates of residue removal by air or water are difficult to predict because they are highly dependent upon rainfall, weather patterns, the soil moisture regime, and microclimatic conditions at the soil or crop surface. Losses of residues in water by surface runoff, movement with eroding soil or downward leaching are discussed elsewhere. It may be noted that, except in special circumstances, the total seasonal losses in runoff rarely exceed 5 or 10% of

[1]University of Maryland Agric. Exp. Stn. Misc. Publ. no. A 4660.

the total applied (Wauchope, 1978; Weber et al., 1980); the fraction removed by leaching is probably much less. In contrast, volatilization losses of 80 to 90% have sometimes been measured within a few days after application, depending on weather conditions and management practices.

## 7-1 THE VOLATILIZATION PROCESS

The dispersal of residues into the atmosphere involves two distinct processes. The first is the evaporation of pesticide molecules into the air from the residues present on soil or plant surfaces. The second is the dispersion of the resulting vapor into the overlying atmosphere by diffusion and turbulent mixing. The two processes are fundamentally different in character and are controlled by different chemical and environmental factors. For the sake of clarity, it is convenient to consider them separately. This does not, however, mean that the two processes are wholly independent and, as will be clear in the later sections discussing experimental measurements and the modelling of volatilization, any general view must regard both processes together as an integrated whole.

In physicochemical terms, the first process represents a phase change into vapor from the liquid or solid state. This can be interpreted in terms of the thermodynamic activity or fugacity of the pesticide chemical. Understanding the factors affecting this change is central to any satisfactory interpretation of the volatilization process. These factors include the vapor pressure (or vapor density) of the active ingredient, the way in which this is affected by adsorption upon the soil or plant surfaces, and how this adsorption changes with the pesticide concentration and moisture status of the surfaces.

The second process, vapor dispersion into the atmosphere, is the same one that controls the transfer of water, $CO_2$, and other gases between soil and plant surfaces and the atmosphere. The principal difference between these gases and pesticides is that the concentration of pesticide vapor in the air more than 2 or 3 m aboveground is virtually zero. Without any appreciable background concentration there is no downward "return flow," and only upward dispersal is of concern.

Mathematically, the vertical flux of vapor, $\hat{I}F$, through any plane at height $z$ above the surface is described by the general equation

$$\hat{I}F = k_Z(dp/dz) \quad [1]$$

where $k_Z$ is the diffusivity coefficient and $dp/dz$ the gradient of vapor pressure at the height $z$. The flux $\hat{I}F$ has dimensions of mol $m^{-2}$ $t^{-1}$.

The flow of air over plant and soil surfaces is invariably turbulent except in a shallow layer with a depth about 1 mm in which the flow may be regarded as laminar and controlled by viscosity (Monteith, 1973). This region of laminar surface flow may be identified as a layer in which the dispersal of pesticide vapor can be described in terms of the molecular diffusion coefficient of the vapor. Outside this zone, where the flow in the overlying atmosphere is turbulent, dispersal must be described in terms of the *eddy diffusion* coefficient. The relationship between these regions is illustrated in Fig. 7-1. Expressing Eq. [1] in terms of the finite depth of the laminar surface layer, the flux through this layer is:

$$\hat{I}F = D(p_s - p_d)/\delta \qquad [2]$$

where $p_s$ and $p_d$ are the vapor pressures at the evaporating surface and at the upper surface of the laminar layer of effective depth $\delta$. This depth is equated with the thickness of the thin surface layer of streamline flow through which the drag or frictional force of the surface is imposed on the airstream-by Newtonian viscosity. Since there is no turbulent mixing in this layer, this means that the effective depth, $\delta$, is defined as the distance above the surface through which the vapor moves by molecular diffusion. The value of $D$ may then be identified with the molecular diffusion coefficient of the particular compound. As Hartley and Graham-Bryce (1980b) pointed out, Eq. [2] really only defines $\delta$ as an effective thickness that cannot actually be measured, but must be expected to vary with speed and surface roughness. The postulate of Newtonian flow implies that there is no single plane of wind shear, but a transition zone across which the flow becomes increasingly turbulent. Despite the usefulness of the interpretation summarized in Eq.

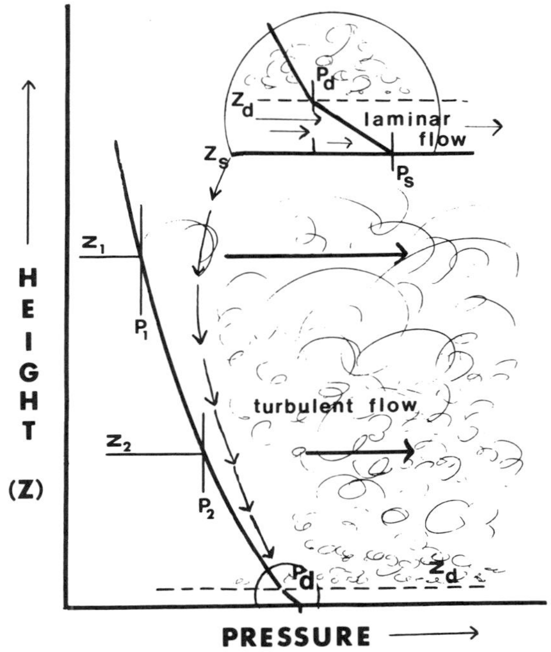


Fig. 7-1. Regions of laminar and turbulent flow in the lower atmosphere and the corresponding profiles of pesticide vapor pressure with height.

[2], it must be recognized that the effective depth is physically indeterminate and the extent to which the diffusivity coefficient in it can be identified with the molecular diffusion coefficient measured in still air is also ambiguous.

In an alternative view, the laminar layer can be regarded as the limiting distance above the surface to which the smallest eddies of the overlying turbulent flow can penetrate. In this picture, the effective depth must however be equally indeterminate because it must fluctuate in response to the pressure variations and varying drag exerted by the turbulence in the overlying flow.

In the turbulent atmosphere, the dispersal of vapor through any layer of finite thickness may be described by

$$\hat{I}F = k_Z(p_1 - p_2)/(z_1 - z_2) \quad [3]$$

where $k_Z$ is the mean eddy diffusivity coefficient between the heights $z_1$ and $z_2$. The depth of the turbulent zone, often described as the *atmospheric boundary layer,* is several orders of magnitude greater than that of the laminar flow layer, being measured in meters rather than the millimetric depths of the latter. The difference between the two diffusion coefficients is also large so that—except when the atmosphere is exceptionally stable—dispersal of the pesticide vapor in the turbulent zone outside the laminar layer is relatively rapid.

Despite these difficulties, the concept of separate regions of dilution by molecular diffusion and by turbulent mixing is of major importance in the theory of the exchange of gases at soil and plant surfaces. The concept is of particular value in the identification of the physical factors that impede the dispersion process under differing soil and microclimate conditions. As Hartley and Graham-Bryce (1980b) noted, Eq. [1] can be rewritten as $\hat{I}F = (f)[(p_1 - p_2/R]$, where $R$ is a resistance term such that $R = (z_1 - z_2)/k_Z$. Stated in this way, the resistance terms are additive and the appropriate forms of Eq. [2] and [3] may be regarded as describing the additive resistances (or impedances) to dispersive flow in the laminar and turbulent layers. If, as Mackay (1984) and Mackay and Paterson (1982) pointed out, the concentration or potential terms are converted to suitable units, the dispersion equations for all the environmental phases are additive and fluxes can be described in terms of similar potential gradients throughout the entire environmental system. This can be done by the use of the fugacity concept. Lewis (1901) originally postulated this concept, then later Mackay and Paterson (1981) reintroduced it as a fundamental principle in environmental chemistry. In using it to interpret the movement of pesticides in the environment, it is necessary to relate the vapor pressure (or vapor density) of the pesticide in the gas phase to the concentration in solution by the use of Henry's law and to characterize residues adsorbed on soil or plant surfaces by using adsorption isotherms stated in terms of vapor pressures. As will be shown later, this is particularly convenient in experimental work with volatile pesticides. The approach provides a powerful tool for relating volatilization to the more general movement and distribution of pesticide residues through the various phases of the general environment. For the purposes of the current discussion, fugacity will be used as a simplifying approach in which pesticide volatilization is regarded as a single flow path through a series of barriers each with a characteristic, but additive, resistance. Since much of the existing literature on volatilization is presented in terms of diffusivities, both approaches will be used interchangeably in this chapter.

### 7–1.1 Vapor Pressures of Pesticide Residues

Since the background vapor pressures of all pesticides in the free atmosphere are extremely low, the overall gradient of vapor pressure or vapor density gradient is established by the value of $p_s$, the vapor pressure of the residues on the plant or soil surface. This is equivalent to a statement that the rate of dispersion will be a function of the fugacity of the residues. This fugacity, which is equivalent to the vapor pressure at the inner surface of the laminar flow layer (Fig. 7–1), will only be that of the pure compound if (i) the residues are present as a continuous exposed layer on the soil or plant surface, (ii) the fugacity is not reduced by chemical or physical adsorption reactions, and (iii) the rate of vapor formation is fast enough to keep the inner surface of the laminar layer saturated.

#### 7–1.1.1 Equilibrium Vapor Pressures

Since the vapor pressure gradient within the laminar surface boundary layer reflects the vapor pressure of the pesticide residues on the surface, it is useful to examine the magnitudes of the vapor pressures of some commonly used pesticides to establish the range of values to be expected. Measured values of the vapor pressures of several commonly used pesticides are presented in Table 7–1. These were selected from the literature to illustrate the range and scale of the values (Beste, 1983; Hartley & Graham-Bryce, 1980d; Jury et al., 1983a). The table also contains the same data expressed in terms of vapor densities, defined as $w/V$, or the mass of pesticide vapor per unit volume, which is frequently used as a convenient unit in pesticide volatility studies. The two sets of units are related by the equation

$$p = (w/V)(RT/M) \quad [4]$$

where $w$ and $M$ are the mass and molecular weight of the pesticide, $V$ is the volume in liters, $T$ is the absolute temperature ($K$), and $R$ is the molecular gas constant.[2]

The values range over about six orders of magnitude from 2800 mPa for EPTC to 0.00074 mPa for picloram. Since the list is not comprehensive, other compounds may be found with values outside this range. The general-scale is, however, extremely low in comparison with more familiar volatile

[2]This equation reduces to $d = w/V = 0.12 \times pM/T$ where $d$ is in $\mu$g/L and $p$ is in mPa. Written in these forms both equations contain the implicit assumption that the pesticide vapor is completely dissociated.

Table 7-1. Saturation vapor pressures and densities of pesticides.

| Pesticide | Molecular wt. | Temperature | Vapor pressure | Vapor density | Reference |
|---|---|---|---|---|---|
| | g | °C | mPa | μg/L | |
| Alachlor | 270 | 25 | 2.9 | 3.2 (−1)† | Beste, 1983 |
| Atrazine | 216 | 25 | 9.0 (−2) | 8.0 (−3) | Jury et al., 1983a |
| Bromacil | 261 | 25 | 2.9 (−2) | 3.0 (−3) | Jury et al., 1983a |
| Carbofuran | 221 | 25 | 1.1 | 1.0 (−1) | Jury et al., 1983a |
| Chlorpropham | 214 | 25 | 1.3 | 1.2 (−1) | Beste, 1983 |
| DDT | 354 | 25 | 4.5 (−2) | 6.0 (−3) | Jury et al., 1983a |
| Diazinon | 304 | 25 | 1.6 (+1) | 2.0 | Jury et al., 1983a |
| Dicamba | 221 | 20 | 5.0 (+2) | 45 | Hartley & Graham-Bryce, 1980d |
| Dieldrin | 381 | 25 | 6.8 (−1) | 1.0 (−1) | Jury et al., 1983a |
| Diuron | 233 | 25 | 2.1 (−2) | 2.0 (−3) | Jury et al., 1983a |
| EPTC | 189 | 20 | 1.1 (+3) | 8.3 (+1) | Hartley & Graham-Bryce, 1980d |
| EPTC | 189 | 25 | 2.8 (+3) | 2.2 (+2) | Jury et al., 1983a |
| Fenitrothion | 277 | 20 | 8.0 (−1) | 9.1 (−2) | Hartley & Graham-Bryce, 1980d |
| Heptachlor | 373 | 20 | 2.2 (+1) | 3.3 | Hartley & Graham-Bryce, 1980d |
| Lindane | 291 | 25 | 8.6 | 1.0 | Jury et al., 1983a |
| Linuron | 249 | 20 | 1.1 | 1.2 (−1) | Hartley & Graham-Bryce, 1980d |
| Malathion | 330 | 20 | 1.3 | 1.8 (−1) | Hartley & Graham-Bruce, 1980d |
| Methyl parathion | 263 | 25 | 2.4 | 2.5 (−1) | Jury et al., 1983a |
| Metolachlor | 284 | 20 | 1.7 | 2.0 (−1) | Hartley & Graham-Bruce, 1980d; Beste, 1983 |
| Monuron | 199 | 25 | 2.3 (−2) | 2.0 (−3) | Jury et al., 1983a |
| Parathion | 291 | 25 | 1.3 | 1.5 (−1) | Jury et al., 1983a |
| Picloram | 241 | 20 | 7.4 (−4) | 7.3 (−5) | Hartley & Graham-Bryce, 1980d |
| Prometon | 225 | 25 | 8.3 (−1) | 1.0 | Spencer et al., 1988 |
| Simazine | 202 | 25 | 2.0 (−3) | 1.7(−4) | Jury et al., 1983a |
| Triallate | 305 | 20 | 2.6 (+1) | 3.2 | Jury et al., 1983a |
| Trifluralin | 335 | 20 | 1.5 (+1) | 2.0 | Jury et al., 1983a |

† Numbers in parentheses represent power of 10 factors.

organic chemicals such as alcohols or ethers that have vapor pressures close to 1 atmos. at normal temperatures. The pressure of a standard atmosphere (760 mm of Hg) is equal to $1 \times 10^5$ Pa, or between about 40 and 100 thousand times that of EPTC. While the tabulated values may seem to suggest that the volatilization rates of the compounds would be low, remember that the background vapor pressures of pesticides in the air are essentially zero so that the residues will evaporate as if they were in a vacuum. This effect is a direct consequence of Dalton's law, which states that the vapor pressure of a single gas is independent of those of any others that may be mixed with it. This absence of any natural background concentration places pesticide vapor dispersion in marked contrast to that of other atmospheric components where a natural background concentration exists. The most striking example is that of water vapor, which has a vapor pressure of about 2300 Pa in saturated air at 20 °C. The limits that this places on the evaporation of water residues are a familiar feature of humid climates. In a more realistic comparison, the evaporation of pesticides at the surface of the Earth should be compared with that of water under the desiccated conditions on the surface of the Moon or Mars.

Comparison of the values in Table 7–1 with others in the literature reveals some uncertainties about the true value for some chemicals. These differences arise from the methods used in their determination. Measurements of very low vapor pressures present considerable experimental difficulties. Plimmer (1976) has reviewed in detail the methods used and the uncertainties these measurements entail. The methods will be briefly described in this chapter because some of the same techniques have been used to measure the vapor pressures of residues in soil and water systems as well as those of the pure compounds.

The methods may be grouped into several distinct categories. The simplest is the direct measurement of the pressure of the vapor in equilibrium with the pure solid or liquid chemical. This is most satisfactory where the vapor pressure is about 130 Pa (1 mm of Hg) or more. This is much greater than that of most pesticides at room temperatures, so that elevated temperatures must be used and the results extrapolated to the normal range by equations based upon the integrated Clapeyron-Clausius equation: $\log P = A - B/(t\,°C + 273.17)$, where $A$ and $B$ are constants, specific for each chemical. Data obtained from such high temperature methods are sometimes uncertain because of errors in both the experimental measurement and in the extrapolation: the error maybe considerable where the extrapolation crosses the temperature of a phase change for the solid or liquid.

Effusion methods measure the rate of escape of vapor through a small orifice into a vacuum. The measurement may be either the weight loss of the container, the torque imposed on a fiber supporting the cell or by collection of the escaping vapor in a liquid N cold trap. Hamaker and Kerlinger (1969) have described in detail several such methods.

The gas saturation method Spencer and Cliath (1969, 1983) described has been used to measure vapor pressures of pesticides both as pure compounds and as residues in soils. This is a direct measurement of the amount of vapor in a given volume (the vapor density) by analysis of a volumetric sample taken at a suitable controlled temperature. The vapor pressure is then calculated from Eq. [4]. Since chromatographic analyses permit accurate determination of nano- and picrogram quantities of pesticides, measurements can be made in the 10 to 40 °C range, and no extrapolations are necessary.

Vapor pressures may also be calculated from gas-liquid chromatographic retention times. This method assumes that column retention times depend only on the vapor pressure: this is then calculated by comparison with the retention time of a standard compound of known vapor pressure. Successful measurements depend on a suitable choice of both column materials and standard substances appropriate to the polarity and chemical character of the particular pesticide (Bidleman, 1984). The advantages of this method are the speed by which data can be obtained and its applicability to mixtures and impure samples. Uncertainties arise from possible specific interactions between particular compounds and column materials and from the need to extrapolate data between column and environmental temperatures. A systematic evaluation of both the gas chromatographic and saturation methods by Kim et al. (1984) using 20 organophosphorus compounds showed that data with a SD of ±10% could be obtained with a suitable choice of temperature and column conditions. Errors were greater for compounds that are solid at room temperature, even after corrections were applied for the effect of the liquid-solid phase change on the extrapolation.

The most consistent results appear to be obtained by the gas saturation and gas-liquid chromatographic methods. Where critical values are sought from the literature, those obtained by these means should be preferred provided that they are supported by an adequate presentation of the original data and description of the experimental conditions. Unfortunately, this is far from the case for many published values. Few critical reviews exist, and where data are required for critical purposes the results should be traced, if possible, to their origin. Spencer and Cliath (1983) have discussed examples of such discrepancies for dieldrin; Plimmer (1976) for several esters of 2,4-D; and Kim et al. (1984) for a number of organophosphorus insecticides.

### 7-1.1.2 Adsorption and Partition Effects

The value of $p_s$ in Eq. [2] is the maximal value that can be established at the inner surface of the laminar flow layer in air moving over a surface. This can only be the saturation value when the surface is covered with pure pesticide residue. Such surfaces are, of course, never found in practice, except for a short time after application, because the pesticide is always adsorbed on soil or plant material, or dissolved in water that is present on the surface or occluded within soil (or leaf) pores. This reduces the vapor pressure below the equilibrium value of the pure compound to an extent that depends upon a number of factors including the soil water content, the soil texture, and organic matter content. Although there is much less information available about the effects of adsorption on plant surfaces, sorption and permeation into cuticular waxes and the epidermal layer of plant leaves are certainly important in reducing the pesticide vapor pressure. Hartley and Graham-Bryce (1980c) have discussed in detail the complexities of herbicide penetration in regard to their effectiveness.

Although the variety of chemicals among the pesticides will show different patterns in their physicochemical behavior in soils, many have similar patterns that are important in controlling their partition between the air, water, and solid phases of the soil. The partitioning between the solid phase and the solution usually follows the Freundlich equation. The vapor density of the pesticide in the air reflects the solution concentration according to Henry's law, defined by the equation

$$d = hc \qquad [5]$$

where $d$ is the vapor density and $c$ the solution concentration, both in the same units (usually mg/L) and $h$ is Henry's law coefficient. Table 7-2 contains data on the water solubilities and Henry's law coefficients for the pesticides listed in Table 7-1. Almost all these values are calculated from vapor densities of saturated vapors and concentrations in saturated solutions, $d_o/c_o$; few measurements have been made under unsaturated conditions.

Table 7-2. Solubility and Henry's law coefficients at 25 °C.

| Pesticide | Solubility | Coefficient | Reference |
|---|---|---|---|
| | mg/L | $h = d_o/c_o$ | |
| Alachlor | 2.4 (+2)† | 1.3 (−6) | Hartley & Graham-Bryce, 1980d; Beste, 1983 |
| Atrazine | 3.2 (+1) | 2.5 (−7) | Jury et al., 1983a; Beste, 1983 |
| Bromacil | 8.2 (+2) | 3.7 (−9) | Jury et al., 1983a; Beste, 1983 |
| Carbofuran | 3.2 (+2) | 3.1 (−7) | Jury et al., 1983a |
| Chlorpropham | 8.9 (+1) | 1.2 (−6) | Hartley & Graham-Bryce, 1980d; Beste, 1983 |
| DDT | 3.0 (−3) | 2.0 (−3) | Jury et al., 1983a |
| Diazinon | 4.0 (+1) | 5.8 (−5) | Hartley & Graham-Bryce, 1980d |
| Dicamba | 4.5 (+3) | 1.0 (−5) | Hartley & Graham-Bryce, 1980d |
| Dieldrin | 1.5 (−1) | 6.7 (−4) | Jury et al., 1983a |
| Diuron | 3.7 (+1) | 5.4 (−8) | Jury et al., 1983a |
| EPTC | 3.7 (+2) | 5.9 (−4) | Jury et al., 1983a; Beste, 1983 |
| Heptachlor | 5.6 (−3) | 5.9 (−2) | Hartley & Graham-Bryce, 1980d |
| Lindane | 7.5 | 1.3 (−4) | Jury et al., 1983a |
| Linuron | 7.5 (+1) | 1.6 (−6) | Hartley & Graham-Bryce, 1980d; Beste, 1983 |
| Malathion | 1.5 (+2) | 1.3 (−6) | Hartley & Graham-Bryce, 1980d; Beste, 1983 |
| Metolachlor | 5.3 (+2) | 3.7 (−7) | Hartley & Graham-Bryce, 1980d; Beste, 1983 |
| Monuron | 2.6 (+2) | 7.7 (−9) | Jury et al., 1983a |
| Parathion | 2.4 (+1) | 6.2 (−6) | Jury et al., 1983a |
| Picloram | 4.3 (+2) | 1.7 (−10) | Hartley & Graham-Bryce, 1980d |
| Prometon | 7.5 (+2) | 1.0 (−7) | Spencer et al., 1988 |
| Simazine | 5.0 | 3.4 (−8) | Jury et al., 1983a |
| Triallate | 4.0 | 8.0 (−4) | Hartley & Graham-Bryce, 1980d; Jury et al., 1983a |
| Trifluralin | 3.0 (−1) | 6.6 (−3) | Jury et al., 1983a; Beste, 1983 |

† Numbers in parentheses indicate power of 10 factors.

Measurements of pesticide concentrations in air equilibrated with treated soils can be made conveniently under controlled conditions in the laboratory using the gas-saturation technique (Spencer & Cliath 1969, 1983). In contrast, measurements of concentrations in the soil solution are much more laborious and less reliable and also require destruction of the sample. The vapor pressure measurements, therefore, offer a direct experimental route to the evaluation of adsorption effects on volatility. Spencer and Cliath (1969, 1970, 1972, 1974) and Spencer et al. (1969) have published data obtained in this way for several major pesticides.

**7-1.1.2.1 Adsorption.** Isotherms of vapor phase desorption of DDT and lindane from moist Gila silt loam (Coarse-loamy, mixed [calcareous], thermic Typic Torrifluvents) soil measured by Spencer and Cliath (1970, 1972) are presented in Fig. 7-2 and 7-3. Figures 7-4 and 7-5 illustrate similar data for dieldrin and trifluralin (Spencer et al., 1969; Spencer & Cliath, 1974). The vapor densities of all the compounds increase with residue concentrations up to values corresponding to those of the pure materials at soil concentrations of 15 mg/kg of p,p 'DDT, between 50 and 60 mg/kg for lindane depending on the temperature, 25 mg/kg for dieldrin and 73 mg/kg for

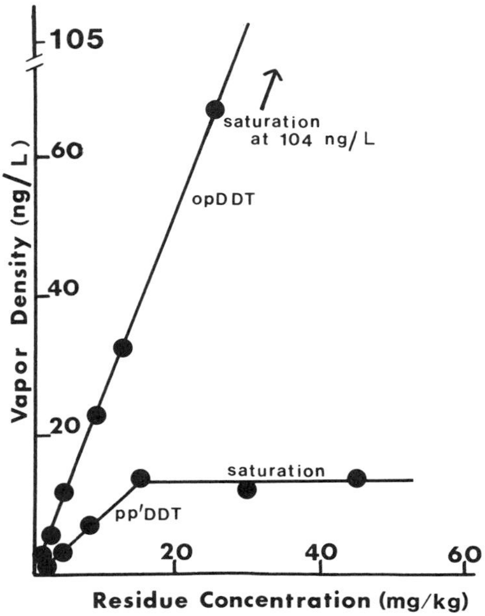


Fig. 7-2. Vapor density of o,p '- and p,p '-DDT in air equilibrated at 30 °C with residues adsorbed on samples of Gila silt loam soil at a moisture content of 0.0394 kg/kg. After Spencer and Cliath (1972).

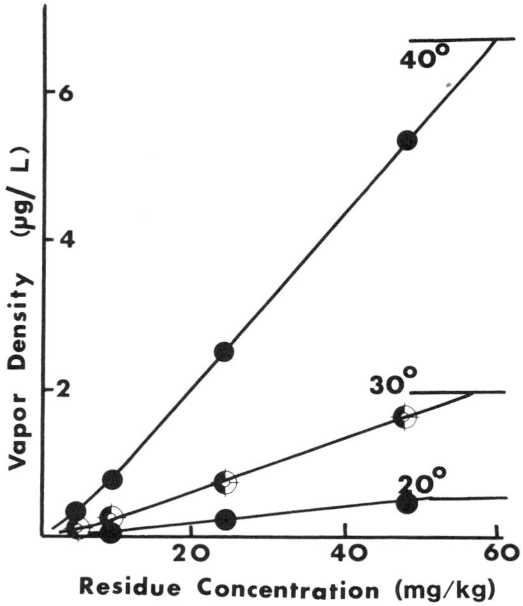


Fig. 7-3. Vapor density of lindane in air equilibrated at 20, 30, and 40 °C with residues adsorbed on Gila silt loam soil at a moisture content of 0.0394 kg/kg. After Spencer and Cliath (1970).

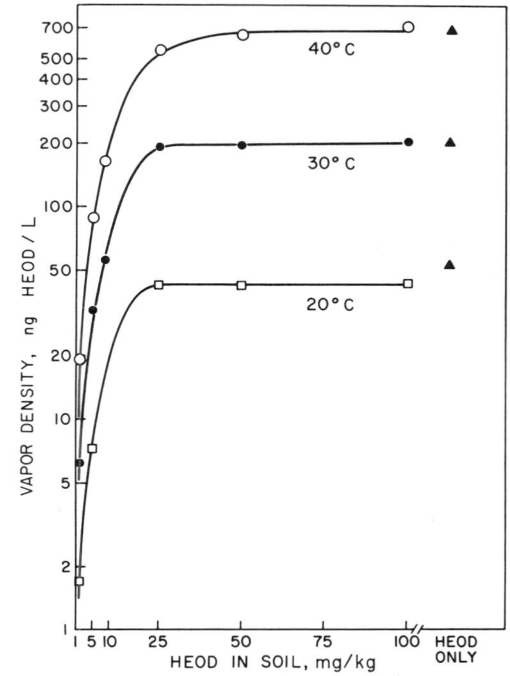


Fig. 7-4. Vapor density of dieldrin (HEOD) in air equilibrated at 20, 30, and 40 °C with residues adsorbed on Gila silt loam soil at a moisture content of 0.10 kg/kg (Spencer et al., 1969).

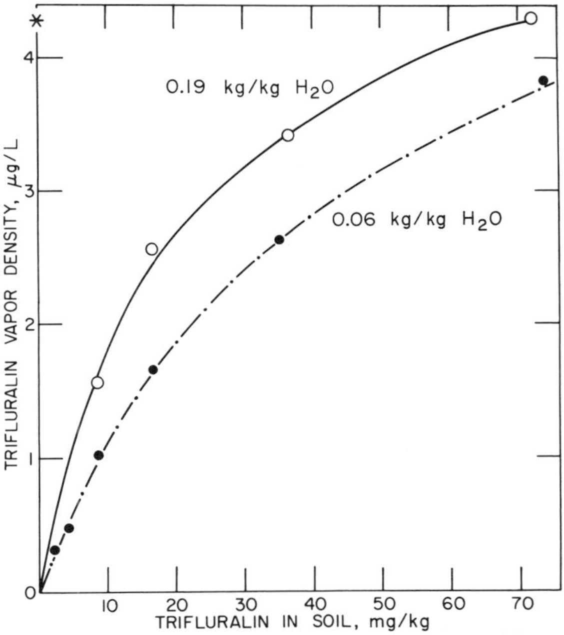


Fig. 7-5. Vapor density of trifluralin in air equilibrated at 30 °C with residues adsorbed on Gila silt loam soil at water contents of 0.06 and 0.19 kg/kg (Spencer & Cliath, 1974).

trifluralin. Above these concentrations, the vapor densities are limited by the pure compounds. The significance of the shapes of these isotherms will be discussed in more detail in section 7–1.1.3.

These results are believed to be typical of many other pesticide residues in soils. To relate them to field conditions, we may assume that if a spray of pesticide applied at a rate of 1.25 kg/ha to the surface of an uncultivated bare soil remains within the top 1-mm layer (bulk density of 1.25), the local concentration may be expected to be about 75 to 100 mg/kg. Figures 7–2, 7–3, 7–4, and 7–5 suggest that at these concentrations the residues will support vapor densities close to those of the parent compounds, provided the soil moisture levels are about 0.04 kg/kg or above. If, however, the pesticide is incorporated by cultivation to a greater depth, e.g., 75 mm, the average soil concentration will be reduced to between 1 or 1.25 mg/kg, and the relative vapor densities are likely to be reduced to 10% or less. Equation [2] then suggests that this will reduce the volatilization rate in proportion. As will be seen from the following discussion, however, other factors, including diffusion characteristics and moisture flow also greatly affect the rates at which incorporated residues move to the open soil surface where they volatilize.

**7–1.1.2.2 Temperature Effects.** Figures 7–3 and 7–4 suggest that where the soil residue concentrations are high, the effect of temperature on the vapor densities will be similar to the effect of temperature on the vapor densities of the pure compounds. Where residue concentrations are below the saturation level (the relative vapor densities ($d/d_0$) are $<$ 1.0), the temperature relations are more complex. In a detailed analysis of the lindane system of Fig. 7–3, Spencer and Cliath (1970) pointed out that below the saturation level, the vapor density increased with temperature more slowly than that of the pure compound. This effect is because of a combination of solubility relationships, Henry's law, (Eq. [5]) and free energy changes during adsorption.

**7–1.1.2.3 Water Solubility.** The validity of the Henry's law relationship in describing the partitioning of the pesticide between the air and water phases of the system is demonstrated by the curve plotted in Fig. 7–6, which

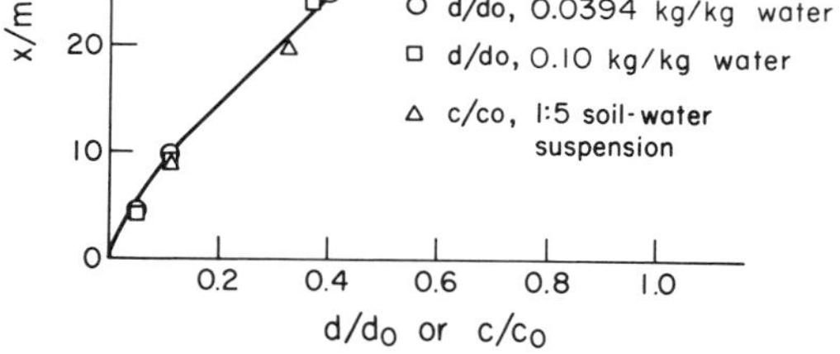

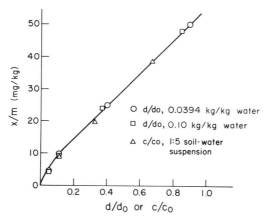


Fig. 7–6. Relative vapor density ($d/d_0$) and relative solution concentration ($c/c_0$) of lindane in air and water equilibrated with residues on Gila silt loam soil at several water contents (Spencer & Cliath, 1970).

shows that, when plotted in terms of $d/d_0$ and the corresponding solubility function $c/c_0$, all the lindane data for a particular temperature (30 °C) fall on the same curve, and are independent of the soil moisture content (Spencer & Cliath, 1970). This result validates the use of vapor measurements in the determination of adsorption (or desorption) isotherms of pesticides in soils. It should be recognized that the isotherms obtained include the effects of concentration and temperature on both the adsorption energy and Henry's law coefficient. A complete calculation of the distribution of residues between soil, water, and air would require independent measurements of these variables, but if the Henry's law coefficient is assumed to be independent of concentration (i.e., $d/d_0 = c/c_0$) the system may be characterized by vapor measurements alone.

**7-1.1.2.4 Soil Moisture Content.** Some investigators (Fang et al., 1961; Harris & Lichtenstein, 1961; Lichtenstein & Schultz, 1961; Deming, 1963; Gray & Weierich, 1965; Parochetti & Warren, 1966; Guenzi & Beard, 1970; Lichtenstein et al., 1970) have noted retardation of the volatilization or disappearance of pesticides and their inactivation in dry soils. Spencer (1970) has discussed in detail the increased adsorption of pesticides in dry soil on the basis of the data presented in Fig. 7-7 and 7-8. These results show that

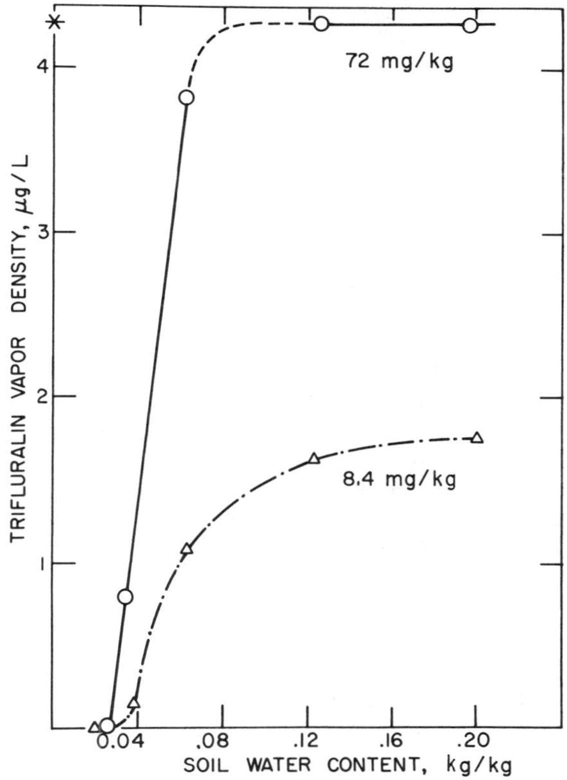


Fig. 7-7. Vapor density of trifluralin in air equilibrated at 30 °C with Gila silt loam containing 8.4 and 72 mg/kg trifluralin and water contents up to 0.20 kg/kg (Spencer & Cliath, 1974).

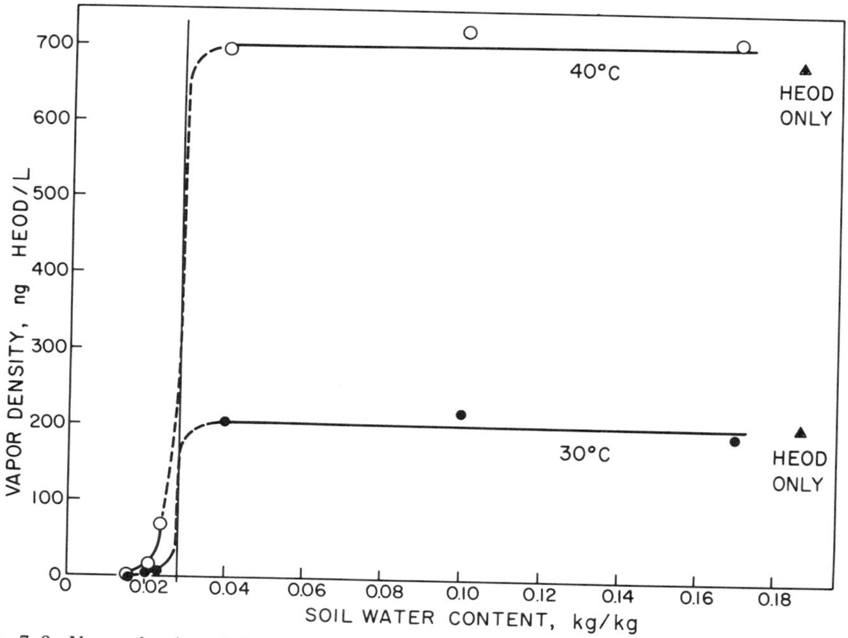


Fig. 7–8. Vapor density of dieldrin (HEOD) in air equilibrated at 30 and 40 °C with Gila silt loam containing 10 mg/kg dieldrin and water contents up to 0.17 kg/kg (Spencer et al., 1969).

even where the concentrations of trifluralin and dieldrin were high enough in moist Gila soil to support vapor densities approaching those of the pure compounds, reduction of the soil water content below 0.03 or 0.04 kg/kg caused a large reduction in the equilibrium vapor densities. At water contents of 0.02 kg/kg or less, the relative vapor densities were below 10%, and at 0.016 kg/kg water content the dieldrin vapor density approached the lower levels of detection. Similar data have been presented for DDT and DDE isomers (Spencer & Cliath, 1972) and lindane (Spencer & Cliath, 1970). In the Gila soil, a water content of 0.028 kg/kg corresponds to a monomolecular layer on the adsorption surfaces. When the moisture content is reduced below this level, adsorption sites on which the pesticide molecules are more strongly bound become active so that their fugacity (and hence their vapor density) is greatly reduced. The effect is reversible in response to changes in water content (Spencer & Cliath, 1974; Spencer et al., 1969). These observations are consistent with the use of clays as carrier materials in wettable powder formulations in which pesticides are stored for long periods in inactive adsorbed forms that are rapidly released on "activation" with water used to prepare the formulation for field use (Van Valkenburg, 1969).

In practical terms, it may be noted that in the Gila soil a moisture content of 0.0394 kg/kg is in equilibrium with air at a relative humidity of 94%. The moisture contents at which the equilibrium vapor densities (and hence the volatilization rates) of the residues are reduced to low values are thus easily reached in surface soils drying in moderately dry air. This dependence of volatilization rate upon atmospheric relative humidity and soil moisture content has been observed for trifluralin in the laboratory by Spencer and Cliath (1974) and in the field by Harper et al. (1976) and Glotfelty et al. (1984). In the field, surface soil temperatures rise rapidly when the surface dries to the point at which heat can no longer be removed by the latent heat of evaporation of water: at this point, pesticide adsorption will increase so that the volatilization rate also falls rapidly. Despite the intuitive expectation that pesticide volatilization should increase with temperature, no simple correlation between volatilization rates and soil temperature can actually be expected. In experiments using a *microagroecosystem* growth chamber to measure the disappearance of trifluralin and several organochlorine insecticides from the surface of bare soil, Nash (1983) found that soil moisture was the most important variable affecting volatilization rate, followed by air temperature.

Vaporization of residues from transpiring leaves, where the water supply is not depleted, may be expected to be independent of the dryness of the soil surface, and to respond to changes in the temperature of the leaf. This will not, however, rise as rapidly as the temperature of dry soil owing to the cooling effect of evapotranspiration.

This interrelation between temperature, soil moisture content, and the fugacity (or activity) of pesticide residues is of major importance in controlling losses of residues of many pesticides from soil surfaces. The effect is of particular importance in controlling the effectiveness of herbicides: the common field observation that herbicides applied to dry soils are not effective until they have been activated by moisture is clearly related to their desorption from the strong adsorption sites as the soil moisture is increased.

Early observations of the interaction between the rates of evaporation of water and DDT were interpreted as *codistillation* (Acree et al., 1963). When correctly used, this term describes a bulk flow process that takes place when the combined vapor pressures of the two components being distilled together equal or exceed the ambient pressure. Since the vapor pressure of water is always $<1$ atmos. under normal environmental temperatures and the pesticide vapor pressures are lower by several orders of magnitude, codistillation is a misnomer for pesticide evaporation. Experimental proof that the volatilization of trifluralin and dieldrin from soils is not directly dependent on water evaporation has been presented by Spencer and Cliath (1974) and Igue et al. (1972) who demonstrated that maximum pesticide volatilization rates were found at 100% relative humidity when the water loss was zero. Any use of the term codistillation as a description of pesticide volatilization is inappropriate and misleading and its use in this context should be discontinued.

**7-1.1.2.5 Soil Organic Matter.** Spencer (1970) investigated the influence of soil organic matter content on the vapor pressure of dieldrin residues in five soils with varied clay and organic matter contents. The results, summarized in Table 7-3, show that the vapor densities over both wet and dry soils were inversely related to the soil organic matter contents. The clay content of the soils, which was unrelated to the organic matter content, had only a minor effect, although the vapor densities were much lower over the dry

Table 7-3. Effect of organic matter and clay content on vapor density of dieldrin (at 10 mg/kg dieldrin and 30 °C) (Spencer, 1970).

| Soil type | Organic matter | Clay | Vapor density, μg/L | |
|---|---|---|---|---|
| | | | Wet† | Dry‡ |
| | g/kg | % | | |
| Rosita very fine sandy loam | 1.9 | 16.3 | 175 | 1.7 |
| Imperial clay | 2.0 | 67.6 | 200 | 0.9 |
| Gila silt loam | 5.8 | 18.4 | 52 | 0.7 |
| Kentwood sandy loam | 16.2 | 10.0 | 32 | 0.4 |
| Linne clay loam | 24.1 | 33.4 | 32 | 0.6 |

† Wet—Approximately 0.2 MPa matrix suction.
‡ Dry—Equilibrated at 50% relative humidity. All soils contain 10 ppm dieldrin: all temperatures 30 °C.

soils. Chiou and Shoup (1985) and Chiou et al. (1983, 1985) have examined in detail the interlinked effects of organic matter and soil moisture on the fugacity and adsorption of nonionic organic compounds in soils. These results show that when the adsorption by clay mineral surfaces is inhibited by the presence of moisture, the adsorption isotherms observed over a wide range of $c/c_0$ values are consistent with partitioning of the residues into the soil organic matter. These isotherms are rectilinear over wide ranges of concentration (and vapor pressure), have low heats of adsorption, and show no competitive effects when two adsorbates are present together. In contrast, adsorption isotherms determined by direct vapor adsorption measurements (Chiou & Shoup, 1985) from solutions with wide ranges of concentration in dry hexane (Chiou et al., 1985) showed large heat of adsorption and competition effects. The curvature of these isotherms was consistent with Brunauer Type II adsorption, characteristic of multilayer formation by vapor condensation.

Except under conditions of intense surface drying, the distribution of residues in field soils will be controlled by the partitioning into the organic matter. Soil organic matter contents and organic matter partition coefficients are thus of primary importance in describing rates of diffusive movement and volatilization in soils and it is, therefore, more useful to express the adsorption coefficients of different soils in terms of their organic contents rather than their textures.

### 7-1.1.3 Fugacity and Vapor Pressure

**7-1.1.3.1 Fugacity and Vapor Flux.** The effects of dissolution in water and adsorption by the solid phases in the soil on the volatility of residues and their partition between the water and solid phases may be brought together in a simplified form by the use of the fugacity approach. As defined by Lewis and Randall (1923), the fugacity of a chemical component is a measure of the escaping tendency of that component in any phase of a system. Since at equilibrium distribution the fugacity of a component is uniform in all the phases of a system, the overall distribution of the component through the whole system can be calculated if the fugacity relationships in each phase are known (Mackay & Paterson, 1982).

Fugacity is closely related to and has the dimensions of vapor pressure (Pa). If we assume ideal gas behavior, the fugacity of the pesticide in air can be equated with its vapor pressure, so that Eq. [1] becomes

$$\hat{I}F = k_Z(df/dz). \quad [1a]$$

If $k'_Z$ is the overall diffusivity coefficient for the entire depth between the surface and any height $z$, $\hat{I}F = k'_Z(f_s - f_z)$, so that when the height is sufficiently large that $p_z \rightarrow 0$,

$$\hat{I}F = k'_Z f_s. \quad [1b]$$

This expression emphasizes the importance of the fugacity of the residues at the surface as the factor controlling the upward flux or evaporation rate.

**7-1.1.3.2 Fugacity in Air and Water.** The relationship between the fugacity of the pesticide vapor and that of the residues dissolved in the soil water or adsorbed on the solid phases has been described by Mackay and Paterson (1981), who point out that, at the low vapor pressures and concentrations at which pesticide residues are present in the environment, we may write, for each phase,

$$C = fZ \quad [6]$$

where $C$ is the concentration of residues in any environmental phase in mol/m$^3$, $f$ is the fugacity in pascals, and $Z$ is the fugacity capacity for that phase.

In air, $C$ is related to the fugacity through the ideal gas law, so that $C_A = n/V = p/RT = fZ_A$, and

$$f = C_A/Z_A = C_ART. \quad [7]$$

It should be noted that $C$ is stated in mol/m$^3$, whereas vapor density, $d$, is stated in mass/volume units, usually milligrams or grams per liter or cubic meter.

In water solution, the appropriate form of Eq. [6] is $C_W = fZ_W$ or $Z_W = C_W/f = C_W/p$, where $p$ is vapor pressure in the atmosphere equilibrated with the solution, and $C_w$ is the concentration in water, again in mol/m$^3$. Thus, $Z_W = 1/H$ where $H$ is the Henry's law coefficient, here expressed in terms of pressure (Pa) and molar concentrations[3], so that

$$f = C_W/Z_W = C_WH. \quad [8]$$

[3]Appropriate values of H may be calculated from the data in Table 7-2, where $H = 2.48 \times 10^3$h: both $H$ ahd $h$ are dimensionless numbers.

This equation states that the fugacity of pesticide vapor in the air over the surface of a water solution will reflect the Henry's law distribution of the residues between those in the air and in those in true molecular solution in the water. This concentration may be less than the total concentration where residues are associated with suspended organic materials or other solid particles or present as emulsions.

**7-1.1.3.3 Fugacity and Adsorption.** Fugacities of adsorbed residues are defined in terms of their adsorption isotherms. Where the amount of adsorbate is a small fraction of the adsorption capacity we may write $C_S = K_p C_W$, where, as above, $C_W$ is the concentration in the water phase, $C_s$ is that in the adsorbent (mol/m$^3$) and $K_p$ is the soil/water partition coefficient.[4] The possible errors that may be incurred by the assumption that adsorption isotherms can be regarded as simple rectilinear relationships have been evaluated by Rao and Davidson (1980).

Since the fugacity of the adsorbed residues is defined by $f_S = C_S/Z_S$ and, under the equilibrium conditions described by the partition coefficient this fugacity must be equal to that in the water phase, $f_S = f_W = C_S/Z_S = HC_W$, and $Z_S = C_S/(HC_W)$, so that $Z_S = K_p/H$ and

$$f_S = C_S(H/K_p). \qquad [9]$$

This equation defines the fugacity capacity of the adsorbed phase in terms of the solid/water and air/water partition coefficients.

In practice, the fugacity of a pesticide in the soil is controlled by its distribution throughout the soil-water-air continuum. This is dominated by the adsorption capacity of the soil so that the correspondingly large fugacity capacity of this phase controls the overall fugacity and hence the volatilization tendency.

The interpretation of adsorption isotherms in terms of fugacity is illustrated in Fig. 7-9, where the curve OM is an isotherm describing adsorption by moist soil in terms of fugacity and adsorbate concentration. Equation [9] shows that the slope of this line at any point is an inverse function of the fugacity capacity of adsorption. Changes in this slope must, therefore, reflect changes in the value of $K_p$ as adsorption sites are filled. The isotherms presented in Fig. 7-2 and 7-3 for desorption of DDT and lindane from moist Gila soil are essentially rectilinear until the vapor pressures are close to saturation with the pure compounds, showing that $Z_s$ remains essentially constant over this range. Dieldrin and trifluralin isotherms on the same soil (Fig. 7-4 and 7-5) are only approximately rectilinear close to the origin: at higher concentrations decreasing slopes indicate that the fugacity capacity increases with pesticide loading.

[4]Classically, partition coefficients are defined by $c_{ad} = x/m = k_p c$, where $c_{ad}$ is in mass concentration units ($x$ gram sorbate/$m$ gram sorbent). In volumetric terms, the sorbed concentration is [($x$ gram)($d_B/m$ gram)], where $d_B$ is the bulk density of the sorbent. Hence $C_S = c_{ad} \times d_B$ and $K_p = k_p \times d_B$. In molar units, the numerical values of the coefficients are unchanged.

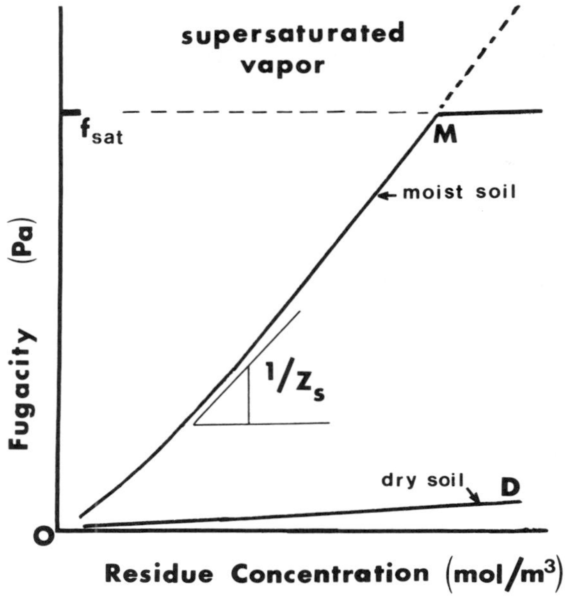


Fig. 7–9. Isotherms describing the fugacity of pesticide vapor in air equilibrated with residues adsorbed on dry and moist soils as a function of the saturation fugacity($f_{sat}$) and the fugacity capacity of the soil ($Z_S$).

Chiou et al. (1985) have suggested that the adsorption of organic compounds by moist soils represents partitioning of the adsorbate into the organic matter with the formation of a "solid solution." The partition between soil and water may thus be written in terms of the partition coefficient $K_{oc} = C_{oc}/C_w$, where $C_{oc}$ is the (volumetric) concentration of pesticide in the organic phase. If the density of the soil organic matter is assumed to be the same as the bulk density of the soil, we may express the soil organic matter content as a volume fraction, and $C_{oc} = C_s/\phi$ where $\phi$ is here the organic matter content (as a decimal fraction). Also, since $K_{oc} = C_s/\phi C_w = K_p/\phi$ and the fugacity in the organic phase is given by $f = C_{oc}/Z_{oc}$, where $Z_{oc}$, the fugacity capacity of the organic phase is given by $H/K_{oc}$. Where the residue concentration is expressed in terms of the entire solid phase, this becomes $f_s = (C_s H)/(\phi K_{oc})$ indicating that the slopes of the isotherms obtained by plotting fugacity as a function of $C_s$ will reflect the partition of the pesticide between vapor, solution, and soil organic matter.[5]

Few values of $K_{oc}$ for pesticides are available in the literature, but they may be approximately estimated from values of the octanol/water distribution coefficients ($K_{ow}$) that have been measured for many compounds, using the regression equation $\log K_{oc} = (1.029 \log K_{ow}) - 0.18$ (Rao & David-

[5] $K_{oc}$ values are calculated in terms of soil organic C content. This is related to soil organic matter fraction by the empirical equation (soil organic matter) = 1.73 × (organic C content). The conversion reflects the N, S, O, and H content of the organic matter.

son, 1980). Values based on the correlation approach should, however, be used with caution: Mingelgrin and Gerstl (1983) have discussed the dangers of using a single correlation to predict values for a wide range of compounds, and point out that the correlations that do exist depend upon the similarity of behavior of substances with similar chemical properties and that the correlations become weaker the wider the group of compounds included.

The effect of reduced soil moisture content on the fugacity of adsorbed residues illustrated in Fig. 7-7 and 7-8 is represented by the isotherm OD in Fig. 7-9. The large decrease in slope is because of a large increase in $Z_s$, the fugacity capacity of the adsorbent, as the adsorbing surfaces become dry. This reflects a corresponding change in the value of $K_p$ because of changes in the free energy and entropy relations of the adsorption reaction as the moisture is lost from the surface. Chiou et al. (1985) have reviewed in detail adsorption on dry mineral surfaces and its sensitivity to moisture. The importance of this for pesticide volatility is that the fugacity of the residues in a dry soil surface is reduced to such levels that volatilization is almost completely suppressed. This, as will be seen below, is a major factor controlling the volatilization from soils in the field.

The inflection in the isotherm OM at the point of intersection with the saturation isotherm of the pure compound (Fig. 7-9) does not reflect any sudden change in the character of the adsorbing surface, but implies that points above this fugacity level are inaccessible because they represent supersaturation with the pure residues. The p,p 'DDT data in Fig. 7-2 show that on the moist Gila soil at 30 °C this point is reached at a fugacity of 0.097 mPa: in contrast, adsorption of the o,p isomer, a similar molecule, continues up to a fugacity of 0.74 mPa. The change in slopes of the lindane isotherms in Fig. 7-3 reflect temperature effects on both $K_p$ and $H$.

## 7-1.2 Residue Distribution Effects

The previous sections have reviewed the principal chemical factors that control the vapor pressures of pesticide residues immediately above soil and plant surfaces. A number of physical factors are equally important because they control the rate at which the supply of vapor is maintained. The distribution of the residues within the soil is the most important of these because it determines the length of the pathway that the residues must travel by diffusion or bulk flow to the evaporating surface. This distribution is in turn controlled by the way the pesticides are applied and, in some degree, by the type of formulation used.

### 7-1.2.1 Surface Residues

Where pesticides are applied as water-based sprays of solutions, dispersed emulsions, or wettable powders, the residues are initially present as a thin layer of material on the exposed plant and soil surface. Where, as on plant leaves, this deposit is a continuous film over the whole surface, the initial volatilization loss (per unit of field area) will be independent of the total amount present because the evaporating area does not change as long as the film remains unbroken. The kinetics of loss during this initial stage of evaporation will, therefore, be that of a zero-order reaction. This rate will not, however, be identical for different compounds even when applied together in the same spray because as indicated by Eq. [2], the rate of volatilization of each compound will depend upon the product of the molecular diffusion coefficient and the saturation vapor density.

Owing to the speed with which water-based formulations are drawn into soil particles by capillary tension, this initial condition may not exist on soil unless the formulation is sprayed on to a wet soil surface. Even on plant leaves it may be short lived and as evaporation continues the surface coverage will become incomplete and break into separate areas as islands of residues. Although the specific volatilization rate (in $g/cm^2$ of covered area) may remain constant, the disappearance of residues in $g/cm^2$ of soil or leaf surface will depend upon the area covered and will decrease as the islands shrink in size. At this stage, the loss rate will be proportional to the amount remaining and the kinetics will, therefore, be those of an apparent first-order reaction. In the last stage of evaporation, where the last residues lie in the interstices of the soil or plant surfaces, the rate may again become independent of the amount remaining. On plant leaves, the last fraction to be volatilized will be that which has been adsorbed into the intercellular material or leaf waxes: here adsorption effects on fugacity will be important.

Thus, disappearance curves of surface applications may be quite complex for purely physical reasons. Where the coverage is initially incomplete, or the initial volatilization rate is large, the initial period of zero-order kinetics may not be present. Taylor et al. (1977) observed this in field measurements of the volatilization of dieldrin and heptachlor residues from a surface application to an orchardgrass meadow. The relative disappearance of the two insecticides did not follow the vapor pressure/molecular weight ratio approximation suggested by Hartley and Graham-Bryce (1980b), according to which the rate of heptachlor loss should be 108 times that of dieldrin, whereas the observed ratio was about fivefold. This difference in the disappearance rates probably caused changes in the distribution of the two insecticides early in the experiment so that the initial differences in disappearance rates because of chemical properties were partly obscured.

A more complex situation arises when spray formulations are applied to dry porous surfaces where the droplets are absorbed into the underlying soil or plant tissue by capillary action. As the residues in the exposed surface are depleted, the volatilization will depend upon the upward diffusion to the open surface as a concentration gradient is established. The volatilization rate will then be controlled by the back-diffusion of the residues. Mayer et al. (1974) analyzed the way that the volatilization rates that are controlled by diffusive flow change with time. This analysis indicates that in moist soil where there is no bulk transport of residues in soil water, the volatilization rates should decrease in approximate proportion to the square root of time, with the exact relationship depending on the boundary conditions. Laboratory measurements of volatilization rates of lindane and dieldrin (Spencer

& Cliath, 1973; Farmer et al., 1972) through 5-mm depths of moist soil verified this prediction: Jury et al. (1980) reported other work with triallate.

Validations of the predictions of such models in the field are difficult owing to the complex effects of residue distribution, sampling errors and the uncertainties about boundary conditions, but the type of relationship predicted was observed by Glotfelty et al. (1984) in measurements of the volatilization of surface-applied trifluralin, heptachlor, chlordane, and lindane from moist soils in the field.

Thus, it is evident that in addition to differences in chemical properties, the volatilization rates of surface-applied pesticides in the field may also be expected to reflect the physical distribution of the residues on and within the target surface in a complex manner depending upon residue penetration, coverage, and the moisture status of the surface.

### 7-1.2.2 Incorporated Residues

Where pesticides are incorporated into the soil by cultivation, the rate at which they volatilize is controlled by their movement through the body of the soil to the overlying air. Incorporated residues differ from surface applications in two respects. First, as noted in section 7-1.1.2.1, the average concentrations in the soil are lower by an order of magnitude or more. As a result, the equilibrium vapor pressure of these residues will be greatly reduced, as illustrated in Fig. 7-4 and 7-5 for dieldrin and trifluralin. Secondly, the average diffusion path length to the surface will be longer by at least the same factor, so that in addition to the decrease in fugacity the resistance to flow will be much greater. Thus, except for the small fraction of exposed material remaining on the surface that is readily lost, the volatilization of incorporated residues will be greatly restricted. The factors controlling their movement to the surface have been discussed in detail by Spencer et al. (1973), who contrasted the volatilization patterns to be expected when the flow to the surface is controlled by a simple diffusion gradient with those expected where pesticide is transported to the surface in the upward flow of soil water (convective flow).

**7-1.2.2.1 Diffusion Controlled Flow.** After initial uniform incorporation into a moist soil, the pesticide residues will partition between the air, water, and solid phases to a uniform fugacity as described in section 7-1.1.3, with the concentration in the soil solution and the soil atmosphere controlled by desorption from the solid phase. This initial state is illustrated in Fig. 7-10a: in the absence of upward evaporative movement to the surface, as when the relative humidity of the air is close to 100%, the pesticide will volatilize from the surface until the diffusive flow becomes limiting (Fig. 7-10b). This situation is described mathematically by model II in the analyses Mayer et al. (1974) presented. This model has been tested and verified in laboratory measurements of lindane and dieldrin volatilization that Igue et al. (1972), Spencer and Cliath (1973), and Spencer et al. (1973) reported. These and other studies by Jury et al. (1980) for triallate and by Spencer and Cliath (1974) for trifluralin suggest that up to 4 weeks are needed for steady-state

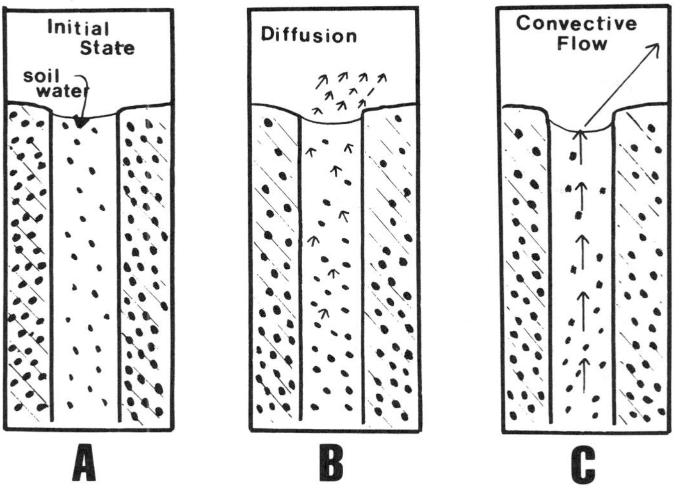


Fig. 7–10. Diffusive and convective (wick effect) mechanisms controlling the volatilization rates of pesticides incorporated into soil.

diffusion gradients to be established. Both the time required for this and the final rate of diffusive flow will reflect the overall diffusion coefficient of the pesticide in the particular soil. Since this coefficient is a complex function of the diffusion pathways through the solid, liquid, and vapor phases in the soil, and these will change with the water content, no single value of such an "effective diffusion coefficient" can be expected to characterize the behavior of even a single pesticide in one soil. Predictions of diffusive flow must rely upon values obtained by calibration observations with the chosen combination of pesticide and soil. Such effective diffusion coefficients may be expected to be sensitive to soil bulk density and temperature in addition to soil water content. Where, however, the soil moisture content is reduced below the critical values so that the fugacity becomes low because of the adsorption in dry soil, diffusive movement will be almost wholly inhibited. These conditions can of course be frequently found in the field, when fallow surface soils—and particularly those of light texture—are exposed to intense drying by wind and sun.

**7–1.2.2.2 Convective Flow and the Wick Effect.** Where continuous upward movement of water is induced by evaporation the resulting upward transport of pesticide may prevent or distort the development of a diffusion gradient, and volatilization at the surface will continue with the water movement (Fig. 7–10c). This flow mechanism has been described as the "wick effect" by analogy with capillary liquid flow to the ends of exposed fibers (Hartley, 1969; Hartley & Graham-Bryce, 1980a). Laboratory data for lindane and dieldrin show that, with these insecticides, volatilization supported by convective flow in moist soil can be up to five times faster than that controlled by diffusion (Spencer & Cliath, 1973). Other experiments with trifluralin (Spencer & Cliath, 1974; Harper et al., 1976) showed much smaller differences reflecting the slower convective transport of trifluralin to the surface because of its lower concentration in the soil water. Jury et al. (1980) also showed that for triallate the relative contribution from convective flow was much less in a silt loam than a sandy loam soil owing to the lower solution concentration established by the greater adsorption in the silt loam, which also had a much higher organic matter content.

Under field conditions, upward movement of moisture to the soil surface is not, however, a continuous process and may cease during rainfall or become slow at high relative humidity or when there is no insolation to evaporate the water. It may, therefore, vary greatly during the day and will almost always be small at night. Convective flow of pesticide to the surface will vary in the same way. This mechanism is illustrated in a number of the field observations discussed in section 7-2.2.

It should, however, be noted that, in common with diffusive flow, convection-controlled volatilization of pesticides will almost totally cease when the surface soil becomes extremely dry.

## 7-2 VOLATILIZATION IN THE FIELD

### 7-2.1 Measurement of Field Volatilization

Volatilization from soils has been measured in both the field and laboratory. Laboratory systems may be divided into experiments using small amounts of soil in confined and highly controlled conditions as Farmer et al. (1972) and Spencer and Cliath (1973) described, and the growth-chamber scale *micro-ecosystems* described by Nash (1983) and Nash and Beall (1979). The data obtained from both kinds of laboratory work differs from field measurements that require observations of a different character taken on a much larger scale. The differences in the results also reflect the much higher degree of experimental control that can be imposed in the laboratory, where temperature, humidity, soil moisture, and air flow speeds can all be kept constant. Laboratory experiments can measure steady-state volatilization rates established when ambient environmental conditions are maintained constant over extended periods or, alternatively, they can measure the changes in rate of loss as the residues are depleted under constant conditions. Such data may be used to compare the behavior of pesticide chemicals of differing character and formulation under similar environmental conditions and to predict the differences between the behavior of different materials in the field.

Field measurement rates are usually based upon upward flux rates calculated from measured gradients of vapor concentration in the air over treated fields. The results reflect the natural uncontrolled soil and weather conditions that can of course themselves show rapid changes. However, since the sampling periods are relatively short, the data can show the effects of such changes on the volatilization rates clearly. The principal disadvantage of field experiments lie in the elaborate instrumentation required and the many samples that must be analyzed.

### 7-2.1.1 Disappearance Methods

When the disappearance of residues in a soil can only be attributed to either chemical degradation or volatilization, the latter can be estimated from the rate of residue disappearance if the rate of chemical degradation is already known. The apparent advantage of this approach lies in its simplicity. If both disappearance and degradation be described by first-order chemical kinetics and both rate constants are uniform during the experiment, the disappearance of the pesticide may be approximated by the kinetic equation describing simultaneous reactions, $R_t = R_0 - \exp(k_c + k_v)t$, where $R_0$ and $R_t$ are the residue concentrations initially and at time $t$, and $k_c$ and $k_v$ are the rate coefficients for degradation and volatilization. Subject to these assumptions, the value of $k_v$ may be obtained by difference from the measured rate of disappearance and the value of $k_c$ where this is known from measurements in closed systems where volatilization is experimentally prevented. In practice, the method can only be expected to give consistent results where the rate of disappearance is measured while the evaporation is not affected by changing temperature and moisture conditions and the residues are not seriously depleted. These restrictions imply that the technique is only likely to be successful with small samples under controlled conditions in the laboratory or ecosystem chambers with elaborate control systems or with highly volatile compounds where measurements can be made over a short time. When used in the field, the method is subject to uncertainty because of sampling variability. This may impose limits of up to ± 30% on individual estimates of $R_t$ (Taylor et al., 1971, 1985). Except where the volatilization loss is much larger than the degradation rate, the value of $k_v$ may be less than the error associated with the measured value of the overall disappearance coefficient ($k_t = k_c + k_v$). Athough the method appears simple in principle, in practice it involves some uncertainties that make it unsuitable except as a means for making comparisons between the volatilization of different compounds from several soil types.

### 7-2.1.2 Vapor Flux Methods

Measurements of pesticide volatilization in the field are made by measurement of the flux of pesticide vapor through a horizontal plane in the turbulent boundary layer of the atmosphere overlying the crop canopy or treated soil. This requires measurement of the gradient of pesticide concentration with height in the air over the field by sampling the air at a series of heights (usually up to about 2 m) above the soil surface for periods of between 30 min and 2 or 3 h, depending upon the weather conditions and the expected pesticide concentrations. The gradients of temperature, water vapor concentration, windspeed, and other supporting meteorological variables must also be measured over the same time. Care must be taken to ensure that the sampling mast is entirely within the atmospheric boundary layer characteristic of the area treated with pesticide. Since the slope of the upper edge of this layer is often as low as 1%, this means that a mast of 2-m height must have a clear fetch of 200 m to the upwind edge of the treated zone to ensure that characteristic profiles are obtained.

**7-2.1.2.1 Theoretical Principles.** Parmele et al. (1972) have presented the principles used in the calculation of the pesticide flux. This calculation is based upon Eq. [1], which becomes, for the pesticide vapor,

$$\hat{I}P = k_p(dp/dz). \qquad [1c]$$

Since there are no sources or sinks for the vapor in the air above the soil or crop canopy, the upward flux through the turbulent air will be uniform with depth and $\hat{I}P$ will be independent of height. (This assumption may not be valid under a crop canopy, where there may be release or adsorption of pesticide by leaves.) The value of $dp/dz$ at the chosen height $z$ is obtained from the profile of vapor concentrations, but the value of $k_p$ is not known and must be measured indirectly.

To this end it must be noted that Eq. [1] is one of a general set of equations of the form $\hat{I}F = k_Z(dx/dz)$ used in micrometeorological studies to describe the vertical transport of water vapor, heat, momentum and $CO_2$ or other vapors through the turbulent boundary layer. Since this transport is outside the laminar layer, molecular diffusion effects are discounted and the dilution of each component is controlled by the mixing in the turbulent air flow. As a first approximation, the values of $k_Z$ in each of the appropriate forms of the equation may then be assumed to be the same: this is called the Similarity Principle. If independent measurements of the flux of a particular component, ($\hat{I}F$), and the corresponding gradient, ($dx/dz$) can be made, the value of $k_p$ may be calculated and used in Eq. [1] to calculate the pesticide flux ($\hat{I}P$).

The validity of the assumption that $k_p$ may be identified with $k_Z$ values for water vapor, heat flow and momentum exchange is not equally good in all cases. Although detailed examination of this topic is beyond the scope of this review, some comment is necessary as a background to the commentary on the field data presented in section 7-2.2. Lemon (1969), Parmele et al. (1972), and Montieth (1973) have published excellent detailed discussions.

Of the three approaches—water vapor flux, energy balance (or heat transport), and aerodynamic (or momentum flux)—the first is probably subject to the fewest assumptions. Since the upward flow of water vapor from the soil or plant surface is a molecular process similar to the dispersion of the pesticide molecules, both may be expected to be diluted in the same way by the turbulent mixing of the air. The principal difference between the two flow patterns is likely to be because of the background concentration of water vapor in the atmosphere that will cause a different gradient in the upper part of the profile as it asymptotically approaches the background level. No similar background exists for the pesticide. Any ambiguities resulting from this difference can be minimized by confining flux and gradient measurements to the lower regions of the boundary layer. The gradient of water vapor concentration is calculated from the humidity measured with wet and dry bulb thermometers mounted at chosen heights. The need for the companion measurement of water vapor flux ($\hat{I}E$) imposes a considerable restriction on the possible use of this method because it requires direct measurements of evapotranspiration from the soil or crop surface. Since this can show rapid fluctuations in response to changes in insolation, it must be measured as accurately as possible over 2 h or less, coincident with the pesticide sampling times. This essentially restricts the use of the method to locations equipped with recording lysimeters: few measurements of pesticide volatilization using this approach have, therefore, been reported. The most notable example is the data Parmele et al. (1972) and Taylor et al. (1976) obtained for dieldrin and heptachlor volatilization at a hydrologic research station in Ohio.

The energy balance, or Bowen ratio method, described by Parmele et al. (1972), Lemon (1969), and Rose (1966), was originally developed to calculate the water vapor flux in the absence of lysimeter data. The flux of water vapor, $\hat{I}E$, is calculated by partitioning the dispersal of energy from the soil surface according to the equation

$$R_n = G + H + (L\hat{I}E) \quad [10]$$

where $R_n$ is the net incident radiation, $G$ the soil heat flux, $H$ the upward transport of sensible heat and $L\hat{I}E$ the heat carried by the dispersing water vapor, $L$ being the latent heat of evaporation. On introducing the Bowen ratio, $H/(L\hat{I}E)$, Eq. [10] transforms to

$$\hat{I}E = (R_n - G)/L(1 + B) \quad [11]$$

where $B = y(\partial T)/(\partial q)$, $(\partial T)$ and $(\partial q)$ being the differences of mean air temperature and water vapor density over a selected height interval within the profile: $y$ is the psychrometric constant. Using the value of $\hat{I}E$ obtained in this way the value of $k_E$ is then estimated from the water vapor flux equation $k_E = \hat{I}E(dq/dz)$. The principal disadvantage of the energy balance method is in the complexity of the supporting data that must include measurements of net radiation, soil heat flux, and both temperature and humidity profiles. This requires a considerable amount of equipment with significant maintenance and calibration costs together with mobile power supplies and shelter in the field. Although a number of sets of field data on pesticide volatilization based upon this method have been published, the need for complex equipment has undoubtedly precluded its wider use.

The aerodynamic method is based on measurement of the gradient of wind speed with height above the surface. This increases with increasing height to the value imposed by the broad-scale meteorological pressure patterns. The lower windspeed in the boundary layer close to the ground reflects the friction or drag exerted by the soil or crop surface as the momentum in the flowing air is transferred to the surface. This downward flux or diffusion of momentum can be described by an equation similar to those for heat, water vapor, and other gases:

$$\hat{I}F_M = - k_M(du/dz) \quad [12]$$

where $k_M$ is the eddy diffusivity coefficient for momentum transfer at the height $z$, and $du/dz$ is the windspeed gradient. In principle it is clear that if we equate the diffusivity coefficients $k_p$ and $k_M$, the pesticide flux $\hat{I}P$ can be derived from measurements of pesticide vapor concentration and wind speed. The conditions under which this equality may be assumed have been discussed by Montieth (1973) and Rose (1966). The assumption implies that momentum, heat, water, and pesticide vapor are all diluted in the same way by the turbulent mixing of the air and that the upward or downward transfers of mass or momentum caused by buoyancy because of heating can be neglected. In turn, this means that the equality can only be expected to hold when the atmosphere is in neutral stability with an adiabatic lapse rate: the profiles of water vapor, pesticide concentration, and momentum above the surface should then all have similar geometrical shapes. Departures from neutral stability occur when the rate of surface heating is greater than the upward flow of heat by turbulent convection, when bubbles of warmer air form and move upward, transporting mass and energy and distorting the wind-induced turbulent flow. In contrast, when the surface layers become dense under radiative cooling the stirring action of the wind is damped out by the increased stability. These effects of buoyancy are corrected for in part by the use of the Richardson number $R_i$, which is defined in terms of the ratio of the temperature lapse rate and the wind velocity gradient (Rose, 1966).

The final form of the equation used in the calculation of the pesticide flux using the aerodynamic method reduces to

$$\hat{I}P = v^2 \frac{(C_1 - C_2)(U_2 - U_1)}{\phi^2 \, [\ln(Z_2 - Z_0)/(Z_1 - Z_0)]^2} \qquad [13]$$

where $C_1$, $U_1$ and $C_2$, $U_2$ are the vapor concentrations and wind speeds at the heights $Z_1$ and $Z_2$ above the surface, $v$ is the Von Karman constant, assumed to be 0.4. Following Parmele et al. (1972), the stability correction term may be defined as $\phi = (1 - bR_i)^{-0.25}$, where $R_i$ is the Richardson number and $b$ is a constant (close to 18) dependent upon the roughness of the surface.

The relative simplicity of the aerodynamic method has made it the method used in many of the published field experiments, despite the possible ambiguities in the assumptions upon which it is based. The extent to which these throw doubt upon the measured volatilization rates must, however, be judged by inspection of the individual experiments. The errors are probably small except where measurements have been made under vigorous thermal heating or stable conditions in the morning and evening: the extent of the possible errors will depend upon the actual measured pesticide volatilization at the particular time.

The choice of the method used in a particular experiment is often dictated by the resources available. The use of the direct water vapor flux method requires an accurate and sensitive lysimeter to measure evapotranspiration from the same crop or soil surface to which the pesticide has been applied. The energy balance method, in which the water vapor flux is calculated from the Bowen ratio, has the theoretical advantage that there probably are less uncertainties in equating the values of $k_E$ and $k_p$ than with the momentum approach, but the equipment needs are much more elaborate. The energy balance method is perhaps to be preferred where fluxes and evapotranspiration rates are likely to be high, but it has the disadvantage that the calculated values of $k_E$ may be indeterminate in the morning and evening or under cool conditions when the net energy input ($R_n - G$) is small. The method can also present difficulties where the energy balance requires terms for advected energy as well as radiation and conduction. This most typically happens in the oasis situation where warm dry air blowing into the experimental area causes water evaporation in excess of that measured as $R_n$ and $G$. These conditions were encountered by Cliath et al. (1980) in measurements of EPTC volatilization from irrigated fields in southern California, when the water evaporation caused stable atmospheric inversions over the plot at midday.

In contrast, the aerodynamic method requires less equipment, but is theoretically less reliable on hot calm days when wind speeds are low and thermal turbulence is dominant. Under calm conditions in the morning and evening—when pesticide volatilization rates may sometimes be large—measurements may become indeterminate, but when the windspeed gradient is maintained outside daylight hours the aerodynamic method may give valuable results when other methods fail.

In the few cases where measurements that permit direct comparisons between the various methods have been made (Harper et al., 1976, 1983; Glotfelty et al., 1984; Grover et al., 1985), the data indicate that the differences between them tend to vary with atmospheric stability and the character of the soil or crop surface.

**7–2.1.2.2 Air-Sampling Techniques.** The profile of pesticide vapor concentration over a treated area is measured by taking samples of air with samplers mounted on vertical masts at a chosen series of heights as illustrated in Fig. 7–11. Each sampler consists of a glass or metal tube containing adsorbent that removes the pesticide vapor from the air drawn through it at a predetermined rate. Since the decrease in pesticide concentration with height is logarithmic, it is convenient to space the sampler heights exponentially to simplify data interpolation. Sampling times can be varied depending upon the experimental objectives, but 30 min is the minimum necessary to integrate the irregularities because of the turbulent vapor flow. Where pesticide concentrations are low, 2 h or more may sometimes be necessary to collect enough residue for accurate analysis. Longer collection times are undesirable because volatilization rates and other factors such as windspeed, temperature, humidity, and soil moisture may be changing so quickly that they cannot be characterized over the longer observation period. Vapor density profiles of trifluralin and heptachlor obtained during 60-min sampling periods at 1 and 24 h after application to a moist silt loam soil are presented in Fig. 7–12 (Turner & Glotfelty, 1977). These vapor concentrations were associated with relatively high volatilization rates (Table 7–4, no. 15 and 27): much lower values are found in many experiments.

A range of materials may be used as the adsorbing substrate in the samplers. In many early experiments, air was pulled through bubblers contain-

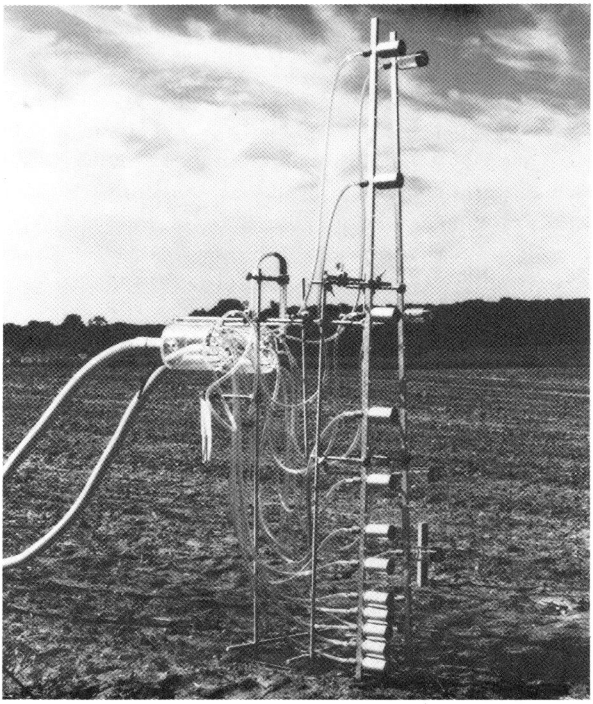

Fig. 7-11. Air sampling system for measurement of profiles of pesticide vapor densities up to 2 m above treated soil or crops in the field.

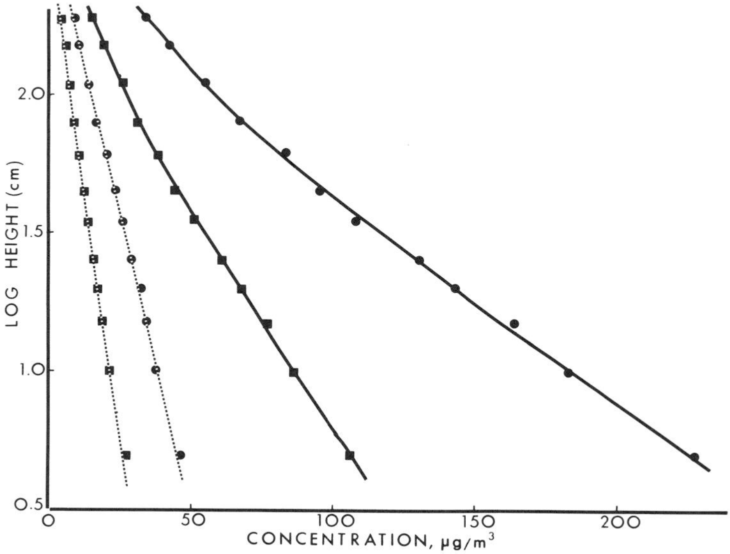


Fig. 7-12. Vapor density profiles of trifluralin (■) and heptachlor (●) during 60-min sampling periods at 1 (—) and 24 h (--) after application to the surface of moist silt loam soil (Turner & Glotfelty, 1977).

Table 7–4. Field measurements of volatilization.

| Pesticide | Exp. no. | Usage | Surface | Season, place | Fraction volatilized in period | Reference |
|---|---|---|---|---|---|---|
| Alachlor | 1 | Surface | Fallow silt loam | May 1981, Maryland | 26% in 24 d | Glotfelty et al., 1989 |
| Atrazine | 2 | Surface | Fallow silt loam | May 1981, Maryland | 2.4% in 24 d | Glotfelty et al., 1989 |
| Chlorpropham | 3 | Surface | Soil under soybean | May 1976, Maryland | 49% in 50 d | Turner et al., 1978 |
| Chlorpropham | 4 | Microencap | Soil under soybean | May 1976, Maryland | 20% in 50 d | Turner et al., 1978 |
| Chlordane | 5 | Surface | Moist fallow silt loam | Aug. 1975, Maryland | 50% in 50 h | Glotfelty, 1981; Glotfelty et al., 1984 |
| Chlordane | 6 | Surface | Dry fallow sandy loam | June 1978, Maryland | 2% in 50 h | Glotfelty, 1981; Glotfelty et al., 1984 |
| Dacthal | 7 | Surface | Moist fallow silt loam | Aug. 1975, Maryland | 2% in 34 h | Glotfelty, 1981; Glotfelty et al., 1984 |
| DDT | 8 | Incorp. to 9 cm, then surface | Moist fallow silt loam | Oct. 1968, Louisiana | 44% in 120 d | Willis et al., 1971 |
| DDT | 9 | Cotton field | Foliage, 45% groundcover | Sept. 1976, Mississippi | 21% in 5 d<br>58% in 11 d | Willis et al., 1980 |
| Dieldrin | 10 | Surface | Moist soil fallow | Sept. 1969, Louisiana | 20% in 50 d | Willis et al., 1972 |
| Dieldrin | 11 | Incorp. to 7.5 cm | Soil under corn | May 1969, Ohio | 4% in 170 d | Taylor et al., 1976 |
| Dieldrin | 12 | Vegetation | Short orchardgrass | July 1973, Maryland | 90% in 30d | Taylor et al., 1977 |
| EPTC | 13 | Surface in irrigation water | Soil under 25 cm alfalfa | May 1977, California | 74% in 52 h | Cliath et al., 1980 |
| Heptachlor | 14 | Incorp. to 7.5 cm | Soil under corn | May 1969, Ohio | 7% in 170 d | Taylor et al., 1976 |
| Heptachlor | 15 | Surface | Moist fallow silt loam | Aug. 1975, Maryland | 50% in 6 h<br>90% in 6 d | Glotfelty, 1981; Glotfelty et al., 1984 |
| Heptachlor | 16 | Surface | Dry fallow sandy loam | June 1978, Maryland | 40% in 50 h | Glotfelty, 1981; Glotfelty et al., 1984 |
| Heptachlor | 17 | Vegetation | Short orchardgrass | July 1973, Maryland | 90% in 7 d | Taylor et al., 1977 |
| Lindane | 18 | Surface | Moist fallow silt loam | June 1977, Maryland | 50% in 6 h<br>90% in 6 d | Glotfelty, 1981; Glotfelty et al., 1984 |
| Lindane | 19 | Surface | Dry fallow sandy loam | June 1978, Maryland | 12% in 50 h | Glotfelty, 1981; Glotfelty et al., 1984 |
| Photodieldrin | 20 | Vegetation | Short orchardgrass | July 1973, Maryland | 5% $d^{-1}$ | Turner et al., 1977 |
| Simazine | 21 | Surface | Fallow silt loam | May 1981, Maryland | 1.1% in 24 d | Glotfelty et al., 1989 |
| Toxaphene | 22 | Cotton crop | Foliage | Sept. 1974, Mississippi | 25% in 5 d | Willis et al., 1980 |
| Toxaphene | 23 | Cotton field | Foliage 45% groundcover | Aug. 1976, Mississippi | 53% in 33 d | Willis et al., 1983 |
| Toxaphene | 24 | Surface | Fallow silt loam | May 1981, Maryland | 33% in 24 d | Glotfelty et al., 1989 |
| Trifluralin | 25 | Incorp. to 2.5 cm | Soybean on sandy loam 0.5% o.m. | June 1973, Georgia | 22% in 120 d | Harper et al., 1976 |
| Trifluralin | 26 | Incorp. to 7.5 cm | Soybean on loam 4% o.m. | May 1973, New York | 3.4% in 90 d | Taylor, 19XX, unpubl. data |
| Trifluralin | 27 | Surface | Moist fallow silt loam | Aug. 1975, Maryland | 50% in 7.5 h<br>90% in 7 d | Glotfelty, 1981; Glotfelty et al., 1984 |
| Trifluralin | 28 | Surface | Moist fallow | June 1977, Maryland | 87% in 50 h | Glotfelty, 1981; Glotfelty et al., 1984 |
| Trifluralin | 29 | Surface | Dry fallow sandy loam | June 1978, Maryland | 25% in 50 h | Glotfelty, 1981; Glotfelty et al., 1984 |
| 2,4-D | 30 | Surface | 20-cm high spring wheat | June 1981, Saskatchewan | 21% in 5 d | Grover et al., 1985 |

ing hexylene or ethylene glycol. This method was discarded in favor of solid adsorbents because of the difficulties of transferring liquids in the field and adsorption of water by the glycol caused changes in liquid volume and flow rate in all but the driest atmospheres. Solid adsorbers such as polyurethane foam offer great convenience in transfer and storage of samples, but require elaborate cleaning and purification before use (Turner & Glotfelty, 1977; Grover & Kerr, 1981). For many purposes, powdered materials such as chromasorb or XAD resin may be used: these require less chemical preparation but, as powders, they require more careful handling in transfer and filling operations and in controlling the density of packing in the sampler, which may affect resistance to air flow. Trapping efficiency and the ease with which the residues may be recovered for analysis are also important considerations in the choice of material. The stability of the residues on the adsorbent during sampling and storage is of major importance: some adsorbents may adsorb significant amounts of water from humid air, which may cause the hydrolysis of some pesticide residues (Grover & Kerr, 1978). Bidleman (1985) has reviewed the range of materials available.

### 7-2.2 Field Experiments

The earlier discussion shows that the properties of both the pesticide and the soil are important in controlling the volatilization in the field. Predictions of their relative effects may, however, require simplifying assumptions in the complex and often transient conditions of the uncontrolled field environment. Field measurements of volatilization, supported by data on usage and management, residue concentrations, soil moisture levels, surface temperatures, and weather conditions are, therefore, essential in understanding how these interact with the chemical properties and soil factors to control volatilization in the field. Without such data no real calibration of laboratory data or modelling projections can be made.

Field experiments are, however, complex and expensive, requiring the collaboration of meteorologists, pesticide chemists, weed scientists, and field staff. Equipment must include special meteorological instrumentation and laboratory facilities for the rapid and accurate analyses of many soil and air samples. Such experiments must yield the maximum amount of data to evaluate the important factors in the particular situation. Flexibility in experimental plans is important because the time span during which the pesticide can be applied is often limited by the crop management practice, and once the application is made the experimenter has no further control over the course of events. Since the results of chemical analyses are never available during the experiment, decisions about changes in experimental plans must often be based solely on judgment aimed to obtain the best information possible under the circumstances.

In recent years, a considerable body of field data has been published despite these difficulties. This information is summarized in Table 7-4. All the original reports contain much data that cannot be presented here and the individual experiments will, therefore, be reviewed only so far as the results relate to the effects of chemical properties, residue distributions, soil conditions, and management practices discussed above.

A preliminary inspection of the data, however, reveals a wide range in the observed volatilization rates extending from losses of 70% within a few hours to small losses over weeks or months (Table 7-4, Exp. no. 13, 15, 18, and 28 and 11, 14, and 26). Some comment is appropriate on the probable precision of the measured loss rates. Where initial volatilization is high and the rates are changing rapidly (Cliath et al., 1980, Table 7-4, no. 13; Glotfelty et al., 1984, no. 5, 15, 16, 18, 27, 28, and 29; Taylor et al., 1976, no. 12, 17) precision is limited by the need to sample for minimal periods of at least 30 min to obtain useful profiles. The calculated flux rates are then integrated values averaged over the sampling time. Measurements immediately after application are often impractical because of the time required to mount sampling equipment that cannot be installed until spraying and cultivation are finished. Direct measurements of the disappearance of residues by soil analyses are also complicated by the time consumed in taking the number of samples needed to obtain statistically meaningful results. Although the extent of uncertainty resulting from these difficulties in both air and soil sampling cannot be generalized, it appears likely that there may be a tendency to underestimate rapidly changing initial volatilization rates. The probable error in each case may be best estimated by a careful inspection of the data.

Where the volatilization rates are low, estimates of long-period losses must be made with interpolated data for the days when observations are not available. The reliability of these interpolations must again be evaluated from the available data. In the experiment by Harper et al. (1976, Table 7-4, no. 25) where 79% of the trifluralin volatilized was lost in the first 18 d, the weather data suggest that the rate showed a steady decline without erratic variations, so that the interpolations can be made with some confidence. Harper et al. (1976) have presented a detailed interpretation of the daily data from this experiment. Data obtained by Taylor et al. (A.W. Taylor, Soil Nitrogen and Environmental Chemistry Lab., USDA-ARS, Beltsville, MD. Unpublished data) (Table 7-4, no. 26) on heavier-textured soil under cooler and wetter conditions in New York indicated much smaller and more uniform trifluralin losses with a total growing-season volatilization of <3.4%.

Data for dieldrin and heptachlor reported by Taylor et al. (1976, Table 7-4, no. 11, 14) showed that estimates based upon linear interpolations of the individual daily data were within 2% of the totals calculated using measured daily evapotranspiration rates, which were also shown to be related to pesticide volatilization on days in which this was measured.

### 7-2.2.1 Usage Patterns

**7-2.2.1.1 Surface Applications.** The results in Table 7-4 for surface-applied residues show a considerable range of volatilization rates, extending from 50% within 6 to 8 h for heptachlor, lindane, and trifluralin on moist soil (Glotfelty et al., 1984, Table 7-4, no. 15, 16, 18, 19, 27, 28, 29) down

Table 7-5. Initial volatilization rates of lindane and trifluralin from surfaces of moist soil at Beltsville, MD, June, 1977 (Glotfelty, 1981).

| Sampling time | Lindane flux | | Trifluralin flux | |
|---|---|---|---|---|
| h, EDT | $g\ ha^{-1}\ h^{-1}$ | %† | $g\ ha^{-1}\ h^{-1}$ | %† |
| | | 10 June | | |
| 1200 | 126 | 11.4 | 195 | 7.0 |
| 1315 | 73 | 6.6 | 121 | 4.3 |
| 1430 | 27 | 2.5 | 56 | 2.0 |
| 1545 | 22 | 2.0 | 55 | 2.0 |
| 1700 | 24 | 2.2 | 49 | 1.8 |
| 1800 | 18 | 1.6 | 38 | 1.3 |
| Day 1, Total§ | 363 | 33 | 643 | 23 |
| | | 11 June | | |
| 0645 | 2.8 | 0.3 | 11.0 | 0.4 |
| 0800 | 5.2 | 0.5 | 18.0 | 0.6 |
| 0915 | 9.0 | 0.8 | 26.0 | 0.9 |
| 1030 | 12.0 | 1.1 | 35.0 | 1.3 |
| 1140 | 12.0 | 1.1 | 35.0 | 1.3 |
| 1300 | 9.1 | 0.8 | 28.0 | 1.0 |
| 1415 | 7.9 | 0.7 | 24.0 | 0.9 |
| 1530 | 5.3 | 0.5 | 11.0 | 0.4 |
| 1645 | 2.7 | 0.4 | 7.3 | 0.3 |
| 1800 | 1.9 | 0.2 | 3.2 | 0.1 |
| 1915 | 1.6 | 0.1 | 3.2 | 0.1 |
| Day 2, Total | 87 | 8 | 252 | 9 |
| Day (1 + 2) Total | 450 | 41 | 895 | 32 |

† Percent fraction of applied (lindane, 1.1 kg/ha; trifluralin, 2.8 kg/ha).
‡ Application 1030 to 1115 EDT 10 June 1977.
§ Rates measured during 60 min of each period; 75 min totals by extrapolation.

to <2% $d^{-1}$ for alachlor, dacthal, and toxaphene, and even lower values for atrazine and simazine (no. 1, 7, 24, 2, and 21). The most rapid losses are found where materials with the highest vapor pressures are exposed on the surface of moist soils. Some typical initial flux rates for volatile compounds from moist soil surfaces are shown by the lindane and trifluralin data presented in Table 7-5; the highest rate for lindane was 11.4% $h^{-1}$ during the second hour after application.

Comparisons of volatilization rates after the initial losses are more complex because of diurnal variations reflecting changes in insolation and soil moisture conditions and differential depletion of residues. In many experiments, volatilization rates from moist soils show marked diurnal changes with maximum rates in the early afternoon. This pattern, which was shown by both lindane and trifluralin on the second day of the above experiment (Table 7-5), is characteristic of pesticide evaporation where the soil moisture level remains above the critical levels at which the fugacity is reduced by strong adsorption on dry mineral surfaces as discussed in section 7-1.1.2.4.

The effect of reduced soil moisture is shown by comparing losses of surface-applied chlordane, heptachlor, trifluralin, and lindane on moist and dry soils presented in Table 7-6 (Glotfelty, 1981). Although the soil surface

Table 7-6. Diurnal change in volatilization rates of surface-applied† trifluralin, heptachlor, and lindane from sandy loam soil at Salisbury, MD, June 1978. (Glotfelty, 1981).

| Time‡ (EDT) | Volatilization rate | | | Ratio§ | |
|---|---|---|---|---|---|
| | Trifluralin | Heptachlor | Lindane | T/H | L/H |
| | $g\ ha^{-1}\ h^{-1}$ | | | | |
| | | 14 June | | | |
| 0900 | 11.0 | 57.0 | 13.0 | 0.29 | 1.4 |
| 1030 | 7.2 | 41.0 | 8.5 | 0.27 | 1.3 |
| 1230 | 4.3 | 32.0 | 6.2 | 0.20 | 1.5 |
| 1515 | 1.0 | 21.0 | 4.0 | 0.07 | 1.1 |
| 1645 | 5.1 | 9.2 | 2.0 | 0.83 | 1.3 |
| | | 15 June | | | |
| 0700 | 5.1 | 32.0 | 5.9 | 0.24 | 1.1 |
| 0815 | 3.6 | 17.0 | 3.9 | 0.32 | 1.4 |
| 0930 | 1.2 | 8.5 | 2.3 | 0.21 | 1.6 |
| 1045 | 0.7 | 5.8 | 1.3 | 0.20 | 1.4 |
| 1200 | 1.1 | 8.6 | 2.1 | 0.19 | 1.5 |
| 1315 | 1.0 | 9.4 | 1.9 | 0.16 | 1.2 |
| 1500 | 0.7 | 6.9 | 1.6 | 0.15 | 1.4 |
| 1615 | 0.7 | 7.1 | 1.8 | 0.15 | 1.5 |
| 1945 | 0.7 | 11.0 | 3.0 | 0.10 | 1.7 |
| 2110 | 1.3 | 24.0 | 7.4 | 0.08 | 1.9 |

† Application rates (kg/ha): trifluralin, 2.50; heptachlor, 2.75; lindane, 0.62. Application from 0610 to 0640, 14 June.
‡ Each time is mid-point of a 1-h sampling period.
§ Ratios of volatilization rates per unit weight of residue—T/H = Trifluralin/heptachlor; L/H = Lindane/heptachlor.

was initially moist the low soil moisture retention capacity of the sandy soil allowed the top 5 mm to become extremely dry during the day, so that the fugacity of the pesticide residues, which were all concentrated in the surface layer, were reduced to a low value by the increased adsorption. Similar results were reported by Turner et al. (1978, Table 7-4, no. 3), from an experiment in which a spray of a waterbased emulsion of chlorpropham was applied to a dry bare soil.

Glotfelty et al. (1989) (Table 7-4, no. 1) observed the effect of moisture on the volatilization of surface residues of alachlor. High volatilization losses of EPTC were measured by Cliath et al. (1980, Table 7-4, no. 13) during flood irrigation of an alfalfa (*Medicago sativa* L.) field in California.

Measurements of the volatilization of the octyl ester of 2,4-D from 20-cm high wheat (*Triticum aestivuum* L.) by Grover et al. (1985, Table 7-4, no. 30) in Saskatchewan in May showed a more complex pattern because of volatilization from the residues on the plant leaves, which had intercepted 52% of the application. There was little or no loss from the soil during the first 2 d, but an increase on the third day was because of desorption of the soil residues following rainfall. These results indicate the complexity that may be encountered in volatilization patterns where surfaces with different characteristics are present at the same time. Other work with insecticides has shown that volatilization from transpiring plant leaves can take place freely during the day even when the underlying soil surface is dry (Taylor et al., 1977, Table 7-4 no. 12 and 17; Willis et al., 1983, no. 23).

Thus, the field observations all confirm the primary importance of the moisture content of the surface soil layer in controlling the volatilization of surface residues. The effect is clearly because of the reduction in the fugacity of the residues by the increased adsorption in dry soil.

Data from two experiments (Glotfelty, 1981; Glotfelty et al., 1989, Table 7-4, no. 2, 7, and 21) indicate that measurable amounts of some pesticides can enter the atmosphere by wind erosion of surface residues. The evidence was obtained with a set of air samplers containing plugs of glass fiber filters mounted alongside the regular polyurethane plug vapor samplers. The difference between the amounts of dacthal and the heptachlor and trifluralin found on the fiber filters reflected the wind erosion of the wettable powder formulation in which the dacthal was applied: all the other compounds were applied as water-based emulsions. In a second field experiment with alachlor, atrazine, and simazine, data from glass fiber samplers again suggested wind erosion of the atrazine and simazine (Glotfelty et al., 1989). The actual amount of pesticide transported in this way appears to be small and may be neglected in comparison to vapor dissipation for all but the least volatile materials. It may, however, be of some significance in the long distance transport and deposition of pesticides of this type (Glotfelty, 1983).

Volatilization of residues from plant leaves appears to follow the same general patterns as that from moist soils except that the effects of changes in moisture content are less evident and the effects of decreasing residue coverage are more prominent. Initial losses are sometimes rapid: 46% of the heptachlor and 12% of the dieldrin were lost within 8 h of application to short orchardgrass (Taylor et al., 1977, Table 7-4, no. 12, 17). These heptachlor losses, which are comparable to those of EPTC from irrigation water under desert conditions found by Cliath et al. (1980, Table 7-4, no. 13), appear to be the highest reported in the literature up to this time.

Toxaphene losses from cotton (*Gossypium hirsutum* L.) leaves (Willis et al., 1983, Table 7-4, no. 23) showed rates consistent with first-order kinetics suggesting that the decrease in rate was controlled by changing residue distribution and fluctuations of specific volatilization rates (kg/kg of residue) with leaf temperature (Harper et al., 1983). Other measurements of the rates of disappearance of toxaphene from cotton plants in California by Seiber et al. (1979) were also consistent with first-order losses, but no measurements of vapor fluxes were made.

A full understanding of the rates of volatilization from plant leaves requires more data than is available, particularly for nonpersistent materials. The reduction in fugacity by adsorption on plant leaves is more complex than in soils, and the picture is further complicated by more complex metabolism kinetics and the possibilities of wash-off from leaves during rainfall or overhead irrigation.

**7-2.2.1.2 Incorporated Residues.** Inspection of the results summarized in Table 7-4 shows that volatilization rates are much less when residues are incorporated into soils. This observation is of course consistent with the reasons for this practice, which is aimed to extend the persistence of compounds that would otherwise be greatly shortened by physical losses. Organochlorine insecticides, whose use is now discontinued in the USA, were incorporated to extend their effectiveness against soil-borne insects: trifluralin, which is sensitive to photodecomposition, is still used in this way to extend its effectiveness as a herbicide. Taylor et al. (1976, Table 7-4, no. 14) showed that incorporation to a depth of 7.4 cm before corn (*Zea mays* L.) planting on 30 April (in Ohio) reduced volatilization losses of heptachlor to $<10\%$ over a growing season of 100 d or more indicating that the persistence depended on the rate of chemical degradation rather than volatilization (Freeman et al., 1975).

The highest daily fluxes measured in this experiment were 5.0 g $ha^{-1}$ $d^{-1}$ of heptachlor and 4.0 g $ha^{-1}$ $d^{-1}$of dieldrin on 26 June 1969. Fluxes showed marked midday maxima on all the days in which the surface soil remained moist and there was free surface water evaporation. These diurnal variations were closely correlated with the evaporation of soil water as measured in the lysimeter, confirming the importance of the wick effect in maintaining the supply of heptachlor at the soil surface, as discussed in section 7-1.2.2.2.

The effects of soil moisture upon trifluralin volatilization from a Cecil sandy loam (clayey, kaolinitic, thermic Typic Hapludult) in Georgia were described by Harper et al. (1976) and White et al. (1977), (Table 7-4, no. 25). In this experiment, 95% of the trifluralin was contained in the top 2.5-cm layer of soil. The water content of this layer was measured throughout the experiment, so that volatilization rates could be directly related to the soil moisture levels. Major effects of soil moisture levels on volatilization rates were observed. As the surface soil dried to a water content $<0.01$ $cm^3/cm^3$ between 0400 and 1200 h on the second day, the volatilization rate fell from 1.7 to below 0.15 g $ha^{-1}$ $d^{-1}$ despite a fourfold increase in the atmospheric diffusivity coefficient, $k_Z$. This soil moisture level corresponds to less than the monomolecular layer level at which the vapor density of trifluralin would be reduced to a low level by adsorption (Spencer & Cliath, 1974).

In another and later field experiment at Hartford, NY (A.W. Taylor, Soil Nitrogen and Environmental Chemistry Lab., USDA-ARS, Beltsville, MD. Unpublished data) (Table 7-4, no. 26) in which trifluralin was applied at 0.7 kg/ha and immediately incorporated to the 7.5-cm depth in a gravelly loam the results suggested that diffusion controlled flow was the limiting mechanism throughout the growing season. Even though the soil was moist on all the days when fluxes were measured, no diurnal variations with marked midday maxima were observed and the rates were uniform from early May to the end of July, with daily losses ranging between 0.22 and 0.4 g $ha^{-1}$ $d^{-1}$. Comparison of this result with the laboratory data Spencer and Cliath (1974) obtained, which indicated that the limiting rate of diffusion controlled flux of trifluralin mixed into Gila silt loam at 14 kg/ha was between 2.5 and 4 g $ha^{-1}$ $d^{-1}$ after 20 d, shows an acceptable level of agreement between the laboratory and field studies when allowance is made for the different application rates and soil conditions.

None of these experiments revealed meaningful correlations between flux rates and air or soil temperature. The results demonstrate the dominant role of soil moisture in controlling the volatilization of soil-incorporated pesticides. Where soil moisture moves continuously to the surface in response to surface evaporation, the volatilization is controlled by the wick effect and reflects the rate at which the residue concentration at the soil surface is replenished by the upward convective movement of the pesticide in the flow of soil moisture. Marked diurnal variations with midday maxima are often observed as the rate of water evaporation changes with insolation or energy advection by wind. When the soil remains moist but there is little water evaporation, volatilization becomes controlled by diffusive flow: both daily and long-term volatilization then tend to be slower and more uniform.

Where the surface soil dries to a moisture content $<0.02$ to 0.05 kg/kg depending upon soil texture, volatilization is reduced to low values or ceases completely because of reduction of the pesticide vapor density by increased adsorption on the dry soil. Since air and surface soil temperatures may rise steeply over dry soil that is not cooled by water evaporation, no meaningful relationship can be expected between volatilization and temperature or windspeed.

All these field experiments indicate clearly that the volatilization rates of incorporated residues are controlled by soil conditions. Meteorological-variables such as temperature, radiation input, and windspeed are important in so far as they control the soil conditions, but soil conditions—and particularly soil moisture levels—are critical in controlling the supply of available pesticide at the surface, which in turn controls the volatilization rate.

**7–2.2.1.3 Volatilization from Water Surfaces.** The volatilization of pesticide residues applied to open water surfaces for the control of insects or aquatic vegetation is of special environmental interest, emphasized by the great importance of rice (*Oryza sativa* L.) culture in many parts of the world. Although the physical chemistry of the distribution of the residues between the air, water, and sediments will not differ fundamentally from that in soils, the large uniform water surface will greatly modify the dynamics of the volatilization process. The rate of loss will also be influenced by the retention of residues by the submerged sediments.

Liss and Slater (1974) have discussed the theoretical approach to volatilization loss from water bodies in terms of the *two resistance-layer model* that Whitman (1923) originally introduced. A more comprehensive interpretation has been presented by Mackay (1984) and by Mackay et al. (1986) in terms of fugacity. The question has also been reviewed by Hartley and Graham-Bryce (1980a) who point out the importance of solar heating, wave action, and wind currents in renewing the concentrations at the surface and reducing the importance of diffusion as the rate-limiting step.

Although a considerable amount of data exists for air-water transfer of PCBs and other volatile toxics, there are no measurements of pesticide volatilization from treated open water bodies currently available in the literature, and no evaluation of the modelling interpretations are therefore possible in this context. This lack of data also represents a major gap in environmental studies of pesticide behavior.

**7-2.2.1.4 Microencapsulation.** The volatilization of chlorpropham from a micro-encapsulated formulation was compared with that of a conventional emulsion in a field experiment reported by Turner et al. (1978, Table 7-4, no. 4). The formulation was a chlorpropham solution contained in 25 $\mu$m nylon capsules suspended in water and applied as a water-based spray. The specific volatilization rate of the encapsulated formulation was between 12 and 26% of the conventional during the first 8 d of the experiment. Over 50 d, about one-half of the disappearance of the conventional formulation and about one-fourth of that of the encapsulated material were due to volatilization, with the remainder due to chemical or biochemical degradation. The similarity in volatilization patterns suggested that the volatilization of the encapsulated formulation was not controlled by direct release from the capsules themselves, but was associated with revolatilization of chlorpropham that had been adsorbed on the soil surface after its release from the capsules. Field evaluations by Gentner and Danielson (1976) showed that the formulation Turner et al. used was herbicidally effective for a significantly longer period than the conventional emulsions.

These results suggest that encapsulation techniques offer real possibilities for the reduction of the environmental impact of some pesticides where there is no reduction in biological activity and the higher cost can be justified. This cost may be greater than that of using the larger amounts of pesticide in conventional formulations needed to offset the volatilization loss. The need for special equipment to apply such formulations—which sometimes cause settling and distribution problems in storage and spray tanks—also represents a further increase in cost. The reduced volatilization from encapsulation must also be balanced against the increased risk of movement of such formulations in surface runoff, since these may be easily detached from soil surfaces with an increased direct impact on water quality in streams or waterways.

### 7-2.2.2 Comparative Behavior

These experiments show that field volatilization rates are highly variable and sensitive to weather, soil conditions, and usage patterns. While the influence of the soil chemistry and chemical properties of individual compounds are often evident in the ways that their volatilization patterns change with the soil conditions in a particular experiment, the variable nature of the field data makes it difficult to compare different compounds in separate experiments done at different times and places. The effects of chemical properties are, therefore, best evaluated in experiments where two or more compounds are applied together and sampled simultaneously in the same experiment. Even where two materials are compared in separate experiments against a third standard compound used in each (e.g., B compared with C from experiments with AB and AC) the results are less satisfactory than direct comparisons unless the weather and usage patterns are similar.

Comparison of the fluxes of compounds sampled together in the same equipment are in effect comparisons of simultaneous vapor density profiles and discount any errors introduced by ambiguities in flux calculations. An example of such an experiment is presented in Table 7–7 which lists the volatilization rates of trifluralin and heptachlor during the first 36 h after a surface application to a moist silt loam (Glotfelty, 1981; Glotfelty et al., 1984). There was no restriction of volatilization by soil drying. The ratios of the specific volatilization rates—(volatilization per unit weight of residue) show that during the first 2 d the heptachlor was lost between 1.6 and 2.1 times more quickly than the trifluralin: the average ratio was 1.73. According to the equation that Hartley (1969) and Hartley and Graham-Bryce (1980b) suggested, the ratios of the fluxes of two pesticides through the laminar layer when the concentrations at the inner surface are $p_1$ and $p_2$ should be described by the equation

$$\hat{I}F_1/\hat{I}F_2 = (p_1/p_2)(M_1/M_2)^{1/2} \quad [14]$$

in which the diffusion coefficients are assumed to be proportional to the square roots of the molecular weights. This equation predicts a flux ratio of 2.8 if the vapor pressures were those of the pure compounds. The observed value of 1.73 suggests that the reduction in the fugacity of the heptachlor by adsorption was about 60% of that of the trifluralin. This estimate is consonant with the respective values of 6 and 4% relative vapor densities of the two compounds in the air close to the surface (Glotfelty et al., 1984). Although there was a considerable decrease in the specific volatilization rates with time, the ratio remained essentially constant suggesting that the common reduction of the volatilization of both compounds probably reflected slow changes in the distribution of the residues as the formulation penetrated the soil and the rate became controlled by movement back toward the soil surface as Glotfelty et al. (1984) noted.

In contrast to this result, where the similar behavior of these compounds in moist soil indicates the importance of physical factors, their behavior in dry soils reveals contrasting adsorption effects. The volatilization rates of trifluralin, heptachlor, and lindane during the first 2 d after application to the surface of a bare sandy loam soil are given in Table 7–6 (Glotfelty, 1981). Here the volatilization of all three compounds was reduced by midday drying of the soil surface. The ratios of the specific volatilization rates of the two organochlorine insecticides remained constant at a value of about 1.4 over the whole experiment. In contrast, the trifluralin/lindane ratios showed a consistent decrease after the reduction of volatilization because of soil drying and did not recover on the second evening. It is clear that during the soil rewetting the fugacity of the trifluralin on the dry soil did not recover at the same rate as that of the organochlorines. This result may be compared with the labora-

Table 7-7. Diurnal change in volatilization rates of surface-applied† trifluralin (T) and heptachlor (H) from moist silt loam soil at Beltsville, MD, August 1975 (Glotfelty, 1981).

| Time | Heptachlor flux | | Trifluralin flux | | Ratio H/T‡ |
|---|---|---|---|---|---|
| h, EDT | g $ha^{-1}$ $h^{-1}$ | %§ | g $ha^{-1}$ $h^{-1}$ | § | |
| | | | 8 Aug. | | |
| 0900 | 76 | 2.15 | 38 | 1.35 | 1.6 |
| 1000 | 117 | 3.30 | 56 | 1.95 | 1.7 |
| 1100 | 117 | 3.30 | 58 | 2.05 | 1.6 |
| 1200 | 71 | 2.00 | 33 | 1.15 | 1.7 |
| 1300 | 83 | 2.35 | 38 | 1.35 | 1.7 |
| 1400 | 65 | 1.85 | 30 | 1.05 | 1.7 |
| 1500 | 62 | 1.75 | 27 | 0.95 | 1.8 |
| 1600 | 35 | 1.00 | 16 | 0.55 | 1.8 |
| 1700 | 21 | 0.59 | 10 | 0.35 | 1.7 |
| 1800 | 20 | 0.56 | 9.4 | 0.33 | 1.7 |
| | | | 9 Aug. | | |
| 1030 | 7.3 | 0.21 | 4.3 | 0.15 | 1.4 |
| 1130 | 12 | 0.34 | 6.8 | 0.24 | 1.4 |
| 1230 | 20 | 0.56 | 9.7 | 0.34 | 1.7 |
| 1330 | 21 | 0.59 | 9.6 | 0.34 | 1.7 |
| 1430 | 24 | 0.67 | 9.7 | 0.34 | 2.0 |
| 1530 | 15 | 0.42 | 5.8 | 0.20 | 2.1 |
| 1630 | 11 | 0.31 | 4.6 | 0.16 | 1.9 |
| 1730 | 10 | 0.28 | 4.2 | 0.14 | 1.9 |

† Application rates—Heptachlor 3.55 kg/ha; trifluralin 2.84 kg/ha.
‡ Ratio of volatilization rates per unit weight of residue.
§ Percent fraction of applied lost per hour.

tory data on the effects of drying on the vapor pressures of trifluralin and organochlorine residues in soil. Spencer and Cliath (1974) showed that in dry Gila soil the vapor pressure of trifluralin residues was 0.02% of that in the same moist soil, whereas the lindane vapor pressure was not reduced below about 13% of the moist value. Owing to the marked difference in the character of the soils used in the field and laboratory studies no direct comparison is possible, but it is clear that the results are mutually consistent. No comparable laboratory data are available for heptachlor.

This result is important because it shows that comparisons of the volatilization rates of two compounds, even under identical experimental conditions, can only be expected to reflect the vapor pressures of the pure materials as long as they are evaporating from surfaces where neither has been strongly adsorbed. Where the adsorption characteristics have been altered as a result of soil drying, their recovery on rewetting may not be wholly reversible.

Comparative volatilization rates of heptachlor and dieldrin residues incorporated to the 7.5 cm depth in the same experiment were reported by Taylor et al. (1976). During the growing season, the average daily heptachlor/dieldrin volatilization ratio varied between 0.7 and 2.9 with an average value of 1.92. This is much smaller than the ratio of their vapor pressures (about 33), and it appears likely that the relative rates of volatilization were being governed by the rates of their movement to the soil surface.

Two sets of field data are available that permit direct comparisons of insecticide residues on plant leaves. Measurements of volatilization of dieldrin and heptachlor applied together to an orchardgrass pasture by Taylor et al. (1977) are summarized in Table 7–8. Rapid initial losses were observed with close to 30% of the heptachlor being lost in the first 2 h after the application. During this time, the dieldrin was volatilized about one-fifth as fast as the heptachlor. Extrapolation of the dieldrin profiles to ground level suggested that during this time the air close to the surface was saturated with dieldrin vapor, whereas the relative vapor density of heptachlor at the surface was 12% or less. It is likely that the heptachlor volatilization was controlled by depletion of the residues soon after application, but this condition was not established for the dieldrin for 2 or 3 h. After the initial conditions had stabilized, the disappearance of the residues of both insecticides followed apparent first-order kinetics for about 1 wk with half-lives of 1.7 d for heptachlor and 2.7 d for dieldrin. The marked noonday maxima typical of volatilization from moist surfaces were observed until after the 9th day. These results indicate that fugacity of residues on moist plant leaves is not reduced by adsorption until some time after application and the difference in the losses of the two insecticides reflects the differing distribution of the residues as discussed in section 7–1.2.1.

Comparative results for the losses of toxaphene and DDT from the leaves of cotton plants in the field for up to 30 d after application were reported by Willis et al. (1983) and Harper et al. (1983). The complex chemical character of toxaphene makes detailed comparison of the results uncertain, but the two compounds showed generally similar behavior with disappearance of the residues again approximating first-order kinetics. Daily evaporative losses were about 6% for both pesticides during the 33-d experiment.

In general, the results obtained from experiments comparing the behavior of two or more pesticides under identical conditions show that such comparative experiments can give information about the chemical and environmental factors controlling volatilization that cannot be obtained in any other way. In view of the cost in time and resources required in field experiments on volatility, it is strongly recommended that such an approach should always be considered in future experimental work in the field.

Table 7–8. Volatilization of heptachlor (H) and dieldrin (D) residues from orchardgrass at Beltsville, MD, 1973 (Taylor et al., 1977).

| Sample period† | Heptachlor flux | | Dieldrin flux | | Ratio H/D |
|---|---|---|---|---|---|
| | $g\ ha^{-1}\ h^{-1}$ | %‡ | $g\ ha^{-1}\ h^{-1}$ | %‡ | |
| 1–2 | 822 | 14.7 | 169 | 3.0 | 4.9 |
| 3–4 | 296 | 5.3 | 80 | 1.4 | 3.8 |
| 5–6 | 128 | 2.3 | 61 | 1.1 | 2.1 |
| 7–8 | 29 | 0.5 | 16 | 0.3 | 1.7 |
| Total loss | 2550 | 46 | 652 | 11.6 | |

† Two-hour sampling periods after application of 5.6 kg/ha.
‡ Percent fraction of applied lost per hour.

#### 7-2.2.3 Rate-Limiting Steps

Data from field experiments can also give much valuable information in identifying the rate-limiting steps in the overall process of residue dispersal into the atmosphere. With one or two exceptions (Taylor et al. 1977), the vapor concentrations at even the lowest heights over treated areas always represent relative vapor densities of 10% or less, even where volatilization is taking place from moist soils where the vapor pressures of the residues are close to those of the pure compounds. This indicates that there is rapid dilution of the vapor by turbulent mixing in the lower atmospheric boundary layer and that the vapor flux rate depends on the rate at which vapor enters the turbulent atmosphere from the underlying surface. This rapid turbulent dispersal of the vapor is exemplified by the value of the residence time of the vapor in the air above the surface, calculated by dividing the amount of pesticide present in a volume of air by the flux through it. Values of 20 to 40 s or less for depths up to about 1 m are typical of turbulent mixing during spring or summer days (Harper et al., 1976; Taylor et al., 1977), with values increasing to between 2 and 10 min under stable atmospheric conditions in the morning and evening. Where these increases are not associated with accumulations of pesticide vapor in the air close to the ground, as would be expected if volatilization continued but atmospheric dispersal ceased, they must clearly reflect reduced vapor input from the soil or plant surfaces, again demonstrating the importance of the surface conditions as the rate-limiting factors in the volatilization process.

## 7-3 PREDICTION AND MODELING

### 7-3.1 General Considerations

Any comprehensive model describing the volatilization of pesticides from soil or plant surfaces must include several sets of factors and describe the ways in which they interact with each other. These sets of factors include:

1. The chemical properties of the pesticide, including the vapor pressure, the Henry's law coefficient, the soil-water partition coefficients in both wet and dry soils and the degradation rate in soil.
2. Management and usage factors, including the mode and rate of application, the resulting residue concentration and distribution in the soil profile or the surface-loading rate in $g/m^2$ of field soil or plant leaf area.
3. The soil environment, including the distribution of soil moisture over the depth containing the pesticide, the rate of flow of soil moisture through this layer and the variations in this flow because of surface water evaporation on both a daily and seasonal basis in soils of differing texture and bulk density.
4. Weather patterns of rainfall, insolation, and windspeed on a daily or shorter basis for the prediction of soil moisture changes because of rainfall, surface evaporation, or soil drying due to crop transpiration.

Currently, no comprehensive model is available that will take into account the complexity of these factors in the field situation.

### 7-3.2 Environmental Models

Existing models may be classified as one of two types, both of limited scope. The first model describes environmental distribution processes on a large scale, using a partitioning approach that predicts the ultimate distribution of a particular chemical between the overall atmosphere, water bodies, sediments, and soils. The fugacity approach Mackay and Paterson (1982) and Mackay et al. (1983) presented also takes into account differing degradation rates in each phase. McCall et al. (1983) have presented an equilibrium distribution approach that also includes partitioning into biota. Mackay et al. (1986) and Bomberger et al. (1983) have discussed the extension of such models to include the kinetics of exchange between phases, but none of these can be used to predict volatilization.

### 7-3.3 Mechanistic and Screening Models

The second type of models provides more detailed descriptions of smaller parts of the overall process. These models are concerned with mechanistic descriptions of the factors controlling the redistribution and movement of residues within soil profiles or from the soil into the atmosphere. The time scale over which these descriptions are valid is often limited, but such models can often be used to analyze the effects of rapid changes in environmental conditions such as diurnal temperature or soil drying or rewetting. Since they deal with small-scale effects, their validity can often be tested in controlled laboratory experiments. Spencer et al. (1982) have presented a review of the factors that must be incorporated into models predicting evaporation rate of pesticides from soil.

A set of five models Mayer et al. (1974) developed described the diffusive movement of pesticide upward after uniform incorporation in moist soil without convective water movement. The first two models assume only upward diffusive movement with the concentration at the soil surface maintained at zero by volatilization: the third model allows for downward diffusion out of the application zone. The fourth model assumes a constant pesticide concentration at the surface as a result of the upward flow, but without any resistance to movement in the overlying air. In the fifth model, upward movement is restricted by limited diffusion through a stagnant layer of air over the surface. The models were tested by comparison with laboratory data on lindane and dieldrin volatilization from soil obtained by Spencer and Cliath (1973) and Farmer et al. (1972). Lindane volatilization was successfully predicted by model IV. For dieldrin, both models II and IV were equally successful at airflow rates of 0.43 cm/s, or greater. This suggested that the diffusive flow maintained a low but constant concentration at the soil surface: at a lower flow rate (0.11 cm/s) model IV gave the most satisfactory prediction, but the observed rates were somewhat less than expected.

Jury et al. (1980) extended the diffusion model and included convective transport to compare model predictions with laboratory measurements of the volatilization of triallate from two soils. This model also included calculation of the distribution of the residues between the air, water, and solid phases of the soil using measured values of the Henry's law coefficient for triallate and adsorption isotherms for the two soils: the pesticide movement was, however, stated in terms of effective diffusion coefficients with no distinction between movement in the air and water phases in the soil. Calculations were made of triallate vapor losses from the soil surface into atmospheres at humidities of 100 and 50% to simulate diffusive and convective transport conditions over 30-d. The model descriptions of diffusive flow were in excellent agreement with the laboratory results for both soils, but the diffusive losses were slow and confined to the immediate soil surface because of the relatively large adsorption of the triallate by both soils. The model tended to over-predict the increase in loss because of convective flow in the silt loam soil. Measurement of the herbicide distribution in the profiles at the end of the experiment revealed a considerable concentration gradient within the top 2 cm. It appeared that the convective flow was insufficient to keep the surface concentration at its initial level because of the low concentration of herbicide in the soil water. In the silt loam, greater adsorption resulted in lower soil water concentrations, and both model and laboratory results showed that the flow was mainly diffusion controlled.

Jury et al. (1983b, 1984a, b, c) further developed this approach to include diffusion, convective flow, leaching, chemical degradation, and gaseous diffusion at the atmospheric boundary layer in assessing the relative volatility, mobility, and persistence of a range of pesticides and other trace organics in soils. This model assumes that the concentrations in the soil air and water are described by Henry's law, that the adsorption isotherms are linear, and that degradation follows first-order rate kinetics. The model assumes a stagnant boundary layer at the soil-air interface through which the vapor must diffuse to reach the atmosphere. The model is based on a mass balance equation for the amount of pesticide contained in a specified volume of soil with a uniform initial concentration $C_T$ (the total mass per unit weight of soil, in $\mu g/cm^3$),

$$(\partial C_T/\partial t) + (\partial J_S/\partial Z) + uC_T = 0 \quad [15]$$

where $C_T = d_B C_S + \theta C_L + aC_G$. Here $C_S$ is the adsorbed concentration (mg/kg soil), $C_L$ the concentration in soil solution ($\mu g/cm^3$ water), and $C_G$ the vapor concentration ($\mu g/cm^3$ of soil air); $d_B$ is the dry soil bulk density ($g/cm^3$); $\theta$ and $a$ are the volumetric water and air contents in $cm^3/cm^3$. Chemical transport through the soil $J_S$, ($\mu g\ cm^{-2}\ d^{-1}$) is described by a solute flux equation containing vapor and liquid diffusion and solute convection terms

$$J_S = -D_G a(\partial C_G/\partial Z) - D_L \Theta(\partial C_L/\partial Z) + J_W C_L \quad [16]$$

in which $D_G$ and $D_L$ are soil vapor and liquid diffusion coefficients ($cm^2$ $d^{-1}$). Chemical degradation is described by a net first-order degradation coefficient $u(d^{-1})$. $J_W$ is the soil water flux through the volume containing the pesticide: the soil is assumed to remain moist with a constant water content and the adsorption isotherms are those appropriate for moist soils. The model does not, therefore, contain any function to predict the effect of soil drying upon volatilization rate.

The equations are solved analytically for an initial soil concentration $C_O$ averaged over an arbitrary finite depth interval $0 < Z < L$. The loss of pesticide to the atmosphere is presumed to be regulated by a stagnant boundary layer through which the vapor must move by diffusion. This process is represented mathematicallyby $J_S(o, t) = -h\ (C_G(o, t) - C_A)$, where $h = D/\delta$, and $C_A$ is the vapor concentration at the upper surface of the boundary layer,analogous to $p_d$. The description of the effects of resistances to vapor loss at the soil surface are introduced into the model by varying the values for $C_A$ and the depth of the stagnant layer.

Basically, the screening model is intended to classify pesticides and other chemicals according to their environmental behavior calculated from their physical and chemical characteristics such as vapor pressure, solubility, Henry's law and organic C partition coefficients and degradation rate. The model can be used to predict the probable consequences of the use of certain types of chemicals, but it cannot be used to calculate the amounts of pesticides that will move from soil to air from specific places or land areas. In one use of the model, Jury et al. (1984b) compared the predicted behavior of 20 pesticides for which the necessary data were available in the literature. Both the relative rates of volatilization and their changes with time depended on the water evaporation rate and the chemical properties of the pesticides. The model indicated that volatilization patterns were mainly controlled by the Henry's law coefficient, which describes the distribution of pesticide between the air boundary layer above the soil and the water at the surface or in the soil pores and consequently indicates if the chemical will volatilize as fast as it is carried to the surface by convection in the water flow. The value of this coefficient (Eq. [5]) can thus be used as a criterion for classifying pesticides depending on whether their volatilization is controlled within the soil or in the boundary layer above the soil surface. The results suggest that pesticides can be divided into three categories depending on whether the Henry's law coefficient (in dimensionless units) is greater than, approximately equal to, or $<2.65 \times 10^{-5}$. Below this the volatilization will be controlled in the atmospheric boundary layer so that if the chemical moves to the surface more quickly than it evaporates it will accumulate at the surface, and the volatilization rate will tend to increase with time. In contrast, the volatilization rates of pesticides with high Henry's law coefficients decrease with time whether water is evaporating or not. The details of the pesticide classification based on the model and the supporting experimental evidence have been presented by Jury et al. (1983a; 1984a, b, c).

Laboratory measurements reported by Spencer et al. (1988) confirmed the screening model predictions with respect to the importance of Henry's law coefficients ($h$) in controlling relative volatilization of pesticides. Volatilization losses and surface distributions of two pesticides with widely differing $h$ values (prometon, $h = 1 \times 10^{-7}$; lindane, $h = 1.3 \times 10^{-4}$) were measured with and without water evaporating. Volatilization of prometon increased with time with water evaporating, whereas volatilization of lindane continually decreased. Prometon accumulated at the soil surface. These data confirm that volatilization of chemicals with low $h$ values (Category III compounds) is controlled by the boundary layer above the soil surface, whereas the control of Category I compounds with high $h$ values, is within the soil. The phenomena of organic chemicals with low $h$ values accumulating at the soil surface following convective movement in evaporating water could enhance their volatilization and increase their availability for photolysis and runoff into surface water.

## 7-4 SURVEY AND CONCLUSION

### 7-4.1 Agronomic and Environmental Significance

#### 7-4.1.1 Agronomic Implications

Volatilization is a major cause of pesticide loss from target areas, particularly when they are applied to the surfaces of soils or plants. The rate of this loss often exceeds that by chemical degradation. Recognition of this fact is of major importance for improved pesticide management. In many instances, the effective life of both herbicides and insecticides is probably limited by their rapid disappearance from the target area. This often results in the need for repeated applications, particularly when weather and soil conditions favor volatilization loss, as in the moist warm conditions of the tropics. Because of the volatile nature of many pesticide chemicals, the possibilities for improved usage focus on preventive measures that avoid use patterns that favor volatilization losses.

The most rapid losses are those from residues on the surfaces of bare moist soils. Serious losses may occur in no-tillage cropping practices where herbicide use is essential: restriction of applications to dry soil is likely to be the best preventive measure. Although subsequent volatilization losses will certainly occur when the soil is moistened by rain, this loss is likely to be less than that occurring immediately after direct application to moist soil surfaces where there may be little tendency for the formulation to diffuse downward into the upper surface soil layer before volatilization begins. Where choices of chemicals are possible, the selection of less-volatile materials is clearly desirable.

Where cultivation is consistent with the soil management practice, incorporation immediately after application, even to a shallow depth, will greatly reduce the loss of even the more volatile compounds. Restriction of application to dry soils will again further decrease the risk of volatilization loss.

Losses from plant surfaces can also be rapid, although residues under the canopy may be in some degree protected by the sheltering action of the leaf cover.

In some circumstances, the volatilization of residues into still or stable air may produce vapor concentrations high enough to have adverse impacts on sensitive nontarget species if the vapor is carried in stable low-lying air or by nonturbulent catabatic flow. Instances have been reported where this has apparently happened even where residues were dispersed over considerable distances in turbulent air. Adsorption of vapor into fog particles that may be deposited on nontarget plant species (Glotfelty et al., 1987) represents another pathway that may cause phytotoxic off-target effects of herbicides on sensitive species or contamination of other crops by insecticides, similar to the effects associated with spray drift (Farwell et al., 1976).

#### 7–4.1.2 Environmental Significance

The most important environmental aspect of pesticide volatilization is the potential for the rapid injection of a large fraction of the applied residue into the atmosphere. The further possibility of the rapid transport of the resulting vapor over long distances makes this a distribution pathway of unique importance. The large distances over which the airborne residues may be carried is offset by the rapid dilution in the atmosphere. This, coupled with the possible degradation by photochemical and oxidative reactions reduces the risk of acute environmental impacts. These risks, however, cannot be entirely discounted, particularly where the chemicals may be reconcentrated by bioaccumulation. The discovery of DDT and other stable organochlorine residues in regions as remote as the polar icecaps is a dramatic demonstration. Bidleman et al. (1981) and Glotfelty (1983) have reported the transport and redeposition of some materials in this way. More examples are likely to be found as analytical methods continue to increase in sensitivity. Reconcentration of vapor by adsorption into rain and fog droplets, with possible redeposition on vegetation, has also been reported (Glotfelty et al., 1987).

In assessing the particular case of the environmental impacts of herbicide use, it should be recognized that there are direct benefits from their use in minimum tillage or no-tillage soil management systems, where there are important long-term benefits in the reduction of soil erosion and improvements in water quality that could not be achieved without their use.

### 7–4.2 Research and Information Needs

#### 7–4.2.1 Theoretical Understanding

There are several weaknesses in our understanding of the volatilization process. All the experimental results confirm that the rate-limiting step is located at the soil surface and the rate is controlled by the vapor pressure of the pesticide at the evaporating surface, the value of $p_s$ of Eq. [2]. The value of this reflects the kinetics of supply and loss at the inner surface of the laminar layer and the fugacity of the residues at or below the surface. The importance of soil moisture content in controlling the adsorption and fugacity of the residues has been demonstrated in both field and laboratory data, but the factors controlling the changes in moisture content in the surface layers of soils are not well understood. An improved understanding of drying and rewetting in surface soils of varied texture under a range of management practices is an essential preliminary step to improved interpretation and prediction of pesticide volatilization rates and their effects on persistence and efficiency.

There are also questions about the nature of the laminar flow layer that is conceived to exist on the surface of soil and plant surfaces. The idea of the existence of this layer is derived from the aerodynamic theory that postulated this as existing as a lamina covering of an airplane wing or other surface exposed to flowing air that transmits the drag forces to the smooth surface of the solid. Surface smoothness is a key element in this concept. Soil and plant surfaces are, however, not smooth on the cellular or micropore scale. It seems possible that turbulent mixing may penetrate within their pore structures so that laminar flow may only exist for short and transient periods punctuated by rapid exchanges or microbursts of external air. The gas and momentum exchange may take place more by a type of pumping action rather than diffusive flow. The question as to which concept is more valid is important in locating the depth at which the dispersion of the pesticide vapor becomes controlled by turbulent mixing rather than by diffusive molecular flow. This is pertinent to the improvement of the models because it defines the level above the soil at which the transition is believed to take place.

Difficulties arise in the interpretation of field measurements of volatilization rates where assumptions are made using the Similarity Principle that equates the diffusivity coefficients of pesticide vapor with those of momentum (or heat transfer) in the atmospheric boundary layer. Detailed comparisons of profiles of different compounds flowing through the same layer at the same time show differences that suggest the diffusivity coefficients for different compounds are not the same (Glotfelty et al., 1983). This throws some doubt on the assumption that they can be identified with the momentum coefficient, an assumption whose main justification is that it is convenient to make. Furthermore, comparison of calculated loss rates with measured residue disappearances in experiments where we may expect to find reasonable agreement when the sampling uncertainties are discounted, suggests that the flux rates calculated using the Similarity Principle tend to underestimate the actual values. The resolution of this question concerning diffusivity coefficients is an old problem in agricultural micrometeorology, which merits more detailed examination using the sensitive analytical techniques that allow us to follow the dilution of pesticides and other organic molecules to concentrations that are not accessible for profiles of water vapor, $CO_2$, and ammonia where atmospheric background levels are present. While the resolution of this issue may not validate the use of the Similarity Principle, it may be expected to improve our knowledge of the atmospheric conditions under which it breaks down and thus evaluate the uncertainties in our current experimental methods.

### 7–4.2.2 Measurements

The lack of field measurements of pesticide volatilization from water surfaces, as in rice culture or lakes treated with herbicides or insecticides, is a major gap in our knowledge. In addition to improving both our understanding of the environmental significance of such use and the possible contributions to improved management, the data will assist in resolving the theoretical issues discussed in the previous section. Although the physicochemical factors governing the distribution of the residues between sediment, water, and air in water bodies are similar to those in the soil system, the dynamics of the delivery of the residues to the evaporating surface are entirely different because of the turbulence and circulation of the water phase. Problems of the interpretation of data obtained from soils because of the uneven distribution of residues on the surface and the changes in fugacity because of the effects of moisture on the adsorption isotherms may be discounted so that the micrometeorological questions concerning the boundary layer and turbulent dispersal of the vapor are likely to be clarified.

The complexity of the experiments is also one of the principal difficulties that limits the amount of data available from field measurements of volatilization. Part of this complexity is because of the number of samples required to obtain the profiles necessary for flux calculations, together with the associated analytical work. There is a clear need for the development of simplified procedures for making reliable flux measurements with fewer pesticide analyses and less meteorological data. Methods using horizontal rather than vertical flux measurements would permit much simpler and more flexible field experimentation.

Further work on the significance of particle transport and identification of the conditions under which wind erosion of surface-applied powder formulations is important, particularly in relation to our knowledge of the conditions under which such applications may act as sources for long distance transport of some of the more widely used but less volatile materials.

### 7–4.2.3 Future Modeling

The development of successful models that will predict the rates of volatilization of pesticides from target areas over extended periods will depend upon improved descriptions of the release of pesticides from soil and plant surfaces. Such descriptions must include the effects of soil drying on the adsorption and reduction in fugacity of residues. The kinetics of changes in surface soil moisture content, including the balance between evaporation and upward movement of moisture will be an essential element in such models, emphasizing the need for an improved understanding of the behavior of the soil water as much as that of the pesticide residues. Models are also needed to predict the effects of a range of soil management practices—including reduced tillage—in soils with a wide range of differing textures and organic matter contents.

Successful development of models that will predict volatilization from target areas over extended periods must predict the effects of the climatic factors such as rainfall, and water evaporation on the behavior of pesticides with a wide range of chemical properties. Such models will require climatic or weather data, and the precision of the output may well depend upon the time scale—daily, weekly, or monthly—on which this can be supplied. Many of the field experiments listed in Table 7–4 show the rapid changes that may occur in volatilization rates, particularly during the first few days after application when volatilization losses are greatest.

Distinctions should continue to be drawn between predictive or simulation models and screening models designed to predict the differences in behavior of a range of chemicals under similar management or environmental conditions. The principal value of screening models is to assess the comparative environmental impact or toxicological risks of different materials, whereas broad-scale predictive models are often designed to predict which phase of the environment will be most affected by a particular material or class of materials.

### 7–4.2.4 Data Base and Soil Parameters

Fundamental weaknesses also exist in our knowledge of the values of basic parameters used in modelling to characterize the chemical properties of pesticide compounds. The values of Henry's law coefficients are a clear example: almost all the data presented in Table 7–2 are constructed from measurements of saturation vapor pressures of pure compounds and their saturation solubilities in water. Few direct measurements of this coefficient, or the way in which it varies with concentration and vapor pressure, are available. The relationships between Henry's law coefficients, vapor pressure, water solubility, and activity coefficients were recently described by Suntio et al. (1988). The current data base on the partition coefficients of pesticides in soils, and particularly soil organic matter is also unsatisfactory. The use of values derived from the general correlation with octanol/water distribution coefficients may lead to substantial errors. While it may be argued that the determination of soil partition coefficients of individual compounds upon many soils is impractical, determinations made using a few standard soil types with a range of textures and organic matter contents would greatly improve the confidence with which predictions of the environmental behavior of important compounds can be made.

## 7–5 SUMMARY

Volatilization is one of the principal pathways by which pesticides are lost from target areas after application. In many cases, the magnitude of the loss is comparable to or greater than that because of chemical degradation. Volatilization losses can exceed 90% of the application within 48 h or less when residues of volatile pesticides are exposed on moist soil or plant surfaces.

The principal factors controlling the rate of volatilization are (i) the vapor pressure of the pesticide; (ii) the distribution of the residues and (iii) the moisture status of the soil or plant surface.

The highest rates of volatilization are found where residues are exposed to the atmosphere after direct application to moist soil or plant surfaces. Such residues of volatile chemicals may have half-lives of a few hours. The rapid initial rates may be reduced to some degree where the residues are applied as oil-in-water emulsions or water-soluble formulations that may be absorbed to shallow depths in porous soil or plant surfaces.

Volatilization is greatly reduced by incorporation into the soil, where the rate becomes dependent upon movement of the residues to the soil surface by diffusion or convective transport by soil water (the wick effect). Both these mechanisms are sensitive to soil moisture conditions. Convective flow is dependent upon continuous upward movement of moisture caused by evaporation at the surface. Changes in moisture flow during the day result in marked diurnal variations in the pesticide volatilization rate. The movement of residues to the surface of moist soil is controlled by diffusion when there is no upward moisture flow.

Most pesticides are strongly adsorbed by dry soil so that the volatilization of both incorporated and surface-exposed residues ceases almost completely from dry soil. This inhibition is essentially unaffected by increases in soil temperature. This reduction in chemical activity is reflected in the reduction of the biological activity of herbicides in dry soils.

Volatilization is a pathway of primary importance for the rapid dispersal of volatile pesticides into the general environment. Although the speed and ease of this dispersal is to some degree offset by rapid dilution in the atmosphere, consequent redeposition as a result of adsorption and reconcentration of the vapor in rain or fog droplets is a known pathway that may represent the first step in the bio-accumulation of those pesticides and their degradation products that are sufficiently stable in air or water. Where organisms that are sensitive to a particular chemical are present in the environmental phase where reconcentration occurs, the possibilities of chronic low-level toxic effects cannot be discounted without evaluation.

## APPENDIX

$a$ = Volumetric content of soil air
$C$ = Solution concentration in mol/m$^3$
$c$ = Concentration in solution (units as stated)
$c_o$ = Equilibrium solubility of the pure compound
$D$ = Molecular diffusion coefficient of vapor
$d$ = Vapor density (units as given)
$d_o$ = Equilibrium vapor density of the pure compound
$d_S$ = Vapor density at evaporating surface
$d_B$ = Bulk density of the sorbent phase
$\delta$ = Depth of surface of laminar flow (Greek delta)
$f$ = Fugacity (pascals: subscripts as with other symbols)
$\phi$ = Stability correction coefficient (Greek phi): or, soil organic matter content (%)
$G$ = Soil heat flux

$h$ = Henry's law coefficient (in concentration terms, $h = d/c = d_o/c_o$)
$H$ = Henry's law coefficient (molar concentration/Pa)
$\hat{I}E$ = Water vapor flux in air
$\hat{I}F$ = Flux per unit area (general, no component or units given)
$\hat{I}H$ = Flux of sensible heat
$\hat{I}P$ = Pesticide vapor flux in air
$k_E$ = Eddy diffusivity coefficient of heat in air
$k_M$ = Eddy diffusivity coefficient of momentum in air
$k_p$ = Eddy diffusivity coefficient of pesticide vapor in air
$k_Z$ = Eddy diffusivity coefficient in air (component undefined)
$K_p$ = Soil/water partition coefficient
$K_{oc}$ = Organic matter/water partition coefficient
$K_{ow}$ = Octanol/water partition coefficient
$L$ = Latent heat of water
$M$ = Molecular weight (grams)
$p$ = Vapor pressure (units as stated)
$p_d$ = Vapor pressure at outer surface of laminar layer
$p_o$ = Equilibrium vapor pressure of the pure compound
$p_S$ = Vapor pressure at evaporating surface
$R$ = Gas constant (molecular)
$R_i$ = Richardson number
$R_n$ = Net incident radiation
$R_t$ = Residues remaining at time, t.
$t$ = Temperature °C
$T$ = Temperature °K
$\theta$ = Volumetric soil water content (Greek theta)
$u$ = First order degradation coefficient
u = Wind velocity (cm/s).
$U$ = Wind speed (m/s)
$v$ = von Karman constant
$V$ = Volume
$w$ = Mass
$y$ = Psychrometric constant
$Z$ = Fugacity capacity (subscripts as defined in text)
$z$ = Vertical distance above (or below) soil surface

## REFERENCES

Acree, F., M. Beroza, and M.C. Bowman. 1963. Codistillation of DDT with water. J. Agric. Food Chem. 11:278–280.

Beste, C.E. (ed.). 1983. Herbicide handbook. 5th ed. Weed Sci. Soc. Am., Champaign, IL.

Bidleman, T.F. 1984. Estimation of vapor pressure for non polar organic compounds by capillary gas chromatography. Anal. Chem. 56:2490–2496.

Bidleman, T.F. 1985. High volume collection of organic vapors using solid adsorbents. p. 51–100. *In* J.F. Lawrence (ed.) Trace analysis. Vol. 4. Academic Press, New York.

Bidleman, T.F., E.J. Christensen, and H.W. Harder. 1981. Aerial deposition of organochlorines in urban and coastal South Carolina. p. 481–508. *In* S.J. Eisenreich (ed.) Atmospheric pollutants in natural waters. Ann Arbor Press, Ann Arbor, MI.

Bomberger, D.C., J.L. Gwinn, W.R. Mabey, D. Tuse, and Tsong Wen Chou. 1983. Environmental fate and transport at the terrestrial-atmosphere interface. p. 197–214. *In* R.L. Swann and A. Eschenroeder (ed.) Fate of chemicals in the environment. ACS Symp. Ser. 225. Am. Chem. Soc., Washington, DC.

Chiou, C.T., and T.D. Shoup. 1985. Soil sorption of organic vapors and effects of humidity on sorptive mechanism and capacity. Environ. Sci. Technol. 19:1196–1200.

Chiou, C.T., P.E. Porter, and D.W. Schmedding. 1983. Partition equilibria of nonionic compounds between soil organic matter and water. Environ. Sci. Technol. 17:227–231.

Chiou, C.T., T.D. Shoup, and P.E. Porter. 1985. Mechanistic roles of soil humus and minerals in the sorption of nonionic organic compounds from aqueous and organic solutions. Org. Geochem. 8:9–14.

Cliath, M.M., W.F. Spencer, W.J. Farmer, T.D. Shoup, and R. Grover. 1980. Volatilization of *s*-ethyl n,n,dipropylthiocarbamate from water and wet soil during and after flood irrigation of an alfalfa field. J. Agric. Food Chem. 28:610–613.

Deming, J.M. 1963. Determination of volatility losses of C14-CDAA from soil samples. Weeds 11:91–95.

Fang, S.C., P. Theisen, and V.H. Freed. 1961. Effects of water evaporation, temperature, and rates of application on the retention of ethyl-N, N-di-n-propylthiocarbamate in various soils. Weeds 9:569–574.

Farmer, W.J., K. Igue, W.F. Spencer, and J.P. Martin. 1972. Volatility of organochlorine insecticides from soil: I. Effect of concentration, temperature, air flow rate, and vapor pressure. Soil Sci. Soc. Am. Proc. 36:443–447.

Farwell, S.O., E. Robinson, W.J. Powell, and D.F. Adams. 1976. Survey of airborne 2,4-D in south-central Washington. J. Air Pollut. Control Assoc. 26:224–230.

Freeman, H.P., A.W. Taylor, and W.M. Edwards. 1975. Heptachlor and dieldrin disappearance from a field soil measured by annual residue determinations. J. Agric. Food Chem. 23:1101–1105.

Gentner, W.A., and L.L. Danielson. 1976. The influence of microencapsulation on the herbicidal performance of chlorpropham. p. 7.26–7.32. *In* Proc. Controlled Release Pesticide Symp. The Univ. of Akron, Akron, OH.

Glotfelty, D.E. 1981. Atmospheric dispersion of pesticides from treated fields. Ph.D. diss. Univ. of Maryland, College Park (Univ. Library Call no. 10879681).

Glotfelty, D.E. 1983. Pathways of pesticide dispersion in the environment. p. 425–435. *In* J.L. Hilton (ed.) Agricultural chemicals of the future. Beltsville Agric. Res. Ctr. Symp. 8. Rowman and Allanheld Publ., Totowa, NJ.

Glotfelty, D.E., J.N. Sieber, and L.A. Liljedahl. 1987. Pesticides in fog. Nature (London) 325:602–605.

Glotfelty, D.E., A.W. Taylor, and W.H. Zoller. 1983. Atmospheric dispersion of vapors: Are molecular properties unimportant? Science 219:843–845.

Glotfelty, D.E., A.W. Taylor, B.C. Turner, and W.H. Zoller. 1984. Volatilization of surface-applied pesticides from fallow soil. J. Agric. Food Chem. 32:638–643.

Glotfelty, D.E., M.M. Leech, J. Jersey, and A.W. Taylor. 1989. Volatilization and wind erosion of soil-surface applied atrazine, simazine, alachlor, and toxaphene. J. Agric. Food Chem. 37:546–555.

Gray, R.A., and A.J. Weierich. 1965. Factors affecting the vapor loss of EPTC from soils. Weeds 13:141–146.

Grover, R., and L.A. Kerr. 1978. Evaluation of silica gel and XAD-4 as adsorbents for herbicides in air. Environ. Sci. Health B13:311–321.

Grover, R., and L.A. Kerr. 1981. Evaluation of polyurethane foam as a trapping medium for herbicide vapor in air monitoring and worker inhalation studies. J. Environ. Sci. Health B16:59–66.

Grover, R., S.R. Shewchuck, A.J. Cessna, A.E. Smith, and J.H. Hunter. 1985. Fate of 2,4-D iso-octyl ester after application to a wheat field. J. Environ. Qual. 14:203–210.

Guenzi, W.D., and W.E. Beard. 1970. Volatilization of lindane and DDT from soils. Soil Sci. Soc. Am. Proc. 34:443–447.

Hamaker, J.W., and H.O. Kerlinger. 1969. Vapor pressure of pesticides. p. 39–54. *In* R.F. Gould (ed.) Pesticidal formulations research: Physical and colloidal chemical aspects. Adv. Chem. Ser. 86. Am. Chem. Soc., Washington, DC.

Harper, L.A., A.W. White, R.R. Bruce, A.W. Thomas, and R.A. Leonard. 1976. Soil and microclimate effects on trifluralin volatilization. J. Environ. Qual. 5:236–242.

Harper, L.A., L.L. McDowell, G.H. Willis, S. Smith, and L.M. Southwick. 1983. Microclimate effects on toxaphene and DDT volatilization from cotton plants. Agron. J. 75:295–302.

Harris, C.H., and E.P. Lichtenstein. 1961. Factors affecting the volatilization of insecticidal residues from soils. J. Econ. Entomol. 54:1038–1045.

Hartley, G.S. 1969. Evaporation of pesticides. p. 115–134. *In* R.F. Gould (ed.) Pesticidal formulations research: Physical and colloidal chemical aspects. Adv. Chem. Ser. 86. Am. Chem. Soc., Washington, DC.

Hartley, G.S., and I.J. Graham-Bryce. 1980a. Principles of diffusion and flow. p. 110–203. *In* Physical principles of pesticide behavior. Vol. 1. Academic Press, New York.

Hartley, G.S., and I.J. Graham-Bryce. 1980b. Behavior of pesticides in air. p. 337–385. *In* Physical principles of pesticide behavior Vol. 1. Academic Press, New York.

Hartley, G.S., and I.J. Graham-Bryce. 1980c. Penetration of pesticides into higher plants. p. 544–657. *In* Physical principles of pesticide behavior. Vol. 2. Academic Press, New York.

Hartley, G.S., and I.J. Graham-Bryce. 1980d. Physical properties of pesticides. p. 896–925. *In* Physical principles of pesticide behavior. Vol. 2. Academic Press, New York.

Igue, K., W.J. Farmer, W.F. Spencer, and J.P. Martin. 1972. Volatility of organochlorine insecticides from soil: II. Effect of relative humidity and soil water content on dieldrin volatility. Soil Sci. Soc. Am. Proc. 36:447–450.

Jury, W.A., W.J. Farmer, and W.F. Spencer. 1984a. Behavior assessment model for trace organics in soil: II. Chemical classification and parameter sensitivity. J. Environ. Qual. 13:567–572.

Jury, W.A., R. Grover, W.F. Spencer, and W.J. Farmer. 1980. Modeling vapor losses of soil-incorporated triallate. Soil Sci. Soc. Am. J. 44:445–450.

Jury, W.A., W.F. Spencer, and W.J. Farmer. 1983a. Use of models for assessing relative volatility, mobility, and persistence of pesticides and other trace organics in soil systems. p. 1–43. *In* J. Saxena (ed.) Hazard assessment of chemicals: Current developments. Vol. 2. Academic Press, New York.

Jury, W.A., W.F. Spencer, and W.J. Farmer. 1983b. Behavior assessment model for trace organics in soil: I. Model description. J. Environ. Qual. 12:558–564.

Jury, W.A., W.F. Spencer, and W.J. Farmer. 1984b. Behavior assessment model for trace organics in soil: III. Application of screening model. J. Environ. Qual. 13:573–579.

Jury, W.A., W.F. Spencer, and W.J. Farmer. 1984c. Behavior assessment model for trace organics in soil: IV. Review of experimental evidence. J. Environ. Qual. 13:580–585.

Kim, Y.H., J.E. Woodrow, and J.N. Seiber. 1984. Evaluation of a gas chromatographic method for calculating vapor pressures with organophosphorus pesticides. J. Chromatogr. 314:37–53.

Lemon, E.R. 1969. Gaseous exchange in crop stands. p. 117–142. *In* J.D. Eastin et al. (ed.) Physiological aspects of crop yield. ASA, Madison, WI.

Lewis, G.N. 1901. The law of physico-chemical change. Proc. Am. Acad. 37:49–96.

Lewis, G.N., and M. Randall. 1923. Thermodynamics and the free energy of chemical substances. McGraw-Hill Book Co., New York.

Lichtenstein, E.P., and K.R. Schultz. 1961. Effect of soil cultivation, soil surface, and water on the persistence of insecticidal residues in soils. J. Econ. Entomol. 54:517–522.

Lichtenstein, E.P., K.R. Schultz, T.W. Fuhremann, and T.T. Liang. 1970. Degradation of aldrin and heptachlor in field soils during a ten-year period: Translocation into crops. J. Agric. Food Chem. 18:100–106.

Liss, P.S., and P.G. Slater. 1974. Flux of gases across the air-sea interface. Nature (London) 247:181–184.

Mackay, D. 1984. Air/water exchange coefficients. p. 91–108. *In* W.B. Neely and G.E. Blau (ed.) Environmental exposure from chemicals. Vol. 1. CRC Press, Boca Raton, FL.

Mackay, D., and S. Paterson. 1981. Calculating fugacity. Environ. Sci. Technol. 5:1006–1014.

Mackay, D., and S. Paterson. 1982. Fugacity revisited. Environ. Sci. Technol. 16:645A–660A.

Mackay, D., S. Paterson, and M. Joy. 1983. Application of fugacity models to the estimation of chemical distribution and persistence in the environment. p. 175–196. *In* R.L. Swann and A. Eschenroeder (ed.) Fate of chemicals in the environment. ACS Symp. Ser. 225. Am. Chem. Soc., Washington, DC.

Mackay, D., S. Paterson, and W.H. Schroeder. 1986. Model describing the rates of transfer processes of organic chemicals between atmosphere and water. Environ. Sci. Technol. 20:810–816.

McCall, P.J., R.L. Swann, and D.A. Laskowski. 1983. Partition models for equilibrium distribution of chemicals in environmental compartments. p. 105–123. *In* R.L. Swann and A. Eschenroeder (ed.) Fate of chemicals in the environment. ACS Symp. Ser. 225. Am. Chem. Soc., Washington, DC.

Mayer, R., J. Letey, and W.J. Farmer. 1974. Models for predicting volatilization of soil-incorporated pesticides. Soil Sci. Soc. Am. Proc. 38:563–568.

Mingelgrin, U., and Z. Gerstl. 1983. Re-evaluation of partitioning as a mechanism of nonionic chemicals adsorption in soils. J. Environ. Qual. 12:1–11.

Monteith, J.L. 1973. Principles of environmental physics. American Elsevier, New York.

Nash, R.G. 1983. Comparative volatilization and dissipation rates of several pesticides from soil. J. Agric. Food Chem. 31:210–217.

Nash, R.G., and M.L. Beall. 1979. A micro-agroecosystem to monitor the environmental fate of pesticides. p. 86–94. *In* J.M. Witt et al. (ed.) Terrestrial microcosms and environmental chemistry. NSF/RA 79-0026. Natl. Sci. Foundation, Washington, DC.

Parmele, L.H., E.R. Lemon, and A.W. Taylor. 1972. Micrometeorological measurement of pesticide vapor flux from bare soil and corn under field conditions. Water Air Soil Pollut. 1:433–451.

Parochetti, J.V., and G.F. Warren. 1966. Vapor losses of IPC and CIPC. Weeds 14:281–285.

Plimmer, J.R. 1976. Volatility. p. 891–934. *In* P.C. Kearney and D.D. Kaufmann (ed.) Herbicides: Chemistry, degradation and mode of action. Vol. 2. Marcel Dekker, New York.

Rao, P.S.C., and J.M. Davidson. 1980. Estimation of pesticide retention and transformation parameters required in nonpoint source pollution models. p. 23–67. *In* M.R. Overcash and J.M. Davidson (ed.) Environmental impact of nonpoint source pollution. Ann Arbor Sci. Publ., Ann Arbor, MI.

Rose, C.W. 1966. The physical environment of agriculture. II. p. 30–68. *In* Agricultural physics. Pergamon Press, New York.

Rose, C.W. 1966. The physical environment of agriculture. III. p. 69–87. *In* Agriculture physics. Pergamon Press, New York.

Seiber, J.N., S.C. Madden, M.M. McChesney, and W.L. Winterlin. 1979. Toxaphene dissipation from treated cotton field measurements: Component residual behavior on leaves and in air, soil, and sediments determined by capillary gas chromatography. J. Agric. Food Chem. 27:284–290.

Spencer, W.F. 1970. Distribution of pesticides between soil, water, and air. p. 120–128. *In* Pesticides in the soil: Ecology, degradation, and movement. Michigan State Univ., East Lansing.

Spencer, W.F., and M.M. Cliath. 1969. Vapor density of dieldrin. Environ. Sci. Technol. 3:670–674.

Spencer, W.F., and M.M. Cliath. 1970. Desorption of lindane from soil as related to vapor density. Soil Sci. Soc. Am. Proc. 34:574–578.

Spencer, W.F., and M.M. Cliath. 1972. Volatility of DDT and related compounds. J. Agric. Food Chem. 20:645–649.

Spencer, W.F., and M.M. Cliath. 1973. Pesticide volatilization as related to water loss from soil. J. Environ. Qual. 2:284–289.

Spencer, W.F., and M.M. Cliath. 1974. Factors affecting vapor loss of trifluralin from soil. J. Agric. Food Chem. 22:987–991.

Spencer, W.F., and M.M. Cliath. 1983. Measurement of pesticide vapor pressures. Residue Rev. 85:57–71.

Spencer, W.F., M.M. Cliath, and W.J. Farmer. 1969. Vapor density of soil-applied dieldrin as related to soil-water content, temperature, and dieldrin concentration. Soil Sci. Soc. Am. Proc. 33:509–511.

Spencer, W.F., M.M. Cliath, W.A. Jury, and L.Z. Zhang. 1988. Volatilization of organic chemicals from soil as related to their Henry's law constants. J. Environ. Qual. 17:504–509.

Spencer, W.F., W.J. Farmer, and M.M. Cliath. 1973. Pesticide volatilization. Residue Rev. 49:1–45.

Spencer, W.F., W.J. Farmer, and W.A. Jury. 1982. Review: Behavior of organic chemicals at soil, air, water interfaces as related to predicting the transport and volatilization of organic pollutants. Environ. Toxicol. Chem. 1:17–26.

Suntio, L.R., W.Y. Shiu, D. Mackay, J.N. Seiber, and D. Glotfelty. 1988. A critical review of Henry's law constants for pesticides. Rev. Environ. Contamin. Toxicol. 103:1–59.

Taylor, A.W., J.H. Caro, H.P. Freeman, and B.C. Turner. 1985. Sampling and variance in measurements of trifluralin disappearance from field soil. p. 25–36. *In* D.A. Kurtz (ed.) Trace residue analysis: Chemometric estimations of amount, uncertainty, and error. ACS Symp. Ser. 284. Am. Chem. Soc., Washington, DC.

Taylor, A.W., H.P. Freeman, and W.M. Edwards. 1971. Sample variability and measurement of the dieldrin content of a soil in the field. J. Agric. Food Chem. 19:832–836.

Taylor, A.W., D.E. Glotfelty, B.L. Glass, H.P. Freeman, and W.M. Edwards. 1976. The volatilization of dieldrin and heptachlor from a maize field. J. Agric. Food Chem. 24:625–630.

Taylor, A.W., D.E. Glotfelty, B.C. Turner, R.E. Silver, H.P. Freeman, and A. Weiss. 1977. Volatilization of dieldrin and heptachlor residues from field vegetation. J. Agric. Food Chem. 25:542–548.

Turner, B.C., and D.E. Glotfelty. 1977. Field air sampling of pesticide vapors with polyurethane foam. Anal. Chem. 49:7–10.

Turner, B.C., D.E. Glotfelty, and A.W. Taylor. 1977. Photodieldrin formation and volatilization from grass. J. Agric. Food Chem. 25:548–550.

Turner, B.C., D.E. Glotfelty, A.W. Taylor, and D.R. Watson. 1978. Volatilization of microencapsulated and conventionally applied chlorpropham in the field. Agron. J. 70:933–937.

Van Valkenburg, J.W. 1969. The physical and colloidal chemical aspects of pesticidal formulations research: A challenge. p. 1–6. *In* R.F. Gould (ed.) Pesticidal formulation research: Physical and colloid chemical aspects. Adv. Chem. Ser. 86. Am. Chem. Soc., Washington, DC.

Wauchope, R.D. 1978. The pesticide content of surface water draining from agricultural fields—A review. J. Environ. Qual. 7:459–472.

Weber, J.B., P.J. Shea, and H.J. Strek. 1980. An evaluation of nonpoint sources of pesticide pollution in runoff. p. 69–98. *In* M.R. Overcash and J.M. Davidson (ed.) Environmental impact of nonpoint source pollution. Ann Arbor Sci. Publ., Ann Arbor, MI.

White, A.W., L.A. Harper, R.A. Leonard, and J.W. Turnbull. 1977. Trifluralin volatilization from a soybean field. J. Environ. Qual. 6:105–110.

Whitman, W.G. 1923. Preliminary experimental confirmation of the two-film theory of gas adsorption. Chem. Metall. Eng. 29:146.

Willis, G.H., J.F. Parr, and S. Smith. 1971. Volatilization of soil-applied DDT and DDD from flooded and nonflooded plots. Pestic. Monit. J. 4:204–208.

Willis, G.H., J.F. Parr, S. Smith, and B.R. Carroll. 1972. Volatilization of dieldrin from fallow soil as affected by different soil water regimes. J. Environ. Qual. 1:193–196.

Willis, G.H., L.L. McDowell, S. Smith, L.M. Southwick, and E.R. Lemon. 1980. Toxaphene volatilization from a mature cotton canopy. Agron. J. 72:627–631.

Willis, G.H., L.L. McDowell, L.A. Harper, L.M. Southwick, and S. Smith. 1983. Seasonal disappearance and volatilization of toxaphene and DDT from a cotton field. J. Environ. Qual. 12:80–85.

# Chapter 8

# Organic Chemical Transport to Groundwater[1]

**C. G. ENFIELD,** *RSKERL, Ada, Oklahoma*

**S. R. YATES,** *University of California, Riverside, California*

The use of pesticides in the production of agricultural commodities is widespread. During 1982, 370 000 t of active ingredients were used in the USA (U.S. EPA, 1982) and the use of pesticides is expected to increase in the future (Knusli, 1979). Since nearly one-half of the U.S. population relies on groundwater as their source for drinking water (McEwen & Stephenson, 1979), contamination potential of groundwater, because of pesticide manufacture and use, must be understood.

The processes of sorption, biotic and abiotic transformation, and vapor transport have been discussed in previous chapters of this book. The objective of this chapter is to integrate the above processes into chemical mass transport models that can be used to forecast environmental exposure.

Almost any modeling activity related to groundwater starts with a water flow model, since, for any significant change to take place in the flow field, flow is an essential ingredient. Van der Heijde et al. (1985) reviewed several hundred groundwater management models from around the world and classified them in a variety of ways. Their review included both saturated and unsaturated flow models along with identifying the source and availability of computer codes for the models. Considering this review and several other reviews (e.g., Rao & Jessup, 1983; Boesten & Leistra, 1983; Addiscott & Wagenet, 1985), the current chapter will emphasize chemical transport rather than mass water flow. For completeness, a brief overview of water flow through saturated and unsaturated soils will be included. Decoupling the water and chemical transport is a major assumption in the following discussion. This means that the influence of the chemical is insignificant on water flow, and properties such as density gradients that can cause fluid movement can

[1] Funded by grants from the Swedish Natural Sci. Res. Counc. No. G-GU-RT 1613-107, the National Swedish Environmental Protection Board No. 5336064-0 and the Robert S. Kerr Environmental Res. Lab., U.S. Environmental Protection Agency, Ada, OK. Although this work was funded in part by the U.S. Environmental Protection Agency, it has not undergone formal agency review. Mention of trade names or commercial products does not constitute endorsement or recommendation for use.

be ignored. This makes it possible to calculate the water flux independent of the chemical and then use the water fluxes in calculating the chemical flux.

## 8-1 MASS FLUX OF WATER

Most recharge of groundwater occurs during the percolation of water through an unsaturated soil. Water movement is controlled by both gravitational and capillary forces.

Capillarity results from two forces: the mutual attraction (cohesion) between water molecules and the molecular attraction (adhesion) between water and solid materials. As a consequence of these forces, water will rise in small diameter glass tubes above the water level in a large container. Most pores in granular material are of capillary size and, as a result, water is pulled upward into a capillary fringe above the water table in the same manner as water would be pulled up into a column of sand whose lower end is immersed in water.

Steady-state water flow in unsaturated soil can be determined from a modified form of Darcy's law (Richards, 1931). Steady state in this context refers to a condition in which the moisture content remains constant, as would be the case under a disposal pond kept at a constant head and separated from the water table by unsaturated soil. The steady-state Darcy flux, $V_d$ $(LT^{-1})$, is proportional to the effective hydraulic conductivity, $K(\psi)$ $(LT^{-1})$, which is a function of the water potential, $\psi$, and gradients because of both capillary and gravitational forces.

$$V_d = -K(\psi) \nabla H \qquad [1]$$

where $\nabla$ is the standard differential operator of vector notation and $H = \psi - z$ is the total potential expressed as total head $(L)$ with the vertical coordinate taken as positive downward. All other components of soil-water potential have been neglected.

If a water-saturated soil could be considered as a bundle of straight and smooth capillary tubes, Poiseuille's and Darcy's equations would be analogous

$$V_d = \frac{-\tau^2 \rho_a g}{8\gamma} \nabla H \qquad [2]$$

where $\tau$ is the radius of a tube or pore (L), $\rho_a$ is the density of water $(ML^{-3})$, $g$ is the gravitational acceleration $(LT^{-2})$, and $\gamma$ is the viscosity of water $(MT^{-1}L^{-1})$.

Similarities between these two equations have been used in the development of equations to describe a functional relation between water content and effective hydraulic conductivity (Millington & Quirk, 1961). Under saturated conditions, all of the soil pores are filled with water; thus, the maximum hydraulic conductivity is observed under saturated conditions (note:

water can be assumed to be an incompressible fluid under most environmental conditions). As the soil desaturates (soil-water pressure head decreases), the cross-sectional area of the soil available to conduct water decreases. The first pores to empty during desaturation of a soil are the largest pores in direct connection with the atmosphere. These large pores offer the least resistance to water flow; thus, as the soil desaturates there is a sharp reduction in hydraulic conductivity. Decreases of three or four orders of magnitude are not uncommon in highly structured, well-drained soils (Davidson et al., 1969).

The above two equations imply that soil-water flux vs. hydraulic gradient is a linear relationship. This is true only for laminar flow conditions. At high fluxes, turbulent conditions may exist creating non-Newtonian fluid properties. Similarly, at low gradients water may become non-Newtonian. For most environmental conditions, it is reasonable to assume that water flow is laminar and water behaves as a Newtonian fluid.

The quantitative application of unsaturated water flow theory to field or laboratory soil systems requires a knowledge of the soil hydraulic conductivity [$K(\psi)$] and soil-water characteristic [$\theta(\psi)$] relationships. The soil-water characteristic describes the relationship between water content and water potential. A typical relationship is shown in Fig. 8-1. There are two curves

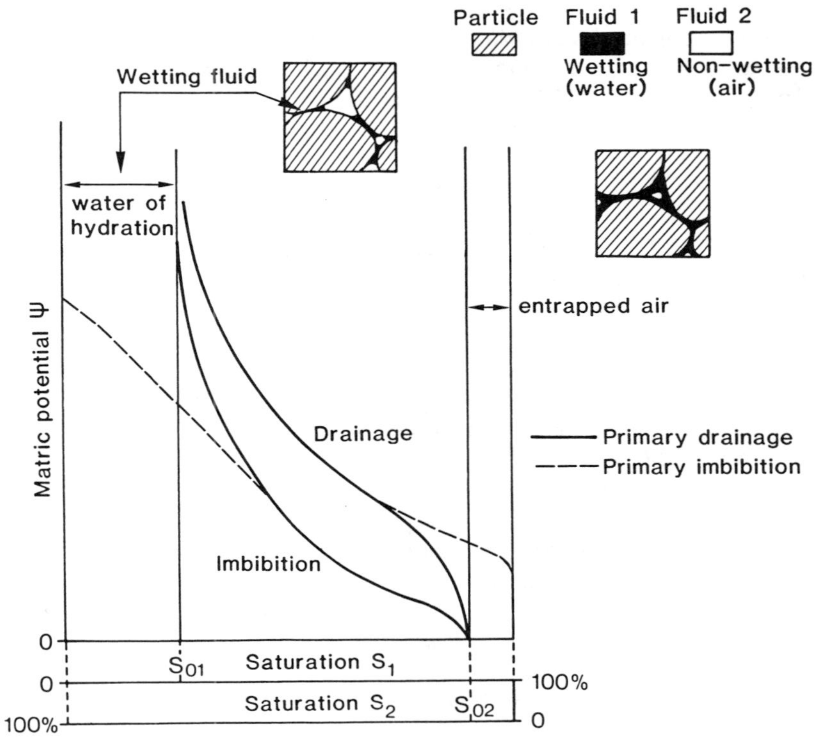


Fig. 8-1. Typical soil-water characteristic curve.

shown: one for drainage and one for imbibition. The soil characteristic can be at any point between the primary drainage and primary imbibition curves depending on past history. In general, flow theory is applied assuming there is a unique function that describes the soil-water characteristic and effects of hysteresis are ignored. Klute (1972) reviewed various methods for measuring hydraulic conductivity and discussed the advantages and disadvantages of each method; methods reviewed included steady-state, unsteady-state, instantaneous profile methods, and those involving the use of soil-water characteristic curves to estimate $K(\psi)$ relationships. Similarly, Heath (1983) reviewed currently used methods in aquifers to determine hydraulic properties.

When evaluating water flow in one dimension under steady-state conditions, once the water flux is known at one of the boundaries, the flow is known throughout the problem domain. Under agricultural conditions, this flux is often approximated from the water balance equation

$$L + P_r = \mathrm{ET} + V_d + \omega. \qquad [3]$$

Precipitation ($P_r$) and irrigation ($L$) records are generally available. The net runoff ($\omega$) can be estimated from existing watershed models (e.g., Donigian et al., 1977). Evapotranspiration (ET) has been the subject of numerous reports (e.g., Jensen, 1973; Slatyer, 1967). Many of the methods require considerable input of climatic data, much of which is difficult to obtain and most are semiempirical. One simple approach is to estimate ET from pan evaporation (Jensen, 1973). The approach, which may be adequate for crude estimates, uses the equation

$$E_g = C_{et} E_{pan} \qquad [4]$$

which gives the ET for a turfgrass. To translate from turf to some other crop, one can use the equation

$$\mathrm{ET} = \kappa\, E_g. \qquad [5]$$

Values for the constants $C_{et}$ and $\kappa$ are reproduced in Tables 8–1 and 8–2. If one assumed there was no runoff, it would be possible to estimate a Darcy flux through the soil.

Many times steady-state conditions are not adequate approximations of water flux. This is particularly true when the temporal distribution of a chemical is wanted near the source of application (near field). As distance and time increase (far field), the variations because of transient climatic activities are minimized. For near-field conditions, Eq. [1] can be combined with the equation of continuity:

$$(\partial\theta/\partial t) = -\nabla V_d \qquad [6]$$

which yields:

$$(\partial\theta/\partial t) = \nabla[K(\psi)\nabla H]. \qquad [7]$$

Table 8-1. Suggested value for $C_{et}$ relating evapotranspiration from a U.S. Class A pan to evapotranspiration from eight 15-cm tall, well-watered grass turf (Jensen, 1973).

| | Pan surrounded by short green crop | | | | Pan surrounded by dry surface ground | | | |
|---|---|---|---|---|---|---|---|---|
| | Upwind fetch of | Relative humidity, %† | | | Upwind fetch of | Relative humidity, %† | | |
| Wind | crop, m | 20–40 | 40–70 | 70 | fallow, m | 20–40 | 40–70 | 70 |
| Light | 0 | 0.55 | 0.65 | 0.75 | 0 | 0.7 | 0.8 | 0.85 |
| $<170$ km d$^{-1}$ | 10 | 0.65 | 0.75 | 0.85 | 10 | 0.6 | 0.7 | 0.8 |
| | 100 | 0.7 | 0.8 | 0.85 | 100 | 0.55 | 0.65 | 0.75 |
| | 1000 | 0.7 | 0.85 | 0.85 | 1000 | 0.5 | 0.6 | 0.7 |
| Moderate | 0 | 0.5 | 0.6 | 0.65 | 0 | 0.65 | 0.75 | 0.8 |
| 170–425 km d$^{-1}$ | 10 | 0.5 | 0.7 | 0.75 | 10 | 0.55 | 0.65 | 0.7 |
| | 100 | 0.65 | 0.75 | 0.8 | 100 | 0.5 | 0.6 | 0.65 |
| | 1000 | 0.7 | 0.8 | 0.8 | 1000 | 0.45 | 0.55 | 0.6 |
| Strong | 0 | 0.45 | 0.5 | 0.6 | 0 | 0.6 | 0.65 | 0.7 |
| 425–700 km d$^{-1}$ | 10 | 0.55 | 0.6 | 0.65 | 10 | 0.5 | 0.55 | 0.65 |
| | 100 | 0.6 | 0.65 | 0.7 | 100 | 0.45 | 0.5 | 0.6 |
| | 1000 | 0.65 | 0.7 | 0.75 | 1000 | 0.4 | 0.45 | 0.55 |
| Very strong | 0 | 0.4 | 0.45 | 0.5 | 0 | 0.5 | 0.6 | 0.65 |
| $>700$ km d$^{-1}$ | 10 | 0.45 | 0.55 | 0.6 | 10 | 0.45 | 0.5 | 0.55 |
| | 100 | 0.5 | 0.6 | 0.65 | 100 | 0.4 | 0.45 | 0.5 |
| | 1000 | 0.55 | 0.6 | 0.65 | 1000 | 0.3 | 0.4 | 0.45 |

† Mean of maximum and minimum relative humidities.

Equation [7] is appropriate for $n$-dimensional water flow in a heterogeneous anisotropic soil. The equation for isotropic soils can be written in two dimensions as:

$$\frac{d\theta}{d\psi}\frac{\partial\psi}{\partial t} = \frac{\partial}{\partial x}\left[K(\psi)\frac{\partial\psi}{\partial x}\right] + \frac{\partial}{\partial z}\left[K(\psi)\frac{\partial\psi}{\partial z}\right] - \frac{\partial K(\psi)}{\partial z}. \qquad [8]$$

An alternate form of Eq. [8] is sometimes used by defining a new variable, the soil water diffusivity $D(\theta) = [K(\theta)(d\psi/d\theta)]$ (Childs & Collis-George, 1950). In any event, the dependence of $D$ or $K$ on $\psi$ or $\theta$ makes the equation nonlinear. Thus, analytical solutions to Eq. [8] are not available except for simple initial and boundary conditions. According to van der Heijde et al. (1985), there are five supported numerical computer codes available that will handle a saturated-unsaturated water flow as listed in Table 8-3. Four of the five codes listed will handle transient flow conditions and some will also handle anisotropic media.

Table 8-2. Crop coefficient for estimating evapotranspiration ($\kappa$) (Jensen, 1973).

| Crop | Period | Coefficient, $\kappa$ |
|---|---|---|
| Alfalfa (*Medicago sativa* L.) | 1 Apr.–10 Oct. | 0.87 |
| Potato (*Solanum tuberosum* L.) | 10 May–15 Sept. | 0.65 |
| Small grain | 1 Apr.–20 July | 0.6 |
| Sugarbeet (*Beta vulgaris* L.) | 10 Apr.–15 Oct. | 0.6 |

Table 8–3. Single-phase saturated-unsaturated flow models. Adapted from van der Heijde et al. (1985).

| Author(s) | Original affiliation | Country | Model name | Completion year (Update) | Model type | Spatial characteristics and dynamics of flow | | System | Process |
|---|---|---|---|---|---|---|---|---|---|
| | | | | | | Saturated | Unsaturated | | |
| P.J.M. De Laat | International institute | Netherlands | SUM-2 | 1972/76 (1980) | Hydraulic | Two-dimensional, horizontal; transient | One-dimensional vertical quasi-steady state | Isotropic heterogeneous porous medium two-layered homogeneous soil | Evapotrans., plant uptake |
| C.R. Amerman | Federal government | USA | STDY-2 | 1976 | Hydrodynamic | Two-dimensional vertical steady state | Two-dimensional vertical steady state | Anisotropic heterogeneous multi-layered soil | |
| J.L. Nieber | University | USA | FEATSMF | 1979 | Hydrodynamic | Two-dimensional vertical transient | Two-dimensional vertical transient | Isotropic homogeneous hill-slope soil | Seepage |
| G.T. Yeh D.S. Ward | Research laboratory | USA | FEMWATER FECWATER | 1981 (1982) | Hydrodynamic | Two-dimensional vertical transient | Two-dimensional vertical transient | Anisotropic heterogeneous porous medium | Ponding |
| T.N. Narismham | Research laboratory | USA | TRUST | 1981 | Hydrodynamic | Multi-dimensional steady state or transient | Multi-dimensional steady state or transient | Anisotropic heterogeneous variable saturated deformable porous medium complex geometry | Drainage dewatering hysteresis |

## 8-2 TRANSPORT OF MISCIBLE NONVOLATILE REACTIVE COMPOUNDS

The most common approach to chemical transport is to consider the contaminant to be nonvolatile and miscible in the liquid phase. A solute could then react with the solid phase by sorption and be transformed while sorbed or be transported and transformed in the liquid phase. The vapor phase could be present, but not directly participate in the transport or transformation process. One way of describing the system is to look at each phase independently and mathematically describe the transfer of chemicals between the phases. The aqueous (liquid) phase could be described in one dimension by the equation

$$\frac{\partial(\theta\rho_a C_a)}{\partial t} = D_a \frac{\partial^2(\theta\rho_a C_a)}{\partial x^2} - V_a \frac{\partial(\theta\rho_a C_a)}{\partial x} - k_{as}\theta\rho_a C_a + k_{sa}(1 - n)\rho_s C_s - k_{ta}\theta\rho_a C_a . \quad [9]$$

Equation [9] describes the change in mass, of a contaminant in a unit volume of soil represented in Fig. 8-2, per unit time. The first term on the right-hand side of the equation describes the dispersive flux into the elemental volume. The physical significance of the hydrodynamic dispersion term ($D_a$) has been the topic of considerable debate. The term is often used to combine the influences of molecular diffusion caused by a concentration gradient

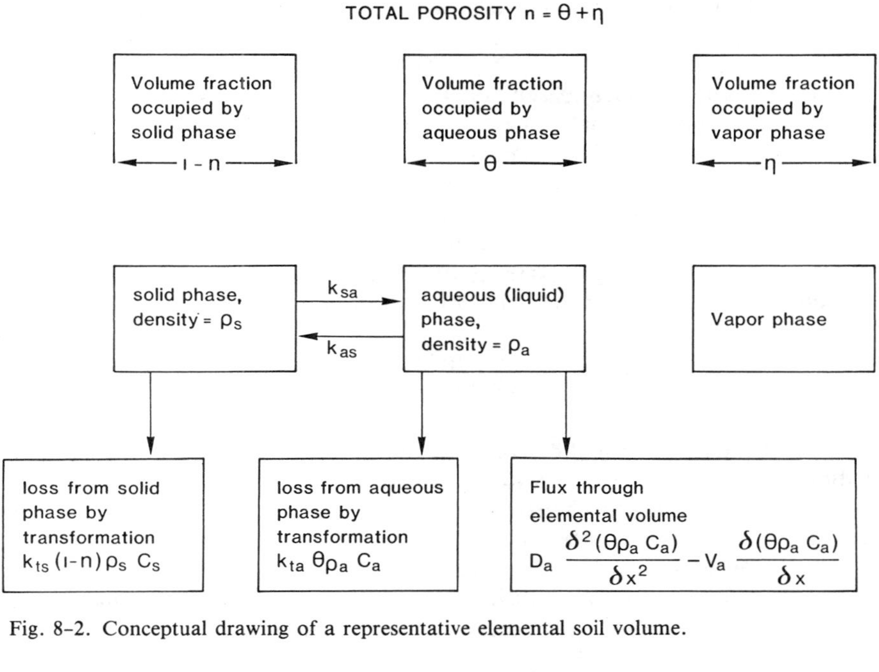


Fig. 8-2. Conceptual drawing of a representative elemental soil volume.

along with dispersion due to mechanical mixing during fluid advection that is a function of the porous media. Many hydrogeologists use the term *dispersivity* ($\alpha$) with the units of length to describe the porous media and then relate the dispersion coefficient to the fluid velocity by the equation

$$D_a = \alpha V_a + \lambda D \qquad [10]$$

where $\lambda$ is a constant related to the tortuosity and $D$ is the molecular diffusion coefficient in water. The tortuosity coefficient, with values typically between 0.5 and 0.01 (Freeze & Cherry, 1979), is frequently estimated from empirical relationships such as Millington and Quirk (1961) presented. The molecular diffusion coefficient for dichlorobenzene in water is approximately 0.5 $cm^2\ d^{-1}$. With typical values of tortuosity the impact of molecular diffusion is small when there is a significant interstitial velocity. Figure 8–3 shows two experimental breakthrough curves for tritiated water movement through a given experimental column at different flow velocities. Although the interstitial velocity ($V_a$) and dispersion coefficient ($D_a$) are significantly different, there is little difference in the dispersivity ($\alpha$) as seen by the overlap in the experimental data. The experimental data shown in Fig. 8–3 are consistent with Eq. [10]. A difficulty with this concept is the dispersivity appears to be scale dependent. The larger the spatial scale from which measurements are taken, the larger the dispersivity (Smith & Schwartz, 1980). Molz et al. (1983) have postulated, in a field study, that the field-scale dispersivity term represents a lack of understanding of the spatial variability of the hydraulic conductivity in the geohydrologic system. The scale dependence of dispersivity used in groundwater movement studies may, thus, be considered a convenient approach to describing our lack of understanding of the flow system. Bresler and Dagan (1981), Amoozegar-Fard et al. (1982), and Parker and van Genuchten (1984) proposed an alternate approach that describes the spatial variability of the flow system while projecting an average concentration for the chemical in the flow field.

The second term on the right-hand side of Eq. [9] describes the mass transfer of the compound out of the representative elemental volume by fluid flow (often called *advective transport*). The velocity term ($V_a$) is usually calculated from a saturated-unsaturated flow model that is assumed to be unaffected by the composition of the aqueous phase (i.e., solute concentration gradients do not affect or cause water flow). The interstitial velocity ($V_a$) is equal to the Darcy velocity ($V_d$) divided by the volumetric water content ($\theta$). The term may either be a variable that is dependent on the transient nature of the flow field, or a constant if the hydraulic conditions are at steady state. If it is desired to make near-field projections of the concentration distribution (i.e., projections near the source of chemical application), the transient nature of water flow generally must be considered. The greater the time of travel to the point of interest, under most field conditions, the less important the transient nature of the water flow problem.

The third term on the right-hand side of Eq. [9] describes the mass loss rate because of a transfer of molecules from the aqueous phase to the solid

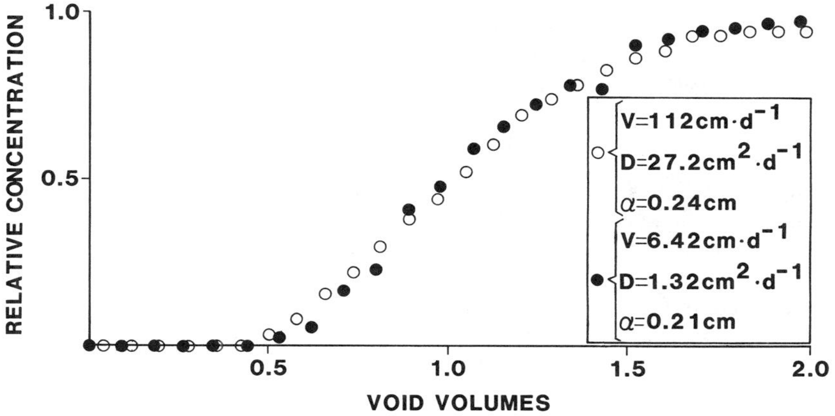


Fig. 8-3. Experimental breakthrough curve for tritium through a sandy soil at two interstitial velocities.

phase. This is written as a first-order kinetic term. It does not attempt to describe the mechanism of the sorption reaction. The term simply states that the higher the activity of a compound in the aqueous phase the higher the probability the molecule can be lost to the solid phase. Similarly, the fourth term describes the mass rate of gain caused by transfer of molecules from the solid phase to the liquid phase.

The final term in Eq. [9] describes the transformation of the compound that takes place in the liquid phase. Overall, the reaction rate is assumed to be a first order. Many different processes can be incorporated in this term. The first-order kinetic model is selected here because it is convenient. When processes other than first order are known to occur, they should be incorporated into transport models.

The contaminant's solid phase concentration can be described by the equation

$$\frac{\partial[(1 - n)\rho_s C_s]}{\partial t} = -k_{sa}(1 - n)\rho_s C_s + k_{as}\theta\rho_a C_a - k_{ts}(1 - n)\rho_s C_s. \qquad [11]$$

As written, Eq. [11] assumes the solid phase is stationary and, therefore, does not include a diffusive or a mass transfer term as included in Eq. [9]. The other three terms have corollaries in Eq. [9]. The change of mass in the total elemental volume is the sum of Eq. [9] and [11] or

$$\frac{\partial(\theta\rho_a C_a)}{\partial t} + \frac{\partial[(1 - n)\rho_s C_s]}{\partial t} = D_a \frac{\partial^2(\theta\rho_a C_a)}{\partial x^2} - V_a \frac{\partial(\theta\rho_a C_a)}{\partial x} - k_{ta}\theta\rho_a C_a - k_{ts}(1 - n)\rho_s C_s. \qquad [12]$$

Equation [12] is written with two dependent variables: the concentration of the contaminant in the aqueous phase ($C_a$) and the concentration of the contaminant in the solid phase ($C_s$).

The net rate of change from an individual phase can be written

$$r = k_{as}\theta\rho_a C_a - k_{sa}(1 - n)\rho_s C_s. \qquad [13]$$

Many researchers find equilibrium takes place rapidly with acceptable reversibility (e.g., Schwarzenbach & Westall, 1981 or McCarthy & Jimenez, 1985) while others find minimum reversibility and relatively slow kinetics (e.g., Di Toro, 1985 or Means et al., 1985). Assuming equilibrium exists, the net rate of change from one phase to another would be zero and the mass ratio would be the equilibrium constant

$$C_s/C_a = (k_{as}\theta\rho_a)/[k_{sa}(1 - n)\rho_s] = k_d. \qquad [14]$$

As developed, Eq. [14] leads to a linear sorption partition coefficient with units $M_a\ M_s^{-1}$, i.e., unitless if phases are not identified. (Note: there are a variety of partition coefficients in the literature; many have associated units. The most common units reported assume that the density of water is 1 Mg $m^{-3}$; this yields the units $m^3\ Mg^{-1}$.)

If we let

$$k_d' = (k_{as}/k_{sa}) \qquad [15]$$

then when local equilibrium exists Eq. [12] can be rewritten in terms of one dependent variable as

$$(1 + k_d')\frac{\partial C_a}{\partial t} = D_a\frac{\partial^2 C_a}{\partial x^2} - V_a\frac{\partial C_a}{\partial x} - (k_{ta} + k_{ts}k_d')\ C_a. \qquad [16]$$

Defining a retardation factor as

$$R = 1 + k_d' \qquad [17]$$

and a lumped transformation term as

$$k_t = k_{ta} + k_d'\ k_{ts} \qquad [18]$$

then Eq. [15] becomes

$$R\frac{\partial C_a}{\partial t} = D_a\frac{\partial^2 C_a}{\partial x} - V_a\frac{\partial C_a}{\partial x} - k_t C_a. \qquad [19]$$

Analytical solutions to Eq. [19] have been developed for a variety of boundary conditions and many of these have been compiled by van Genuchten and Alves (1982). In addition to the analytical solutions that are available, several numerical solutions have been presented in the literature. Two numerical codes are available and supported by the International Ground Water Modeling Center, Butler Univ., Indianapolis, IN. These two codes are listed in Table 8–4.

Table 8-4. Distributed nonconservative mass transport models: Miscible (convection and dispersion)—unsaturated and saturated. Adapted from van der Heijde et al. (1985).

| Author(s) | Original affiliation | Country | Model name | Completion year (Update) | System | Flow characteristics | Solute transport process | Method |
|---|---|---|---|---|---|---|---|---|
| Kaszeta<br>Simmons<br>Cole<br>Ahistrom<br>Foote<br>Serne | Research laboratory | USA | MMT-DPRW (1D/3D) | 1976 (1980) | Isotropic, heterogeneous porous medium; confined, semiconfined or water-table aquifer; multi-aquifer system | One- or two-dimensional, horizontal or vertical, or three-dimensional; steady state or transient | Convection, dispersion, adsorption, ion-exchange, radioactive decay, chemical reactions | Discrete parcel random walk |
| G.T. Yeh<br>D.S. Ward | Research laboratory | USA | Femwaste/ FECWASTE | 1981 | Anisotropic, heterogeneous porous medium | Two-dimensional, horizontal or vertical; steady state or transient | Convection, dispersion, diffusion, adsorption, first-order decay | Finite element |

To illustrate the movement of a retarded compound, where biological transformation is inhibited by sodium azide (0.02%), consider the movement of hexachlorobenzene (HCB) through a soil column. In this experiment, groundwater amended with HCB at 0.013 $\mu M\ L^{-1}$ (3.6 $\mu g\ L^{-1}$) was passed through a sandy soil (same soil as illustrated in Fig. 8-3). The porosity of the column was determined from $^3H$ breakthrough curves and assumed to be equal to the water content ($\theta$). The experimental breakthrough curves for HCB through replicate columns are shown in Fig. 8-4. A solution to Eq. [11], assuming $k_t = 0$ with boundary conditions

$$\left(D\frac{\partial C}{\partial x} + V_a C\right)\Bigg|_{x=0} = \begin{cases} V C_o & 0 < t \le t_o \\ 0 & t > t_o \end{cases} \qquad [20]$$

following the development of van Genuchten and Alves (1982), is

$$\frac{C(x,t)}{C_o} = \begin{cases} A(x,t) & 0 < t \le t_o \\ A(x,t) - A(x,t - t_o) & t > t_o \end{cases} \qquad [21]$$

where

$$A(x,t) = \frac{1}{2}\,\text{erfc}\left[\frac{R\,x - V_a t}{2(D_a R\,t)^{1/2}}\right] + \left[\frac{V_a^2\,t}{\pi D_a R}\right]^{1/2} \exp\left[-\frac{(R\,x - V_a t)^2}{4D_a R\,t}\right]$$
$$- \frac{1}{2}\left[1 + \frac{V_a x}{D_a} + \frac{V_a^2\,t}{D_a R}\right] \exp\left[\frac{V_a x}{D_a}\right] \text{erfc}\left[\frac{R\,x + V_a t}{2(D_a R\,t)^{1/2}}\right] \qquad [22]$$

is also presented in Fig. 8-4. The curve presented is a least squares best fit to the experimental data with regression coefficients listed on the figure. Although an attempt was made to maintain a constant water flux through the column throughout the experiment, variations did occur. The projected curve used mean interstitial velocity measured prior to the sampling time. The variation in velocity caused the projected curve to have minor perturbations in the shape of the curve. The same $k_d$ was used to describe the ascending and descending portion of the curve. The agreement between experimental and regressed function indicate the assumption of reversibility is adequate for HCB. The column results are consistent with batch studies shown in Fig. 8-5 where $^{14}C$ labeled HCB in 30 mL of groundwater was equilibrated with 2 g of soil. The soil partition coefficient obtained from the batch study is not significantly different than the one in the column study. Under appropriate conditions, the analytic solutions currently available in the literature can be a valuable asset to projecting the movement of chemicals through soils and thus contribute to exposure assessments. When the flow conditions are significantly dynamic to make it necessary to consider the temporal and spatial variability in water flow, it is possible to apply one of the numerical models listed in Table 8-4 or other appropriate model to project chemical concentration distributions in environmental situations.

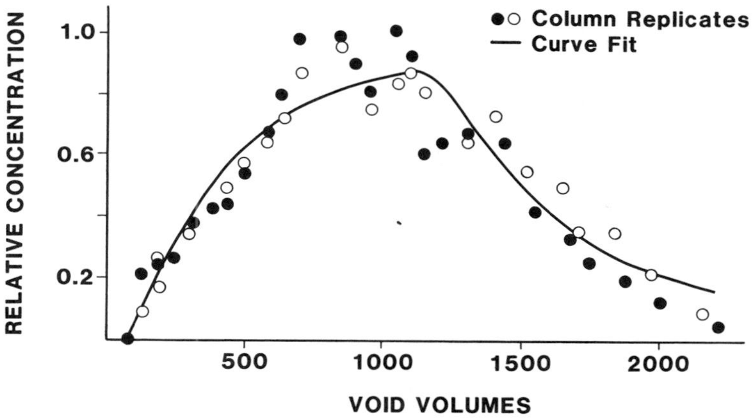


Fig. 8–4. Hexachlorobenzene breakthrough curve for duplicate columns. For col. 1 the following physical parameters apply: $\theta = 0.501$, mean interstitial velocity 112 m $d^{-1}$. For col. 2, the following physical parameters apply: $\theta = 0.469$, mean interstitial velocity 112 m $d^{-1}$. The parameters obtained from the least squares analysis are $k_d = 97$ and $\alpha = 2$ m.

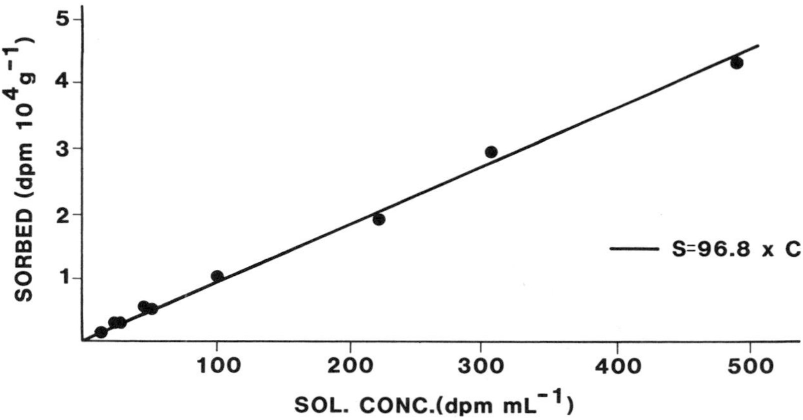


Fig. 8–5. Hexachlorobenzene sorption isotherm.

## 8–3 TRANSPORT FACILITATED BY COMPLEX FLUIDS

The fluid passing through the soil is not pure water with a single chemical dissolved in the water as assumed in the previous section. There has been an increasing awareness that there are synergistic and antagonistic reactions taking place within the soil system because of the complex nature of the fluid as well as the soil. Two approaches have been proposed to modify the miscible displacement theory of dilute systems to permit consideration of more complex fluids. Nkedi-Kizza et al. (1985) have considered the possibility of mixed miscible solvents. They have shown that it is possible to predict the

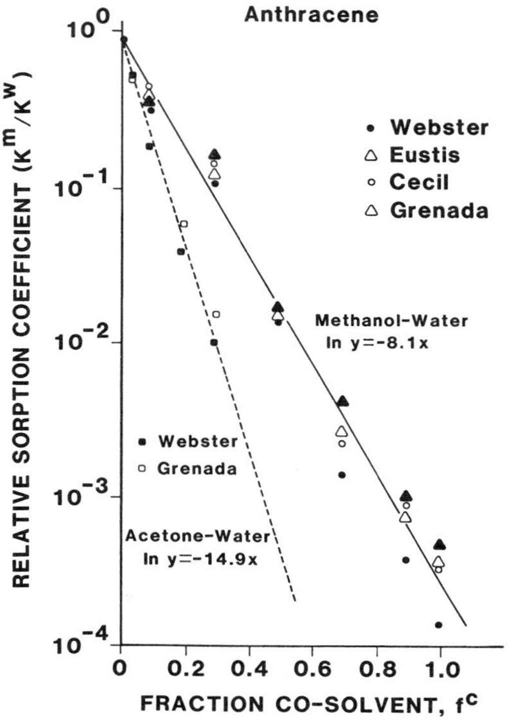


Fig. 8–6. Influence of mixed solvents on solubility in fluid phase. Adapted from Nkedi-Kizza et al. (1985).

enhanced solubility of a chemical because of the chemical properties of the solvents involved. Figure 8–6 shows the impact of the mixed solvents on the retardation factor. The co-solvent can potentially reduce the retardation factor several orders of magnitude. The importance of this phenomenon is particularly significant at municipal or industrial waste sites where there is a potential of co-disposed wastes enhancing the movement of toxic chemicals to groundwater. Nkedi-Kizza et al. (1985) also demonstrated that the impact of the co-solvent is greater the more hydrophobic the chemical. Enfield (1985) proposed that the movement of chemicals can be enhanced by the presence of macromolecules or immiscible fluids that may be moving with the water and acting as surfactants in the soil water system. Enfield (1985) considered the mobile phase to consist of two liquids plus a vapor moving through the soil system as shown in Fig. 8–7. Through a transformation of variables, the form of the equation was shown to be the same as those already solved for a variety of boundary conditions. In the following paragraphs, a similar development is shown with an emphasis on the movement of macromolecules. Change in the aqueous phase follows the same form as Eq. [9] except there are terms to describe the exchange between the aqueous phase and the macromolecule. This results in the one-dimensional equation

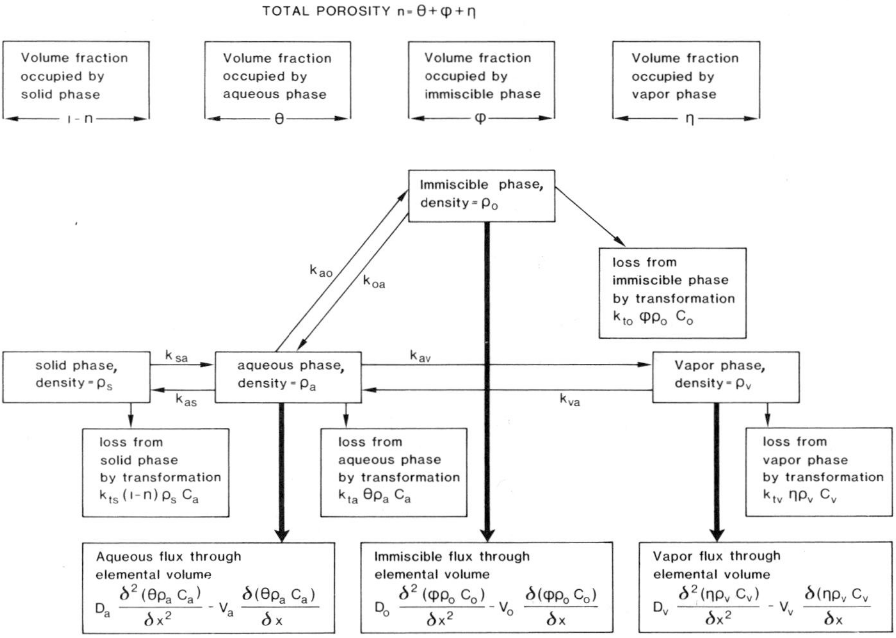


Fig. 8-7. Conceptual model for a representative elemental volume of soil with multiple fluid phases.

$$\frac{\partial(\theta\rho_a C_a)}{\partial t} = D_a \frac{\partial^2(\theta\rho_a C_a)}{\partial x^2} - V_a \frac{\partial(\theta\rho_a C_a)}{\partial x} - k_{ao}\theta\rho_a C_a + k_{oa}\phi\rho_o C_o$$

$$-k_{av}\theta\rho_a C_a - k_{va}\eta\rho_v C_v - k_{as}\theta\rho_a C_a + k_{sa}(1 - n)\rho_s C_s - k_{ta}\theta\rho_a C_a. \quad [23]$$

A second equation must be developed to describe the rate of change in concentration associated with the macromolecules.

$$\frac{\partial(\phi\rho_o C_o)}{\partial t} = D_o \frac{\partial^2(\phi\rho_o C_o)}{\partial x^2} - V_a \frac{\partial(\phi\rho_o C_o)}{\partial x}$$

$$- k_{oa}\phi\rho_o C_o + k_{ao}\theta\rho_a C_a - k_{to}\phi\rho_o C_o. \quad [24]$$

The change in the solid phase in this example is identical to Eq. [11]. The vapor phase follows the equation

$$\frac{\partial(\eta\rho_v C_v)}{\partial t} = D_v \frac{\partial^2(\eta\rho_v C_v)}{\partial x^2} - V_v \frac{\partial(\eta\rho_v C_v)}{\partial x}$$

$$- k_{va}\eta\rho_v C_v + k_{av}\theta\rho_a C_a - k_{tv}\eta\rho_v C_v. \quad [25]$$

Equation [25] allows transformations to take place in the vapor phase according to first-order kinetics and partitioning between the liquid and vapor phase. When local equilibrium is assumed,

$$k_{ao}\theta\rho_a C_a = k_{oa}\phi\rho_o C_o \qquad [26]$$

$$k_{av}\theta\rho_a C_a = k_{va}\eta\rho_v C_v \qquad [27]$$

$$k_{as}\theta\rho_a C_a = k_{sa}(1 - n)\rho_s C_s . \qquad [28]$$

By defining a new variable $C^*$ as the total mass concentration of the mobile contaminant over all phases,

$$C^* = \theta\rho_a C_a + \phi\rho_o C_o + \eta\rho_v C_v \qquad [29]$$

or

$$C^* = \left(1 + \frac{k_{ao}}{k_{oa}} + \frac{k_{av}}{k_{va}}\right) \theta\rho_a C_a \qquad [30]$$

and by defining an additional variable

$$\beta = 1 + \frac{k_{ao}}{k_{oa}} + \frac{k_{av}}{k_{va}} \qquad [31]$$

such that

$$C_a = C^*/\beta\theta\rho_a \qquad C_o = k_{ao}C^*/\beta k_{oa}\phi\rho_o$$

$$C_v = k_{av}C^*/\beta k_{va}\eta\rho_v \qquad C_s = k_{as}C^*/\beta k_{sa}(1 - n)\rho_s \qquad [32]$$

then the total change in an elemental volume can be shown to follow the equation

$$\left[1 + \frac{k_{ao}}{k_{oa}} + \frac{k_{av}}{k_{va}} + \frac{k_{as}}{k_{sa}}\right] \frac{\partial C^*}{\partial t} = \left[D_a + \frac{k_{ao}}{k_{oa}} D_o + \frac{k_{av}}{k_{va}} D_v\right] \frac{\partial^2 C^*}{\partial x^2}$$

$$- \left[V_a + \frac{k_{ao}}{k_{oa}} V_o + \frac{k_{av}}{k_{va}} V_v\right] \frac{\partial C^*}{\partial x}$$

$$- \left[k_{ta} + \frac{k_{ao}}{k_{oa}} k_{to} + \frac{k_{av}}{k_{va}} k_{tv} \frac{k_{as}}{k_{sa}} k_{ts}\right] C^* . \qquad [33]$$

By defining the variables

$$R^* = 1 + \frac{k_{ao}}{k_{oa}} + \frac{k_{av}}{k_{va}} + \frac{k_{as}}{k_{sa}} \quad [34]$$

$$D^* = D_a + \frac{k_{ao}}{k_{oa}} D_o + \frac{k_{av}}{k_{va}} D_v \quad [35]$$

$$V^* = V_a + \frac{k_{ao}}{k_{oa}} V_o + \frac{k_{av}}{k_{va}} V_v \quad [36]$$

$$K_t^* = k_{ta} + \frac{k_{ao}}{k_{oa}} k_{to} + \frac{k_{av}}{k_{va}} k_{tv} + \frac{k_{as}}{k_{sa}} k_{ts}. \quad [37]$$

Eq. [33] becomes

$$R^* \frac{\partial C^*}{\partial t} = D^* \frac{\partial^2 C^*}{\partial x^2} + V^* \frac{\partial C^*}{\partial x} - K_t^* C^* \quad [38]$$

which is the same form as Eq. [19] already discussed. Enfield and Bengtsson (1988) discuss the impact of the macromolecules on the mobility of chemicals. They note that large hydrophilic macromolecules may be excluded by smaller pores in the soil and thus, the macromolecules may actually move more rapidly than the water.

Two theoretical figures (8–8 and 8–9) were developed to elucidate the significance of both the presence of the macromolecule and the impact of differences in interstitial velocities between fluid phases. Several simplifying assumptions were made in the development of the figures. First, dispersion and transformation in Eq. [38] were ignored which yields

$$R^* \frac{\partial C^*}{\partial t} = V^* \frac{\partial C^*}{\partial x}. \quad [39]$$

Further, the following assumptions are applied to generate theoretical figures (8–8 and 8–9): (i) the density of the macromolecule equals the density of the water; (ii) the partition coefficient to the macromolecule's organic C is the same as the partition coefficient to the soil's organic C ($k_{oc}$), $k_p = k_{oc}$ times the weight fraction C occupied by the macromolecules; (iii) the fraction of organic C of the soil (F) is 0.02; (iv) the soil partition coefficient can be described by the equation of Briggs (1981) as

$$\log k_d = 0.52 \log k_{ow} + 0.65 + \log F \quad [40]$$

(v) the total porosity (n) of the soil is 0.5; and (vi) the particle density of the soil is 2.65.

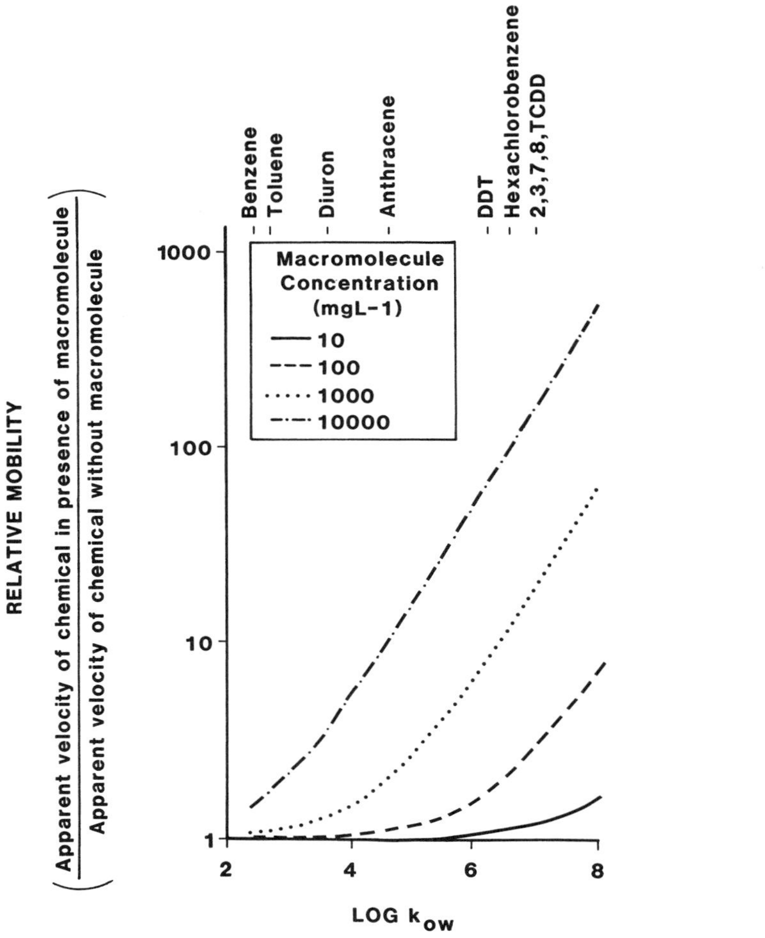


Fig. 8-8. Mobility of a hydrophobic compound relative to the mobility of the same compound without the presence of a macromolecule as a function of octanol-water partition coefficient (theoretical computation).

The impact of the macromolecule on the mobility of hydrophobic compounds is demonstrated in Fig. 8-8. When one assumes that the interstitial velocity of the macromolecule and the water are the same ($V_a = V_o$), the octanol/water partition coefficient vs. the relative chemical mobility (mobility in the presence of macromolecule/mobility without macromolecule) can be presented. The macromolecules in the mobile phase can significantly alter the relative mobility of extremely hydrophobic compounds even when the amount of macromolecule is in concentrations typical of groundwater or agricultural soil solution. This might be one explanation why hydrophobic pesticides, such as DDT, have been reported to move farther under field conditions than model projections (e.g., Enfield et al., 1982). The importance

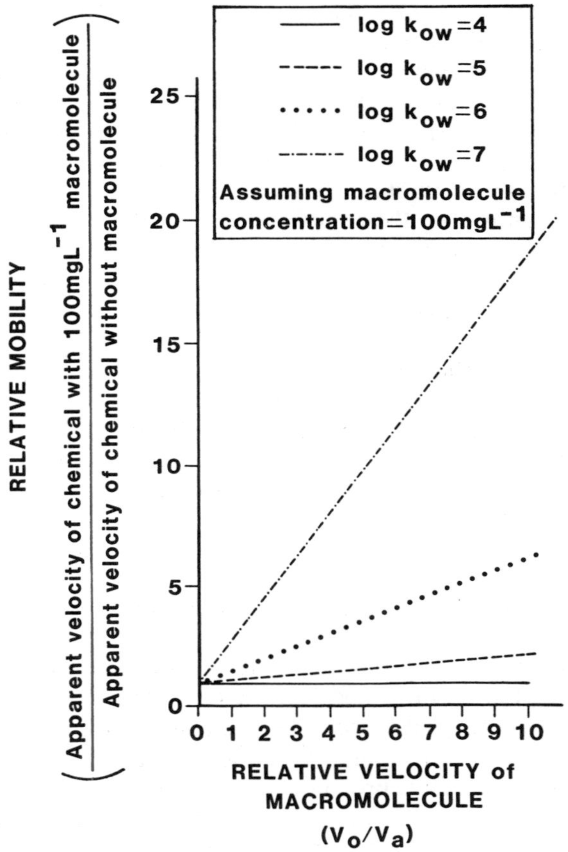


Fig. 8–9. Mobility of a hydrophobic compound relative to the mobility of the same compound without the presence of a macromolecule vs. the velocity of the macromolecule relative to the velocity of the water where the concentration of the macromolecule is 100 mg $L^{-1}$ (theoretical computation).

of the macromolecule rapidly diminishes as the octanol/water partition coefficient goes down. At municipal or industrial waste sites where relatively high concentrations of macromolecule might exist (1 or more g/L of total organic C in the fluid [Williams et al., 1984]) the importance of the macromolecule to chemical mobility could be significant even for chemicals with log $k_{ow}$ of 3 or less. Hydrophilic macromolecules with C concentrations of 100 to 500 mg/L may change the relative mobility of hydrophobic compounds by an order of magnitude in low C soils if the partition coefficient to the macromolecule's C is the same as the partition coefficient to the soil's organic C.

The significance of differences in interstitial velocities of the two mobile phases are shown in Fig. 8–9; it was assumed that the C concentration of the macromolecule was 100 mg/L. By assuming differences in the interstitial velocities of the mobile phases, it is possible to plot relative mobility of the macromolecule ($V_o/V_a$) vs. the relative mobility of the compound. As the relative mobility of the macromolecule goes down, the importance goes down. By the time the relative velocity of the macromolecule reaches 10% of the velocity of the water, the macromolecule is no longer significant in facilitating chemical transport. Therefore, consideration of a mobile organic phase for accelerating chemical transport is limited to macromolecules that behave as hydrophilic compounds but have the capacity to sorb hydrophobic solutes in groundwater. Examples of this type of macromolecule would be the naturally occurring dissolved organic C (DOC) found in groundwater as well as some of the nonionic or anionic biologically or chemically produced micelle-forming surfactants. The importance of the macromolecule is much more important for chemicals with a high $k_{ow}$ than chemicals with a low $k_{ow}$. This is similar to the observations Nkedi-Kizza et al. (1985) presented for the presence of co-solvents. For concentrations of dissolved organic C commonly found in most agricultural soils only the hydrophobic compounds will be influenced by the DOC, but in cases where the DOC is greater than normally observed the infuence of the DOC may extend to less hydrophobic chemicals.

## 8-4 MODEL APPLICATION

The selection and application of a model to an environmental situation depends on the objective of the modeling exercise. There is no one best model for all purposes. Users need to consider their objectives as well as the assumptions in a model prior to implementation of the model. Models might be classified into two or three broad categories (research, screening, and/or educational models). Research models try to accurately describe a portion of the transport processes at a specific site. The models divide the problem domain into elements. Each element is assumed to accurately represent a continuum in the flow field. The models overlook the microscopic flow patterns and attempt to describe some fictious average flow. Model elements are generally much larger than a minimum representative elementary volume defined by Bear (1979), but will be considered the model's representative elemental volume (REV). To calculate the water fluxes in the flow field, the hydraulic properties, which describe the soil-water characteristic and hydraulic conductivity function, are required input for each REV. This information is supplied by either tabular or defined functional relationships. For each REV, the partition coefficient between the solid and liquid phase as well as the transformation rate must be specified to calculate the chemical flux. The user needs to pay particular attention to the units used by the model developer and the assumptions on where the transformation takes place. Some model developers have assumed that the only place transformation will take place is in the bulk liquid. Others have assumed that transformation takes place only when the chemical is sorbed to the soil. Either of these assumptions can be incorrect descriptions of the processes involved. Nevertheless, it is possible to transform the data such that either of the methods gives a numerically correct presentation of the result using a procedure similar to that described in section 8–3. In addition to describing the properties of each of the elements of the problem domain, boundary conditions must be described either as a potential or flux. Realistic description of the boundary conditions is often the most difficult task for the user of an existing model. This is particularly true for transient flow conditions where boundary conditions change as a function of time.

Research models are used to answer a variety of questions. They are often used in legal actions at hazardous waste sites to help forecast future environmental insult and potential impact of proposed remedial actions. Research models are used in the design of containment systems for landfills and to see if transport processes are understood.

Screening and educational models require much less input data than the research models. As a result, the screening models compromise in the way they describe one or more of the transport processes. Most screening models assume steady-state water flux throughout the problem domain. This is a definite limitation, particularly when the analysis is describing the spatial and temporal distribution of a chemical near the point of application. Other screening models route the water through compartments such that the output of one compartment is the input to the next compartment without maintaining the flow system as a continuum. Even considering the simplifying assumptions in the screening models, they are often as good as the available input data. Screening models are often used where the hydrogeologic system is not well defined. For example, in the past, the potential behavior of a new pesticide has been inferred from a combination of laboratory data and a few field monitoring studies. A major limitation of field studies is that each study represents only one combination of field, climatic, and management factors. Modeling as a preliminary screening and forecasting tool permits extrapolating the limited field data to a variety of conditions. Screening models are often used in developing sampling schemes to minimize the number of compounds quantitatively determined in a sampling train. In this case, even with their inaccurate description of the actual flow process, screening models will often be adequate to determine which chemicals should first arrive at sampling wells and suggest how the sampling wells should be placed to obtain samples of a chemical plume. Examples of models suitable for this purpose are Plume 2D and Plume 3D two- or three-dimensional plumes in uniform groundwater flow presented by Wagner et al. (1984a, b). The model in three dimensions analytically solves the equation:

$$R\frac{\partial C}{\partial t} + V\frac{\partial C}{\partial x} = D_x\frac{\partial^2 C}{\partial x^2} + D_y\frac{\partial^2 C}{\partial y^2} + D_z\frac{\partial^2 C}{\partial z^2} - Rk_tC. \quad [41]$$

Equation [41] is similar to Eq. [19] except for the way the transformation rate coefficient is defined, but assumes dispersion is in three dimensions rather than in one dimension. The two models assume:

1. The groundwater flow regime is completely saturated.
2. All aquifer properties are constant and uniform throughout the aquifer.
3. The groundwater flow is horizontal, continuous, and uniform throughout the aquifer.
4. The chemical input is a point at the origin of the coordinate system.
5. Mass flow rate of the chemical input is constant.
6. At zero time, the concentration of chemical in the aquifer is zero.
7. The aquifer is infinite in extent.

The models use the principle of superposition to overcome the problem associated with the assumption of an infinite aquifer and to permit multiple sources and variable source rate terms. Input data requirements are listed in Table 8–5. Output from the model is the spatial and temporal distribution for the chemical in tabular format. The models were written in FORTRAN, for interactive use on a microcomputer.

Enfield et al. (1986) presented, with the acronym SWAG (simulated waste access to groundwater), a one-dimensional compartmental screening model developed for use in land treatment of wastewater, where there is an excess of water creating significant amounts of recharge. The compartmental model was developed to describe the movement of volatile or nonvolatile transformable or recalcitrant organic chemicals in rapid-infiltration wastewater treatment systems. The rapid infiltration wastewater treatment system is assumed

Table 8–5. Input data requirements for the screening model Plume 3D.

| |
|---|
| Title |
| Units for length |
| Units for time |
| Units for concentration |
| Saturated thickness of aquifer (L) |
| Aquifer porosity |
| Seepage (interstitial) velocity (L $t^{-1}$) |
| Retardation factor |
| X dispersion coefficient, (sq. L $t^{-1}$) |
| Y dispersion coefficient, (sq. L $t^{-1}$) |
| Z dispersion coefficient, (sq. L $t^{-1}$) |
| Transformation rate (1 $t^{-1}$) |
| Select transient or steady-state solution |
| Number of sources |
| X, Y, and Z coordinates of sources |
| Starting time, ending time, and mass rates of sources† |
| Steady-state mass rate of source‡ |
| The remaining input data define the matrix where the concentration is evaluated. |
| First X, Last X, and Delta X (L) |
| First Y, Last Y, and Delta Y (L) |
| First Z, Last Z, and Delta Z (L) |
| First t, Last t, and Delta t (t) |

† Input data required for transient solution.
‡ Input data required for steady-state solution.

to operate in a cyclic manner with periods of flooding where the soil surface is covered with water and periods of drying where the soil profile is allowed to dry and re-oxidize. The first compartment describes chemical loss from the infiltration basin when the infiltration basin is flooded with water. The second compartment considers losses because of volatilization and transformation in near-surface soils during periods of drying. The third compartment describes the transport and transformation of the remaining chemical to groundwater. The model was designed to consider an accidental spill of chemical into a treatment system or the constant input of chemical into a system. The theoretical basis for the model is given in brief below.

The model assumes the volume of flow applied to the infiltration basins all passes through the soil profile to groundwater. Precipitation is assumed equal to ET and there is no runoff. The first compartment of the model was considered to be a perfectly mixed container with constant depth with no mass flow. The chemical was allowed to transform by first-order kinetics, and loss to the atmosphere was shown to follow first-order kinetics, based on the Lewis and Whitman (1924) two film model. The chemical remaining after losses in the pond was placed at the bottom of the succeeding compartment that is an unsaturated soil layer. The soil layer acted as a restrictive layer for volatile losses. It was assumed that there was no convective flow through this soil layer and the layer was at a quasi-steady-state condition. First-order transformation was permitted in the soil layer and volatile losses followed Fick's law where the concentration of the chemical in the atmosphere was assumed to be zero. The concentration remaining for input into the third compartment ($C_i$) was shown to follow the equation:

$$C_i = \frac{CV_d \sinh(\sqrt{\beta}\,\iota)}{(1 - \mathrm{FTF})\,D^* \sqrt{\beta} \cosh(\sqrt{\beta}\,\iota) + V_d \sinh(\sqrt{\beta}\,\iota)}. \qquad [42]$$

Several terms in Eq. [42] have not been defined. The FTF is the fraction of time flooded. This was included to show that volatile losses could take place only during the portion of time the pond was not present. Beta is a temporary variable related to the transformation rate and effective dispersion coefficient. Iota is the thickness of the soil layer restricting volatile losses and is estimated based on the distance a chemical should move during one cycle (wetting and drying) of system operation. The third and final compartment was described in a manner similar to the development in section 8–3 of this chapter ignoring the possibility of an immiscible organic phase.

The output of the model was compared to observations from a replicated laboratory microcosm for 18 organic compounds. Input data requirements are given in Table 8–6 along with the units used in the model. The saturated hydraulic conductivity, total porosity, and Clapp and Hornburger (1978) curve coefficient ($b$) were used along with the water application rate to describe the hydraulic regime. Given the total porosity ($n$), the saturated

Table 8-6. Input data requirements for the model SWAG.

| Input parameter | 1,1,1-Trichloroethane | Bis-(2-chloroethyl)ether |
|---|---|---|
| Sat. hydraulic conductivity, m $d^{-1}$ | 0.83 | 0.83 |
| Total porosity, $m^3$ $m^{-3}$ | 0.4 | 0.4 |
| Clapp & Hornburger curve coefficient | 0.8 | 0.8 |
| Dispersivity, m | 0.2 | 0.2 |
| Molecular wt., g | 133.41 | 143.02 |
| Vapor pressure, kPa | 16.5 | 0.19 |
| Solubility in water, mol $m^{-3}$ | 5.40 | 71.32 |
| Temperature, K | 293 | 293 |
| Average water application rate, m $d^{-1}$ | 0.044 | 0.044 |
| Depth of water in pond, m | 0.003 | 0.003 |
| Fraction time flooded | 0.1 | 0.1 |
| Repeat cycle time, d | 0.167 | 0.167 |
| Bulk density of solid, Mg $m^{-3}$ | 1.6 | 1.6 |
| Density of liquid, Mg $m^{-3}$ | 1.0 | 1.0 |
| Density of air at temperature, Mg $m^{-3}$ | $1.3 \times 10^{-3}$ | $1.3 \times 10^{-3}$ |
| Ten meter wind speed, m $s^{-1}$ | $10^{-6}$ | $10^{-6}$ |
| Diffusion coefficient in water, $m^2$ $d^{-1}$ | $5.2 \times 10^{-5}$ | $5.2 \times 10^{-5}$ |
| Diffusion coefficient in air, $m^2$ $d^{-1}$ at 273 | 0.6 | 0.6 |
| Partition coefficient liquid/solid phase | 0.052 | 1.58 |
| Duration of chemical pulse, d | ∞ | ∞ |
| Transformation rate liquid phase, $d^{-1}$ | 0 | 0.06 |
| Transformation rate solid phase, $d^{-1}$ | 0 | 0.2 |
| Constant input conc., mol $m^{-3}$ | 1 | 1 |
| Initial distance, m . . . . . . . . . . . . . . . . . . . . . . | | |
| Distance increment, m . . . . . . . . . . . . . . . . . . . . | | |
| Maximum distance, m . . . . . . . . . . . . . . . . . . . . | | |
| Initial time, d . . . . . . . . . . . . . . . . . . . . . . . . . . | | |
| Time increment, d . . . . . . . . . . . . . . . . . . . . . . | | |
| Maximum time, d . . . . . . . . . . . . . . . . . . . . . . . | | |
| Depth to groundwater, m | 1.5 | 1.5 |

hydraulic conductivity ($K_s$), the water application rate equivalent to the Darcy velocity ($V_d$), and a curve coefficient ($b$), the water content ($\theta$) is calculated from the equation:

$$V_d = K_s\ (\theta/n)^{2b+3}. \qquad [43]$$

The air-filled porosity ($\eta$) is then $n - \theta$, and the interstitial velocity for water $V_a = V_d/\theta$. The effective diffusion coefficient for the vapor phase was approximated from Millington and Quirk's (1961) empirical relationship.

$$D_v = D_o\ (\eta^{10/3}/\theta_s) \qquad [44]$$

with a knowledge of the diffusion coefficient in air. Projections for volatile losses were within a factor of 2 for slowly transformable compounds. Volatile losses for compounds that would degrade would be over-estimated when degradation is ignored. Independent measurements of transformation rates were not available for most of the compounds used in the study. Although the model could be made to fit the observations by adjusting the transfor-

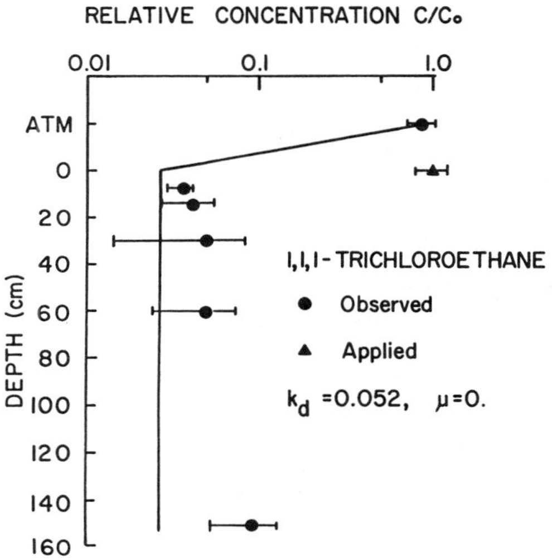


Fig. 8–10. Quasi-steady-state distribution of 1,1,1-trichloroethane in a 1.5-m deep microcosm. The coefficient $\mu$ in the figure is the same as $K_t$* defined by Eq. [37] where $k_{to} = k_{tv} = 0$.

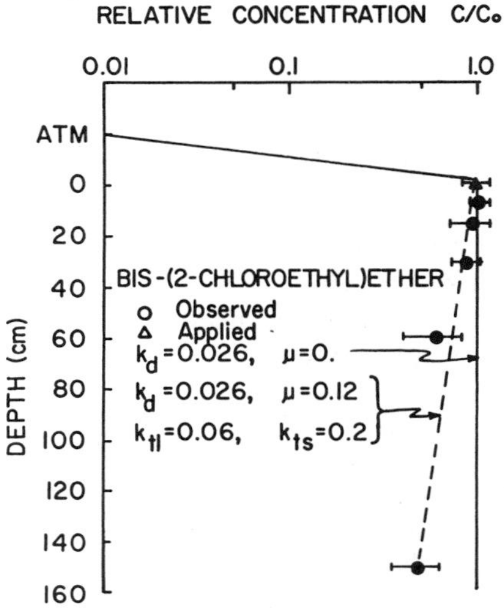


Fig. 8–11. Quasi-steady-state distribution of bis-(2-chloroethyl)ether in a 1.5-m microcosm. The coefficient $\mu$ in the figure is the same as $K_t$* defined in Eq. [37] where $k_{to} = k_{tv} = 0$.

mation parameters, this does not mean the model adequately describes the processes involved. To illustrate the model's output, results of the output and the measurements from the microcosm are shown in Fig. 8–10 and 8–11. 1,1,1-Trichloroethane was selected because it is a volatile, recalcitrant compound showing little transformation. The figures show the relative concentration applied to the microcosm as a triangle and the relative concentration in the soil solution as a function of depth in the soil profile. The computed and measured relative amounts volatilized are also shown. Bis-(2-chloroethyl)ether (Fig. 8–11) was selected as a transformable, nonvolatile compound. No attempt was made to measure volatile losses from the experimental column. The model projected $<1\%$ of the compound would be lost by volatilization. Bis-(2-chloroethyl)ether is known to hydrolyze, and the transformation rate that best fits the experimental data is close to the hydrolysis rate reported in the literature. The examples presented show conditions where the model estimates are close to the observations. First-order kinetics assumed in the model are not always satisfactory. When processes other than first order are known, appropriate models should be used.

## 8–5 NEW DEVELOPMENTS AND OUTLOOK FOR USE OF MODELS IN ASSESSING GROUNDWATER CONTAMINATION PROBLEMS

In recent years, researchers have become aware that modeling chemical transport through soil as an isotropic medium is not always adequate (e.g., Bouma et al., 1982; Smettem & Collis-George, 1985; White, 1985a, b). Many have attempted to consider the medium as anisotropic (Bear, 1979) taking into consideration the stratification of the sediments as they were deposited or developed in layers. This was proposed because when using relatively large REVs the horizontal hydraulic conductivity was generally much greater than the vertical hydraulic conductivity. This approach has been particularly popular in groundwater models. Several researchers have also been trying to integrate soil taxonomy with water and chemical transport. As soils develop a structure, much of the water flow takes place in macropores along the ped faces or worm (*Lumbricus* spp.) channels bypassing the bulk of the soil. Analytic solutions to chemical transport, considering the soil to be a fractured media where water flows through the macropores and then diffuses into the peds where it may react with the soil particles, have been developed for a variety of boundary conditions (White, 1985a, b; van Genuchten, 1985; Parker & Valocchi, 1986) and shown to fit experimental field data. These are definite advances over the approaches mentioned earlier in this chapter. When using these approaches, the boundary conditions can be difficult to adequately describe, particularly under unsaturated transient flow conditions. For example, if, during a given climatic event, the rainfall or irrigation intensity is less than the saturated hydraulic conductivity of the individual soil peds, flow will not take place along the ped faces as the macropores will remain unsaturated. When the intensity of the climatic event is greater than the saturated hydraulic conductivity of the ped then excess water will flow along the ped faces. This leads to a new problem of requiring a knowledge of how a chemical is applied, when the chemical was applied and a history of climatic events prior to an event where water flows along the ped faces. Dekker and Bouma (1984) studied the flow problem with nitrate ($NO_3$) and demonstrated large differences in the amount of N lost from a soil profile based on the history of application. White et al. (1986) made similar observations with the movement of the pesticides bromacil and napropamide through undisturbed cores of a structured clay soil. The macropores will also have a dramatic influence on movement of volatile compounds in the vapor phase. Diffusion in the vapor phase is many times greater than diffusion in the liquid phase (see Table 8–6) and the presence of macropores will present a ready conduit for vapor phase movement to the atmosphere. As a result, much of a volatile substance incorporated in surface soils may be lost to the atmosphere. The loss of volatile materials is significant even through several meters of unsaturated soils that were not directly contaminated with the substance. Plumes of chlorinated solvents in groundwater have been monitored by sampling the gas in soil pores several meters above a contaminant plume.

In the presence of macropores, even when the problems associated with the boundary conditions are solved, there will still be problems with describing the number and size of macropores. Soil taxonomists are evaluating geostatistical methods as a possible tool to help describe the macropores. Possibly a stochastic representation of the macropores in a deterministic model will be effective for forecasting the spatial and temporal distribution of chemicals in the environment.

All of the models discussed assume transformation is pseudo first order. This approach is probably adequate as a first approximation. However, many of the reactions will not be first order. For example, oxidation processes in groundwater will likely be second order dependent not only on the concentration of the chemical but on the concentration of $O_2$. Models have been developed to handle this type of situation (Borden & Bedient, 1986; Yates & Enfield, 1989) and have received limited testing in environmental situations (Borden et al., 1986). Additional work is needed to fully understand the processes and develop rate coefficients as a function of other environmental variables such as $O_2$, pH, temperature, and nutrients required by the biological population before the implications of second-order reactions can be fully understood.

At the beginning of this chapter, decoupling water flow from chemical transport was assumed. This assumption is not always adequate. For example, in geothermal regions there may be significant temperature gradients such that density variations will have a significant influence on mass flow. Lindstrom and Piver (1985) presented a model for organic transport through unsaturated/saturated soils under nonisothermal conditions addressing such problems. As more is learned about the processes involved and the significance of these processes, approaches similar to that of Lindstrom and Piver will become useful.

The models currently in use have definite limitations and shortcomings. Nevertheless, models are the only rational way to forecast the spatial and temporal distribution of a chemical in the environment. In the registration of pesticides for use in agriculture, the use of field studies gives an excellent way of evaluating the fate of a chemical under a given set of climatic conditions for a given soil. However, to evaluate the fate of a chemical under a variety of climatic conditions for a variety of soils, models must be used with an understanding of their limitations.

In addition to using models for forecasting prior to an environmental problem, models are useful in evaluating alternative reclamation approaches to existing environmental contamination. In the case of an accidental spill, there is a question of how much time is available before a segment of the population is exposed. This type of information can be useful to an emergency response team faced with the responsibility of warning the public of potential exposure. The next question addresses whether human intervention is desirable or if natural physical, chemical, and biological processes will be adequate to sufficiently mediate the problem. Modeling in a forecasting mode is the only current approach to address these problems. When forecasting, the accuracy of model estimates is not good. The modeler rarely has the luxury of calibrating one's model with some past history at the site. The modeler must rely on information in the literature and, hopefully, the experience of a soil taxonomist and hydrogeologist familiar with the area and chemists and microbiologists familiar with the chemicals in question. Even with the limitations on accuracy, the modeler can give a range for the expected spatial and temporal distribution of a chemical in the environment.

Cleanup of soils and geologic materials is an expensive proposition. Often the cleanup procedure is to remove the material and place it in a contained system, potentially, to become a problem at a future date. Models can be used for evaluating in situ reclamation activities such as well placement to capture a contaminant plume, answering such questions as: How many wells are needed? How should each of the wells be pumped? How long will it be necessary to pump a well field to "clean" an aquifer? Other in situ reclamation approaches require the stimulation of naturally occurring organisms in the aquifer. Studies have shown that there is a reasonable population of microorganisms in aquifer materials but their activity is low (Hutchins et al., 1985). Chemical additions may be possible to stimulate the activity. One difficulty with chemical amendments is that they often increase the biological activity within the well and plug the well screen and aquifer near the well limiting the useful life of the injection system. Models can be used to study alternate schemes of dosing the well, creating "bubbles" of nutrient and giving the well itself a chance to recover its hydraulic properties.

Models are useful in the evaluation of the design of containment systems in both saturated and unsaturated environments. In constructed landfills, barriers to water flow (soils or other materials with low hydraulic conductivity) layered with highly permeable layers are used to divert the water around the waste material. The philosophy is that if water doesn't pass through the material, there will not be contamination of the groundwater. As long as the constructed system remains intact, in theory the system should work and give a long residence time for the chemicals, giving time for chemical and biological transformations.

## APPENDIX

| Symbol | Description | Units |
|---|---|---|
| $b$ | Clapp and Hornburger curve coefficient | |
| $C^*$ | Concentration in the mobile phase | $M\ M^{-1}$ |
| $C_i$ | Dependent concentration variable | $M\ M^{-1}$ |
| $C_a$ | Concentration in aqueous phase | $M\ M^{-1}$ |
| $C_{et}$ | Evaporation pan coefficient | |
| $C_o$ | Concentration in immiscible phase | $M\ M^{-1}$ |
| $C_s$ | Concentration in solid phase | $M\ M^{-1}$ |
| $C_v$ | Concentration in vapor phase | $M\ M^{-1}$ |
| $D$ | Soil water diffusivity | $L^2\ T^{-1}$ |
| $d$ | Molecular diffusion coefficient in water | $L^2\ T^{-1}$ |
| $D^*$ | Lumped dispersion variable of mobile phase | $L^2\ T^{-1}$ |
| $D_a$ | Dispersion in aqueous phase | $L^2\ T^{-1}$ |
| $D_o$ | Dispersion in immiscible phase | $L^2\ T^{-1}$ |
| $D_v$ | Dispersion in vapor phase | $L^2\ T^{-1}$ |
| $E_g$ | Evapotranspiration from grass | $L\ T^{-1}$ |
| $E_{pan}$ | Pan evaporation | $L\ T^{-1}$ |
| ET | Evapotranspiration | $L\ T^{-1}$ |
| $F$ | Fraction organic C associated with solid phase | |
| $g$ | Gravitational acceleration | $L\ T^{-2}$ |
| $H$ | Total potential head | L |
| $K$ | Darcy velocity (hydraulic conductivity) | $L\ T^{-1}$ |
| $k_{ao}$ | First-order transfer coefficient aqueous to immiscible phase | $T^{-1}$ |
| $k_{as}$ | First-order transfer coefficient aqueous to solid phase | $T^{-1}$ |
| $k_{av}$ | First-order transfer coefficient aqueous to vapor phase | $T^{-1}$ |
| $k_d$ | Unitless soil/water partition coefficient | |
| $k_d'$ | Unitless soil/water distribution coefficient | |
| $k_{oa}$ | First-order transfer coefficient immiscible to aqueous phase | $T^{-1}$ |
| $k_{ow}$ | Octanol/water partition coefficient | |
| $K_s$ | Saturated hydraulic conductivity | $L\ T^{-1}$ |
| $k_{sa}$ | First-order transfer coefficient solid to aqueous phase | $T^{-1}$ |
| $k_t$ | Lumped transformation rate | $T^{-1}$ |
| $K_t^*$ | Lumped transformation variable | $T^{-1}$ |
| $k_{ta}$ | First-order transformation rate aqueous phase | $T^{-1}$ |
| $k_{to}$ | First-order transformation rate immiscible phase | $T^{-1}$ |
| $k_{ts}$ | First-order transformation rate solid phase | $T^{-1}$ |
| $k_{tv}$ | First-order transformation rate vapor phase | $T^{-1}$ |
| $k_{va}$ | First-order transfer coefficient vapor to aqueous phase | $T^{-1}$ |
| $L$ | Hydraulic loading from synthetic sources | $L\ T^{-1}$ |
| $n$ | Porosity of the soil | $L^3\ L^{-3}$ |

(continued on next page)

| Symbol | Description | Units |
|---|---|---|
| $P_r$ | Precipitation intensity | L $T^{-1}$ |
| $r$ | Net rate of change | M $L^{-3}$ $T^{-1}$ |
| $R$ | Retardation factor | |
| $R^*$ | Lumped retardation variable | |
| $t$ | Time | T |
| $V^*$ | Lumped velocity variable of mobile phase | L $T^{-1}$ |
| $V_a$ | Interstitial velocity of aqueous phase | L $T^{-1}$ |
| $V_d$ | Darcy velocity | L $T^{-1}$ |
| $V_o$ | Interstitial velocity of immiscible phase | L $T^{-1}$ |
| $V_v$ | Interstitial velocity of vapor phase | L $T^{-1}$ |
| $x$ | Distance along flow path in one-dimensional flow or horizontal distance in two-dimensional flow | L |
| $z$ | Vertical distance | L |
| $\alpha$ | Dispersivity | L |
| $\beta$ | Temporary variable | |
| $\rho_a$ | Density of aqueous phase | M $L^{-3}$ |
| $\rho_o$ | Density of immiscible phase | M $L^{-3}$ |
| $\rho_s$ | Particle density of the soil | M $L^{-3}$ |
| $\rho_v$ | Density of the vapor phase | M $L^{-3}$ |
| $\phi$ | Volume fraction occupied by the immiscible phase | $L^3$ $L^{-3}$ |
| $\theta$ | Volume fraction occupied by the aqueous phase | $L^3$ $L^{-3}$ |
| $\eta$ | Volume fraction occupied by the vapor phase | $L^3$ $L^{-3}$ |
| $\gamma$ | Viscosity of water | |
| $\psi$ | Water potential | L |
| $\tau$ | Radius of capillary tube or pore | L |
| $\kappa$ | Crop coefficient | |
| $\lambda$ | Tortuosity constant | |
| $\omega$ | Runoff | L $T^{-1}$ |
| $\iota$ | Thickness of soil restricting volatilization | L |

## REFERENCES


Addiscott, T.M., and R.J. Wagenet. 1985. Concepts of solute leaching in soils: A review of modeling approaches. J. Soil Sci. 36:411–424.

Amoozegar-Fard, A., D.R. Nielsen, and A.W. Warrick. 1982. Soil solute concentration distributions for spatially varying pore water velocities and apparent diffusion coefficients. Soil Sci. Soc. Am. J. 46:3–9.

Bear, J. 1979. Hydraulics of groundwater. McGraw-Hill, New York.

Boesten, J.J.T.I., and M. Leistra. 1983. Models of the behavior of pesticides in the soil-plant system. p. 35–64. *In* S.E. Jorgensen and W.J. Mitsch (ed.) Application of ecological modeling in environmental management, Part B. Elsevier Sci., Amsterdam.

Borden, R.C., and P.B. Bedient. 1986. Transport of dissolved hydrocarbons influenced by oxygen-limited biodegradation 1. Theoretical development. Water Resour. Res. 22:1973–1982.

Borden, R.C., P.B. Bedient, M.D. Lee, C.H. Ward, and J.T. Wilson. 1986. Transport of dissolved hydrocarbons influenced by oxygen-limited biodegradation 2. Field application. Water Resour. Res. 22:1983–1990.

Bouma, J., C.F.M. Belmans, and L.W. Dekker. 1982. Water infiltration and redistribution in a silt loam subsoil with vertical worm channels. Soil Sci. Soc. Am. J. 46:917–921.

Bresler, E., and G. Dagan. 1981. Convective and pore scale dispersive solute transport in unsaturated heterogeneous fields. Water Resour. Res. 17:1683–1693.

Briggs, G.G. 1981. Theoretical and experimental relationships between soil adsorption, octanol-water partition coefficients, water solubilities, bioconcentration factors, and the parachor. J. Agric. Food Chem. 29:1050–1059.

Carsel, R.F., C.N. Smith, L.S. Mulky, D. Dean, and P. Jowise. 1984. User's manual for the pesticide root zone model (PRZM). Release 1. Environ. Res. Lab., USEPA, Athens, GA.

Clapp, R.B., and G.M. Hornburger. 1978. Empirical equations for some soil hydraulic properties. Water Resour. Res. 14:601–604.

Childs, E.C., and C.N. Collis-George. 1950. The permeability of porous materials. Proc. R. Soc. 201A:392–405.

Davidson, J.M., L.R. Stone, D.R. Nielsen, and M.E. LaRue. 1969. Field measurement and use of soil water properties. Water Resour. Res. 5:1312–1321.

Dekker, L.W., and J. Bouma. 1984. Nitrogen leaching during sprinkler irrigation of a dutch clay soil. Agric. Water Manage. 9:37–45.

Di Toro, D.M. 1985. A particle interaction model of reversible organic chemical sorption. Chemosphere 14:1503–1538.

Donigian, A.S., D.C. Beyerlein, H.H. Davis, and N.H. Crawford. 1977. Agricultural runoff management (ARM) Model. Version II, Refinement and Testing. EPA-600/3-77-098. U.S. Gov. Print. Office, Washington, DC.

Enfield, C.G. 1985. Chemical transport facilitated by multiphase flow systems. Water Sci. Technol. 17:1–12.

Enfield, C.G., and G. Bengtsson. 1988. Macromolecule transport of hydrophobic contaminants in aqueous environments. Ground Water 26:64–70.

Enfield, C.G., R.F. Carsel, S.Z. Cohen, T. Phan, and D.M. Walters. 1982. Approximating pollutant transport to ground water. Ground Water 20:711–722.

Enfield, C.G., D.M. Walters, J.T. Wilson, and M.D. Piwoni. 1986. Behavior of organic pollutants during rapid-infiltration of wastewater into soil. Hazard. Wastes Hazard. Mater. 3:57–76.

Freeze, R.A., and J.A. Cherry. 1979. Groundwater. Prentice-Hall, Englewood Cliffs, NJ.

Heath, R.C. 1983. Basic ground-water hydrology. Geol. Surv. water-supply Paper 2220. U.S. Geol. Surv., Alexandria, VA.

Hutchins, S.R., M.B. Tomson, P.B. Bedient, and C.H. Ward. 1985. Fate of trace organics during land application of municipal wastewater. CRC Crit. Rev. Environ. Contro. 15:355–416.

Jury, W.A., W.F. Spencer, and W.J. Farmer. 1983. Behavior assessment model for trace organics in soil: I. Model description. J. Environ. Qual. 12:558–564.

Klute, A. 1972. The determination of the hydraulic conductivity and diffusivity of unsaturated soils. Soil Sci. 113:264–276.

Knusli, E. 1979. Advances in pesticide science, Part 1. Pergamon Press, Oxford, England.

Lewis, W.K., and W.G. Whitman. 1924. Principles of gas absorption. Ind. Eng. Chem. 16:1215–1220.

Lindstrom, F.T., and W.T. Piver. 1985. A mathematical model for the transport and fate of organic chemicals in unsaturated/saturated soils. Environ. Health Perspectives 60:11–28.

McCarthy, J.F., and B.D. Jimenez. 1985. Interactions between polycyclic aromatic hydrocarbons and dissolved humic material: Binding and dissociation. Environ. Sci. Technol. 19:1072–1076.

McEwen, F.L., and G.R. Stephenson. 1979. The use and significance of pesticides in the environment. John Wiley and Sons, New York.

Means, J.C., R.W. Walters, M.M. Rao, and A. Guiseppi-Elie. 1985. Effect of cosolvents on the sorption of 2,3,7,8-tetrachlorodibenzodioxin in soil. p. 231. Proc. 189th Natl. Meet.—ACS, Div. Environ. Chem., Miami, FL. 29 Apr.-3 May. Am. Chem. Soc., Washington, DC.

Millington, R.J., and J.P. Quirk. 1961. Permeability of porous solids. Trans. Faraday Soc. 57:1200–1207.

Molz, F.J., O. Guven, and J.G. Melville. 1983. An examination of scale-dependent, dispersion coefficients. Ground Water 21:715–725.

Nkedi-Kizza, P., P.S.C. Rao, and A.G. Hornsby. 1985. Influence of organic cosolvents on sorption of hydrophobic organic chemicals by soil. Environ. Sci. Technol. 19:975–979.

Parker, J.C., and M. Th. van Genuchten. 1984. Determining transport parameters from laboratory and field tracer experiments. Virginia Agric. Exp. Stn. Blacksburg Bull. 84-3.

Parker, J.C., and A.J. Valocchi. 1986. Constraints on the validity of equilibrium and first-order kinetic transport models in structured soils. Water Resour. Res. 22:399–408.

Rao, P.S.C., and R.E. Jessup. 1983. Sorption and movement of pesticides and other toxic organic substances in soils. p. 183–201. *In* D.W. Nelson (ed.) Chemical mobility and reactivity in soil systems. SSSA, Madison, WI.

Richards, L.A. 1931. Capillary conductivity of liquids through porous media. Physics. 1:318–333.

Slatyer, R.O. 1967. Plant-water relationships. Academic Press, London.

Smettem, K.R.J., and N. Collis-George. 1985. The influence of cylindrical macropores on steady-state infiltration in a soil under pasture. J. Hydrol. 79:107–114.

Smith, L., and F.W. Schwartz. 1980. Mass transport, 1: A stochastic analysis of macroscopic dispersion. Water Resour. Res. 16:303–313.

Schwarzenbach, R.P., and J. Westall. 1981. Transport of nonpolar organic compounds from surface water to groundwater. Laboratory sorption studies. Environ. Sci. Technol. 15:1360–1367.

Technical Committee on Irrigation Water Requirements. 1973. Sources of evapotranspiration data. p. 63–111. *In* M.E. Jensen (ed.) Consumptive use of water and irrigation water requirements. Am. Soc. Civ. Eng., New York.

U.S. Environmental Protection Agency. 1982. Pesticide industrial sales and usage—1982 market estimates. USEPA Inhouse Report.

van der Heijde, P., Y. Bachmat, J. Bredeheoft, B. Andrews, D. Holtz, and S. Sebastian. 1985. Groundwater management: The use of numerical models. Water Resour. Monogr. 5. Am. Geophys. Union, Washington, DC.

van Genuchten, M. Th. 1985. A general approach for modeling solute transport in structured soils. p. 513–526. *In* Hydrology of rocks of low permeability. Memoires, Vol. 17, Part 2. Proc. Int. Assoc. Hydrogeologists, Tucson, AZ. 7–12 Jan. Int. Assoc. Hydrogeologists.

van Genuchten, M.Th., and W.J. Alves. 1982. Analytical solutions of the one-dimensional convective-dispersive solute transport equation. Tech. Bull. 1661. USDA, Washington, DC.

Wagner, J., S.A. Watts, and D.C. Kent. 1984a. Plume 3D three-dimensional plumes in uniform ground water flow. Project Rep. CR811142. EPA 600/2-85-067, PB85-214443. Robert S. Kerr Environ. Res. Lab., USEPA, Ada, OK.

Wagner, J., S.A. Watts, and D.C. Kent. 1984b. Plume 2D two-dimensional plumes in uniform ground water flow. Project Rep. CR811142. EPA 600/2-85-065, PB85-214450. Robert S. Kerr Environ. Res. Lab., USEPA, Ada, OK.

White, R.E. 1985a. The analysis of solute breakthrough curves to predict water redistribution during unsteady flow through undisturbed structured soil. J. Hydrol. 79:21–35.

White, R.E. 1985b. A model for nitrate leaching in undisturbed structured clay soil during unsteady flow. J. Hydrol. 79:37–51.

White, R.E., J.S. Dyson, A. Gerstl, and B. Yaron. 1986. Leaching of herbicides through undisturbed cores of a structured clay soil. Soil Sci. Soc. Am. J. 50:277–283.

Williams, G.M., C.A.M. Ross, A. Stuart, S.P. Hitchman, and L.S. Alexander. 1984. Controls on contaminant migration at the Villa Farm Lagoons. Q. J. Eng. Geol. London 17:39–55.

Yates, S.R., and C.G. Enfield. 1989. Transport of dissolved substances with second-order reaction. Water Resour. Res. 25:1757–1762.

# Chapter 9

# Movement of Pesticides into Surface Waters

**R. A. LEONARD,** *USDA-ARS-AR, Tifton, Georgia*

Dispersion of pesticide residues into the environment by surface runoff from agricultural lands has been a major concern for about the last 20 yr. Early concerns centered primarily on the persistent chlorinated hydrocarbon insecticides (Nicholson et al., 1964, 1966) that were transported by both water and air far from their application site. Because of their adverse ecological impacts, most of these compounds have been banned or restricted from extensive use. Many pesticides including herbicides, however, remain in use as essential tools of modern agriculture and these span a wide range of behavorial and toxicological properties.

An analysis of pesticide impacts is beyond the scope of this chapter, but a few observations are useful for perspective. Public sensitivity to the pesticide-in-water problem is somewhat a carry-over of concerns that were generated as a result of problems with the persistent organochlorine insecticides. Aquatic life forms were acutely affected at low concentrations and both chronic and acute effects were associated with biological magnification in food chains. Insecticides in current use are much less persistent and many are significantly less toxic. Some of the insecticides, however, are quite toxic to both mammals and aquatic organisms (Stewart et al., 1975). Significant transport of these materials in runoff from application sites could, therefore, constitute a hazard. Because most currently used pesticides are nonpersistent and do not concentrate in the food chain, any problem created is likely to be of an acute nature, highly transitory, and occur near the source of runoff.

Currently, herbicides used in agriculture constitute the greatest tonnage of all pesticides. Although some herbicides are acutely toxic to fish (Stewart et al., 1975), the hazard of herbicides to the aquatic environment is small compared to insecticides (Caro, 1976). With a few noticeable exceptions, herbicides have little toxicity to humans, wildlife, and livestock. Herbicides may destroy or suppress aquatic vegetation, producing both desirable and undesirable effects. Kemp et al. (1982) and Forney and Davis (1981) report growth inhibition of submerged aquatic vegetation in the presence of 60 to 1040 $\mu$g/L atrazine depending on species. No effects were observed at con-

 *Pesticides in the Soil Environment*—SSSA Book Series, no. 2.

centrations of a few micrograms per liter (about 10 or less). Some species may show minor effects around 10 μg/L. The presence of herbicide residues such as picloram may also impact the water quality for irrigation of sensitive crops (Merkle & Bovey, 1974).

Pesticides in groundwater are of particular concern since much of the population depends on groundwater for drinking (CAST, 1985; Natl. Res. Counc., 1986). While concerns for surface water quality are often separated from those of groundwater quality, the hydrologic cycle provides direct connection in many geologic regions. Depending on hydraulic gradients, surface water may recharge groundwater or be replenished by groundwater (Hicks et al., 1987). Therefore, levels of pesticides in surface water may affect groundwater or be affected by groundwater.

In general, much more data is available on the presence of pesticide residues in surface water than on their environmental significance. As an example, we have amassed data on transitory concentration pulses in streams and in storm hydrographs from plots and watersheds from times of minutes to a few hours. These data have been necessary to understand transport processes and controlling variables. Toxicity tests are usually based on 24, 48, and even 96-h exposure to static concentrations. Perhaps we oversimplify when we compare pesticide concentration from runoff plots and small watersheds to these toxicity values. However, in many instances runoff concentrations do not exceed concentrations determined to be toxic by these tests, and here the potentiality of an acute problem can be dismissed. State-of-the-art methods can detect pesticide residues at concentrations as low as 1 ng/L or less. When pesticide residues are present in water at any detectable levels, some concern as to what may be their unknown long-term effects are inevitable. Therefore, society has continued to promote and encourage practices and agricultural production methods that minimize or reduce off-site pesticide transport, and much research and education is, therefore, devoted to this end.

This chapter covers processes and factors that determine quantities and distribution of pesticides in surface waters emanating from treated lands and modeling technology for assessing and predicting pesticide runoff. A similar chapter published elsewhere (Leonard, 1988) dealing only with herbicides has been updated and extended to pesticides in general.

For the purpose of this chapter, *runoff* is defined as water and any dissolved or suspended matter it contains that leaves a plot, field, or small single-cover watershed in surface drainage. Specifically, pesticide runoff includes dissolved, suspended particulate, and sediment-adsorbed pesticide that is transported by water from a treated land surface. At the field-scale, direct surface runoff (overland flow) is the major component of runoff. That is, return of subsurface water to the soil surface by seepage as prolonged interflow at this scale is minimal. In some landscapes, larger watersheds will have streams with significant subsurface or interflow components as well as overland flow components. In these larger, more complex drainage areas, the term *streamflow* will be used instead of runoff in reference to flowing water at the land surface.

# 9-1 PRINCIPLES GOVERNING PESTICIDE ENTRAINMENT AND TRANSPORT IN RUNOFF

## 9-1.1 Pesticide Extraction into Runoff

Processes of pesticide entrainment and transport in runoff may be visualized as in Fig. 9-1. At the microscale or interrill scale, pesticide extraction

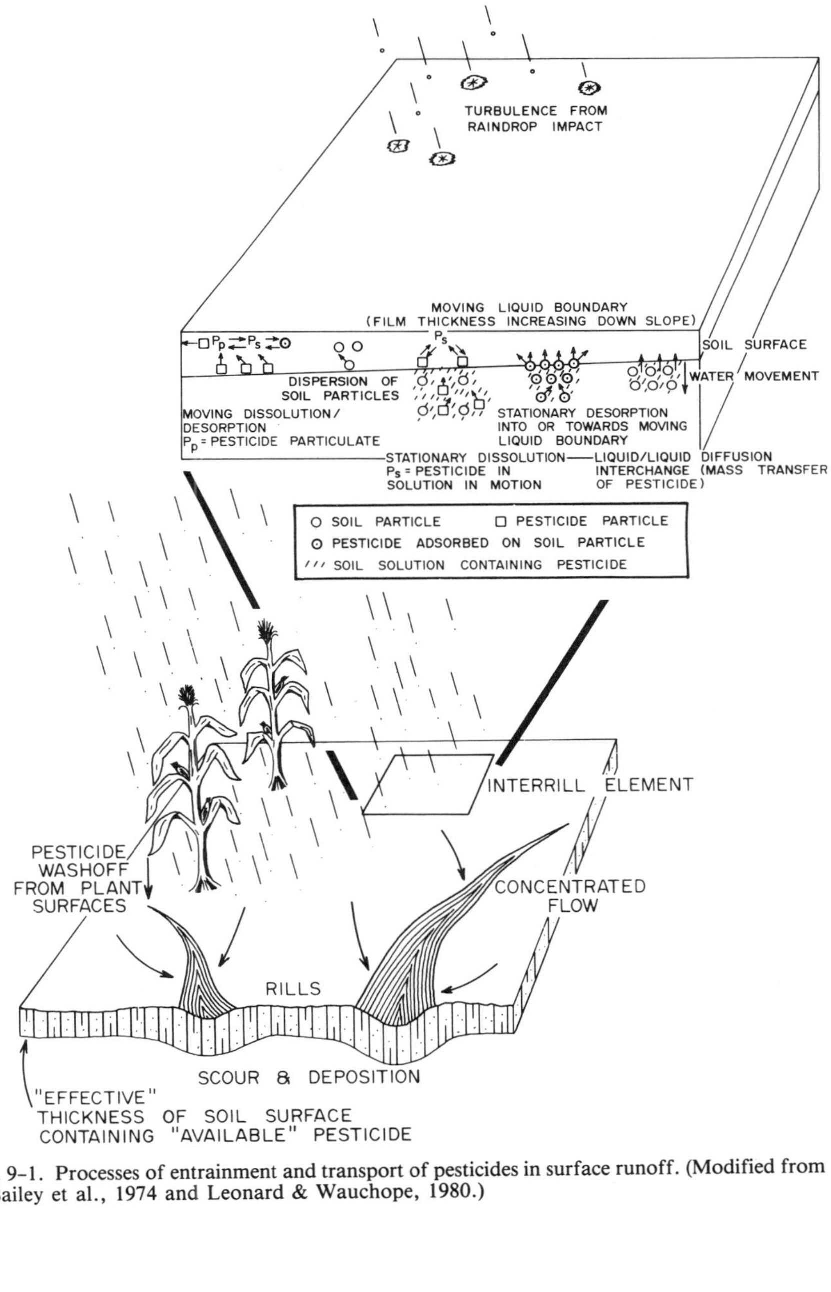


Fig. 9-1. Processes of entrainment and transport of pesticides in surface runoff. (Modified from Bailey et al., 1974 and Leonard & Wauchope, 1980.)

into runoff may be described as mechanisms of (i) diffusion and turbulent transport of dissolved pesticide from soil pores to the runoff stream; (ii) desorption from soil particles into the moving liquid boundary; (iii) dissolution of stationary pesticide particulates; (iv) scouring of pesticide particulates and their subsequent dissolution in the moving water (Bailey et al., 1974). Pesticide is also entrained in runoff attached to suspended soil particles. Raindrop impact in addition to suspending and dislodging soil particles, produces instantaneous pressure gradients affecting the turbulent interchange between the soil pore solution and the runoff stream (Ahuja et al., 1981; Ahuja & Lehman, 1983).

As flows within a field become concentrated because of topography, tillage patterns, or other irregularities affecting flow patterns, rills often develop. Rills are formed by concentrated flows detaching and transporting soil from a limited part of the land surface (Meyer et al., 1975). Rills vary in depth and size, but by definition are limited to depths that can be crossed by farm machinery and filled in by tillage. In contrast to interrill chemical extraction, rill erosion exposes a much deeper layer of the soil surface. Depending on chemical distribution with depth, this soil material may contribute chemicals to the runoff stream or remove chemicals by adsorption.

Shallow interflow may also contribute dissolved chemicals to the runoff stream (Ahuja et al., 1981a; Donigian et al., 1977). Shallow interflow is defined as water that has infiltrated the soil surface, but returns to the surface as seepage downslope or into rills, furrows, and other surface depressions.

The process of extraction of chemicals into runoff is obviously complex, with the degree of complexity increasing as the complexity of slope, drainage, and management increases within a catchment. In development of mathematical models, modelers have sought a definition of the "effective" surface soil mass or depth that interacts with runoff. Early models assumed the entrainment process to be dominated by soil erosion and chemical extraction was described as equivalent to a completely mixed reactor, runoff water equilibrating with transported sediment (Huff & Kruger, 1967; Crawford & Donigian, 1973; Bruce et al., 1975). Steenhuis and Walter (1980), Haith and Tubbs (1981), and Williams and Hann (1978) also used the concept of a "mixing zone" at the soil surface in which chemicals in this zone were assumed to be equal to that in surface runoff. The depth of this mixing zone was conveniently chosen to be 10 mm. Frere et al. (1975) did not define a finite thickness of the mixing zone in their model, but did assume complete mixing of soil pore water with runoff water. In their early model, Crawford and Donigian (1973) assumed a surface active depth of 3 mm consistent with later findings of Ahuja et al. (1981b). Conducting runoff studies using soil boxes and $^{32}P$ tracers placed at different depths, they concluded that the "effective average" zone of interaction ranged from 2 to 3 mm. However, the degree of interaction decreased exponentially with depth. Additional studies (Sharpley et al., 1981) showed that the effective depth of interaction was related to the degree of soil aggregation and increased with soil slope, kinetic energy of raindrops, and to a lesser extent, rainfall intensity.

Leonard et al. (1979) examined pesticide data from a number of watersheds (Smith et al., 1978) and found runoff concentrations over a wide range of storm conditions to be strongly correlated with pesticide concentrations in the surface 10 mm of watershed soil (Fig. 9–2). They, therefore, concluded that this 10-mm surface zone could be used as a descriptor or predictor of the amount of pesticide available for runoff. They made no inferences, however, as to the depth of the layer actually affected by runoff. In development of the CREAMS model, Leonard and Wauchope (1980) chose to use the 10-mm depth as a predictor of pesticide available for runoff, primarily because it is about the minimum depth that can be delineated and sampled under actual field conditions considering normal surface roughness. Also, their model is a "lumped" model where interrill extraction and rill processes are not conceptually or mathematically separated and, therefore, the surface depth chosen is an "effective average" in both vertical and horizontal dimensions. Steenhuis and Walter (1980) and Haith (1980) apparently used a similar rationale in defining the effective surface depth. In

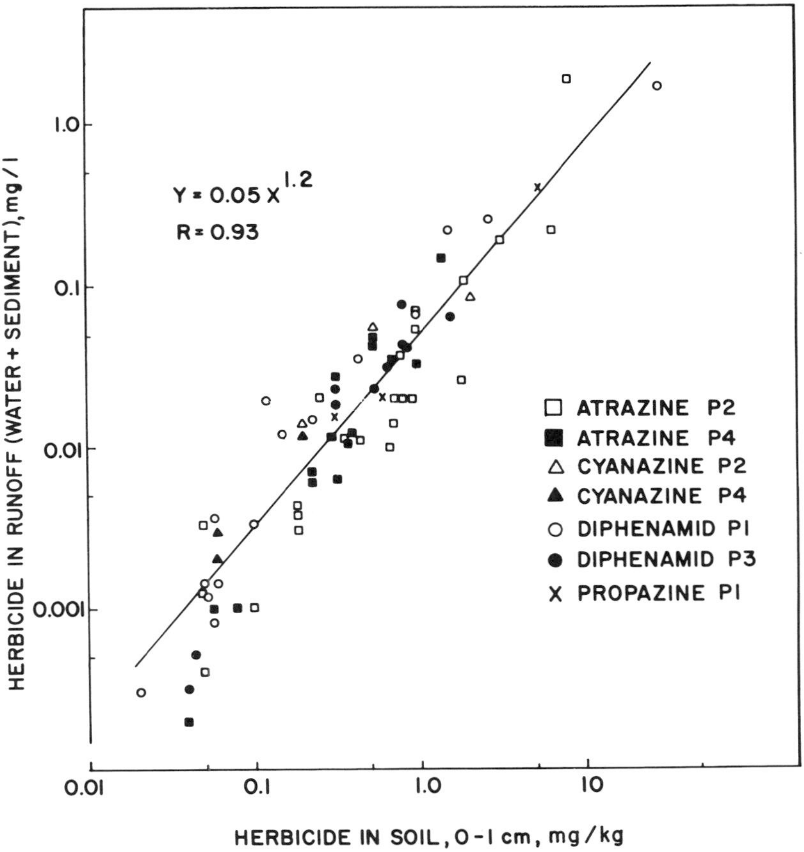


Fig. 9–2. Relationship between pesticide concentrations in 0 to 1 cm surface soil layer and concentrations in surface runoff (Leonard et al., 1979).

contrast, however, Leonard and Wauchope (1980) did not assume complete mixing throughout the source zone, but defined an additional parameter they called an *extraction ratio* which was assumed to approximate the effective mass ratio of the interacting soil to water in the runoff stream. Numerically, a value of 0.05 to 0.2 for the extraction ratio was required to provide reasonable model fit to a wide range of observed watershed and plot data with a value of 0.1 giving adequate predictions in most situations (Leonard & Nowlin, 1980). Conceptually, this parameter was envisioned as a variable depending on soil properties, rainfall intensity, and runoff rate, much as observed for the depth of interactions Sharpley et al. (1981) reported. However, insufficient data sets have been available for statistical evaluation.

Ingram (1979) and Ingram and Woolhiser (1980) derived and evaluated a chemical extraction model in which incomplete mixing of pore water and runoff water was assumed. Dispersion of sediment as a source of extracted chemical was not considered. Their work showed that the concentration of a chemical in overland flow is much lower than in soil pore water and, therefore, assuming complete mixing of the surface zone with runoff is a gross oversimplification.

### 9–1.2 Role of the Adsorbed Phase and Sediment Transport

Some pesticides are preferentially transported adsorbed to entrained sediments. Wauchope (1978), using published data, related mode of transport to pesticide solubility. He found only those pesticides with solubilities below 2 mg/L transported primarily by sediment. Notable exceptions were soluble salts that are strongly adsorbed by clay surfaces, e.g., paraquat, MSMA. Sediment pesticide transport may also be described using adsorption partition coefficients ($K_d$) (Mulkey & Falco, 1977; Pionke, 1977).

Adsorbed pesticide contributes to runoff as sediment-transported pesticide and by pesticide desorption into the flowing water from either moving or stationary soil particles. The mixing zone is a dynamic system affected by raindrop impact and by the turbulent nature of the overland flow process. In spite of this, most runoff models assume instantaneous and reversible equilibrium (Crawford & Donigian, 1973; Leonard & Wauchope, 1980). The distribution of pesticide between solution and adsorbed phases can then be assumed to follow simple relationships such as the Freundlich equation: $S = KC^N$, where $S$ is the adsorbed-phase concentration, $C$ is the solution-phase concentration, and $K$ and $N$ are constants specific to a pesticide-soil combination. Furthermore, if the concentration ranges of interest are in the nearly linear portion of the adsorption isotherm, $N = 1$, and the relationship reduces to $S = K_dC$. Rao et al. (1983) and Green et al. (1980) describe in detail equations used for expressing pesticide sorption in soil and methods of estimating sorption constants. Since soil organic matter is the primary soil constituent responsible for adsorption of nonionic organic compounds, the sorption constant may be based on only the organic matter present such that

$K_{oc} = K_d/oc$, where oc is the fraction of organic C present in the soil or sediment. A sorption constant expressed in this manner is dependent only on the pesticide and is independent of soil type.

For many soil-pesticide systems, adsorption-desorption relationships are extremely complex and isotherms are often nonsingular and not completely reversible. However, estimated errors in predicting pesticide distribution assuming reversible, singular isotherms are within factors of 2 or 3 for concentration ranges observed in agricultural soil (Rao & Davidson, 1980). Assuming instantaneous equilibrium may also introduce errors in prediction. However, these errors may not be much greater than the uncertainty of 2 or 3 present as a result of isotherm nonsingularity. Sorption kinetics are considered in P entrainment models of Novotny et al. (1978) and Sharpley et al. (1981). Similar approaches should apply for pesticides and any adsorbed constituent in runoff (Wauchope & Sharpley, 1984); however, rate constants would be difficult to evaluate and would probably depend on soil (sediment)/water ratios, pesticide type and concentrations, organic matter contents, and relative ages of the organic matter pesticide complexes. In spite of these uncertainties, models assuming equilibrium sorption relations appear to give adequate predictions (Lorber & Mulkey, 1982; Nutter et al., 1984). Leonard and Wauchope (1980), however, noted that in some situations the apparent sorption coefficient ($K_d$) increased with time during a growing season, that is, pesticides appeared to become more difficult to desorb with time of contact in soil. A possible explanation is pesticide diffusion into the soil organic matter particles with subsequent desorption becoming kinetically inhibited.

### 9–1.3 Relationships between Pesticide Runoff and Distribution in the Soil-Runoff Zone

Pesticides leaving a treated area in runoff constitute only a small percentage of that applied since most of the applied pesticide is dissipated by other processes. These processes are discussed at length in other chapters of this publication. They are mentioned briefly here for perspective. Relationships between pesticide runoff and distribution and persistence in the runoff active zone are depicted in Fig. 9–3. An application of pesticide, depending on crop stage, formulation, intended target, application method, and weather conditions is distributed between soil, plant foliage or crop residue, and off-catchment losses. Herbicides applied by ground equipment mainly reach the soil surface if that is the intended target. Depending on the density of the weed and crop canopy and application methods, contact herbicides are distributed between the soil surface and crop foliage. Newer methods such as by rope-wick applications are efficient and deposit essentially all of the herbicide on foliage (Dale, 1981). In no-till or conservation tillage systems, herbicides may be intercepted by crop residues (Banks & Robinson, 1982; Wagenet, 1987). Aerially applied insecticides may suffer significant drift and volatilization losses and be distributed between both plant and soil surface depend-

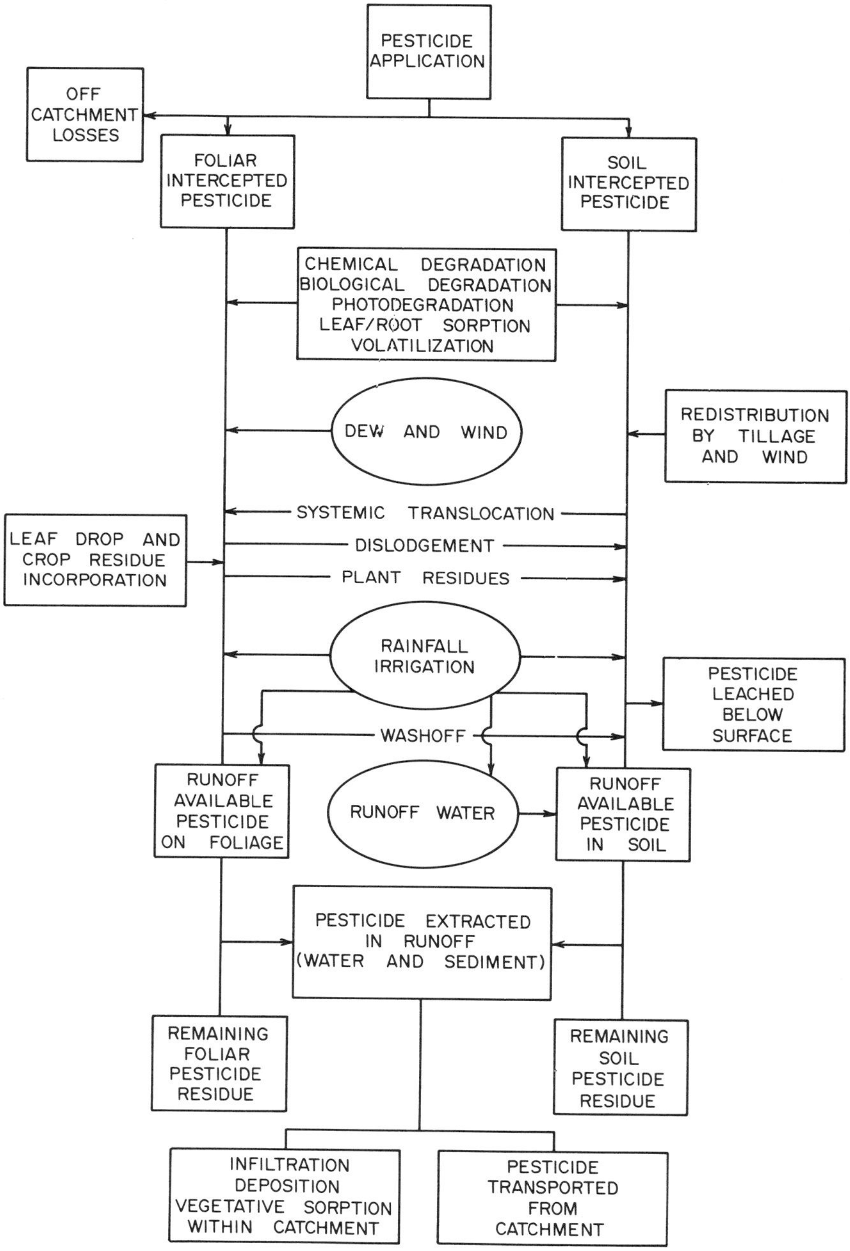


Fig. 9–3. Relationship between pesticide runoff and pesticide distribution and persistence in the runoff active zone.

ing on the degree of canopy closure (Willis et al., 1980). Once a pesticide reaches either the plant or soil surface, it is degraded or transformed by chemical, biological, and photochemical processes and subjected to volatilization losses. Although dissipation is by a combination of processes, pesticide persistence with time on foliage and soil may be approximated by an exponential function mimicking a first-order process (Nash, 1980; Willis et al., 1980). Pesticide half-lives or half-concentration times based on an assumed first-order function vary from a few hours to months depending on the pesticide and the local environment. Pesticides deposited on surfaces are subjected to environmental extremes, and therefore, may dissipate more rapidly than in the soil bulk (Nash, 1980). For rapidly dissipating pesticides, rainfall timing with respect to application is obviously the most critical factor affecting runoff potential.

In a given rainfall/runoff event, the contribution of foliar and plant residue to surface runoff is highly variable, depending on pesticide chemistry and formulation, rainfall timing with respect to application, rainfall intensity and duration, initial soil water content, and other factors that affect water infiltration and runoff. Presumably, only a fraction of the pesticide residing on foliage and crop residue at the time of a rainfall event has the potential for washoff. Insecticide washoff and persistence on foliage has been investigated more extensively than herbicide washoff and foliar persistence (Willis et al., 1980). Lipophillic compounds such as organochlorine pesticides penetrate waxes at the leaf surface and become difficult to dislodge by rainfall. Conversely, polar compounds do not penetrate the leaf surface. As a result, only 5 to 10% of a compound such as toxaphene is considered available for rainfall washoff, whereas 70 to 80% of the organophosphates may be removed by rainfall (Willis et al., 1980). Some triazines (Martin et al., 1978) and other soluble herbicides (Bovey et al., 1974; Trichell et al., 1968; Cohen & Steinmetz, 1986) are apparently readily removed from crop surfaces. Banks and Robinson (1982), however, found that only 45% of applied metribuzin could be washed from wheat (*Triticum aestivum* L.) straw. Regardless of the fraction of pesticide potentially dislodged, the amount washed off is a function of rainfall volume and independent of rainfall intensity (McDowell et al., 1984; Cohen & Steinmetz, 1986) assuming the same pattern of washoff holds for herbicides as observed with insecticides. During a rainstorm, pesticide concentrations in foliar washoff decline rapidly as hyperbolic functions of rainfall volume. That portion reaching the soil surface before runoff begins may be moved into the soil surface by infiltrating water or become absorbed by the soil; or in case of dusts and particulates, become intimately mixed with soil. However, pesticides reaching the soil surface before runoff begins may be more susceptible to runoff than soil-applied pesticides. Water dripping from plant canopy and residue may concentrate pesticides into drip lines, rills, and row furrows that are part of the surface drainage network. After runoff begins, pesticides washed from foliage and plant residue may fall directly into flowing water and be rapidly transported as surface runoff. Runoff begins when rainfall rates exceed infiltration rates. Therefore, even though washoff from the plant canopy is not dependent on rainfall intensity, time to runoff inception may be highly dependent on rainfall intensity, thus affecting the net contribution of foliar and residue washoff to runoff. Conditions most favorable for runoff would be where pesticides are applied to foliage or crop residue immediately before an intense rainstorm when the soil is nearly saturated from a previous rain. Contributions of plant residue washoff to runoff are discussed in more detail in a later section on conservation tillage.

Pesticide residues may redistribute within and between the soil and plant compartment as a result of some other processes. Foliar pesticide may be dislodged to the soil surface by the action of dew and wind as well as by rainfall. Leaf drop and incorporation of crop residue may also add pesticide residues remaining in and on plants to the soil. Some pesticides, particularly systemics, are translocated from the root/soil interface to aboveground plant parts although in concentrations that should not contribute materially to runoff. Soil tillage after pesticide application may incorporate pesticide to lower depths and reduce the concentration at the immediate soil surface. In situations where the surface has been depleted, tillage may bring additional pesticide residues to the soil surface. Mobile pesticides are readily leached below the soil surface before runoff begins. Conversely, mobile pesticides brought to the soil surface by capillary forces are deposited as the soil solution evaporates.

Factors that affect runoff potential most are those that affect pesticide persistence in the runoff active zone at the soil surface. These are factors that affect overall pesticide resistance or susceptibility to degradation and volatilization, leaching below the soil surface, and physical redistribution of residues. Most of the factors, affecting runoff potential interact in a complex manner. Table 9–1 lists climatic, soil, pesticide, and management factors that are known to affect pesticide runoff and references to selected works that illustrate their importance.

### 9–1.4 Temporal and Spatial Relationships

If a small unit of the soil surface is viewed as a simple reactor providing entrained chemicals to runoff in proportion to the mass of chemicals present, pesticide concentrations in runoff should decline during the runoff event as an approximate first-order function. The rate of decline would be affected by the efficiency of the mixing/extraction process and the rate of movement of chemicals through the mixing zone with infiltrating water. Factors in the preceding table such as pesticide solubility, sorption properties, infiltration rates, and rainfall intensity will affect the rate at which a given chemical leaves the mixing zone. On small runoff plots, particularly when using controlled or simulated rainfall, an ideal behavior is observed (Fig. 9–4). That is, runoff concentrations are high initially, and decline exponentially with time in the event. Concentrations of pesticides from small plots receiving simulated rainfall are usually higher than those observed from natural systems because

Table 9-1. Factors affecting pesticide entrainment and transport in runoff. After Leonard (1988).

| Factors | Comment | Selected references |
|---|---|---|
| | 1. Climatic | |
| A. Rainfall/runoff timing with respect to pesticide appliction | Highest concentration of pesticide in runoff occurs in first significant runoff event after application. Pesticide concentration and availability at the soil and foliar surfaces dissipate with time thereafter. | White et al., 1967; Bovey et al., 1975; Bradley et al., 1972; Wauchope & Leonard, 1980a, b; Baker & Johnson, 1979; Edwards et al., 1980; Smith et al., 1983; Triplett et al., 1978 |
| B. Rainfall intensity | Surface runoff occurs when rainfall rate exceeds infiltration rate. Increasing intensity increases runoff rate and energy available for pesticide extraction and transport. May also affect depth of surface interaction. Increasing intensity reduces time to runoff within storm. | Skaggs & Khaleel, 1982; Sharpley et al., 1981; Sharpley, 1985a |
| C. Rainfall duration/ amount | Affects total runoff volumes; pesticide washoff from foliage related to total rainfall amount; leaching below soil surface also affected. | White et al., 1967; Bovey et al., 1975; Baker et al., 1981; Willis et al., 1980 |
| D. Time to runoff after inception of rainfall | Runoff concentrations increased as time to runoff decreased. Pesticide concentrations and availability are greater in first part of the event before significant reduction occurs as a result of leaching and incorporation by raindrop impact. | Baker & Laflen, 1979; Gaynor & Volk, 1981; Baker et al., 1982; Barnett et al., 1967 |
| E. Water temperature | Little data available, but increasing temperature normally increases pesticide solubility and decreases physical adsorption. | Barnett et al., 1967; Bailey et al., 1974 |
| | 2. Soil | |
| A. Soil texture and organic matter contents | Affects infiltration rates; runoff is usually higher on finer-textured soils. Time to runoff is greater on sandy soils reducing initial runoff concentrations of soluble pesticides. Organic matter content affects pesticide adsorption and mobility. Soil texture also affects soil erodibility, particle transport potential, and chemical enrichment factors. | Rawls & Brakensiek, 1982; Rao & Davidson, 1980; Wischmeier & Smith, 1978; Foster et al., 1980 |
| B. Surface crusting and compaction | Crusting and compaction decreases infiltration rates, reduces time to runoff, and increases initial concentrations of soluble pesticides | Baker & Laflen, 1979 |

(continued on next page)

Table 9-1. Continued.

| Factors | Comment | Selected references |
|---|---|---|
| C. Water content | Initial soil water content at beginning of a rainstorm may increase runoff potential, reduce time to runoff, and reduce leaching of soluble chemicals below soil surface before runoff inception. | Knisel & Baird, 1969; Davidson et al., 1975; Barnett et al., 1967 |
| D. Slope | Increasing slope may increase runoff rate, soil detachment and transport, and increase effective surface depth for chemical extraction. | Wauchope, 1978; Sharpley et al., 1981; Foster et al., 1980 |
| E. Degree of aggregation and stability | Soil particle aggregation and stability affects infiltration rates, crusting potential, effective depth for chemical entrainment, sediment transport potential, and adsorbed chemical enrichment in sediment. | Sharpley et al., 1981; Foster et al., 1980; Ahuja et al., 1981 |
| | 3. Pesticide | |
| A. Solubility | Soluble pesticides may be more readily removed from crop residue and foliage during the initial rainfall or be leached into the soil. However, when time to runoff is short, runoff concentration may be enhanced by increasing solubility. | Barnett et al., 1967; Trichill et al., 1968; Baker et al., 1978; Baker & Johnson, 1979; Willis et al., 1980; Baker et al., 1982 |
| B. Sorption properties | Pesticides strongly adsorbed in soil will be retained near application site, i.e., possibly at soil surface and be more susceptible to runoff. Amounts of runoff when dependent on amount of soil erosion and sediment transport. | McDowell et al., 1981; Willis et al., 1983 |
| C. Polarity/ionic nature | Adsorption of nonpolar compounds determined by soil organic matter; ionized compounds, and weak acids/bases affected more by mineral surface and soil pH. Lyophillic compounds retained on foliage by leaf surface and waxes, whereas polar compounds more easily removed from foliage by rainfall. | Rao & Davidson, 1980; Willis et al., 1980; Wauchope & Leonard, 1980a, b |
| D. Persistance | Pesticides that remain at the soil surface for longer periods of time because of their resistance to volatilization, chemical, photochemical, and biological degradation have higher probability of runoff. | Wauchope, 1978; Mills & Leonard, 1984; Leonard & Knisel, 1986 |

(continued on next page)

Table 9-1. Continued.

| Factors | Comment | Selected references |
|---|---|---|
| E. Formulation | Wettable powders are particularly susceptible to entrainment and transport. Liquid forms may be more readily transported than granular. Esters less soluble than salts produced higher runoff concentrations under conditions where initial leaching into soil surface is important. | Wauchope, 1978; Rohde et al., 1979; Wauchope & Leonard, 1980a, b; Wauchope, 1987b |
| F. Application rate | Runoff concentrations are proportional to amounts of pesticide present in runoff zone. At usual rates of application for pest control, pathways and processes (e.g., sorption and degradation rates) are not affected by initial amounts present, therefore, runoff potential is in proportion to amounts applied. | Barnett et al., 1967; Hall, 1974; Leonard et al., 1976 |
| G. Placement | Pesticide incorporation or any placement below the soil surface reduces concentrations exposed to runoff process. | Leonard et al., 1979; Wauchope, 1978; Rohde et al., 1979; Wauchope & Leonard, 1980a, b |
| | 4. Management | |
| A. Erosion control practices | Reduces transport of adsorbed/insoluble compounds. Also reduces transport of soluble compounds if runoff volumes are also reduced during critical times after pesticide application. | Caro, 1976; McDowell & Grissinger, 1976; Pionke, 1977; McDowell et al., 1981; Willis et al., 1983 |
| B. Residue management | Crop residues can reduce pesticide runoff by increasing time to runoff, decreasing runoff volumes, and decreasing erosion and sediment transport. However, pesticide runoff may be increased under conditions where pesticides are washed from the crop residue directly into runoff water (high initial soil water, clay soil, intense rainfall immediately after pesticide application). | Triplett et al., 1978; Baker et al., 1978; Baker & Johnson, 1979; Edwards et al., 1980; Baker et al., 1982; Hall et al., 1984 |
| C. Vegetative buffer strips | Buffer strips around treated fields may reduce transport of some pesticides by secondary infiltration, sediment deposition and sorption on plant surfaces and debris. | Asmussen et al., 1977; Rohde et al., 1980 |
| D. Irrigation | Chemical application by sprinkler irrigation may move soluble pesticides into soil surface and reduce runoff potential. Aerial application of pesticides during periods of flood irrigation greatly increases pesticide runoff in surface drainage. | Dowler et al., 1982; Spencer et al., 1985 |

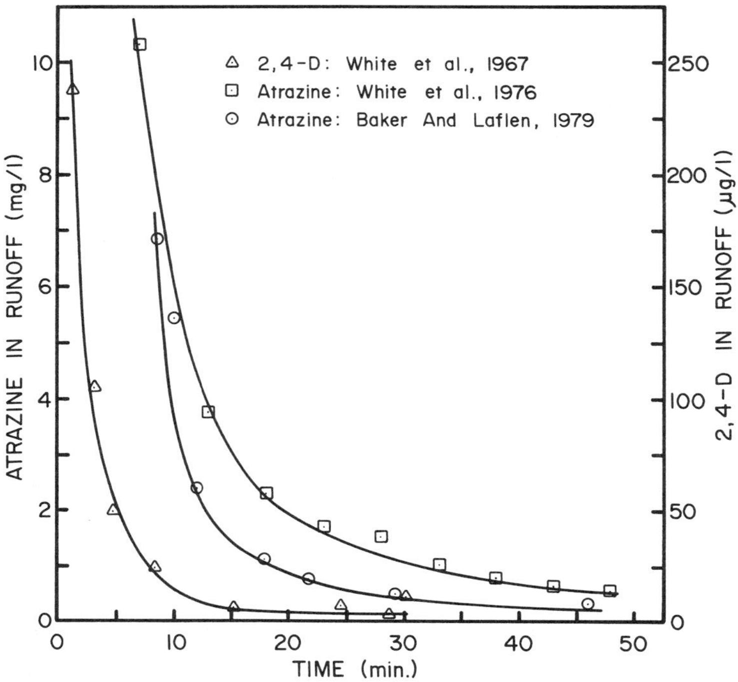


Fig. 9-4. Runoff concentration of pesticides from small plots as related to time after start of runoff. From Leonard (1988).

the runoff extraction processes are often maximized. The ideal behavior is more difficult to observe as the complexity of the runoff system increases in both time and space. Studies conducted on field-size watersheds or larger plots under natural rainfall conditions often give pesticide runoff concentrations that vary more than an order of magnitude within a single event, sometimes in a seemingly random manner (Smith et al., 1978). Wauchope (1978a) concluded that within-event concentrations for a given chemical are almost unique for each situation, depending on application rate, storm intensity and timing, and field size. Some rational trends are usually observed, however, with pesticide runoff concentrations being higher in the early portion of runoff events (Fig. 9-5). Rohde et al. (1979), however, observed an induction period resulting in higher storm concentrations of the nematicide ethoprop 5-10 min after initiation of runoff. In their studies, early runoff could have originated in untreated field borders. Results from small plots are controlled primarily by those processes illustrated in Fig. 9-1 as interrill or microscale processes. As the size and complexity of the drainage area increases, runoff water becomes concentrated in rills, furrows, waterways, and field ditches.

If a watershed is conceptually divided into some small elements, each giving the ideal response, the watershed response at the outlet would be an integration of the individual responses. The shape of the concentration vs.

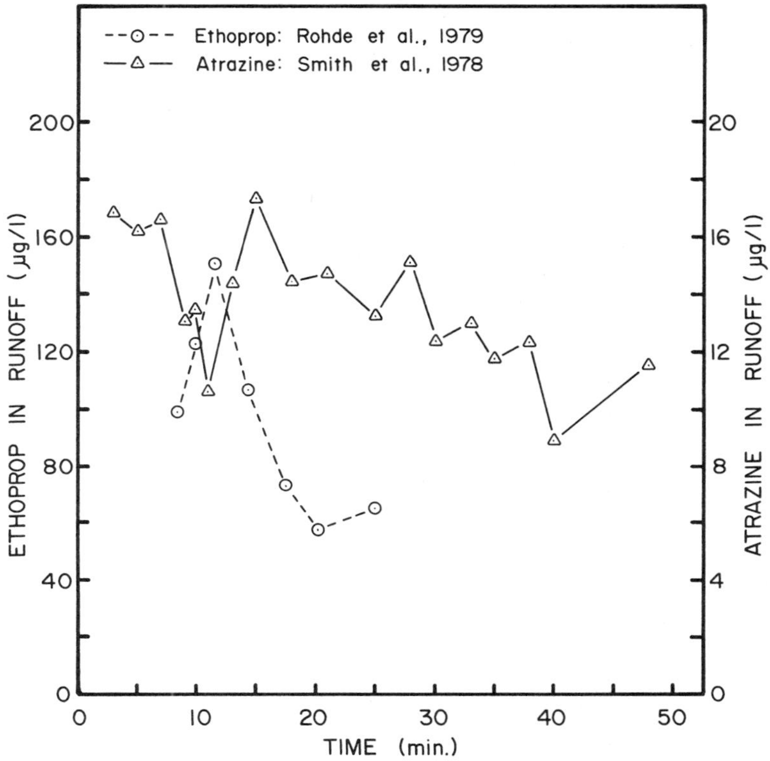


Fig. 9–5. Runoff concentration of pesticides from field-sized watersheds as related to time after start of runoff. From Leonard (1988).

time curve at the watershed outlet would depend on times-of-travel of runoff from each element, the degree of mixing during transport, and effects of channel processes such as deposition and reentrainment. Therefore, considering the complexity of the runoff system, it is not surprising that concentration patterns within single storm events are difficult to interpret. A difficulty is also introduced in sampling storm hydrographs. Infrequent instantaneous samples may give erroneous data. Wauchope et al. (1977) discuss in detail experimental procedures necessary to obtain acceptable data on pesticide runoff. Runoff models predicting within storm pesticide runoff concentrations need to consider how to route and combine flows from each subelement of the catchment. To avoid this complexity, several of the published pesticide runoff models are lumped models predicting only event-average concentrations and total pesticide yields in runoff by event or day (Steenhuis & Walter, 1980; Haith, 1980; Leonard & Wauchope, 1980).

Event average pesticide concentrations over a growing season or period of pesticide persistence are much more predictable. Event average concentrations decrease exponentially with time after pesticide application as illustrated in Fig. 9–6 (also see Wauchope, 1978; Leonard et al., 1979; Wauchope

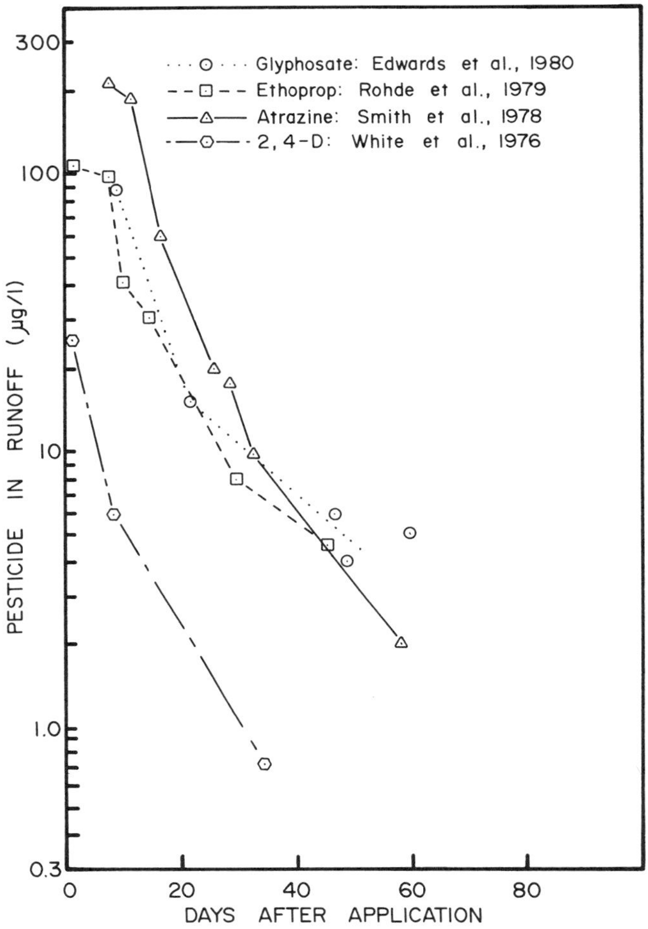


Fig. 9–6. Concentration of pesticides in runoff in relation to time after application. From Leonard (1988).

& Leonard, 1980a, b). This is a reasonable expectation since pesticides are known to dissipate from soil and foliage by processes that can be approximated by first-order rate equations (Nash, 1980; Willis et al., 1980). Since runoff concentrations correlate strongly with pesticide concentrations in soil (Leonard et al., 1979) (Fig. 9–2), we must assume that the decline in runoff concentrations with time primarily reflects this decline at the soil surface. However, runoff available pesticide residues at the soil surface may dissipate more rapidly by a combination of volatilization, photolysis, leaching, and other processes leading to surface depletion more rapidly than expected based on conventional estimates of persistence. Using semilog plots of runoff concentration vs. time, similar to those in Fig. 9–6, Wauchope and Leonard (1980a, b) concluded that the half-lives of available residues for a wide range of pesticides were remarkably similar, but depended on the period after application used for calculations. That is, the plots were only approximated by straight-line functions (true first-order equation). The curvilinear nature of the plots show that half-lives of the runoff available residues tend to increase over the period of persistence. Stated in another way, the initial pesticide residues soon after application that are available for runoff dissipate more rapidly than do remaining residues later in the season. Volatilization, photolysis, and other processes that are driven by the environmental extremes at the soil/air interface presumably contribute to this rapid decline of available pesticide residues. Although dissipation from the soil bulk is approximated by first-order equations, Nash (1980) also recognized that dissipation curves could be best described by polynomial functions. Apparently pesticide half-lives, particularly after one to three half-lives, increase with time in bulk soil somewhat analogous to the observations on runoff available residue, although to varying degrees. With time of contact in soil, pesticides may become somewhat protected from degradation processes by adsorption, occlusion, or both.

Physical processes active during runoff may also contribute to the decline in runoff available residues. During the first runoff event after pesticide application, the surface drainage network developed and composed of rills and furrows may be selectively depleted of pesticide residues. The soil surface may also become armored by sand or gravel as the finer silts and clays are removed. Some pesticide formulations (e.g., wettable powders) may be easily transported initially because of direct transport of the pesticide carrier (Wauchope, 1987b). Ease of transport may then decrease as the carrier is mixed into the soil by raindrop impact and overland flow.

### 9–1.5 Pesticide Extraction by Overland Water Flow

The preceding discussion emphasized pesticide extraction from the soil surface by the combination of raindrop impact and overland flow. In irrigated areas of the western USA, pesticide residues have been detected in large drains; surface runoff from irrigated cropland suspected as the main contributor of these residues (Bailey & Hannum, 1967; Johnson et al., 1967; Eccles, 1979). For furrow irrigation, the land surface is often graded to prevent standing irrigation water that permits surface runoff into agricultural drains during irrigation (Spencer et al., 1985). Spencer et al. (1985) conducted an extensive study on pesticides in irrigation return flows and found that within limits of the analytical methods, surface runoff was the sole source of pesticide residues in drainage water. Magnitudes of annual pesticide losses reported were similar to those reported for rain-driven runoff. As observed in rain-driven runoff, pesticide concentrations were highest in the first events after application. This is when soil surface pesticide concentrations are highest and most susceptible to extraction in runoff.

Statistical correlations by Spencer et al. (1985) between soil surface pesticide concentrations and pesticide runoff concentrations showed, as did Leonard et al. (1979), that soil surface pesticide concentrations are good indicators of expected pesticide concentrations in runoff. Magnitudes of coefficients on equations derived by Spencer et al. (1985) compared to those of Leonard et al. (1979) suggest that at the same soil pesticide concentration, rain-driven processes are more efficient in extracting pesticides into runoff. This could be expected because of the influence of raindrop splash, however, the differences observed may also be because of soil differences. The effective depth of interaction between the soil surface and runoff is affected by soil aggregation and increases with increasing rainfall and runoff energy (Sharpley, 1985a).

In the work of Spencer et al. (1985), most of the pesticide residues in runoff except those of the pyrethroids were transported in the water phase. Since the mass of sediment transported in runoff was low ($\leq 5$ g/L), a large pesticide $K_d$ would be required to shift the primary mode of transport to sediment. Even though water transport was dominant, the apparent $K_d$ determining partitioning between water and sediment in irrigation runoff was numerically greater than normally reported from laboratory studies. This observation could not be explained based on enrichment of finer particles and organic matter in sediment since the sediment was found to have a composition similar to the residual soil. Leonard and Wauchope (1980) also noted apparent increases in $K_d$ with time of pesticide contact with field soils.

## 9–2 PESTICIDE RUNOFF AND LOADS IN SURFACE WATERS

### 9–2.1 Losses from Cropland

Wauchope (1978) conducted an extensive review of available data and proposed a rule-of-thumb pesticide classification with estimates of average or reasonably expected edge-of-field runoff losses. Estimated seasonal losses were 1% of the amount applied for foliar-applied organochlorine insecticides, 2 to 5% for wettable powders, depending on slope and hydrologic response, and 0.3% or less for the remaining pesticides. Recognizing that in many studies, the bulk of measured runoff occurred in a single storm soon after pesticide application, he defined these as critical events; those occurring within 2 wk of pesticide application with at least 10 mm of rainfall, 50% of which becomes runoff. Noting that runoff losses in the range of 1 to 2% are not uncommon for a wide range of pesticides, he defined "catastrophic" events as those in which runoff losses exceed 2% of the application. These events, almost without exception, are first events after application. Exceptions might be with the most persistent compounds and those such as paraquat that are strongly adsorbed and totally dependent on sediment transport.

Wauchope and Leonard (1980a, b) expanded Wauchope's earlier (1978) generalizations to derive a simple predictive formula for maximum expected concentrations in edge-of-field runoff. Prediction by use of the formula required classification of pesticides into four availability classes based on susceptibility to runoff. Wettable powder formulations applied to the soil surface were among the most runoff-susceptible pesticides and soil incorporated emulsifiable concentrates were among the least susceptible.

Since 1978, field research has, in general, confirmed the above generalizations, and has additionally provided information on effects of management, particularly conservation tillage. Data on newer pesticide material has also been provided and extensive data based for model development and testing have been accumulated. As an example, Spencer et al. (1985) compiled a comprehensive data base on pesticides in irrigation runoff. Recent studies on pesticide runoff from small plots and single-cover fields and watersheds are summarized in Table 9–2. Specific references to these works are included in the various sections of this chapter. As stated earlier, pesticide runoff losses from simulated rainfall may be much higher than from natural rainfall. For this reason, studies within simulated rainfall are reported separately in Table 9–2.

### 9–2.2 Pesticide Transport from Forests and Rangelands

Data on pesticide runoff losses from forested areas and rangelands as well as that from cropland are included in Tables 9–2 and 9–3. Most pesticides applied to forests and rangelands are herbicides used for control of understory, brush, or for forest specie conversion. Concentrations and amounts of pesticides in runoff from forested or brushland areas are generally lower than for pesticides applied to cropland at the same rate (Newton & Norgren, 1977; Norris, 1969; Wauchope, 1978; Nutter et al., 1984; Norris et al., 1982; Davis et al., 1968). Important factors that contribute to these differences include: (i) partial area contribution to runoff, that is, all areas of the watershed do not contribute runoff equally (Neary et al., 1983); (ii) herbicides tend to be sufficiently mobile to penetrate the soil surface before runoff begins and be translocated below the surface that is in contact with runoff (Neary et al., 1979, 1985; Bouchard et al., 1985). The forest floor is covered with detritus contributing to high infiltration rates such that baseflow or interflow may be the main contribution to watershed runoff (Neary et al., 1979, 1983; Nutter et al., 1984). In spite of these differences, concepts and models developed primarily for cropland application have been adapted and are proving useful in forest application (Nutter et al., 1984).

### 9–2.3 Pesticides in Runoff from Nonagricultural Areas

About 30% of pesticides sold in the USA in 1983 were for nonagricultural purposes (CAST, 1985). Nonagricultural uses cited were in industry and commerce, by governments, and for home and garden purposes. Some of these pesticides used for purposes such as rights-of-way clearance, gardens, and turf will be subject to transport in runoff. Principles controlling pesticide runoff from these treated areas are basically the same as those controlling pesticide runoff from agricultural areas, however, little specific information is available.

Urban runoff is an important contributor of nonpoint source pollution (Novotny & Chesters, 1981). Principal nonpoint pollutants of interest have been N, P, toxic organic chemicals including pesticides, biological oxygen demand (BOD), and metals. Little data on specific pesticides in urban runoff has been published. One potential source of pesticides in urban runoff is by wash-off of atmospheric dusts containing pesticide residues that have settled on streets and other impervious urban surfaces (Cohen & Pinkerton, 1966).

Table 9-2. Runoff losses of pesticides from small plots and single cover watersheds (1978-1985).

| Compound | Rate, kg/ha | Location | Crop/Cover | Conc. in runoff, μg/kg† | Total seasonal losses, % of application | Comments | Reference |
|---|---|---|---|---|---|---|---|
| | | | Runoff from natural rainfall or snowmelt | | | | |
| Atrazine | Various | Ohio | No-till and conventionally tilled corn | 480 | 0-5.7 | No-till reduced herbicide runoff primarily by reducing runoff volume. Time of runoff event relative to application most important factor | Triplett et al., 1978 |
| Simazine | Various | Ohio | No-till and conventionally tilled corn | 1200 | 0-5.4 | | |
| Simazine | Various | Various | Irrigation canals | 250 | -- | Applied to canals with and without flowing water | Anderson et al., 1978 |
| 2,4-D | 3.7 | Florida | Citrus | 75 | -- | Concentrations in tile outflow generally in range of 20-25 μg/L | Wheeler et al., 1978 |
| Atrazine | 1.45-4.03 | Georgia | Corn | 1 900 (water) | 0.2-1.9 | Results from comprehensive studies to provide data for model development and testing | Leonard et al., 1979 |
| Paraquat | 1.53-15.3 | Georgia | Corn | 980 000 (sed.) | 3.4-10.9 | | |
| Trifluralin | 1.12 | Georgia | Soybean | 21 (water) | 0.1-0.3 | | |
| Propazine | 1.66 | Georgia | Grain sorghum | 400 (water) | 6.7 | | |
| Cyanazine | 1.35-1.61 | Georgia | Corn | 180 (water) | 0.07-1.0 | | |
| Diphenamid | 2.31-3.52 | Georgia | Soybean | 2 070 (water) | 0.1-7.2 | | |
| Alachlor | 2.24 | Iowa | Corn | | 0.96 (avg.) | Losses depend on time between application and runoff; decreased runoff and erosion decreased pesticide losses, but not in proportion because conc. in water and sediment were higher for conservation systems | Baker & Johnson, 1979 |
| Atrazine | 2.24 | Iowa | Corn | | 2.1 (avg.) | | |
| Cyanazine | 2.24 | Iowa | Corn | | 2.1 (avg.) | | |
| Fonofos | 1.12 | Iowa | Corn | | 0.36 (avg.) | | |
| Ethoprop (liquid) | | Georgia | Soybean | 283 | 0.1 | Major differences in runoff and dissipation of ethoprop observed between liquid and granular formulations | Rohde et al., 1979 |
| Ethoprop (granular) | | Georgia | Soybean | 45 | 0.01 | | |

(continued on next page)

Table 9-2. Continued.

| Compound | Rate, kg/ha | Location | Crop/Cover | Conc. in runoff, μg/kg† | Total seasonal losses, % of application | Comments | Reference |
|---|---|---|---|---|---|---|---|
| Atrazine | | Maryland | Corn | 16.9 (avg.) | 1 | Field soil sampling indicated both vertical and lateral movement of atrazine, but not of alachlor | Wu, 1980 |
| Alachlor | | Maryland | Corn | 0.6 (avg.) | 0.16 | | |
| Picloram | 2.8 | Arizona | Pinyon-Juniper | 320 | 1.1 | Highest runoff conc. in initial runoff event after application | Johnsen, 980 |
| Glyphosate | 1.10–8.96 | Ohio | No-till corn | 5200 one event others ≤100 | 1.85 (extreme yr, <1 other yr and watersheds | Abnormally high conc. because of high application rate and runoff occurring 1 d after application | Edwards et al., 1980 |
| Trifluralin | 1.12 | Georgia | Soybean | 38 | 0.17 | Trifluralin detected in surface runoff for 16 wk after application; none in tile outflow except trace 16 wk after application | Rohde et al., 1980 |
| Permethrin | 0.112 (10 applications) | Louisiana | Cotton | <1 | <1 | Runoff losses low even under extreme runoff conditions | Carrol et al., 1981 |
| Toxaphene | | Mississippi | Cotton | -- | 1–0.5 | Linear relationships observed between sediment yields and toxaphene yields in runoff. 93% of toxaphene in runoff attached to sediment; 7% in solution | McDowell et al., 1981 |
| 2,4-D | | Oregon | Rangeland | -- | 0.014 | Nearly all herbicide runoff observed resulted from direct deposits in stream channels and streambanks | Norris et al., 1982 |
| Picloram | | Oregon | Rangeland | -- | 0.35 | | |
| 2,4-D | | Saskatchewan, Canada | Wheat stubble, fallow | 31 (avg.) | 4.1 (6-yr avg.) | Snowmelt runoff | Nicholaichuk & Grover, 1983 |
| | | | Fallow | 3 (avg.) | 0.3 | | |

(continued on next page)

Table 9-2. Continued.

| Compound | Rate, kg/ha | Location | Crop/Cover | Conc. in runoff, µg/kg† | Total seasonal losses, % of application | Comments | Reference |
|---|---|---|---|---|---|---|---|
| Toxaphene<br>DDT<br>DDE<br>Trifluralin | | Mississippi<br>Mississippi<br>Mississippi<br>Mississippi | Cotton<br>Cotton<br>Cotton<br>Cotton | | | Pesticide conc. in sediment were directly proportional to sediment clay and organic matter conc. Storm and yield of pesticides were linear functions of storm sediment yields in years where no new applications made. In those years, correlations required separation into similar tillage-application regimes | Willis et al., 1983 |
| Azinophos-Methyl<br>Fenvalerate | | Louisiana<br>Louisiana | Sugarcane<br>Sugarcane | 250 | 0.55<br>0.56 | 1981 losses shown were twice that in 1980, mainly because of rainfall timing relative to application. Fenvalerate might cause problems for aquatic habitats immediately surrounding application sites | Smith et al., 1983 |
| Cyanazine | 1.1–1.7 | Pennsylvania | No-till and conventionally tilled corn | | 0.73–5.7 conventional<br><0.01–0.75 no-till | Herbicide runoff reduction accomplished primarily by reduction in volume of runoff | Hall et al., 1984 |
| Picloram | | Texas | Bermudagrass | 250 | 6.3 | Conditions during study strongly conducive to herbicide transport from treated source area. Studies additionally traced transport through larger watershed systems | Mayeaux et al., 1984 |

(continued on next page)

Table 9-2. Continued.

| Compound | Rate, kg/ha | Location | Crop/Cover | Conc. in runoff, μg/kg† | Total seasonal losses, % of application | Comments | Reference |
|---|---|---|---|---|---|---|---|
| | | | | Runoff from simulated rainfall | | | |
| Cyanazine | 2.24 | Iowa | Corn with various treatments | 1 330 (water)<br>5 140–420 (sed.) | 11.0 avg. all treatments | Herbicide losses under conservation tillage greater than under conventional tillage; effects of reduced runoff volumes offset by higher conc. Total losses of Fonophos related to sediment transport | Baker et al., 1978 |
| Alachlor | 2.24 | Iowa | Same as above | 610–60 (water)<br>3 590–510 (sed.) | 7.9 avg. all treatments | | |
| Fonofos | 1.2 | Iowa | Same as above | 19–41 (water) | 1.8 avg. all treatments | | |
| Propachlor | 2.5 | Iowa | Fallow with wheel-track and pesticide incorporation variables | 3 800 (water)<br>7 000 (sed.) | 0.8–12.7 | Runoff losses from surface applications compared to incorporated applications; runoff losses enhanced by wheel tracks because of increased runoff volumes and shorter times to start of runoff | Baker & Laflen, 1979 |
| Atrazine | 2.5 | Iowa | | 6 800 (water)<br>28 000 (sed.) | | | |
| Alachlor | 2.5 | Iowa | | 5 000 (water)<br>22 000 (sed.) | 1.7–22.1 | | |
| Fluometuron | 4.4 | Various | -- | 0.87<br>0.30 (avg.) | <1 avg. | Major emphasis placed on first event | Wiese et al., 1980 |
| Atrazine | | | Limed and unlimed soil | | 3.7 | Greater sediment transport of terbutryne. Liming significantly reduced runoff volumes. | Gaynor & Volk, 1981 |
| Terbutryne | | | Same as above | | 0.3 | | |
| Propachlor | 2.09 | Iowa | Fallow and plots with corn residue | 59–173 (water; avg.)<br>370–840 (sed.; avg.) | 0.76–6.1 | Values given are for range of averages across treatments. Herbicide conc. not affected by placement above or below residue, but were negatively correlated with time to runoff which was increased by presence of residue | Baker et al., 1982 |
| Atrazine | 2.09 | Iowa | | 83–141 (water; avg.)<br>600–1110 (sed.; avg.) | 0.97–5.7 | | |
| Alachlor | 2.09 | Iowa | | 78–220 (water; avg.)<br>880–2240 (sed.; avg.) | 1.0–8.6 | | |

(continued on next page)

Table 9-2. Continued.

| Compound | Rate, kg/ha | Location | Crop/Cover | Conc. in runoff, μg/kg† | Total seasonal losses, % of application | Comments | Reference |
|---|---|---|---|---|---|---|---|
| | | | | Runoff from irrigated fields | | | |
| Cycloate | 2.9 | California | Sugarbeets | 6.2 | 0.03 | Irrigation runoff | Spencer et al., 1985 |
| DCPA | 3.4-7.6 | California | Cotton | 189 | 1.22-1.40 | Irrigation runoff | |
| Dinitramine | 1.3 | California | Cotton | 34 | 1.32 | Irrigation runoff | |
| EPTC | 2.8-13.8 | California | Sugarbeet and alfalfa | 1630 | 6.4-7.2 | EPTC applied in irrigation water at 2 mg/L | |
| Prometryn | 1.3-3.1 | California | Cotton | 1408 | 0.95-5.0 | Irrigation runoff | |
| Trifluralin | 0.96-1.1 | California | Cotton | 19 | 0.14-0.29 | Irrigation runoff | |
| Azinphosmethyl | 0.52 | California | Cotton | <0.5 | -- | Irrigation runoff | |
| Chlorpyrifos | 0.98-2.88 | California | Cotton | 480 | 0.02-0.24 | Significant proportion of losses because of aerial application during irrigation | |
| Diazinon | 0.48-2.69 | California | Sugarbeet, mellon | 22 | 0.04-0.07 | Irrigation runoff | |
| Malathion | 1.17-4.46 | California | Cotton, sugarbeet, alfalfa, lettuce | 21 | 0.0003-0.09 | Irrigation runoff | |
| Methidathion | 1.02-1.12 | California | Cotton | 473 | 0.16-2.0 | Significant proportion of losses because of aerial application during irrigation | |
| Mevinphos | 0.14 | California | Alfalfa | 26 | 0.27 | Irrigation runoff | |
| Ethyl Parathion | 0.28-4.20 | California | Lettuce, sugarbeet, alfalfa | 77 | 0.02-0.51 | Irrigation runoff | |
| Methyl Parathion | 0.14-2.1 | California | Lettuce, sugarbeet | 27 | 0.003-0.32 | Irrigation runoff | |
| Sulprophos | 4.55-6.00 | California | Cotton | 0.32 | 0.001-0.007 | Irrigation runoff | |
| Methomyl | 0.4-5.82 | California | Cotton, lettuce, sugarbeet, alfalfa | 223 | 0.13-1.73 | Irrigation runoff | |
| Endosulfan | 1.66-5.96 | California | Lettuce, melon | 104 | 0.19-0.62 | Irrigaiton runoff | |
| Ethylan | 3.2 | California | Lettuce | 8.0 | 0.008 | Irrigation runoff | |
| Fenvalerate | 0.27-1.23 | California | Cotton | 133 | 0.05-0.21 | Significant proportion of losses because of aerial application during irrigation | |
| Permethrin | 0.10-0.71 | California | Cotton | 71 | 0-0.16 | Significant proportion of losses because of aerial application during application | |

† Maximum reported concentrations unless specified otherwise.

Table 9-3. Pesticides in streams and water bodies resulting from agricultural applications. Selected examples (1978-1985).

| Watershed stream systems | Location | Pesticide residues found | Conc. ranges ($\mu$g/kg) | Loads | Comments | Reference |
|---|---|---|---|---|---|---|
| Grassed watershed (53 ha); with partial treatment by herbicide. Stream system draining 1772 ha of perennial pasture watershed | Riesel, TX | Picloram | 250 maximum from treated 8-ha area | Most of herbicide leaving treated area was transported through the 53-ha watershed-system, but at reduced concentrations | Experiments conducted to study patterns of dilution, transport, and dissipation of herbicides in complex watersheds | Mayeux et al., 1984 |
| | | | 13 720 injected directly into stream | Only a small fraction of injection detected to pass 5400-m point. Apparent loss because of concentration decreasing to below limit of detection | | |
| Rivers and agricultural drainage | Japan | CNP | 0.01-16.67 | | Highest levels found above 4 wk after rice planting and when flood waters released from paddies. | Suzuki et al., 1978 |
| 2025 ha of agricultural watershed; corn and soybean major crop | Lincoln, NE | Atrazine<br>Alachlor<br>Propachlor | 14-24<br>0-1.4<br>0.58-3.0 | | Results based on limited sampling during drought years | Schepers et al., 1980 |
| Black Creek Watershed | Allen Co., IN | 2,4,5-T | 0.2-7.7 | | Atrazine, alachlor, carbofuran, and malathion not detected | Dudley & Karr, 1980 |

(continued on next page)

Table 9–3. Continued.

| Watershed stream systems | Location | Pesticide residues found | Conc. ranges ($\mu$g/kg) | Loads | Comments | Reference |
|---|---|---|---|---|---|---|
| Honey Creek Watershed and rivers of Northwest Ohio | Northwestern Ohio | Atrazine<br>Simazine<br>Metribuzin<br>Alachlor<br>Metolachlor<br>Butylate<br>Phorate<br>Terbufos<br>Fonofos<br>Carbofuran | 87†<br>7.4†<br>2.3†<br>105†<br>140†<br>0.49†<br>0.24†<br>0.54†<br>1.0†<br>45† | 7.5% of applied atrazine exported from Honey Creek Watershed in 1981 | Study reports higher concentration values than expected from other published sources; however, rainfall during study was two to three times above long-term average. Author concludes that concentrations observed not acutely toxic to fish and invertebrates, but may produce inhibitory growth effects on plants and algae | Baker et al., 1981 |
| Parana River, 600 km upstream from mouth | Argentina | Lindane<br>Parathion<br>Alpha-BHC | 0.009‡<br>0.022‡<br>0.009‡ | | Sediment transported pesticides were positively correlated with discharge as was sediment concentrations | Lenardon et al., 1984 |
| Well water, surface water (lakes, ponds, and rivers), and municipal water | 205 sites in South Carolina | DBCP | 0–0.4 | | In area of high use, 37% of surface water samples exceeded background (0.05 $\mu$g/L) levels, but none above 0.4 | Carter & Riley, 1981 |
| Forested watersheds | Central Tennessee | None | -- | -- | 1.68 kg/ha a.i. hexazinone applied as pellets. No detectable residue in streamflow for 28-wk after application | Neary, 1983 |
| Forested watershed; 104 ha containing four 1-ha watersheds | Upper Piedmont, GA | Hexazinone | 442 ± 53 first event maximum for treated area. 0–40 hexazinone + metabolite in streamflow for 104-ha watershed | 0.53% of application (avg. from watersheds) | Authors conclude that residue not high enough for aquatic damage | Neary et al., 1983 |

(continued on next page)

Table 9-3. Continued.

| Watershed stream systems | Location | Pesticide residues found | Conc. ranges (μg/kg) | Loads | Comments | Reference |
|---|---|---|---|---|---|---|
| Wye River Estuary | Eastern Chesapeake Bay region | Atrazine | 0–300 edge of field; 15 maximum in estuary; <3 avg. in estuary at peak loading | >3% moved in estuary | Herbicide level rarely approached levels that would reduce aquatic photosynthetic rate | Glotfelty et al., 1984 |
| | | Simazine | Simazine concentration in estuary significantly lower than for atrazine | | | |
| Various drainage basins surrounding Chesapeake Bay | Chesapeake Bay area | Linuron | <10 (sed.)<br><0.2 water | | No apparent accumulation of linuron in estuary | Zahnow & Riggleman, 1980 |
| Rhode River Watershed | Chesapeake Bay, MD | Atrazine | 0–40 | | Little correlation between use and hericide loading rates in water; herbicides in runoff from nontreated areas suggested aerial or subsurface transport in addition to surface runoff | Wu et al., 1983 |
| | | Alachlor | 0–6 | | | |
| 11 agricultural watersheds ranging in size from 20–79 $km^2$ | Canadian Great Lakes Basin, Co. Ontario | 18 parent compounds plus isomers and metabolites found in drainage waters. Organochlorines not in current use also detected. Only atrazine, endosulfan, and simazine appeared year round | Atrazine 1.1, 1.6§<br>Endosulfan 0.0037, 0.002§<br>Simazine 0.02, 0.06§ | 2.02 g $ha^{-1}$ yr | Report summarizes extensive studies on pesticides in streamflow; concentrations of four pesticides (organochlorines consistently greater than established water quality criteria. No herbicide concentrations exceeded criteria | Frank et al., 1982 |

† Maximum values observed. ‡ Mean values. § Overall mean for 11 watersheds 1976 and 1977, respectively.

### 9–2.4 Pesticides in Streams and Water Bodies

When runoff from a treated area containing pesticide enters a water course or body of water, concentrations are rapidly diluted and are also partitioned among various components of the hydrosphere. Pesticides in current use are considered nonpersistent compared to the organochlorine insecticides. Early studies such as those by Nicholson et al. (1964, 1966) concentrated almost totally on defining the problem relative to organochlorine insecticides that were in prevalent use at that time. Results of these and other studies were summarized in reviews by Pionke and Chesters (1973), Leonard et al. (1976), and Caro (1976). In general, widespread contamination by organochlorine compounds such as DDT and dieldrin had occurred, although at low concentrations ranging from a few parts per trillion in water to a few parts per billion in sediments and streambed material. These concentrations, however low, were deemed environmentally significant because of biological magnification through food chains and resulted in eventual banning of most uses of the organochlorine compounds. Since the mid- to late 1960s, the number of reported occurrences and levels of organochlorines in water has continued to decline (Lichtenberg et al., 1970; Schafer et al., 1969; Schulze et al., 1973). Extremely low levels of three herbicides, 2,4-D, 2,4,5-T, and silvex were also occasionally found in rivers and lakes, but at concentrations rarely exceeding a few parts per trillion.

Examples of more recent reports of pesticides in surface waters are summarized in Table 9–3. Organochlorine pesticide residues from past use are still being found at low, but environmentally significant concentrations (Frank et al., 1982). Concerns have been voiced about the possible effects of herbicide residues in surface water on aquatic vegetation. Of particular concern are possible links between atrazine residues and die-back of submerged aquatic vegetation in estuarine systems such as the Chesapeake Bay (Glotfelty et al., 1984; Wu et al., 1983). However, concentrations observed in the Wye River estuary, in a portion of the river receiving high loadings from agricultural runoff, have rarely approached levels thought to produce even minor effects (Glotfelty et al., 1984) (Table 9–3).

Movement of atrazine in complex watershed systems has been studied extensively in Ontario. Frank and Sirons (1979) and Frank et al. (1982) reported that between 0.3 and 1.9% of the applied atrazine was transported in streamflow from watersheds ranging in size from 1860 to 7913 ha with peak concentrations of about 33 $\mu$g/L. Average concentrations were 1.1 and 1.6 $\mu$g/L in 1976 and 1977, respectively. Atrazine and simazine were two of only three pesticides detected in streamflow throughout the year, but concentrations never reached levels thought to be harmful to vegetation. Baker et al. (1981), however, found concentrations of atrazine and other herbicides in streamflow from a watershed in northwestern Ohio, that could have been sufficiently high to produce inhibitory effects on plants and algae (87 $\mu$g/L atrazine, see Table 9–3). Rainfall during their study period was reported to be two to three times normal and total atrazine transported was 7.5% of that applied. Apparently, runoff conditions during their study met the criteria of "catastrophic" events as Wauchope (1978) defined which must occur rarely on a basin-wide scale.

Atrazine was also reported in streamflow in Nebraska by Schepers et al. (1980) and Wu et al. (1983) in Maryland, but in much lower concentrations and loads. Residues of 2,4,5-T were found in drainage from the Black Creek Watershed in Indiana, but residues of atrazine and alachlor both of which were used in the watershed, were not detected (Dudley & Karr, 1980).

### 9-2.5 Attenuation in the Transport System

As summarized above, runoff losses at the edge-of-the field may reach several percent of the application and concentrations may reach several milligrams per liter if runoff occurs soon after application. As evident from observations in large diverse watershed systems, these runoff concentrations are rapidly attenuated in the transport system by dilution, deposition and trapping of sediments, adsorption by bottom and bank materials, and infiltration along the various flow paths. Another factor to consider in large diverse watersheds is application timing. Not all the watershed will be treated on any one day. Consider the hypothetical situation depicted in Fig. 9-7.

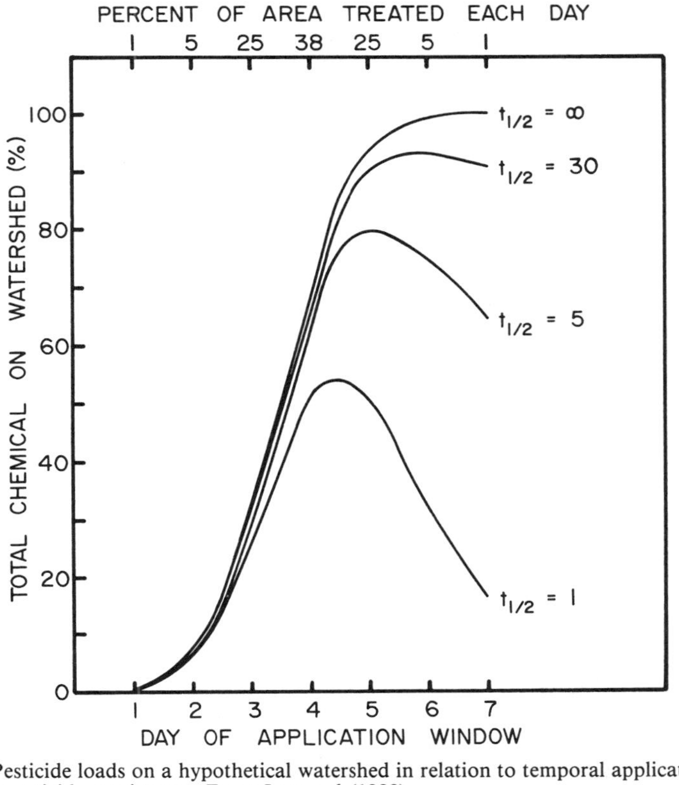


Fig. 9-7. Pesticide loads on a hypothetical watershed in relation to temporal application pattern and pesticide persistence. From Leonard (1988).

Assume that a given chemical is applied to the entire watershed sometime during a 7-d period or in a 7-d window distributed normally about the fourth day. The amounts on the watershed each day potentially available for runoff is represented by a family of curves, depending on the persistence of the chemical applied. The width of the application window could also be varied and it would be obvious that the runoff potential would be reduced as the half-life of the chemical is decreased or the width of the window is increased. These concepts may be combined with probability theory as Mills and Leonard (1985) used for a quantiative expression of these effects. Also, in large watersheds, rainfall and runoff is usually distributed in time and space such that an additional attenuation of pesticide loads in runoff at the watershed outlet could be expected. Mayeux et al. (1984) demonstrated that a mobile material such as picloram may be transported through a watershed system in storm runoff with little loss in actual mass transported, but at concentrations significantly reduced by dilution. In an injection experiment in a stream system, they showed that dilution could reduce concentrations below detectable limits and net losses in transported mass could be inferred if these limits were not considered. Glotfelty et al. (1984) reported a direct proportionality between atrazine concentrations in the Wye River estuary and salinity, and used this relationship to describe dilution and to estimate the average fresh water atrazine concentration (about 23 $\mu$g/L) by extrapolation. The work of Trichell et al. (1968) illustrates how loss of a mobile pesticide from runoff may occur as a result of infiltration and other processes occurring in overland flow. They treated a plot uniformly with picloram and lost 5.5% of the application in runoff during a simulated rainstorm. By treating the upper one-half of the plot with twice the rate, only 3.2% of the pesticide was lost in runoff. Rohde et al. (1980) showed that transport of pesticides such as trifluralin may be significantly reduced by infiltration, sedimentation, and filtration processes occurring in grassed waterways and buffer strips.

## 9-3 EFFECTS OF MANAGEMENT ON PESTICIDE RUNOFF

### 9-3.1 Relationships between Runoff and Erosion Control and Pesticide Runoff

Numerous management practices have been suggested to reduce potential nonpoint source pollution from pesticides (Caro, 1976; Novotny & Chesters, 1981). These range from using alternative pesticides, crop rotations, integrated pest management systems to soil conservation, substitution of crops and use of mechanical procedures. Discussions here are limited to interrelationships between soil and water management and pesticide runoff potential. Soil and water management systems may be designed to reduce pesticide runoff where specific problems have been identified, such as runoff from soils containing high levels of persistent organochlorine residues. Management systems, however, are usually developed and employed to accomplish other objectives such as reduction of soil erosion by reduced tillage. Questions then are what are the important tradeoffs or impacts such as possibly increased pesticide runoff potential.

As previously mentioned, the amount of pesticide in the active zone at the soil surface at the time of runoff is perhaps the most important variable affecting amounts and concentrations in runoff. Certainly, changes in rates, formulation/application methods, and chemical type will all affect maximum concentrations and persistence of pesticides available for runoff. For example, Wauchope (1987b) using small runoff beds and simulated rainfall found that pesticide concentrations in runoff events immediately after pesticide application were strongly formulation dependent. Whether these formulation effects carryover to subsequent events is not known. Secondary effects are related to changes in infiltration, runoff timing, and soil water content (Table 9-1). Finally, the effect of management depends on the mode of pesticide transport in runoff and whether or not a given management practice affects runoff volume, sediment yields, or both.

The effects of soil erosion control practices on pesticide runoff depends on the adsorption characteristics of the pesticide and degree of reduction of fine sediment transport. As sediment yield is reduced, adsorbed pesticide in runoff is reduced, but not necessarily in proportion because control practices tend to reduce transport of coarse particles more than fine particles, and therefore, the capacity for adsorbed chemical transport per unit sediment mass is increased. Sediment enrichment ratios, the ratio of sediment adsorption surface to that of the residual soil, have been reported to range from about 1 to 5 with about 2 to 3 being most common (Smith et al., 1978; McDowell et al., 1981; Menzel, 1980; Sharpley, 1985b). Ignoring these changes in sediment composition and assuming a liner form of the Freundlich equation, the distribution of pesticide between water and sediment in hypothetical runoff events is illustrated in Fig. 9-8. Relationships between $K_d$ and the fraction of the total pesticide transported by sediment are shown. The family of curves results from different sediment concentrations. As can be seen, whether a pesticide is transported primarily dissolved in water or attached to sediment depends both on $K_d$ and sediment concentrations. When sediment concentrations are low, solution transport in terms of percent of the total is dominant even for strongly adsorbed pesticides because volumes of runoff water greatly outweigh masses of sediment. Conversely, when sediment concentrations are high, sediment pesticide transport will be significant even for those pesticides with intermediate $K_d$. As examples, MSMA and paraquat are transported almost entirely by sediment, whereas, the primary mode of transport of compounds such as trifluralin depends strongly on sediment concentrations (Wauchope, 1978; Willis et al., 1983; Smith et al., 1978).

Smith et al. (1978) compared pesticide runoff from terraced watersheds to runoff from watersheds with no planned conservation practices. Paraquat which was strongly bound to sediment was reduced in proportion to sediment reduction. Terraces did not reduce runoff volumes, and therefore, losses of atrazine, diphenamid, cyanazine, propazine, and 2,4-D were not affected since these were transported primarily in the aqueous phase. Ritter et al. (1974) showed that conservation practices that reduce runoff volumes also reduce losses of atrazine and propachlor. Baker and Johnson (1979) and

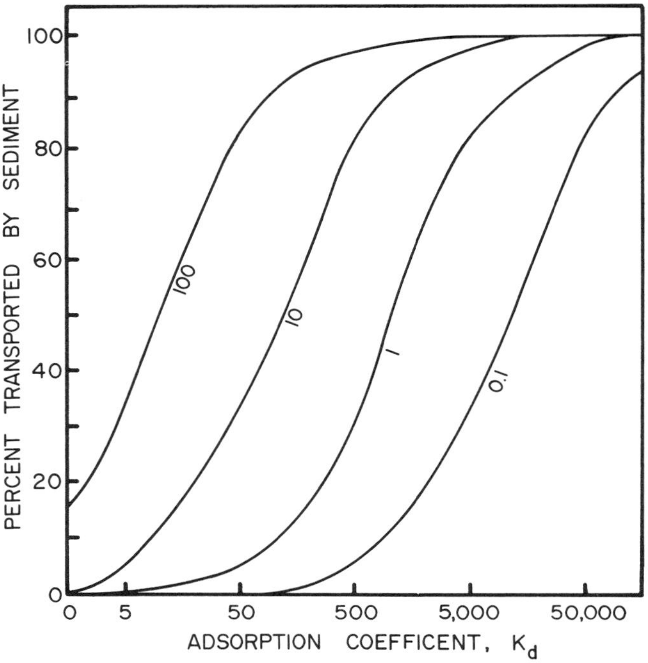


Fig. 9-8. Percentage of pesticide transported by sediment as function of pesticide $K_d$ and sediment concentrations in runoff. Numbers on curves are sediment concentrations in gram per liter. From Leonard (1988).

Baker et al. (1978) related runoff and soil loss to crop residues in some tillage practices. Crop residue reduced runoff volumes on some soils, but not losses of alachlor and cyanazine because concentrations tended to increase with increasing crop residue.

### 9-3.2 Conservation Tillage and Pesticide Runoff

Considerable interest has been expressed in conservation tillage systems (Wauchope et al., 1984; Wagenet, 1987; Wauchope, 1987a) and whether pesticide losses in runoff may be enhanced or reduced. Probably both situations exist depending on a combination of conditions. Triazines (Martin et al., 1978) and other soluble herbicides (Bovey et al., 1974; Trichell et al., 1968) are easily removed from crop surfaces by rainfall and runoff and this washoff may be a source of enhanced concentrations in runoff as observed by Baker and Johnson (1979) and Baker et al. (1978). However, Baker et al. (1982) reported that concentrations in runoff were not affected by herbicide placement above or below crop residue, but were negatively correlated with time to runoff. Baker and Laflen (1979) had earlier shown that wheel tracks reduced time to runoff, increased initial herbicide concentration in runoff, total runoff volumes, and therefore, total herbicide losses. Triplett et al. (1978) and Edwards et al. (1980), in their studies of no-till systems, emphasized the importance of rainfall/runoff timing relative to herbicide application and reported reduced seasonal herbicide runoff losses as a result of reduced runoff volumes, as did Hall et al. (1984). Although pesticides in runoff were not measured, Langdale and Leonard (1983) reported significant reductions in growing season runoff under no-till systems in the Georgia Piedmont. Kramer (1984) in a comprehensive analysis found little difference in runoff from conservation tillage vs. conventional tillage systems in the Central Claypan Region of Missouri.

Considering the above observations, pesticide losses in runoff from no-till systems may be enhanced under conditions where soluble pesticides are applied to crop residues and runoff occurs soon after application, and quickly after rainfall begins. These conditions may occur under a combination of high initial soil water content, intense rainfall, and low infiltration rates (e.g., compacted soil or high clay soil). Conversely, no-till or the presence of crop residues may reduce herbicide runoff under conditions where crop residue reduces surface sealing and maintains higher infiltration rates, and therefore, increases time to runoff and reduces total runoff volumes. Here most of the soluble pesticide will be leached from the residue to the soil before runoff begins.

Runoff of relatively insoluble pesticides adsorbed to sediment will be reduced under no-till because sediment yields are reduced (Langdale & Leonard, 1983). Some pesticides may also be adsorbed or retained by vegetative surface such that their runoff concentrations may be reduced (Asmussen et al., 1977).

### 9–3.3 Irrigation Management and Pesticide Runoff

Supplemental irrigation is often used in humid production regions to augment soil water during periods of rainfall deficiency. Most of this irrigation is by overhead sprinklers and water is normally applied at rates that do not produce significant surface runoff. However, irrigation management can affect pesticide runoff potential. Runoff potential, particularly of pesticides that are mobile in soil can be reduced by irrigation soon after pesticide application. Mobile pesticides will be translocated from the immediate soil surface potentially in contact with runoff to lower soil layers. Immobile pesticides would not be affected. Application of pesticides in irrigation water (Dowler et al., 1982), depending on pesticide formulation and mobility in soils, also may distribute pesticides to deeper soil depths and reduce pesticide runoff potential compared to conventional surface applications. Since irrigation maintains higher levels of soil water, probabilities will be greater for higher antecedent soil water levels for antecedent runoff events in response to rainfall. Higher levels of soil water would, however, have greater effects on runoff from clay or heavy-textured soils with relatively low infiltration rates compared to sandy soils (Knisel & Leonard, 1986). DelVecchio et al. (1983) using model simulations showed that supplemental irrigation has little effect on pesticide runoff from a loamy sand soil.

Masses of pesticides lost via surface runoff from furrow irrigation are determined to a large extent by volumes of drainage water (Spencer et al.,

1985). Therefore, quantities of pesticides lost, particularly those applied in the irrigation water itself, can be reduced by increased irrigation efficiency and preventing runoff during actual application. In the study of Spencer et al. (1985), losses of insecticides were greatly increased when aerial application took place during irrigation periods.

## 9-4 MODELING AND COMPUTER SIMULATION OF PESTICIDE RUNOFF

### 9-4.1 Model Selection and Use

Collection of field data on pesticide runoff is an expensive and time consuming process, and therefore, methods have been sought to predict or simulate expected results under the different scenarios of pesticide application of interests. High-speed computers perform repetitive, complex computations allowing representations of physical processes in mathematical form. As a result, numerous computer models have been developed to simulate pesticide runoff and fate in soils. Because pesticide runoff is driven by rainfall-induced losses of surface water and sediments, pesticide runoff models must incorporate hydrology and erosion/sediment transport models. In addition, pesticide models must incorporate algorithms or expressions that describe pesticide state in the runoff active zone from the day of application through the period of persistence. Basically, pesticide runoff models must consider all states and processes depicted in Fig. 9-3. Models may be similar in structure, but all differ, each reflecting the model builder's concepts of the physical processes, simplifications employed, methods of representation and programming, and the intended use of the model. Models vary greatly in temporal and spatial resolution. The burden is on the model user to select a model appropriate to his particular application. Comprehensive discussions on model selection and use are found elsewhere (Knisel & Leonard, 1989; Leonard & Knisel, 1986, 1988; Knisel, 1985; Bailey & Swank, 1983). Model users should carefully consider attributes of different models to determine which can provide the desired results. Some considerations include intended model purpose, data required, data available, ease of parameter estimation and use, and cost of simulation.

Some models were developed to simulate response for a single, design-type storm. This is entirely adequate when considering only surface runoff and sediment yield, but a design storm may not be one that results in a maximum pesticide loss. For nonpoint source pollutants, continuous (or daily) simulation over a relatively long climatic record is recommended to examine risk analysis (Knisel & Leonard, 1989).

Often, modelers think that physical processes are the same regardless of scale, and a model can be applied equally as well for fields as for basins. While the processes are the same, the sensitivity of processes vary with size of area. For example, the infiltration process and the simulation time increment in the model are much more significant for small (plot size) areas than for large basins. Therefore, a user should consider the scope of a model when making a selection. Another consideration is whether or not a model is sensitive to changes in managment practices. If not, there is little chance of success in selecting among alternate practices for nonpoint source pollution control.

### 9–4.2 Runoff Models Available

Crawford and Donigian (1973) developed one of the first complete, continuous simulation models they called the Pesticide Runoff Transport (PRT) model to estimate runoff, erosion, and pesticide losses from field-sized areas. The hydrologic component of the PRT model is the Stanford Watershed Model (Crawford & Linsley, 1966) and the erosion component that Negev (1967) developed. The Stanford Watershed model was one of the first computer simulation models for hydrology. Donigian and Crawford (1976) later incorporated other water-quality components into a model called the Agricultural Runoff Model (ARM), which retained the essential pesticide features of the PRT model. Continued development of these models and linkages with other water-quality models has led to a comprehensive program (HSPF, Hydrologic Simulation Program—Fortran) for modeling runoff, sediment, pesticides, nutrients, and other water-quality constituents from urban, agricultural, and other land uses (Donigian et al., 1983). The HSPF allows detailed simulation of stream hydraulics, water-quality processes, pesticide and nutrient behavior in soils and water bodies, and sediment contaminant transport. All these modeling programs require extensive data for calibration.

Bruce et al. (1975) developed an event model to estimate runoff, erosion, and pesticide losses from field-sized areas for single runoff-producing events. This model requires data for calibration to the specific site and conditions.

Frere et al. (1975) developed an Agricultural Chemical Transport Model (ACTMO) to estimate runoff, sediment yield, and chemicals from field- and basin-sized areas. The hydrology component is the USDA Hydrograph Laboratory Model (Holtan & Lopez, 1971), based on an infiltration concept. The erosion component is based on the rill and interrill erosion concepts and the Universal Soil Loss Equation (USLE) modification that Foster et al. (1977) developed. The ACTMO model requires no calibration with historical data.

The pesticide runoff model of Haith and Tubbs (1981) is structured to couple with any hydrologic and soil loss model. In application, they used the USDA-Soil Conservation Service (SCS) (1972) curve number (CN) method for simulation of daily runoff and the modified USLE as Williams (1975) proposed. They assumed the available pesticide for runoff to be that in the surface 10-mm soil layer. This pesticide was allowed to decay exponentially with time. A single-valued linear equilibrium adsorption process was assumed for partitioning pesticide between adsorbed and aqueous phases. Since most of the required data inputs are readily available, no calibration is necessary in applying this model.

Steenhuis (1979) and Steenhuis and Walter (1980) described a model, the Cornell Pesticide Model (CPM) which is similar to the model of Haith and Tubbs (1981) in that the USDA-SCS CN method and USLE submodels are used as options. Additional hydrology options, however, are included for application to snowmelt and shallow soils. Their pesticide submodel has three components: a first-order degradation component, a pesticide-leaching component (Rao et al., 1976), and a pesticide runoff function. The model uses a mixing depth of about 10 mm, but this depth can be adjusted by calibration, if necessary, using measured pesticide losses. The Continuous Pesticide Simulation (CPS) model tested by Lorber and Mulkey (1982) is based on the Cornell model.

In 1980, the USDA-ARS, in a nationally coordinated project developed and published the Chemicals, Runoff, and Erosion from Agricultural Management Systems (CREAMS) model (Knisel, 1980). CREAMS is a relatively simple, computer-efficient, physically based model for field-sized areas. CREAMS does not require observed data for parameter calibration, although observed data are beneficial to adjust the sensitive parameters such that good agreement can be obtained with simulation results. This model is especially useful in making relative comparisons of pollutant loads from alternate management practices. Compared with other pesticide models, unique features are the ability to simulate foliar-applied chemicals and sediment enrichment ratios. Also, complete mixing of the surface layer with runoff water is not assumed.

The Pesticide Root Zone Model (PRZM) (Carsel et al., 1984) was developed primarily for use in simulating vertical pesticide movement in the root zone and in the unsaturated zone above a water table. However, runoff losses are also simulated. The hydrology component of PRZM is basically the daily hydrology option (CN) as in CREAMS, with modified water balance equations. The soil erosion component is based on the Modified Universal Soil Loss Equation (MUSLE) (Williams, 1975).

The CREAMS model has been modified to also address vertical pesticide movement in the root zone (Leonard et al., 1987; Knisel & Leonard, 1986). This modified version (Groundwater Loading Effects of Agricultural Management Systems, GLEAMS) retains the basic structure of the CREAMS model as to the surface runoff components, but links the hydrology, erosion, and pesticide submodels in one program for greater efficiency in computer operation. Other changes include simplified input files, increased maximum simulation up to 50 yr, and an irrigation option.

Application of the above models with the exception of HSPF, is limited primarily to field-size areas. The HSPF was initiated to provide capabilities of continuously simulating dynamics of river basins with generation of detailed information on aquatic impacts and system responses (Donigian et al., 1983). Other basin and complex-watershed models containing pesticide components are Small Watershed Agricultural Model (SWAM) (DeCoursey, 1982), and the Pesticide Runoff Simulator (Computer Sciences Corp., 1980). Both of these models use pesticide concepts similar to that in CREAMS with additional algorithms for routing the field outputs through the watershed system. Hydrology and erosion/sediment transport submodels in the Pesticide Runoff Simulator were adapted from Simulator for Water Resources in Rural Basins (SWRRB) (Williams et al., 1985).

Long records of 20 to 50 yr on pesticide runoff from a given management system are nonexistent. Much more data is available on long-term rainfall. Models may then be used in long-term simulations to generate data on pesticide runoff for frequency analysis (Knisel & Leonard, 1989) and assessment of risks in probabilistic terms. Often however, even long records of rainfall are unavailable. One solution is to use stochastic rainfall generation models (Nicks, 1974; Haith, 1987) to create these records. Another approach in probability analysis is to use a model such as Mills and Leonard (1985) developed. They developed a stochastic pesticide state model assuming that the state of the pesticide, and the amount available for runoff at the time of a significant rainfall event, is a random variable and the primary variable affecting runoff losses and demonstrated how probability density functions are generated specific to given pesticides, expected climatic patterns, and application methods and rates.

In developing the stochastic model, the amount of pesticide on foliage or soil available for runoff was assumed to be largely dependent upon the amount of pesticide applied, the percentage of the applied amount that is susceptible to runoff, the persistence or half-life of the pesticide, and the time from application to the occurrence of runoff producing rainfall (Mills & Leonard, 1985). The amount of pesticide applied, the pesticide half-life and its susceptibility to runoff are not random variables, but are determined by pesticide chemistry, management decisions, and other conditions. Time from application to occurrence of significant rainfall is, however, highly subject to random variation. Coupling stochastic pesticide state models with stochastic models of rainfall characteristics and watershed state and deterministic models for event runoff of water, sediment, and chemicals and environmental damage functions are possible (Mills & Leonard, 1985).

### 9–4.3 Model Application

Use of models is illustrated by selected examples as follows. Donigian et al. (1983) gives a thorough discussion on how the HSPF model is applied to the Four-Mile Creek Watershed in east central Iowa. An excellent example of procedures necessary in calibration and verification of this model is provided. Once calibration was obtained under base-line conditions of conventional corn/soybean [*Zea mays* L./*Glycine max* (L.) Merr.] tillage (plow, disk, and plant), selected parameters were adjusted to reflect the effects of Best Management Practice (BMP) scenario which included conservation tillage. Their simulations showed that for a 2.5 yr of simulation for the BMP, runoff was reduced by an average of 8% and sediment production was reduced by 42% with a subsequent reduction in alachlor transport of 32%. Knisel et al. (1983) demonstrated use of CREAMS on this same watershed. The CREAMS model has also been applied to production systems in the southeast Coastal Plain for evaluation of effects of management practices on potential pesticide runoff (DelVecchio & Knisel, 1982; DelVecchio et al., 1983).

Effects of management practices on pesticide runoff from a Tifton loamy sand (fine-loamy, siliceous thermic Plinthic Paleudult) planted to corn in the southeast Coastal Plain as judged from model simulation are shown in Fig. 9–9 (Knisel, 1985). Management practice (1) was continuous corn, conventional moldboard tillage where a plow pan had developed limiting rooting depth to 20 cm. Practice (2) was continuous corn, chisel 30-cm deep, residue management, producing a rooting depth of 34 cm. Practice (3) was the same as (2), but with a winter cover crop of small grains. Two pesticides, each surface applied at planting at rates of 2.8 kg/ha were assumed. Pesticide A represented a water-transported chemical and was assigned a $K_d$ of 0.2 whereas pesticide B, representing a sediment-transported chemical, was assigned a $K_d$ of 200. Both were assumed to have other properties in common including a half-life of 30 d. Compared to practice (1), practices (2) and (3) reduced runoff volumes. This is because of increased hydraulic conductivity of the root zone, increased rooting depth, and increased evapotranspiration particularly in winter months for practice (3).

Reduced runoff and residue management result in a significant reduction in sediment yield. Surface runoff losses of the water transported pesticide (A) were negligible for all practices (<1 g/ha cumulative for 20 yr). This mobile pesticide moved below the active runoff surface rapidly with rainfall in the loamy sand soil such that the major pathway of loss was leaching through the root zone (Fig. 9–9B). For the sediment-transported pesticide, runoff losses were reduced in approximate proportion to sediment reduction. However, because of the 30-d half-life, pesticide losses were only affected by sediment losses occurring a few weeks after pesticide application. For about 8-yr (1965–1972), rainfall distribution was such that little sediment production occurred from any management practice. If only this period had been simulated, little differences between practices would have been shown, emphasizing the need for long-term simulation when assessing effects of management.

The CREAMS model has been used in evaluation of management practices for watershed land treatment projects (Nicks et al., 1984) and development of a field planning guide for recommendation of practices to minimize pesticide runoff (Burt et al., 1984). The CREAMS model has also been adapted for use in pesticide management in forestry seed orchards (Nutter et al., 1984).

No model has yet been developed that can be proven to give accurate predictions of pesticide runoff on an absolute basis, i.e., predicting pesticide concentrations in runoff consistently to within a few percent (Leonard & Knisel, 1989). Therefore, in model application, decisions should be made based on relative comparisons rather than absolute numbers. That is, does one practice reduce pesticide loss compared to another, or is one pesticide more susceptible to runoff than some other? Model users must remember that models are approximations and simplifications of reality and hope that their results reflect at least proper directions and relative effects on which, along with only information and judgment, proper decisions can be made.

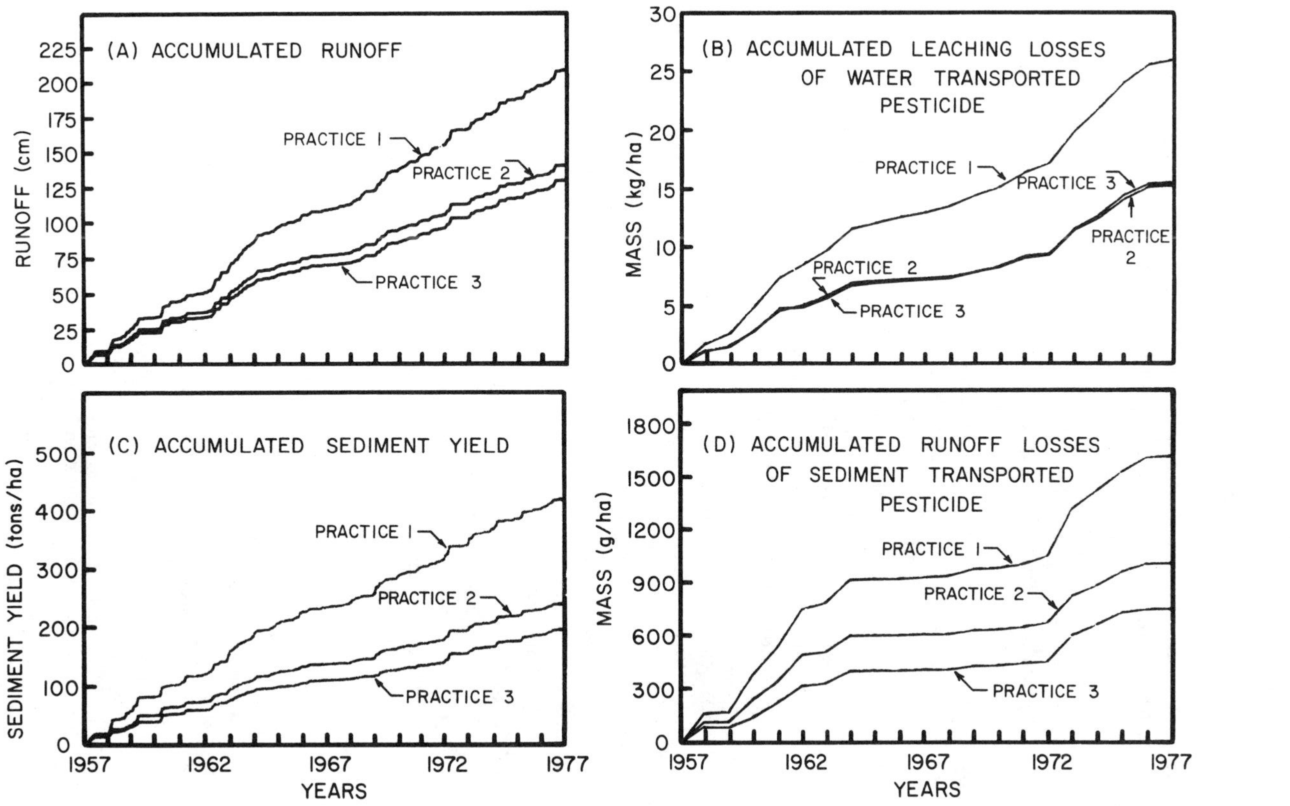


Fig. 9–9. Accumulated runoff, sediment yield, and pesticide runoff and leaching simulated with the CREAMS/GLEAMS model, 1958–1977. Tifton loamy sand, continuous corn. Practice (1) is conventional tillage; (2) is chisel plow, residue management; and (3) is the same as (2), but with winter small grain (Knisel, 1985).

## 9-5 SUMMARY AND CONCLUSION

Chemicals applied to cropland, forests, and other areas for pest control may enter surface waters by rainfall or irrigation runoff. Amounts of runoff losses are highly variable depending on pesticide rates, formulation, and application methods; runoff timing with respect to application timing; and pesticide persistence and chemical properties. Annual runoff losses for a large group of surface soil-applied pesticides average about 2% of the application with occasional losses reaching the range of about 5% when runoff occurs soon after application. Runoff losses exceeding 10% have been reported. Most of these reports, however, are from studies using simulated rainfall where measured pesticide losses may be higher than under conditions of natural rainfall. Under natural conditions, pesticide runoff losses in the 10% range would be rare. For soil-incorporated pesticides or foliar-applied pesticides, runoff losses are commonly <1% and often as low as 0.1% or less. Incorporation into the soil reduces pesticide concentrations in the immediate soil surface in contact with rainfall and overland flow, thereby reducing the potential for pesticide runoff. Pesticides applied to foliage may dissipate rapidly or may not be readily dislodged by rainfall.

Pesticide concentrations in streams and significant water bodies are much lower than those measured in edge-of-field runoff because of processes of dilution, sedimentation, vegetative trapping, and degradation in transport. Where detectable, concentrations are commonly in the low microgram per liter to milligram per liter range. Modern analytical techniques allow detection of pesticide residues in water at levels below which clear toxicological interpretations can be made.

Computer models have been developed to simulate pesticide runoff losses for a wide range of pesticide type, management, soils, and climate conditions. These models do not provide predictions in terms of absolute values to a high degree of accuracy, but do provide information for relative comparisons between management alternatives, weather patterns, and other variables.

## REFERENCES

Ahuja, L.R., and D.R. Lehman. 1983. The extent and nature of rainfall-soil interaction in the release of soluble chemicals to runoff. J. Environ. Qual. 12:34–40.

Ahuja, L.R., J.D. Ross, and O.R. Lehman. 1981a. A theoretical analysis of interflow of water through surface soil horizons with implications for movement of chemicals in field runoff. Water Resour. Res. 17:65–72.

Ahuja, L.R., A.N. Sharpley, M. Yamamoto, and R.G. Menzel. 1981b. The depth of rainfall-runoff-soil interaction as determined by $^{32}P$. Water Resour. Res. 17:969–974.

Anderson, L.W.J., J.C. Pringle, R.W. Raines, and D.A. Cisneros. 1978. Simazine residue levels in irrigation water after ditchbank application for weed control. J. Environ. Qual. 7:575–579.

Asmussen, L.E., A.W. White, E.W. Hauser, and J.M. Sheridan. 1977. Movement of 2,4-D in a vegetated waterway. J. Environ. Qual. 6:159–162.

Bailey, G.W., A.P. Barnett, W.R. Payne, Jr., and C.N. Smith. 1974. Herbicide runoff from four Coastal Plain soil types. USEPA, Environ. Prot. Technol. Ser. EPA-660/2-74-017. U.S. Gov. Print. Office, Washington, DC.

Bailey, G.W., A.R. Swank, Jr., and H.P. Nicholson. 1974. Predicting pesticide runoff from agricultural land: A conceptual model. J. Environ. Qual. 3:95–102.

Bailey, G.W., and R.R. Swank, Jr. 1983. Modeling agricultural nonpoint source pollution: A research perspective. p. 27–47. *In* F.W. Schaller and G.W. Bailey (ed.) Agricultural management and water quality. Iowa State Univ. Press, Ames.

Bailey, T.E., and J.H. Hannum. 1967. Distribution of pesticides in California. J. of Sanitary Engineering Division. Proc. Am. Soc. Civ. Eng. 93(SA5):27–43.

Baker, D.B., K.A. Kruger, and J.V. Setzler. 1981. The concentrations and transport of pesticides in northwestern Ohio rivers—1981. Report on contract CACW 49-81-C-0028. Tech. Rep. Ser. 20. Lake Erie Wastewater Management Study. U.S. Army Corps of Engineers, Buffalo, NY.

Baker, J.L., and J.M. Laflen. 1979. Runoff losses of surface applied herbicides as affected by wheel tracks and incorporation. J. Environ. Qual. 8(4):602–607.

Baker, J.L., J.M. Laflen, and H.P. Johnson. 1978. Effect of tillage systems on runoff losses of pesticides. A rainfall simulation study. Trans. ASAE 21:886–892.

Baker, J.L., and H.P. Johnson. 1979. The effect of tillage systems on pesticides in runoff from small watersheds. Trans. ASAE 22:554–559.

Baker, J.L., J.M. Laflen, and R.O. Hartwig. 1982. Effects of corn residue and herbicide placement on herbicide runoff losses. Trans. ASAE 25:340–343.

Baldwin, F.L., P.W. Santelmann, and J.M. Davidson. 1975. Movement of fluometuron across and through the soil. J. Environ. Qual. 4:191–194.

Banks, P.A., and E.L. Robinson. 1982. The influence of straw mulch on the soil reception and persistence of metribuzin. Weed Sci. 30:164–168.

Barnett, A.P., E.W. Hauser, A.W. White, and J.H. Holladay. 1967. Loss of 2,4-D in washoff from cultivated fallow land. Weeds 15:133–137.

Bouchard, D.C., T.L. Lavy, and E.R. Lawson. 1985. Mobility and persistence of hexazinone in a forest watershed. J. Environ. Qual. 14:229–233.

Bovey, R.W., E. Burnett, C. Richardson, M.G. Merkle, J.R. Baur, and W.G. Knisel. 1974. Occurrence of 2,4,5-T and picloram in surface runoff water in the Blacklands of Texas. J. Environ. Qual. 3:61–64.

Bradley, J.R., Jr., T.J. Sheets, and M.D. Jackson. 1972. DDT and toxaphene movement in surface water from cotton plots. J. Environ. Qual. 1:102–105.

Braun, H.E., and R. Frank. 1980. Organochlorine and organophosphorus insecticides. Their use in eleven agricultural watersheds and their loss to stream waters in southern Ontario, Canada 1975-1977. Sci. Total Environ. 15:169–192.

Bruce, R.R., L.A. Harper, R.A. Leonard, W.M. Snyder, and A.W. Thomas. 1975. A model for runoff of pesticides from small upland watersheds. J. Environ. Qual. 4:541–548.

Burt, J.P., A.D. Nicks, and G.A. Gander. 1984. Development of a field planning guide for the Mississippi Delta using CREAMS. *In* Proc. of Winter Meet., New Orleans, LA. 11–14 Dec. Paper 84-2636. ASAE, St. Joseph, MI.

Caro, J.H. 1976. Pesticides in agricultural runoff. p. 91–119. *In* B.A. Steward (ed.) Control of water pollution from cropland. Vol. 2. An overview. USEPA EPA-600/2-75-026b or USDA ARS-H-5-2. U.S. Gov. Print. Office, Washington, DC.

Carrol, B.R., G.H. Willis, and J.B. Graves. 1981. Permethrin concentration on cotton plants, persistence in soil, and loss in runoff. J. Environ. Qual. 10:497–500.

Carsel, R.F., C.N. Smith, L.A. Mulkey, J.D. Dean, and P. Jowise. 1984. Users' manual for the pesticide root zone model (PRZM) Release 1. USEPA EPA-600/3-84-109. U.S. Gov. Print. Office, Washington, DC.

Carter, G.E., Jr., and M.B. Riley. 1981. 1,2,dibromo-3-chloropropane residues in water in South Carolina, 1979-1980. Pestic. Monit. J. 15:139–142.

Cohen, J.M., and C. Pinkerton. 1966. Widespread translocation of pesticides by air transport and rainout. p. 163–176. *In* R.F. Gould (ed.) Organic pesticides in the environment. Adv. Chem. Ser. 60. Am. Chem. Soc., Washington, DC.

Cohen, M.L., and W.D. Steinmetz. 1986. Foliar washoff of pesticides by rainfall. Environ. Sci. Technol. 20:521–523.

Computer Sciences Corporation. 1980. Pesticide runoff simulator, user's manual. USEPA, Falls Church, VA.

Council for Agricultural Science and Technology 1985. Agriculture and groundwater quality. Rep. 103 (ISSN b194-4088) CAST, Ames, IA.

Crawford, N.H., and A.S. Donigian. 1973. Pesticide transport and runoff model for agricultural lands. USEPA EPA-660/2-74-013. U.S. Gov. Print. Office, Washington, DC.

Crawford, N.H., and R.K. Linsley. 1966. Digital simulation in hydrology: Stanford watershed model IV. Stanford Univ. Tech. Rep. 39.

Dale, J.E. 1981. Control of Johnsongrass (*Sorghum halepense*) and volunteer corn (*Zea mays*) in soybeans (*Glycine max*). Weed Sci. 29:708–711.

Davidson, J.M., G.H. Brusewitz, D.R.Baker, and A.L. Wood. 1975. Use of soil parameters for describing pesticide movement through soils. USEPA EPA-660/2-75-009. U.S. Gov. Print. Office, Washington, DC.

Davis, E.A., P.A. Ingebo, and P.C. Page. 1969. Effect of a watershed treatment with picloram on water quality. U.S. Forest Serv., Rocky Mountain For. Range Exp. Stn. Res. Notes RM-100.

DeCoursey, D.G. 1982. ARS small watershed model. p. 1–33. *In* Proc. Summer Meet. Madison, WI. 14–18 June. Paper No. 82-2094. ASAE, St. Joseph, MI.

DelVecchio, J.R., and W.G. Knisel. 1982. Application of a field-scale nonpoint pollution model. p. 227–236. *In* E.G. Kruse et al. (ed.) Proc. Am. Soc. Civil Eng., Irrigation and Drainage Specialty Conf., Orlando, FL. 20–23 July. ASCE, New York.

DelVecchio, J.R., W.G. Knisel, and V.A. Ferreira. 1983. The impact of irrigation on pollutant loads. p. 113–123. *In* J. Borrelli et al. (ed.) Proc. Am. Soc. Civil Eng., Irrigation and Drainage Specialty Conf., Jackson, WY. 20–22 July. ASCE, New York.

Donigian, A.S., D.C. Beyerlein, H.H. Davis, and N.H. Crawford. 1977. Agricultural runoff management model, Version II: Refinement and testing. USEPA EPA-600/3-77-098. U.S. Gov. Print. Office, Washington, DC.

Donigian, A.S., Jr., and N.H. Crawford. 1976. Modeling pesticides and nutrients on agricultural lands. USEPA EPA-600/2-7-76-043. U.S. Gov. Print. Office, Washington, DC.

Donigian, A.S., Jr., J.C. Imhoff, B.R. Bichnell. 1983. Predicting water quality resulting from agricultural nonpoint source pollution via simulation—HSPF. p. 200–249. *In* F.W. Schaller and G.W. Bailey (ed.) Agricultural management and water quality. Iowa State Univ. Press, Ames.

Dowler, C.C., W.A. Rohde, L.E. Fetzer, D.E. Scott, T.E. Sklaney, and C.W. Swann. 1982. The effect of sprinkler irrigation on herbicide efficacy, distribution, and penetration in some coastal plain soils. Georgia Agric. Exp. Stn. Res. Bull. 281.

Dudley, D.R., and J.R. Karr. 1980. Pesticides and polychlorinated biphenyl residues in the Black Creek watershed, Allen County, Indiana—1977–1978. Pestic. Monit. J. 13:155–157.

Eccles, L.A. 1979. Pesticide residues in agricultural drains, southeastern desert area, California. U.S. Geol. Surv. Water Resourc. Investigation 79-16. U.S. Gov. Print. Office, Washington, DC.

Edwards, W.M., C.G. Triplett, Jr., and R.M. Kramer. 1980. A watershed study of glyphosate transport in runoff. J. Environ. Qual. 9:661–665.

Forney, D.R., and D.E. Davis. 1981. Effect of low concentrations of herbicides on submerged aquatic plants. Weed Sci. 29:677–685.

Foster, G.R., L.J. Lane, J.D. Nowlin, J.M. Laflen, and R.A. Young. 1980. A model to estimate sediment yield from field sized areas: Development of model. Vol. 1. p. 36–64. *In* W.G. Knisel (ed.) CREAMS: A field size model for chemicals, runoff, and erosion from agricultural management systems. USDA Conserv. Res. Rep. 26. U.S. Gov. Print. Office, Washington, DC.

Foster, G.R., D.L. Meyer, and C.A. Onstad. 1977. A runoff erosivity factor and variable slope length exponents for soil loss estimates. Trans. ASAE 20:683–687.

Frank, R., H.E. Braun, M.V. Holdrinet, G.J. Sirons, and B.D. Ripley. 1982. Agriculture and water quality in the Canadian Great Lakes Basin 5. Pesticide use in eleven agricultural watersheds and presence in stream water—1975–1977. J. Environ. Qual. 11:497–505.

Frank, R., and G.J. Sirons. 1979. Atrazine—its use in corn production and its loss to stream waters in southern Ontario, Canada—1975–1977. Sci. Total Environ. 12:223–240.

Frere, M.H., C.A. Onstad, and H.N. Holtan. 1975. ACTMO: An agricultural chemical transport model. USDA-ARS ARS-H-3. U.S. Gov. Print. Office, Washington, DC.

Gaynor, J.D., and V.V. Volk. 1981. Runoff losses of atrazine and terbutryne from unlimed and limed soil. Environ. Sci. Technol. 15:440–443.

Glotfelty, D.E., A.W. Taylor, A.R. Isensee, J. Jersey, and S. Glenn. 1984. Atrazine and simazine movement to Wye River estuary. J. Environ. Qual. 13:115–121.

Green, R.E., J.M. Davidson, and J.W. Biggar. 1980. An assessment of methods for determining adsorption-desorption of organic chemicals. p. 73–82. *In* A. Banin and U. Kafafi (ed.) Agrochemicals in soils. Pergamon Press, New York.

Haith, D.A. 1980. A mathematical model for estimating pesticide losses in runoff. J. Environ. Qual. 9:428–433.

Haith, D.A. 1987. Extreme event analysis of pesticide loads to surface waters. J. Water Pollut. Control Fed. 59:284–288.

Haith, D.A., and L.J. Tubbs. 1981. Operational methods for analysis of agricultural nonpoint source pollution. Search: Agriculture. Cornell Univ., Ithaca, NY.

Hall, J.K. 1974. Erosional losses of *s*-triazine herbicides. J. Environ. Qual. 3:174–180.

Hall, J.K., N.L. Hartwig, and L.D. Hoffman. 1984. Cyanazine losses in runoff from no-tillage corn in "living" and dead mulches vs. unmulched conventional tillage (herbicide, *Zea mays*). J. Environ. Qual. 13:105–110.

Hicks, D.W., L.E. Asmussen, and H.F. Perkins. 1987. Soil and geohydrologic relations on a southern Coastal Plain watershed. p. 27. *In* Agronomy abstract. ASA, Madison, WI.

Holtan, H.W., and N.C. Lopez. 1971. USDAHL-70 model of watershed hydrology. USDA Tech. Bull. 1435. U.S. Gov. Print. Office, Washington, DC.

Huff, D.D., and P. Kruger. 1967. The chemical and physical parameters in a hydrologic transport model for radioactive aerosols. Proc. Int. Hydrol. Symp. 1:128–135.

Ingram, J.J. 1979. Chemical transfer from a saturated soil into overland flow. M.S. thesis. Colorado State Univ., Ft. Collins.

Ingram, J.J., and D.A. Woolhiser. 1980. Chemical transfer into overland flow. Vol. 1. p. 40–53. *In* Proc. Symp. Watershed Management, Boise, ID. 21–23 July. ASCE, Boise, ID.

Johnsen, T.N., Jr. 1980. Picloram in water and soil from a semiarid pinyon-juniper watershed. J. Environ. Qual. 9:601–605.

Johnson, W.R., F.T. Ittihadiek, K.R. Craig, and A.F. Pillsbury. 1967. Insecticides in tile drainage effluents. Water Resour. Res. 3:525–537.

Kemp, W.M., J.C. Means, T.W. Jones, and J.C. Stevenson. 1982. Herbicides in Chesapeake Bay and their effect on submerged aquatic vegetation. p. 503–567. *In* Chesapeake Bay program technical studies: A synthesis, Part IV. U.S. Gov. Print. Office, Washington, DC.

Knisel, W.G. (ed.). 1980. CREAMS: A field-scale model for chemicals, runoff, and erosion from agricultural management systems. USDA-SEA Conserv. Res. Rep. 26. U.S. Gov. Print. Office, Washington, DC.

Knisel, W.G. 1985. Use of computer models in managing nonpoint pollution from agriculture. p. 1–18, section KV. *In* V. Novotny (ed.) Nonpoint Pollution Abatement Symp. 23–25 Apr. Marquette Univ., Milwaukee, WI.

Knisel, W.G., and R.W. Baird. 1969. Runoff volume prediction using daily climatic data. Water Resour. Res. 5:84–89.

Knisel, W.G., G.R. Foster, and R.A. Leonard. 1983. CREAMS: A system for evaluating management practices. p. 178–199. *In* F.W. Schaller and G.W. Bailey (ed.) Agricultural management and water quality. Iowa State Univ. Press, Ames.

Knisel, W.G., and R.A. Leonard. 1986. Impact of irrigation on groundwater quality in humid areas. Water Forum 86. p. 1508–1515. *In* World Water Issues in Evolution. Proc. of ASCE Spec. Conf., Long Beach CA. 4–6 Aug. Am. Soc. of Civil Eng., Boise, ID.

Knisel, W.G., and R.A. Leonard. 1989. Representative climatic record for pesticide runoff and leaching simulations. J. Environ. Health Sci., Part B. Pesticides, Food contaminants, and agricultural wastes. Marcel Dekker, New York. (In press.)

Kramer, L.A. 1984. Seasonal runoff and soil loss from conventional and conservation tilled corn. Winter Meet., New Orleans, LA. Paper 84-2554. ASAE, St. Joseph, MI.

Langdale, G.W., and R.A. Leonard. 1983. Nutrient and sediment losses associated with conventional and reduced tillage agricultural practices. p. 457–467. *In* R. Lowrance et al. (ed.) Nutrient cycling in agricultural ecosystems. Spec. Publ. 23. Univ. of Georgia College of Agriculture, Athens.

Lenardon, A.M., M.I.M. Hevia, J.A. Fuse, C.B. DeNochetto, and P.J. Depetris. 1984. Organochlorine and organophosphorus pesticides in the Paran River, Argentina. Sci. Total Environ. 34:289–298.

Leonard, R.A. 1988. Herbicides in surface waters. p. 45–89. *In* R. Grover (ed.) Environmental chemistry of herbicides. CRC Publ. Co., Boca Raton, FL.

Leonard, R.A., G.W. Bailey, and R.R. Swank, Jr. 1976. Transport, detoxification, fate, and effects of pesticides in soil, water, and aquatic environments. p. 48–78. *In* Land application of waste material. Soil Conserv. Soc. Am., Ankeny, IA.

Leonard, R.A., and W.G. Knisel. 1986. Model selection for nonpoint source pollution and resource conservation. p. 213–229. *In* A. Giorgini and F. Zingales (ed.) Developments in environmental modelling, 10. Agricultural nonpoint source pollution: Model selection and application. Elsevier, Amsterdam.

Leonard, R.A., and W.G. Knisel. 1989. Can pesticide transport models be validated using field data: Now and in the future? J. Environ. Health Sci., Part B. Pesticides, food contaminants, and agricultural wastes. Marcel Dekker, New York. (In press.)

Leonard, R.A., W.G. Knisel, and D.A. Still. 1987. GLEAMS: Groundwater loading effects of agricultural management systems. Trans. ASAE 30:1403–1418.

Leonard, R.A., and J.D. Nowlin. 1980. The pesticide submodel. p. 304–329. *In* W.G. Knisel (ed.) User's manual. Vol. 2. CREAMS: A field scale model for chemicals, runoff, and erosion from agricultural management systems. USDA Conserv. Res. Rep. 26. U.S. Gov. Print. Office, Washington, DC.

Leonard, R.A., G.W. Langdale, and W.G. Fleming. 1979. Herbicide runoff from upland piedmont watersheds—data and implications for modeling pesticide transport. J. Environ. Qual. 8:223–229.

Leonard, R.A., and R.D. Wauchope. 1980. The pesticide submodel. p. 88–112. *In* W.G. Knisel (ed.) Model documentation. Vol. 1. CREAMS: A field scale model for chemicals, runoff, and erosion from agricultural management systems. USDA Conserv. Res. Rep. 26. U.S. Gov. Print. Office, Washington, DC.

Lichtenberg, J.J., J.W. Eichelberger, R.C. Dressman, and J.E. Longbottom. 1970. Pesticides in surface waters of the United States—a 5-year summary, 1964–1968. Pestic. Monit. J. 4:71–86.

Lorber, M.N., and L.A. Mulkey. 982. An evaluation of three pesticide runoff loading models. J. Environ. Qual. 11:519–529.

Martin, C.D., J.L. Baker, D.C. Erbach, and H.P. Johnson. 1978. Washoff of herbicides applied to corn residue. Trans. ASAE 21:1164–1168.

Mayeux, H.S., Jr., C.W. Richardson, C.W. Bovey, R.W. Burnett, and E. Merkle. 1984. Dissipation of picloram in storm runoff. J. Environ. Qual. 13:44–49.

McDowell, L.L., and E.H. Grissingir. 1976. Erosion and water quality. p. 41–56. *In* D.A. Fletcher (ed.) Proc. 23rd Natl. Watershed Congr. Natl. Assoc. Conserv. Districts, Washington, DC.

McDowell, L.L., G.H. Willis, C.E. Murphree, L.M. Southwick, and S. Smith. 1981. Toxaphene and sediment yields in runoff from a Mississippi USA delta watershed. J. Environ. Qual. 10:120–125.

McDowell, L.L., G.H. Willis, L.M. Southwick, and S. Smith. 1984. Methyl parathion and EPN washoff from cotton plants by simulated rainfall. Environ. Sci. Technol. 18:423–427.

Menzel, R.G. 1980. Enrichment ratios for water quality modeling. p. 486–492. *In* W.G. Knisel (ed.) Supporting documentation. Vol. 3. CREAMS: A field scale model for chemical, runoff, and erosion from agricultural management systems. USDA Conserv. Res. Rep. 26. U.S. Gov. Print. Office, Washington, DC.

Merkle, M.G., and R.W. Bovey. 1974. Movement of pesticides in surface water. p. 99–106. *In* W.D. Guenzi (ed.) Pesticides in soil and water. SSSA, Madison, WI.

Meyer, L.D., G.R. Foster, and M.J.M. Romkens. 1975. Source of soil eroded by water from upland slopes. p. 177–189. *In* A.R. Robinson (ed.) Present and prospective technology for predicting sediment yields and sources. ARS-S-40. USDA-ARS, U.S. Gov. Print. Office, Washington, DC.

Mills, W.C., and R.A. Leonard. 1985. Pesticide pollution probabilities. Trans. ASAE 27:1704–1710.

Mulkey, L.A., and J.W. Falco. 1983. Methodology for predicting exposure and fate of pesticides in aquatic environments. p. 250–266. *In* F.W. Schaller and G.W. Bailey. Agricultural management and water quality. Iowa State Univ. Press, Ames.

Nash, R.G. 1980. Dissipation rate of pesticides from soils. p. 560–594. *In* W.G. Knisel (ed.) Supporting documentation. Vol. 3. CREAMS: A field scale model for chemical, runoff, and erosion from agricultural management systems. USDA Conserv. Res. Rep. 26. U.S. Gov. Print. Office, Washington, DC.

National Research Council. 1986. Pesticides and groundwater quality. Issues and problems in four states. Natl. Acad. Press, Washington, DC.

Neary, D.G. 1983. Monitoring herbicide residues in spring flow after an operational application of hexazinone. South. J. Appl. For. 7:217–223.

Neary, D.G., P.B. Bush, and J.E. Douglass. 1983. Off-site movement of hexazinone in stormflow and baseflow from forested watersheds. Weed Sci. 31:543–551.

Neary, D.G., P.B. Bush, J.E. Douglass, and R.L. Todd. 1985. Picloram movement in an Appalachian hardwood forest watershed. J. Environ. Qual. 14:585–592.

Neary, D.G., J.E. Douglass, and W. Fox. 1979. Low picloram concentrations in streamflow resulting from forest application of Tordon 10k. Proc. So. Weed Sci. Soc. 32:182–197.

Negev, M.A. 1967. A sediment model on a digital computer. Stanford Univ. Tech. Rep. 76.

Newton, M.A., and J.A. Norgren. 1977. Silvicultural chemicals and protection of water quality. U.S. Gov. Print. Office, Washington, DC.

Nicholaichuk, W., and R. Grover. 1983. Loss of fall applied 2,4-D in spring runoff from a small agricultural watershed. J. Environ.Qual. 12:412–414.

Nicholson, H.P., A.R. Grzenda, G.J. Lauer, W.S. Cox, and J.I. Teasley. 1964. Water pollution by insecticides in an agricultural river basin, Part 1. Occurrences of insecticides in river and treated municipal water. Limnol. Ocean. 9:310–317.

Nicholson, H.P., A.R. Grzenda, and J.T. Teasley. 1966. Water pollution by insecticides. A six and one-half year study of a watershed. Proc. Symp. Agric. Waste Water. Water Resour. Cent., Univ. Calif., Rep. 10:132–141.

Nicks, A.D. 1974. Stochastic generation of the occurrence, pattern, and location of maximum amount of daily rainfall. p. 154–171. *In* Proc. of Symp. on Statistical Hydrology. USDA-ARS Misc. Publ. 1275. U.S. Gov. Print. Office, Washington, DC.

Nicks, A.D., J.B. Burt, W.G. Knisel, and G.A. Gander. 1984. Application of a field-scale water quality model for evaluation of watershed land treatment project. p. 1–33. *In* Proc. of Winter Meet., New Orleans, LA. Dec. 11–14. Paper no. 84-2633. ASAE, St. Joseph, MI.

Norris, L.A. 1969. Herbicide runoff from forest lands sprayed in summer. p. 24–26. *In* Research program report. Western Soc. Weed Sci., Las Vegas.

Norris, L.A., M.L. Montgomery, L.E. Warren, and W.D. Mosher. 1982. Brush control with herbicides on hill pasture sites in southern Oregon. J. Range Manage. 35:75–80.

Novotny, V., and G. Chesters. 1981. Handbook of nonpoint pollution: Sources and management. van Nostrand Reinhold Co., New York.

Nutter, W.L., T. Tkaus, P.B. Bush, and D.G. Neary. 1984. Simulation of herbicide concentrations in stormflow from forested watersheds. Water Resourc. Bull. 20:851–857.

Pionke, H.B. 1977. Form and sediment associations of nutrients (C, N, and P) and pesticides. p. 199–216. *In* H. Shear and A.E.P. Watson (ed.) The fluvial transport of sediment-associated nutrients and contaminants. Workshop Proc. 22–25 Oct. 1976. Int. Joint Comm., Kitchener, ON.

Pionke, H.B., and G. Chesters. 1973. Pesticide-sediment-water interactions. J. Environ. Qual. 2:29–45.

Rao, P.S.C., and J.M. Davidson (ed.). 1980. Retention and transformation of selected pesticides and phosphorus in soil-water systems: A critical review. USEPA EPA-600/3-82-060. NTIS PB82, 256, 884. U.S. Gov. Print. Office, Washington, DC.

Rao, P.S.C., J.M. Davidson, and L.C. Hammond. 1976. Estimation of nonreactive and reactive solute front locations in soils. *In* Proc. of Hazardous Waste Res. Symp., Tucson. USEPA EPA-600/9-76-015. U.S. Gov. Print. Office, Washington, DC.

Rao, P.S.C., P. Nkedi-Kizza, J.M. Davidson, and L.T. Ou. 1983. Retention and transformation of pesticides in relation to nonpoint source pollution for croplands. p. 126–140. *In* F.W. Schaller and G.W. Bailey (ed.) Agricultural management and water quality. Iowa State Univ. Press, Ames.

Rawls, W.J., and D.L. Brakensiek. 1982. Estimating soil water retention from soil properties. Proc. Am. Soc. Civ. Eng. 108(IR2):161–171.

Ritter, W.F., H.P. Johnson, W.G. Lovely, and M. Molnau. 1974. Atrazine, propachlor, and diazinon residues on small agricultural watersheds. Environ. Sci. Technol. 8:37–42.

Rohde, W.A., L.A. Asmussen, E.W. Hauser, and A.W. Johnson. 1979. Concentrations of ethoprop in the soil and runoff water of a small agricultural watershed. USDA-SEA Agric. Res. Results ARR-S-2. U.S. Gov. Print. Office, Washington, DC.

Rohde, W.A., L.E. Asmussen, E.W. Hauser, R.D. Wauchope, and H.D. Allison. 1980. Trifluralin movement in runoff from a small agricultural watershed. J. Environ. Qual. 9:37–42.

Schafer, M.L., J.T. Peeler, W.S. Gardner, and J.E. Campbell. 1969. Pesticides in drinking water. Waters from the Mississippi and Missouri Rivers. Environ. Sci. Technol. 3:1261–1269.

Schepers, J.S., E.J. Vavricka, D.R. Anderson, H.D. Wittmuss, and G.E. Schuman. 1980. Agricultural runoff during a drought period. J. Water Pollut. Control Fed. 52:711–719.

Schulze, J.A., D.B. Manigold, and F.L. Andrews. 1973. Pesticides in selected western streams—1968–1971. Pestic. Monit. J. 7:73–84.

Sharpley, A.N. 1985a. Depth of surface soil-runoff interaction as affected by rainfall, soil slope, and management. Soil Sci. Soc. Am. J. 49:1010–1015.

Sharpley, A.N. 1985b. The selective erosion of plant nutrients in runoff. Soil Sci. Soc. Am. J. 49:1527–1534.

Sharpley, A.N., L.R. Ahuja, and R.G. Menzel. 1981. The release of soil phosphorus to runoff in relation to the kinetics of desorption. J. Environ. Qual. 10:386–391.

Skaggs, R.W., and R. Khaleel. 1982. Infiltration. p. 121–166. *In* C.T. Haan et al. (ed.) Hydrologic modeling of small watersheds. ASAE, St. Joseph, MI.

Smith, C.N., R.A. Leonard, G.W. Langdale, and G.W. Bailey. 1978. Transport of agricultural chemicals from small upland Piedmont watersheds. USEPA EPA-600/3-78-056. U.S. Gov. Print. Office, Washington, DC.

Smith, S., T.E. Reagan, J.L. Flynn, and G.H. Willis. 1983. Azinphos-methyl and fenvalerate runoff loss from a sugarcane saccharum-officinarum and insect integrated pest management system. J. Environ.Qual. 12:534–537.

Smith, C.N., W.R. Payne, Jr., L.A. Mulkey, J.E. Benner, R.S. Parrish, and M.C. Smith. 1981. The persistence and disappearance by washoff and dryfall of methoxychlor from soybeans. J. Environ. Sci. Health b16:777–794.

Spencer, W.F., M.M. Cliath, J. Blair, and R.A. LeMest. 1985. Transport of pesticides from irrigated fields in surface runoff and tile drain waters. USDA-ARS Conserv. Res. Rep. 31. U.S. Gov. Print. Office, Washington, DC.

Steenhuis, T.S. 1979. Simulation of the action of soil and water conservation practices in controlling pesticides. p. 106–146. *In* D.A. Haith and R.C. Loehr (ed.) Effectiveness of soil and water conservation practices for pollution control. USEPA EPA-600/3-79-106. U.S. Gov. Print. Office, Washington, DC.

Steenhuis, T.S., and M.R. Walter. 1980. Closed form solution for pesticide loss in runoff water. Trans. ASAE 23:615–620, 628.

Stewart, B.A., D.A. Woolhiser, W.H. Wischmeier, J.H. Caro, and M.H. Frere. 1975. Control of water pollution from cropland. Vol. 1. A manual for guideline development. USDA ARS-H-5-1 and USEPA EPA-600/2-75-026a. U.S. Gov. Print. Office, Washington, DC.

Suzuki, M., Y. Yamato, and T. Akiyama. 1978. Fate of herbicide 2,4,6-trichlorophenyl-4-nitrophenyl ether in rivers and agricultural drainages. Water Res. 12:777–782.

Trichell, D.W., H.L. Morton, and M.G. Merkle. 1968. Loss of herbicides in runoff water. Weed Sci. 16:447–449.

Triplett, G.B., Jr., B.J. Conner, and W.M. Edwards. 1978. Transport of atrazine and simazine in runoff from conventional and no-tillage corn. J. Environ. Qual. 7:77–84.

U.S. Department of Agriculture-Soil Conservation Service. 1972. Natl. Eng. Handb., Sec. 4. Hydrology. U.S. Gov. Print. Office, Washington, DC.

Wagenet, R.J. 1987. Processes influencing pesticide loss with water under conservation tillage. p. 189–204. *In* T.J. Logan et al. (ed.) Effects of conservation tillage on groundwater quality, nitrates and pesticides. Lewis Publ., Chelsea, MI.

Wauchope, R.D. 1987a. Effects of conservation tillage on pesticide loss with water. p. 203–215. *In* T.J. Logan et al. (ed.) Effects of conservation tillage on groundwater quality, nitrates and pesticides. Lewis Publ., Chelsea, MI.

Wauchope, R.D. 1987b. Tilted-bed simulations of erosion and chemical runoff from agricultural fields: II. Effects of formulation on atrazine runoff. J. Environ. Qual. 16:212–216.

Wauchope, R.D. 1978. The pesticide content of surface water drainage from agricultural fields: A review. J. Environ. qual. 7:459–472.

Wauchope, R.D., and R.A. Leonard. 1980a. Maximum pesticide concentrations in agricultural runoff. A semiempirical prediction formula. J. Environ. Qual. 9:665–672.

Wauchope, R.D., and R.A. Leonard. 1980b. Pesticide concentration in agricultural runoff: Available data and an approximation formula. p. 546–559. *In* W.G. Knisel (ed.) Supporting documentation. Vol. III. CREAMS: A field scale model for chemical, runoff, and erosion from agricultural management systems. USDA Conserv. Res. Rep. 26. U.S. Gov. Print. Office, Washington, DC.

Wauchope, R.D., L.L. McDowell, and L.J. Hagen. 1985. Environmental effects of limited tillage. p. 266–281. *In* A.F. Wiese (ed.) Weed control in limited tillage systems. Weed Sci. Soc. Am., Champaign, IL.

Wauchope, R.D., K.E. Savage, and D.G. DeCoursey. 1977. Measurement of herbicides in runoff from agricultural areas. p. 49–58. *In* B. Truelove (ed.) Research methods in weed science. 2nd ed. So. Weed Sci. Soc., Champaign, IL.

Wauchope, R.D., and A.N. Sharpley. 1984. Nonpoint pollution of surface waters by chemicals: Kinetic aspects of desorption of pollutants by runoff water. p. 47. *In* Weed Science Society of America abstracts, Weed Sci. Soc. of Am., Champaign, IL.

Wheeler, W.B., R.S. Mansell, D.V. Calvert, and E.H. Stewart. 1978. Movement of 2,4-D in drainage waters from a citrus grove in a Florida USA flatwood soil. Proc. Soil Crop Sci. Soc. Fla 37:180–183.

White, A.W., L.E. Asmussen, E.W. Hauser, and J.W. Turnbull. 1976. Loss of 2,4-D in runoff from plots receiving simulated rainfall and from a small agricultural watershed. J. Environ. Qual. 5:487–490.

White, A.W., A.D. Barnett, B.G. Wright, and J.H. Holladay. 1967. Atrazine losses from fallow land caused by runoff and erosion. Environ. Sci. Technol. 1:740–744.

Wiese, A.F., E.K. Savage, J.M. Chandler, L.C. Liu, L.S. Jeffrey, J.B. Weber, and K.S. LaFleur. 1980. Loss of fluometuron in runoff water. J. Environ. Qual. 9:1–5.

Williams, J.R. 1975. Sediment yield predictions with universal equation using runoff energy factor. p. 244–252. *In* Present and prospective technology for predicting sediment yields and sources. USDA-ARS, U.S. Gov. Print. Office, Washington, DC.

Williams, J.R., and R.W. Hann. 1978. Optimal operation of large agricultural watersheds with water quality constraints. Texas Water Resourc. Inst. Tech. Rep. 96.

Williams, J.R., A.D. Nicks, and J.G. Arnold. 1985. Simulation for water resources in rural basins. ASCE J. Hydraul. 111:970–986.

Willis, G.H., W.F. Spencer, and L.L. McDowell. 1980. The interception of applied pesticide by foliage and their persistence and washoff potential. p. 595–606. *In* W.G. Knisel (ed.) Supporting documentation. Vol. 3. CREAMS: A field scale model for chemicals, runoff, and erosion from agricultural management systems. USDA Conserv. Res. Rep. 26. U.S. Gov. Print. Office, Washington, DC.

Willis, G.H., L.L. McDowell, C.E. Murphree, L.M. Southwick, and M. Smith. 1983. Pesticide concentrations and yields in runoff from silty soils in the lower Mississippi Valley, USA. J. Agric. Food Chem. 31(6):1171–1177.

Wischmeier, W.H., and D.D. Smith. 1978. Predicting rainfall erosion losses—A guide to conservation planning. USDA Agric. Handb. 537. U.S. Gov. Print. Office, Washington, DC.

Wu, T.L. 1980. Dissipation of the herbicides atrazine and alachlor in a Maryland USA corn (*Zea mays*) field. J. Environ. Qual. 9:459–465.

Wu, T.L., D.L. Correll, and H.E.H. Remenapp. 1983. Herbicide runoff from experimental watersheds. J. Environ. Qual. 12:330–336.

Zahnow, E.W., and J.D. Riggleman. 1980. Search for linuron residues in tributaries of the Chesapeake Bay, USA. J. Agric. Food Chem. 28:974–978.

# Chapter 10

# Modeling Pesticide Fate in Soils[1]

**R. J. WAGENET,** *Cornell University, Ithaca, New York*

**P. S. C. RAO,** *University of Florida, Gainesville, Florida*

Predicting the fate of pesticides released into the environment is necessary to anticipate, and thereby minimize, adverse impacts away from the point of application. This means that we must understand what happens to a pesticide once it has been applied to an agricultural field, and we must be able to forecast its behavior in the environment. Using this information, the probable adverse impacts on the environment (e.g., groundwater and surface water contamination) and on human health can then be estimated.

A variety of physical, chemical, and biological processes determine the environmental fate of pesticides. The rates at which these processes operate will determine the mobility and the persistence of a pesticide. The rates and pathways of each of the processes, in turn, are modified to a considerable extent by various environmental factors as well as the properties of the pesticide itself. Microbiologists, chemists, soil scientists, and hydrologists study these processes in isolation. A more holistic approach is required, however, to understand and forecast the interactions among the processes and their integrated impact on the environmental behavior of pesticides.

## 10-1 DEVELOPMENT OF SIMULATION MODELS

The development of simulation models for forecasting pesticide behavior is an attractive way of evaluating solutions to some agricultural and environmental problems. For example, guidance is needed in determining which pesticides should be used for a particular soil-crop-weather combination and to estimate the rate and timing of pesticide application such that crop protection is maximized and adverse environmental impacts are minimized. Similarly, we need to identify pesticides that should not be introduced because of their high potential to contaminate a specific environmental compartment (e.g., groundwater).

[1] Joint contribution from Cornell Univ. and the Univ. of Florida. Approved for publication as New York Agric. Exp. Stn. Journal Paper 1700.

Processes identified as being the primary determinants of pesticide behavior in soils include: (i) degradation of a pesticide by soil microorganisms that use it as a substrate for their growth and maintenance; (ii) chemical degradation, a process that results from the participation of a pesticide in common chemical reactions (e.g., hydrolysis) that can occur in the aqueous phase; (iii) pesticide sorption by the mineral and organic constituents of the solid matrix, which leads to reduced mobility of the pesticide; (iv) uptake of the pesticide by plant roots; (v) volatilization or loss of the pesticide by its evaporation (with or without water) from the soil; and (vi) the diluting effects of water flow processes that act to disperse and distribute the pesticide during its passage through the unsaturated zone towards groundwater. The interrelationships among these processes are depicted schematically in Fig. 10–1.

## 10–2 USES OF SIMULATION MODELS

During the last 10 yr, substantial advances have been made in the use of computers as tools to integrate the environmental processes for predicting pesticide behavior in soil-water-crop systems. Conceptual models, constructed by coupling mathematical representations of the environmental processes, are now being used (or are being considered for use) by the (USEPA) as well as several state regulatory agencies. These models permit an assessment of the time required for the soil-applied pesticide to be dissipated to some acceptable, regulated level before entering groundwater. Simulation models also provide guidance for design of remedial measures for contaminated soils and groundwater and for regulations required to control pesticide use.

Another use for simulation models is as a tool for predicting the mobility and persistence in soils of pesticides that are currently under development. It is hoped that by using a model to project the likely environmental impact of a new pesticide (or a new use of an existing pesticide) before its actual use, we can identify persistent and mobile pesticides that may pose significant risks to the environment and human health. Such a proactive regulatory use of simulation models to screen candidate pesticides is currently being evaluated; we will present here some of the approaches that are being used (section 10–9).

A third major use of pesticide-fate models is to assist farmers and growers in designing effective crop, soil, and chemical management strategies. The short-term economic benefit is the greatest when a maximum crop-yield response can be obtained from the minimum amount of pesticide application required for effective pest control. Such a use, however, must also ensure minimum adverse environmental impact. Models can also play a role in aiding farmers in resolving such apparent conflicting interests. Simulation models may be used to: (i) evaluate alternate management practices (e.g., rates and timings of pesticide applications); (ii) select among alternate pesticides that may have different environmental behavior but offer equivalent pest control; and (iii) design optimum water and chemical management in soils with a wide range in physical and chemical properties.

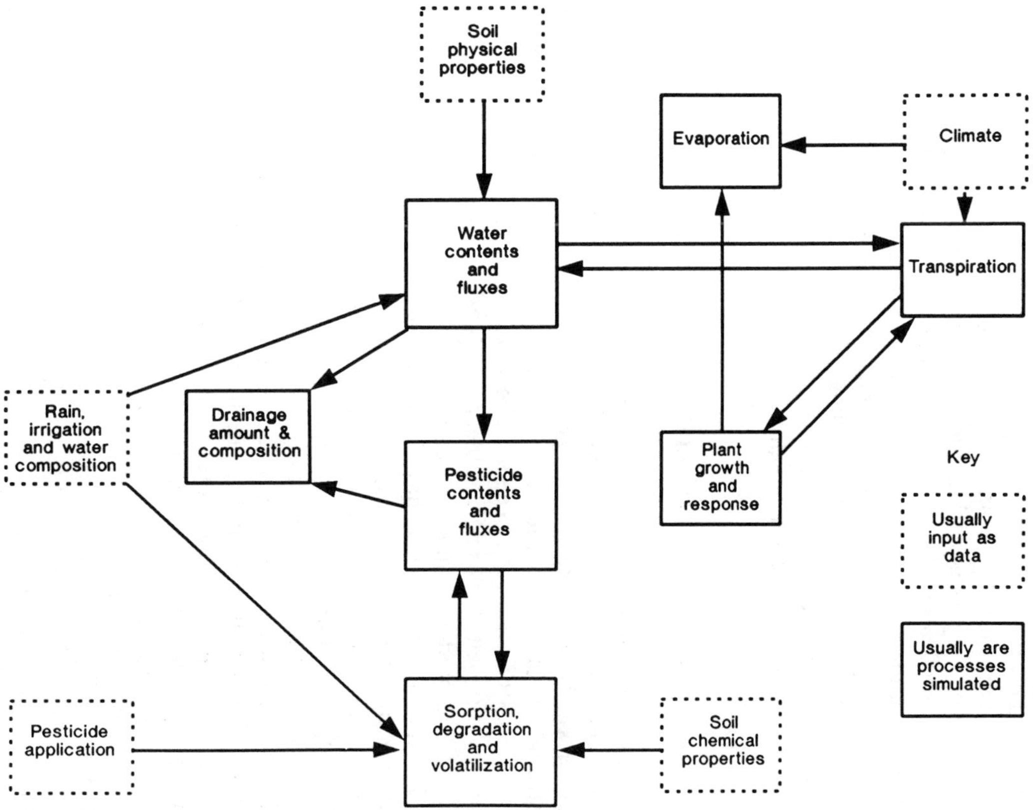


Fig. 10-1. Schematic representation of the interrelationships among the processes that influence pesticide fate in soils.

For the above reasons, modeling is increasingly being used as a tool for evaluating the fate of pesticides in soil-water systems. Sorption, leaching, degradation, volatilization, and other processes are being integrated through the use of simulation modeling techniques. Because of the extreme complexity of these systems, it is obvious that the use of simulation models will continue to be the most expeditious, reliable, and cost-effective means of integrating the various interacting processes that simultaneously determine pesticide fate in soils. For example, modeling may help to summarize and interpret herbicide efficacy trials and can provide the vehicle for transferring experimental results to unstudied situations, such as the probable environmental impact of an applied herbicide. However, proper development, testing, and responsible use of a modeling approach must be based upon a thorough, comprehensive understanding of interdependent and dynamic natural processes, and as we shall see in the following section, upon a clear definition of the purpose for which the model will be used.

## 10-3 TYPES OF SIMULATION MODELS

Recent years have seen a variety of approaches used to describe water and pesticide movement in field soils. Some new models have been proposed that vary widely in their conceptual approach and degree of complexity, and are strongly influenced by the environment, training, and biases of their developers. Few of them have been used beyond their initial development and testing by the scientists responsible for publishing them. In particular, pesticide-fate models have not often been used for management purposes, and, in general, the ones that have been used are either incompletely tested or have been developed for only limited cases.

One major limitation and constraint to the widespread use of pesticide-fate models is the apparent lack of recognition that different types of models are required for different purposes. The quantity of required input data, the level of detail at which the basic processes are described, and sensitivity and accuracy of model outputs all depend upon whether the modeler intends to approach the simulation from a research or management perspective. Lack of appreciation for different modeling approaches has produced confusion and disarray in efforts to apply models for problem-solving purposes. A short review of the range of models developed to describe water and solute movement in soil will place existing pesticide models into the proper perspective. A more detailed discussion of different types of pesticide-fate models will follow in subsequent sections.

A recent review of approaches used to predict solute leaching in soils (Addiscott & Wagenet, 1985) identified a number of research, management, and screening models that have been reported in the scientific literature. These are the categories into which the existing pesticide-fate models will be placed. Research models were identified as generally intended to provide quantitative estimates of water flow and pesticide behavior, but wih comprehensive and often substantial data demands regarding the system to be simulated.

Management models, while less data-intensive, were identified as commensurately less quantitative in their ability to predict water and solute movement under transient field conditions. There are only a few cases in which either type of model has been tested against field data, and little attention has been given to the use of the so-called *management models* for the actual purposes of managing pesticide applications to soil. A third model category, the screening model, was identified as an analytic solution intended for evaluation and comparison of pesticide behavior under constrained and limited conditions, but with commensurately few data requirements. This type of model is not intended for environmental fate studies, but instead categorizes chemicals into broad behavioral classes.

This chapter focuses on the strengths and weaknesses of the pesticide-fate models, presents a few examples of how they have been used, and suggests areas in which future research emphasis should be placed to improve these models.

## 10–4 OVERVIEW OF PESTICIDE FATE MODELS

Deterministic, mechanistic models of solute transport, based upon miscible displacement theory (Nielsen & Biggar, 1962), have been the most widely used transport modeling approaches in soil science since the 1960s. These models presume that soil-water flow and solute transport processes operate such that the occurrence of a given set of physical and chemical events leads to a uniquely definable water or solute distribution in the soil profile. Based on Darcy's law, water flow is assumed to be describable as the product of a hydrualic gradient and soil hydraulic conductivity, which varies with the soil-water content. It is assumed that physical convection (i.e., mass flow) and molecular diffusion combine to displace a solute in porous media. This type of solute transport model is summarized in the convection-dispersion equation (CDE), which has been derived in detail in some places (e.g., Kirkham & Powers, 1972; Wagenet, 1984), and solved analytically for a variety of initial and boundary conditions (reviewed by van Genuchten & Alves, 1982). These analytical solutions, representing particular research models of water flow and solute transport, have been extensively and successfully applied to the results obtained in carefully controlled laboratory studies in which water flow is at a constant rate (i.e., steady-state flow). As a result, the CDE is now a well-established solute modeling approach for such cases (van Genuchten & Cleary, 1979). The pesticide screening model BAM, discussed below, has evolved from the use of analytical solutions for describing solute behavior under carefully defined situations.

Transient field conditions, in which the movement of both water and solute vary with depth and time, require numerical (rather than analytical) solution of the CDE to properly represent the influence of changing soil-water contents and fluxes on solute concentrations and fluxes. Additionally, the transient water-flow equation (commonly referred to as the Richards equation) must be numerically solved to describe the water flow regime.

Research models constructed on such a basis are numerous, including those that simulate plant growth in response to transient water regimes (Nimah & Hanks, 1973), N movement and transformations (Watts & Hanks, 1978; Tillotson & Wagenet, 1982), the movement and chemical reactions of inorganic salts (Bresler, 1973; Childs & Hanks, 1975; Robbins et al., 1980), and disposal of wastewater (Iskander & Selim, 1981). The pesticide model Leaching Estimation and CHemistry Model (LEACHM) (Wagenet & Hutson, 1989) discussed in a later section, is similar in structure to these models and has evolved from such modeling experiences since the early 1970s.

In the absence of comprehensive and reliable management models, research models have been employed for management purposes with variable success. The deterministic nature of these models, coupled with their data-intensive demands, often make them cumbersome for use on a routine management basis. Additionally, most of these models employ numerical differencing or iterative solution techniques that until only recently have necessitated the use of mainframe computers, not the desktop microcomputers that are now readily available. Therefore, while these types of models are still used on a research basis, they are presently useful for management applications only if implemented on a mainframe computer or a computationally efficient microcomputer (e.g., those based on 80386 processors).

Recognizing both the constraints of deterministic models developed from analytical and numerical solutions of the CDE and the need to offer management guidance, at least two alternative paths have recently been followed. This separation of approach has been catalyzed by numerous observations of the spatial and temporal variability of water and solute transport at the field-scale. A variety of stochastic models (Bresler & Dagan, 1981; Dagan & Bresler, 1979; Jury et al., 1982; Knighton & Wagenet, 1987a, b) are based upon the assumption that processes in soil-water systems and their outcome are uncertain, and therefore definable only in statistical terms. These models include a description of spatially and temporally variable processes, and are relatively untested for any but a limited set of field conditions. These models focus on a particular case for predicting the mean values for fluxes of water and solutes, and the expected deviations from the mean values. Conceptuall, these models are a step in the right direction given the uncertainty in model parameter values arising from the spatial heterogeneity of field soils; practically, they have not yet been condensed to a form usable as routine management tools.

Some effort has also been expended in developing simpler deterministic representations useful in pesticide management (Nofziger & Hornsby, 1986, 1987; Pacenka & Steenhuis, 1984). These models often use simplified representations of the fundamental processes of transport, but result in a model that is computationally efficient and can be used without a substantial amount of input data. These models, developed for general management guidance or instructional purposes, are intended to illustrate basic principles and to demonstratethe interactions among the major processes affecting pesticide fate in soils. Given such objectives, these models require as inputs only approximate estimates of the system variables. Such models have not yet been proven to provide other than *qualitative* estimates of water and solute movement. They are strictly deterministic, although, when stochastically formulated, they may infact prove to be as good descriptors of field-scale fluxes of water and solutes as the stochastic research models are expected to be. As yet, the available simple, deterministic models have been used mainly for educational purposes, with management use delayed until they are tested and validated under field conditions.

Jury et al. (1983) have proposed screening models useful in estimating pesticide fate in soils. These models are based on analytical solutions of the CDE formulated to include the relevant pesticide-fate processes of sorption, degradation, leaching, and volatilization. These models can be used only to categorize the *relative behavior* of pesticides under the specific set of conditions and assumptions used in developing the analytical solution. The attractiveness of these models lies in the relatively few data demands needed to characterize pesticide behavior, and the relative ease of programming of an analytical solution. The screening models, however, have frequently been incorrectly used to estimate environmental fate under field conditions, a use for which they were never intended. When correctly used, however, screening models provide the opportunity to group pesticides into behavioral classes where their behavior in the field may be inferred from similarly classed, but better studied pesticides. Such an approach has obvious relevance in the pesticide registration process, as the screening models can be used to estimate the general transport and transformations of an unstudied pesticide.

There is currently substantial interest within the agricultural community in using models to guide the application of water and pesticides to soils and crops, and to predict pesticide fate in soils and groundwater. The wide availability of microcomputers and increased recognition (through educational programs) of the complexity of soil-plant-water systems has accelerated the demand for accurate, yet usable, management models of these systems.

It follows that the next step in modeling water flow and pesticide fate under field conditions is to condense the comprehensive descriptions provided by research models (both deterministic and stochastic) into management models useful in providing quantitative guidance under field conditions. One method of accomplishing this is to use a research model that has been constructed in such a manner that data inputs are both easily understood and not excessively demanding in terms of field measurements, and which has model output organized so that it can easily be interpreted. This is the philosophy behind the pesticide model Pesticide Root Zone Model (PRZM) (Carsel et al., 1984) and the LEACHM model (Wagenet & Hutson, 1989). Such models still suffer from being deterministic representations of a spatially and temporally variable system. However, it has been shown that use of such models in an ad hoc stochastic manner is possible, in which repeated executions of the model with different input values of possible field conditions can provide accurate description of the variability of water and pesticide in the field (Bresler et al., 1979; Wagenet & Rao, 1983; Wagenet & Hutson, 1989). Such use of numerical models, if carefully accomplished, offers the opportunity to use research models for management purposes, and also as tools in the design and interpretation of further field research. It is important to understand the fundamental assumptions and inherent strengths and weaknesses of such models.

## 10–5 PESTICIDE-FATE PROCESSES

### 10–5.1 Transport

Most of the process-based models that describe pesticide fate in the unsaturated zone have been developed from basic theory of the physical and chemical processes affecting solute transport in soils. As such, these models are deterministic, with some examples in the literature (Leistra, 1978, 1979; Lindstrom et al., 1968; Rao et al., 1979; van Genuchten et al., 1974; van Genuchten & Wierenga, 1976).

It is usually assumed that pesticide movement is a result of three processes: (i) pesticide diffusion in the aqueous phase along a solute-concentration gradient; (ii) diffusion in the gas phase in response to a gradient in gas-phase concentration, if the pesticide is volatile; and (iii) convection (mass flow) of the pesticide because of movement of the bulk fluid phase (water or gas) in which the pesticide is dissolved. Convective transport in the liquid phase results in a physical dispersion of the solute between domains of varying pore-water velocities. This process is termed *hydrodynamic dispersion* and is conceptually and mathematically combined with the expression for molecular diffusion (see Eq. [1]). Note that convective flux in the gas phase, resulting from pressure gradients, is ignored in most of the models. It has recently been considered in more detail by Wagenet et al. (1989), and found to, in some cases, be of consequence in the transport of extremely volatile chemicals. In addition to the transport processes, sources or sinks of the chemical may be present.

Expressed mathematically, the solute flux ($J$; $M/L^2T$) terms are:

$$J_S = J_{DL} + J_{CL} + J_{DG} \qquad [1]$$

where $J_s$ is total solute flux, $J_{DL}$ is the diffusive flux in the liquid phase, $J_{CL}$ is the convective flux in the liquid phase, and $J_{DG}$ is the diffusive flux in the gas phase.

Diffusive flux is represented in general terms by Fick's law,

$$J_D = -D_o(\partial c/\partial z) \qquad [2]$$

where $D_o$ is the molecular or ionic diffusion coefficient ($L^2/T$) in a pure solution (of air or water), and $c$ is the pesticide concentration ($M/L^3$).

For diffusion in the liquid phase in porous media, Eq. [2] becomes

$$J_{DL} = -D_{OL}\,\theta(L/L_e)^2\,\alpha(\partial c_L/\partial z) = -D_p\,(\theta)\,(\partial c_L/\partial z) \qquad [3]$$

where $D_{OL}$ is the diffusion coefficient ($L^2/T$) in a pure liquid phase, $c_L$ is the solute concentration ($M/L^3$) in the liquid phase, $\theta$ is the volumetric water content ($L^3/L^3$), $D_p(\theta)$ is the effecive diffusion coefficient ($L^2/T$), $(L/L_e)^2$ accounts for the tortuous nature of the pore sequences, and $\alpha$ represents the effects of a solute on the liquid properties. The reader is referred to Wagenet (1984) for further details of the derivation of Eq. [3].

For diffusion in the gas phase in a porous medium, Eq. [2] can be represented as

$$J_{DG} = -D_{OG}\,(\partial c_G/\partial z) \qquad [4]$$

where $D_{OG}$ is the vapor diffusion coefficient ($L^2/T$) in soil air, and can be estimated from (Jury et al., 1983),

$$D_{OG}(\epsilon) = D_o\; T(\epsilon) \qquad [5]$$

where $D_o$ is the pesticide diffusion coefficient ($L^2/T$) in air in the absence of a porous medium, and $T(\epsilon)$ is a tortuosity factor calculated using the Millington-Quirk model (Shearer et al., 1973) as follows:

$$T(\epsilon) = \epsilon^{10/3}\,\theta_s^{-2}. \qquad [6]$$

The convective flux of pesticide is usually represented as (Scheidegger, 1960):

$$J_{CL} = -\left[\theta\, D_M(q)\frac{\partial c_L}{\partial z}\right] + [qc_L] \qquad [7]$$

where $q$ is the macroscopic soil-water flux ($L/T$), and $D_M(q)$ is the mechanical dispersion coefficient ($L^2/T$) that describes mixing between large and small pores as the result of local variations in mean water-flow velocity.

Combining Eq. [3], [4], and [7] into Eq. [1] gives,

$$J_S = -D_p\,(\theta)\,\frac{\partial c_L}{\partial z} - \theta D_M(q)\,\frac{\partial c_L}{\partial z} + qc_L - D_{OG}\,\frac{\partial c_G}{\partial z}. \qquad [8]$$

The first two terms on the right-hand side of Eq. [8] are usually combined to give the general equation,

$$J_S = -\theta D(\theta,\, q)\,\frac{\partial c_L}{\partial z} + qc_L - D_{OG}\,\frac{\partial c_G}{\partial z} \qquad [9]$$

where $D(\theta, q)$ is the hydrodynamic dispersion coefficient ($L^2/T$) that incorporated the effects of mechanical (i.e., flow-induced) dispersion and molecular diffusion upon pesticide movement in the liquid phase, and is defined as

$$D(\theta, q) = [D_p(\theta)/\theta] + D_M(q). \quad [10]$$

The value of $D_M(q)$ can be estimated from (e.g., Ogata, 1970),

$$D_M(q) = \lambda |v| \quad [11]$$

where $v = (q/\theta)$ and $\lambda$ is the dispersivity ($L$), with values that range from a few millimeters to several meters depending on the nature of the porous medium itself as well as the scale of observation (Sillman & Simpson, 1987; Gelhar et al., 1985). For example, an approximate rule of thumb is that $\lambda$ is about 1/10th the scale of observation (Gelhar et al., 1985).

The value of $D_p(\theta)$ can be estimated from (Kemper & van Schaik, 1966):

$$D_p(\theta) = D_O\, a \exp(b\theta) \quad [12]$$

where $a$ and $b$ are empirical constants reported (Olsen & Kemper, 1968) to be approximately $b = 10$ and $0.005 < a < 0.01$. It should be noted, however, that when $q \gg 0$, contributions of molecular diffusion to the observed hydrodynamic dispersion are negligible.

Almost all situations of pesticide transport in the field occur under non-steady (i.e., transient) water flow conditions. For such cases, both the soil-water content ($\theta$) and soil-water flux ($q$) vary with soil depth and time, resulting in pesticide solution concentrations and fluxes that also vary with soil depth and time. The principle of conservation of mass for a solute in a unit volume of soil may be expressed as:

$$\partial c_T/\partial t = -\partial J_S/\partial z \pm \phi \quad [13]$$

where $c_T$ is the total solute concentration ($M/L^3$) in all phases (liquid, gas, and solid), $t$ is time ($T$), and $\phi$ represents as yet unspecified sources or sinks ($M/L^3T$) for the pesticide.

The total pesticide concentration, $c_T$ ($M/L^3$), in a unit volume element of the soil can be partitioned as:

$$c_T = \rho\, c_S + \theta\, c_L + \epsilon\, c_G \quad [14]$$

where $\rho$ is soil bulk density ($M/L^3$), $c_S$ is sorbed-phase concentration ($M/M$), and $\epsilon$ is the volumetric air content [note that $\epsilon = (\theta_s - \theta)$, where $\theta_s$ is the saturated volumetric soil-water content]. Combining Eq. [9], [13], and [14] gives

$$\frac{\partial}{\partial t}(\rho c_S + \theta c_L + \epsilon c_G) = \frac{\partial}{\partial z}\theta D(\theta, q)\frac{\partial c_L}{\partial z} - q c_L - D_{OG}\frac{\partial c_G}{\partial z} \pm \phi. \quad [15]$$

Equation [15] is the expression most often used to describe pesticide fate in a field soil, and includes the effects of sorption, liquid-, and vapor-phase transport; degradation; and plant uptake. Each of these processes will be considered in further detail in the following subsections.

### 10–5.2 Sorption

The sorption of pesticide on the soil retards its movement through the root zone, with the extent of retardation dependent upon the physical and chemical properties of the soil as well as the molecular characteristics of the pesticide (Helling & Dragun, 1981). Extensive compilations of available data on pesticide sorption by soils may be found in recent literature (Karickhoff, 1981; Kenaga & Goring, 1980; Rao & Davidson, 1980).

The linear and Freundlich isotherm equations have been most often used to describe pesticide sorption on soils. These equations are given by,

$$c_S = K_d \, c_L \qquad [16]$$

and

$$c_S = K_f \, c_L^N; \; N < 1 \qquad [17]$$

where $K_d$ and $K_f$ are the sorption coefficients, and $N$ is an empirical constant. Values of $K_d$, $K_f$, and $N$ are usually determined by curve-fitting sorption data obtained from batch equilibrium studies. Rao and Davison (1980) and Green et al. (1980) presented reviews of methods to experimentally measure and to estimate the values of the sorption coefficients.

The value of $K_d$ or $K_f$ is a measure of the extent of pesticide sorption by the soil. The soil organic C (OC) content is apparently the single best predictor of the sorption coefficient for nonionic, hydrophobic pesticides (c.f., Karickhoff, 1981; Rao & Nkedi-Kizza, 1984; Rao et al., 1983). It has been reported that the sorption coefficient for a particular pesticide, when normalized with respect to soil OC, is essentially independent of soil type. This has led to the definition of the OC-normalized sorption coefficient, $K_{oc}$, as

$$K_{oc} = (K_d \text{ or } K_f/\%\text{OC}) \times 100. \qquad [18]$$

Compilations of $K_{oc}$ values are presented in several papers (Karickhoff, 1981; Kenaga & Goring, 1980; Rao & Davidson, 1980). While $K_{oc}$ is useful in studies of nonionic, hydrophobic pesticides in a broad range of soils, it is important to realize that in soils with extremely low or high organic C and for ionizable or ionic pesticides the use of $K_{oc}$ to estimate $K_d$ or $K_f$ may be subject to large error (Hamaker & Thompson, 1972). More recent studies have evaluated some factors, such as variations in chemical and physical composition of the soil organic matter, contributing to variability in $K_{oc}$. Garbarini and Lion (1986) have shown that empirical relationships that include both C and $O_2$ contents of organic matter yielded better estimates of $K_d$ values than those based only on organic C. This was attributed to a better accounting of the variations in the chemical nature of the soil organic matter.

When measured values of $K_d$ or $K_f$ are not available, the $K_{oc}$ value may be estimated as follows if only the melting point and aqueous solubility of the pesticide are given (Karickhoff, 1981):

$$\log K_{oc} = -0.921 \log X_{sol} - 0.00953\,(MP - 25) - 1.405 \quad [19]$$

where

$$X_{sol} = [(C_{sol}/MW)]/55.56. \quad [20]$$

In Eq. [19] and [20], the aqueous solubility is expressed as a concentration, $C_{sol}$ $(M/L^3)$, or as a mole-fraction ($X_{sol}$), MW is the molecular weight (M/mol), and MP is the melting point (°C) which is set equal to 25 °C for pesticides that are liquids at temperatures ≤25 °C. The validity of this expression has been demonstrated for some nonionic pesticides (Karickhoff, 1981; Rao & Nkedi-Kizza, 1984).

Models of pesticide movement use values of $K_d$, $K_f$, or $K_{oc}$. These values depend greatly upon the pesticide, the soil characteristics, and the method used to estimate them. Further efforts to measure $K_{oc}$, $K_f$, and $K_d$ and more accurate methods to estimate them, such as Eq. [19] and [20], are needed. Although more complicated expressions of sorption phenomena may better represent fundamental relationships, such resolution is not required in most management models, given the simplifying assumptions made about other components of the system (water flow, degradation, and plant effects). For a particular field study, it will be important for the modeler to know (i) the value $K_d$ or $K_f$ for each soil horizon, (ii) the value of $K_{oc}$ and the depth-variations in soil organic C or (iii) the aqueous solubility and melting point of the pesticide and depth-variation in soil organic C. Lyman (1982) compared several approaches for indirect estimation of pesticide sorption coefficients. Essentially all of the pesticide-fate models (research, management, and screening) available at present represent sorption as a linear, instantaneous, reversible process.

Since the late 1960s, it has been appreciated that the assumption of describing sorption as a linear, equilibrium, reversible process is not always a valid representation. Most early studies that used such a representation (Kay & Elrick, 1967; Davidson et al., 1968; Davidson & McDougal, 1973) showed substantial deviation between calculated and measured effluent curves obtained from the study of pesticide transport in laboratory soil columns. It is now clear that the sorption process is much more complicated for many pesticides and some other models have been introduced. These include nonlinear, equilibrium isotherms (van Genuchten et al., 1974), as well as nonequilibrium sorption models (Hornsby & Davidson, 1973; Mansell et al., 1977; Selim et al., 1976; Cameron & Klute, 1977; Rao et al., 1979; van Genuchten and Wagenet, 1989).

In a limited number of experimental studies (e.g., Rao et al., 1979; Lee et al., 1988; Gamerdinger et al., 1989), it has been shown that the two-site models (Selim et al., 1976; Cameron & Klute, 1977; Parker & van Genuchten, 1984) better describe much of the asymmetry in the effluent breakthrough curves obtained for pesticide displacement through soil columns. These models assume that the solid phase of the soil is composed of different constituents (soil minerals, oxides, and organic matter), and that the solute will react with the sorbent at different rates and intensities. The model divides the possible sorption domains into two fractions: Type-1 on which sorption is assumed to be instantaneous, and Type-2 on which the process is assumed to be time-dependent. Brusseau and Rao (1988) have critically reviewed the literature dealing with three factors contributing to sorption nonideality: non-linear isotherms, nonsingular isotherms (hysteresis), and nonequilibrium.

There has been little consideration of the implications of the sorption nonideality as estimates are made of the environmental fate of pesticides. Most approximate estimates are still made using $K_d$, $K_f$, and $N$. However, as it becomes necessary to more accurately predict the transport of pesticides in the environment, it may become necessary to revise the approaches currently used to characterize sorption. Several models, all research type, are presently considering this as will be discussed below.

### 10–5.3 Volatilization

The description of volatilization begins by relating solution- and gas-phase pesticide concentrations. Recognizing that the liquid-vapor partitioning can be represented by a modified Henry's law, Jury et al. (1983) proposed that,

$$c_G = K_H^* c_L \qquad [21]$$

where

$$K_H^* = c_G^*/c_L^*. \qquad [22]$$

$K_H^*$ is the modified Henry's law constant (dimensionless), defined here as the ratio of the saturated vapor density ($c_G^*$) and the aqueous solubility ($c_L^*$) of the pesticide, both in expressed units of mass/volume.

The importance of volatilization as a dissipative process makes it necessary to carefully define the boundary condition at the soil-air interface. If we asume that a boundary layer exists through which diffusion of volatilized pesticide occurs at the soil-air interface, then it can be assumed that at some distance from the ground surface,

$$c_G(\delta, t) = 0 \qquad [23]$$

and

$$J_T(0, t) = \{[c_G(0, t) - c_G(\delta, t)]/\delta\} D_O \qquad [24]$$

Using Eq. [19], Eq. [24] can be restated as,

$$J_T(0, t) = D_O K_H^* c_L(0, t)/\delta. \qquad [25]$$

The value of $\delta$ in Eq. [25] has been assumed in some studies to be 5 mm (Wagenet et al., 1988), although it can be calculated (Jury et al., 1983). Given knowledge of $c_L$ (0, $t$), Wagenet et al. (1989) used Eq. [25] to calculate the volatilization flux at the soil surface for the neamticide dibromochloropropane (DBCP). It is interesting to note that with constant $D_O$, $K_H^*$, and $\delta$, Eq. [25] reduces to a pseudo-first-order expression. This has led to the use of a dissipation rate constant that characterizes the combined losses through volatilization and degradation in the upper segment of the soil profile. This approach, often incorrectly applied, has resulted in an underestimation of the actual value for the half-life for pesticide degradation.

### 10–5.4 Degradation

Losses of pesticides via microbiological and chemical pathways of transformation are collectively termed *degradation*. In most current pesticide-fate models, it is usually assumed that microbiological processes are predominant in the root zone. Once the pesticide leaches past the biologically active root zone, and migrates well beyond the upper soil layers or enters into the saturated zone (groundwater), however, chemical degradation pathways, such as hydrolysis, are likely to be relatively more important. The term degradation should not be confused with dissipation, which is a collective and more empirical term relating to disappearance of pesticide by any number of unquantified and unspecified pathways.

For the same reasons that pesticide sorption is simply represented by linear- or Freundlich-type isotherm equations, pesticide degradation is usually represented simply by first-order kinetics. Some experimental studies have established this to be approximately true, or at least true enough to represent little loss of accuracy for most required simulations. Hamaker (1972) presented a comprehensive discussion of pesticide degradation in soil based upon two basic types of rate models.

$$\partial c/\partial t = -\mu c^{\mathrm{n}} \qquad [26]$$

$$\partial c/\partial t = -V_{\max}\,[c/(\alpha + c)] \qquad [27]$$

where $c$ and $t$ have been previously defined, $\mu$ = degradation rate coefficient ($1/T$), n = reaction order, $V_{\max}$ = maximum degradation rate ($M/L^3T$), and $\alpha$ is a constant. When $n = 1$ in Eq. [26], an expression for first-order kinetics is obtained, which upon integration and with $c/c_0 = 0.5$ [where $c_0$ is $c(t = 0)$] gives the value of the half-life, $t_{0.5}$ ($T$). When $c \gg \alpha$, Eq. [27] reduces to an expression for zero-order kinetics. Pesticide degradation using these two rate equations has been compared by Goring et al. (1975).

Rao and Davidson (1980) have compiled the values of first-order degradation rate coefficients for some pesticides from several published sources. Values of $\mu$ and $t_{0.5}$ measured in the field and laboratory for some pesticides exhibited coefficients of variation generally $<100\%$, a relatively narrow range considering the diverse soil and environmental conditions in which the studies were done. Additionally, it should be recognized that field-measured half-lives are generally shorter than those measured under controlled laboratory conditions. This is a consequence of (i) multiple degradation pathways operating under field conditions, which results in a more rapid degradation and (ii) losses by such unquantified pathways as volatilization and incomplete recovery in sampling or analysis. Therefore, the use of laboratory-derived values for degradation rate constants may tend to overestimate the persistence of pesticides under field conditions. Nash and Osborne (1980) have also compiled available data for pesticide dissipation in soils. These data are available from the authors as an "Environmental Chemical Dissipation File." Nash (1988) presented selected examples from this data base in a recent review article on pesticide degradation. Unpublished information on the health and environmental effects as well as environmental fate of toxic organics, submitted to the USEPA's Office of Toxic Substances, can be retrieved using a computerized data base called TSCATS, developed and maintained by the Syracuse Research Corporation (Santodonato et al., 1987).

It is important for the modeler to have a range of $\mu$ values for use in the simulation exercises. It is probable that degradation rates will be both temporally and spatially variable, yet data for characterizing spatial and temporal variability are almost nonexistant. The first-order degradation rate coefficient is a function of environmental factors affecting the microbiological system. The dependence of $\mu$ on such factors as soil-water content, temperature, salinity, and substrate characteristics has been studied (e.g., Ou et al., 1982; Walker, 1976a, b) and used to develop several degradation models. Nash (1988) reviewed methods currently available for estimating pesticide dissipation rate constants.

The present approach used in most modeling studies is to estimate $\mu$ value from laboratory studies, and to adjust this value to account for variations in soil and environmental factors. Some modelers execute their models with a range of $\mu$ values to bracket the possible variations in degradation rates. Some empirical approaches have been proposed to describe the relationships between $\mu$ and various soil and environmental variables (Nash, 1988). While much progress has been made in estimating sorption coefficients using physical-chemical properties of the pesticides, similar developments in relating microbial degradation rates to chemical structure are in their infancy. Howard et al. (1987) reviewed the available data and the literature on various attempts to relate chemical structure to biodegradability. They concluded that the available data are inadequate to develop *quantitative* structure-biodegradability relationships. Howard et al. (1987) proposed a system for collecting and evaluating biodegradability data to be compiled in a computer data-base BIODEG.

### 10–5.5 Plant Uptake

Pesticide uptake by plants has been ignored in most modeling efforts. This is primarily because of an almost total lack of quantitative information available to the modeler on the uptake process. Recognizing this limitation, most modelers have apparently assumed that any inaccuracy in simulation of pesticide fate in the root zone that results from not considering plant uptake is within the noise of inaccuracies produced by other assumptions about the chemical-physical-biological processes operating in the system. It seems prudent to begin to provide estimates of plant uptake as it affects pesticide leaching. In the management model PRZM (Carsel et al., 1984), two options for modeling pesticide plant uptake are provided: passive uptake along with transpiration, or no uptake. The adequacy of the passive uptake assumption remains to be tested. Only recently have researchers begun to examine the quantitative relationships between pesticide chemical properties, such as the octanol-water partition coefficient, and plant uptake (Briggs et al., 1982). Further work along these lines will be extremely valuable in modeling pesticide fate.

Plant uptake of water also has a direct impact upon the movement of pesticides in soils. As soil water is extracted, the flux of water passing through the root zone decreases and might conceivably become less spatially variable (Wagenet & Rao, 1983). This will lead to similar effects on pesticide movement. Accurate simulation modeling of pesticide fate in plant environments then becomes quite dependent upon the consideration of patterns of water extraction by plants. Information needed by the modeler includes: rooting depth at different times of the season, root density distribution with depth and time, and, if possible, the relationship between the soil-water stress and plant extraction. Although several simple, promising approaches are presently available for modeling plant uptake of water, the widespread application of these methods awaits further confirmation of their accuracy.

### 10–5.6 Other Processes and Factors

Some additional processes and factors also influence pesticide behavior in soils, but these have generally not been included in current versions of pesticide-fate models. Some examples are: (i) photodecomposition, which is usually lumped along with other degradation processes; (ii) preferential flow along macropores, which may have a significant impact on rapid movement of small amounts of pesticides to greater depths than that anticipated; and (iii) the influence of pesticide formulations and dissolved organics on sorption and degradation. Other factors or processes addressed inadequately in present models are: the synergistic/antagonistic effects of a mixture of pesticides on sorption and degradation of each component and the effects of soil salinity and cultivation practices on soil hydraulic properties, which, in turn, impact the rates of water flow and pesticide leaching. As more is learned about these factors, enough to develop quantitative relationships, they will need to be accommodated in future versions of the existing pesticide-fate models.

### 10–5.7 Integration of Pesticide-Fate Processes

Input data required by pesticide-fate models and output provided by these models vary according to the modeling purpose (Tables 10–1 and 10–2) and the detail at which the various processes have been represented in the model. Models intended primarily for research purposes are quite data-demanding (Table 10–2), but also have the capability to provide simulations

Table 10-1. Major outputs of four types of pesticide simulation models.

| | Type of model | | | |
|---|---|---|---|---|
| Output | Research, LEACHM | Screening, BAM | Management, PRZM | Instructional, CMLS |
| $c(z, t = t_i)$ | +† | + | + | − |
| $c(t, z = z_i)$ | + | + | + | − |
| Depth of solute peak concentration | + | + | + | + |
| Maximum depth of solute penetration | + | + | + | − |
| $q\ (z, t = t_i)$ | + | Input | − | − |
| $q\ (z, z = z_i)$ | + | Input | + | − |
| $\theta\ (z, t = t_i)$ | + | Input | + | − |
| $\theta\ (t, z = z_i)$ | + | Input | + | − |
| Phase partitioning of chemical mass | + | + | + | − |
| Temperature | * | − | − | − |
| Water uptake | + | − | + | − |
| Pesticide uptake | * | − | * | − |
| Volatile losses | + | + | − | − |
| Runoff | − | − | + | − |

† A plus sign indicates that the parameter is required and a negative sign indicates that the parameter is not required. An asterisk indicates that the parameter can be considered, but usually is not because of insufficient data.

of pesticide behavior under a wide range of conditions. Management models (e.g., PRZM), constructed to include a range of soil processes, can also require substantial amounts of input data. Management and screening models usually require less information to execute the model, but also provide commensurately fewer, and less quantitative, estimates of the system's behavior (Table 10–1). The following section will illustrate the details of this linkage between model complexity, data requirements, and simulation using as examples four models: LEACHM, Behavior Assessment Model (BAM), PRZM, and Chemical Movement in Layered Soil (CMLS) (see Table 10–3).

## 10–6 RESEARCH AND SCREENING MODELS

Models of pesticide fate can be developed from Eq. [15] for two general cases. These cases differ depending upon the nature of water flow. If water flows through the soil at a constant rate which does not change with time, the system is at steady state with respect to water flow and the resulting simplifications in Eq. [15] are possible. If water flow changes with time, the system is not at steady state, or is transient with respect to water, and simplifications in Eq. [15] are not possible. These two water-flow scenarios form the basis for grouping pesticide research models into the two classes listed below. In fact, one of these classes incorporates the screening model approach of Jury et al. (1983), and contains some other useful formulations of the processes included in Eq. [15].

Table 10-2. Major characterizing parameters required by four types of pesticide simulation models.

| Parameter | Type of model: Research, LEACHM | Screening, BAM | Management, PRZM | Instructional, CMLS |
|---|---|---|---|---|
| **Pesticide** | | | | |
| Sorption | | | | |
| Normalized sorption coefficient ($K_{oc}$) | +† | + | + | + |
| Distribution coefficient ($K_d$) | − | − | − | − |
| Volatilization | | | | |
| Henry's constant ($K_H$) | − | + | − | − |
| Saturated vapor density | + | − | − | − |
| Aqueous solubility | + | − | − | − |
| Gas phase diffusion coefficient | + | + | − | − |
| Degradation | | | | |
| Lumped half-life ($t_{0.5}$) | + | + | + | + |
| Hydrolysis half-live | * | − | − | − |
| Oxidation half-live | * | − | − | − |
| Daughter products | * | − | − | − |
| Temperature-dependent degradation | * | − | − | − |
| Foliar decay rate | − | − | + | − |
| **Soil** | | | | |
| Apparent diffusion coefficient | + | + | + | − |
| Saturated water content ($\theta_s$) | + | − | + | + |
| Field capacity water content ($\theta_{fc}$) | − | − | + | + |
| Wilting point water content ($\theta_{wp}$) | − | − | + | + |
| Hydraulic properties (K-$\psi$-$\theta$) | + | − | − | − |
| Bottom boundary flexibility | + | − | + | − |
| Bulk density ($\rho$) | + | + | + | + |
| Organic C content | + | + | + | + |
| pH | − | − | − | − |
| Cation exchange capacity | − | − | − | − |
| Layering | + | − | + | + |
| SCS curve number | − | − | + | + |
| Heat flow parameters | * | − | − | − |
| **Crop** | | | | |
| Root density distribution | + | − | + | − |
| Maximum rooting depth | + | − | + | + |
| Pesticide uptake relationship | * | − | * | − |
| **Climatological** | | | | |
| Daily rainfall or irrigation amounts | + | − | + | + |
| Rainfall or irrigation rates | + | − | − | − |
| Daily pan evaporation | + | − | + | − |
| Daily max./min. temperature | * | − | − | − |
| Daily actual evapotranspiration | − | − | − | + |
| Snowmelt | − | − | + | − |
| Hours of sunlight | − | − | + | − |
| **Management** | | | | |
| Pesticide | | | | |
| Application date | + | − | + | + |
| Application rate | + | + | + | − |
| Depth of incorporation | + | + | + | + |
| Multiple applications | + | − | + | − |
| Crop (production system) | * | − | * | − |
| Soil tillage effects | * | − | * | − |

† A plus sign indicates that the parameter is required and a negative sign indicates that the parameter is not required. An asterisk indicates that the parameter can be considered, but usually is not because of insufficient data.

Table 10–3. Example simulation models developed or useful for pesticide fate assessment.

| Model | Acronym | Purpose | Reference |
|---|---|---|---|
| Behavior Assessment Model | BAM | Screening | Jury et al., 1983; 1987 |
| Leaching Estimation And Chemistry Model | LEACHM | Research | Wagenet & Hutson, 1986; 1989 |
| Pesticide Root Zone Model | PRZM | Management | Carsell et al., 1984 |
| Chemical Movement in Layered Soil | CMLS | Instructional | Nofziger & Hornsby, 1986 |
| Method Of Saturated Zone Solute Estimation | MOUSE | Instructional | Pacenka & Steenhuis, 1984 |
| PESTicide ANalytical Solution | PESTAN | Screening | Enfield et al., 1982 |
| Chemicals, Runoff and Erosion from Agricultural Management Systems | CREAMS | Management | USDA, 1980; Leonard & Ferreria, 1983 |
| Numerical Solution of CDE (untitled) | -- | Research | Leistra, 1979 |
| SEasonal SOIL compartment model | SESOIL | Management | Bonazountas & Wagner, 1984 |

### 10–6.1 Steady-State Water Flow Models

Soil column experiments have been widely used to study pesticide displacement in soil. These experiments often use carefully controlled conditions of constant water flux and uniformly packed soil columns with appropriate initial and boundary conditions. In such cases, analytical solutions can be obtained, which are useful in studying sorption, degradation, and transport. For steady water flow, Eq. [15] becomes:

$$\rho \frac{\partial c_S}{\partial t} + \theta \frac{\partial c_L}{\partial t} = \theta D \frac{\partial^2 c_L}{\partial z^2} - q \frac{\partial c_L}{\partial z} \pm \phi . \qquad [28]$$

When the focus is on sorption, $\phi$ is often assumed to be zero. Classical approaches represent sorption according to Eq. [16], in which case Eq. [28] reduces to:

$$\frac{\partial c_L}{\partial t} = D^* \frac{\partial^2 c_L}{\partial z} - V^* \frac{\partial c_L}{\partial z} \qquad [29]$$

where $D^* = [D/R]$, $V^* = [v/R]$, $v = (q/\theta)$ is the average pore-water velocity $(L/T)$, and $R = [1 + (\rho\ K_d/\theta)]$ is the retardation factor.

When sorption is defined as other than a linear, equilibrium process, a wide range of models is possible. For example, in the case of the two-site nonequilibrium model, the concentration in the sorbed phase $c_S$, is partitioned into Type-1 and Type-2 fractions, so that $c_S = (c_{S1} + c_{S2})$. If we assume that F is the fraction of Type-1 sites, then at equilibrium (Type-1 sites are always at equilibrium):

$$c_{S1} = k_1\ c_L = F\ K_d\ c_L \qquad [30a]$$

$$c_{S2} = k_2\ c_L = (1 - F)\ k_d\ c_L . \qquad [30b]$$

The sorption rate for the kinetic nonequilibrium (Type-2) sites is given by

$$\frac{\partial c_{S2}}{\partial t} = \alpha\ (k_2\ c_L - c_{S2}) \qquad [31]$$

where $\alpha$ is a first-order transfer rate coefficient $(1/T)$. Combining Eq. [28], [30], and [31] and maintaining mass balance considerations for the two sorption sites leads to (Selim et al., 1976; Cameron & Klute, 1977; van Genuchten, 1981):

$$\left[1 + \frac{F\rho\ K_d}{\theta}\right] \frac{\partial c_L}{\partial t} + \frac{\rho}{\theta} \frac{\partial c_{S2}}{\partial t} = D \frac{\partial^2 c_L}{\partial z^2} - v \frac{\partial c_L}{\partial z} \qquad [32]$$

$$\frac{\partial c_{S2}}{\partial t} = \alpha \, [(1 - F) \, K_d \, c_L - c_{S2}] . \qquad [33]$$

Analytical solutions to these equations have been used to extend the understanding of pesticide sorption. Nonlinear, least-squares inversion methods have been developed (van Genuchten, 1981; Parker & van Genuchten, 1984) for estimating the values for the parameters included in Eq. [32] and [33] using data obtained from soil column experiments. Recently, van Genuchten and Wagenet (1989) and Gamerdinger et al. (1989) have incorporated a term in Eq. [32] to account for first-order degradation and have used the resulting solutions to study simultaneous sorption and degradation of pesticides during transport through soils. Lemley et al. (1988) incorporated a first-order degradation term in Eq. [29] to investigate the sequential degradation and transport of aldicarb nematicide and its two oxidative metabolites during steady water flow through soil columns.

Enfield et al. (1982) proposed a screening model, PESTAN, based on Eq. [28] in which steady-state, one-dimensional water flow conditions in a uniform soil profile are assumed. They presented three versions of their model based on the following simplifications:

1. Linear, reversible, equilibrium sorption isotherm; first-order kinetics for degradation; convective solute transport; and hydrodynamic dispersion neglected.
2. Linear, reversible, equilibrium sorption isotherm; degradation neglected; and convective-dispersive solute transport.
3. Nonlinear (Freundlich), reversible, equilibrium sorption isotherm; first-order kinetics for degradation; convective solute transport; and hydrodynamic dispersion neglected.

Enfield et al. (1982) used analytical solutions for model versions 1 and 2, and a numerical solution for version 3; note that none of these versions account for vapor-phase transport.

Three key parameters required in PESTAN are: the steady-state soil-water flux ($q$); the degradation rate constant ($\mu$); and the sorption retardation factor ($R$). The soil-water flux may be estimated from simple water balance computations using long-term weather records (e.g., $q$ can be set equal to the annual groundwater recharge rate at the site), while values of $\mu$ and $R$ specific to a soil-pesticide combination would be needed. This model has been used by the USEPA for an initial screening of pesticide data submitted during the registration process.

Jury et al. (1983) developed a screening model, also based on the assumption of steady-state water flow. Assuming that dispersion was negligible, the steady-state solute flux was represented as:

$$J_S = -D_{OG} \frac{\partial c_G}{\partial z} - D_p(\theta) \frac{\partial c_L}{\partial z} + q \, c_L \qquad [34]$$

$$\frac{\partial c_T}{\partial t} = D_E \frac{\partial^2 c_T}{\partial z^2} - v_E \frac{\partial c_T}{\partial z} - \mu c_T \quad [35]$$

where $D_E = [(K_H D_{OG} + D_p(\theta)/R_L]$, $v_E = (q/R_L)$, and $R_L = (\rho K_d + \theta + \epsilon K_H)$.

Equation [35] was solved by Jury and co-workers with initial and boundary conditions appropriate for the following scenario:

1. Uniform incorporation of a quantity of chemical to a specified depth below the surface.
2. Volatilization through a stagnant air boundary layer at the soil surface.
3. Convection by a steady water flux, $q = \pm J$ or 0.
4. Infinite depth of uniform soil below the depth of incorporation.

The resulting analytical solution, as corrected by Jury et al. (1987b), is thereby constrained to conditions of steady water flux and constant soil-water content over time. The pesticide degradation rate coefficient, $\mu$, in Eq. [35] was assumed to be constant with soil depth.

Jury et al. (1987a) have proposed a variation to their screening model, where $\mu$ was allowed to be depth-dependent. The value of $\mu$ was assumed proportional to depth variations in soil microbial activity, which was divided into the three intervals: surface zone with a constant $\mu$ value ($=\mu_o$); a transition zone in which $\mu$ decreases exponentially from $\mu_o$ to $\mu_r$; and a residual zone where $\mu = \mu_r$.

The assumption of steady water flow prevents the screening models from being used for estimating environmental fate under transient water flow conditions; models appropriate for transient water flow are discussed in section 10–6.2. However, the screening models are quite useful for grouping pesticides into general behavior classes for an initial screening of many chemicals and to assess the relative risks of potential groundwater contamination from pesticides. The models Jury et al. (1983, 1987a) proposed are the only analytical solutions to incorporate volatilization into such an assessment.

### 10–6.2 Transient-State Water Flow Models

When water moves through soil at a rate that changes with time, the system is defined as transient with respect to water flow. It will also be often transient with respect to resulting solute movement, as the soil-water flux and the soil-water content included in Eq. [15] are now functions of both depth and time. Such conditions characterize nearly every field case of pesticide movement in the unsaturated zone, and also pertain to many lysimeter studies. Given the need to consider such important situations, pesticide models have been developed from Eq. [15] that include consideration of the depth and time changes in water and resulting solute movement.

Equation [15] can be solved using numerical methods to provide a pesticide model useful for describing field conditions. These models presume a knowledge (or a simulation) of soil-water flux and soil-water content as a function of depth and time. These variables have been calculated using either a mechanistic interpretation of water flow, or can be estimated using empirical methods. It is important to recognize the difference in approach, as an accurate description of water flow underpins the pesticide models based on Eq. [15], and to a large degree determines the usefulness of the simulations and the situations to which the models can be potentially applied.

One approach for estimating water movement uses the following equation, derived by combining Darcy's law and the equation of continuity, for one-dimensional, transient, vertical water flow:

$$\frac{\partial \theta}{\partial t} = \frac{\partial}{\partial z}\left[K(\theta)\,\frac{\partial H}{\partial z}\right] - A(z,\,t) \qquad [36]$$

where $H$ is hydraulic potential energy ($L$), equal in this case to the sum of matric ($h$) and gravitational ($g$) potential energy; $K(\theta)$ is soil hydraulic conductivity ($L/T$) as a function of soil-water content; $A(z, t)$ is the sink term to account for plant uptake of water; and other terms are as defined earlier. Note that if plants are not present, $A(z, t) = 0$. Equation [36] can be solved numerically to give $H(z, t)$ as a function of the initial and boundary conditions, and soil-water propteties. Estimation of $h(z, t)$ automatically gives $\theta(z, t)$, provided the soil-water characteristic curve $[\theta(h)]$ is known. Knowledge of $\theta(z, t)$ allows the soil-water flux, $q$, in Eq. [15], to be calculated from:

$$q(z,\,t) = \int_{L_1}^{L_2} [\partial\theta/\partial t]\,\partial z \qquad [37]$$

where finite increments of time ($t$) and depth ($z$) are used as defined in the numerical model, and $L_1$ and $L_2$ refer to specific coordinate locations in the soil profile. The pesticide model LEACHM (Wagenet & Hutson, 1987) employs such a method of estimating water flow. The technique is classical, and has been employed in most of the deterministic, numerical models cited above.

The use of such models can perhaps be better appreciated by describing the model LEACHM in more detail. This model has been organized on a modular basis (Fig. 10–2). A main program reads input data, initializes variables, calls subroutines, performs mass balance calculations, and prints output tables. Subroutines deal with time-step calculation, evapotranspiration, water flow, pesticide movement, application, degradation, volatilization, leaf and root growth, temperature, and (if quantifiable) pesticide uptake by plants. Segregation of each of these processes into subroutines called by the main program enables any subroutine in the model to be replaced by an improved or different formulation if desired.

Simulations in LEACHM begin at time zero, and a set of initial conditions are required. The soil need not be homogeneous in the vertical direc-

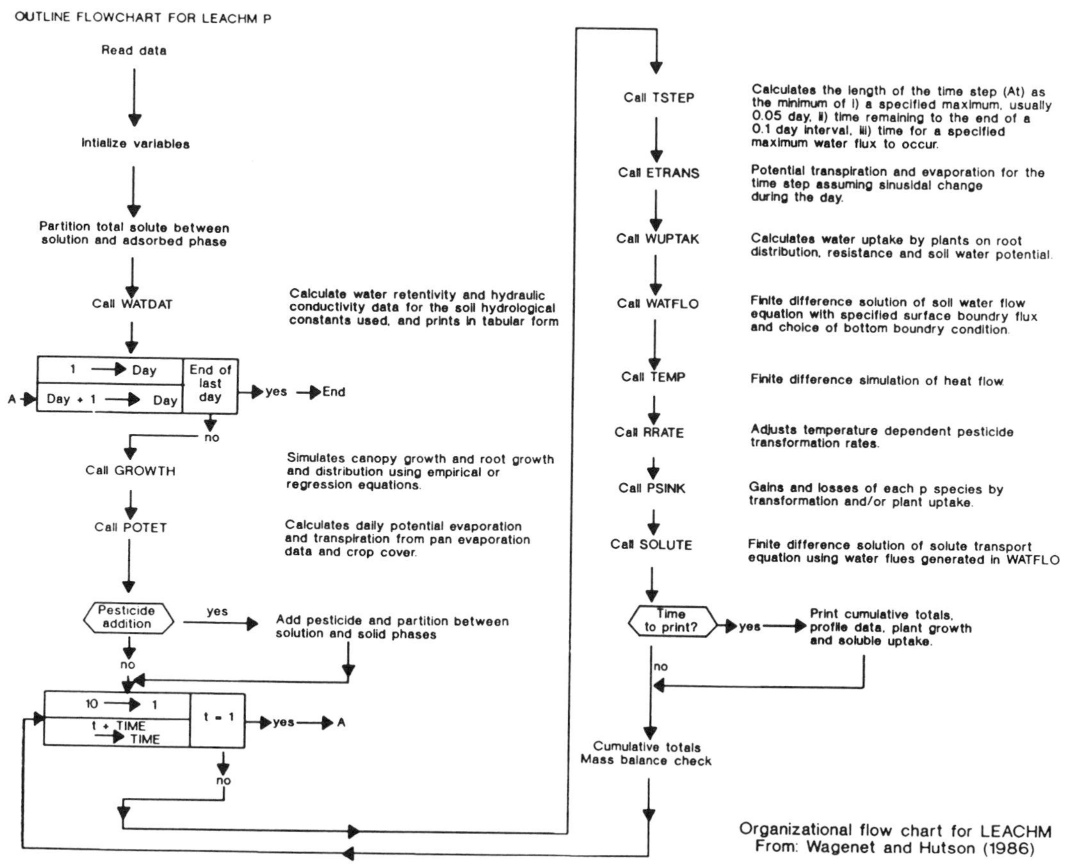


Fig. 10-2. Flow chart for the model LEACHM. Taken from Wagenet and Hutson (1986).

tion. Plants can be present or absent, but if present, germination and root expansion can be simulated, or a static, established root system can be defined. The simulation program requires the following inputs, which are read from a data file:

1. Soil properties and initial conditions with depth of: (a) depth boundaries, (b) soil-water content or matric potential, and (c) hydrological constants for calculating soil-water characteristic curves and hydraulic conductivity.
2. Soil surface boundary conditions of: (a) irrigation frequency, duration, and rate (on a daily basis); (b) rainfall frequency, duration, and rate (on a daily basis); and (c) an evaporation (on a weekly basis).
3. Upper and lower soil-water potentials for water extraction by plants.
4. Crop factors, i.e., days from planting to crop maturity, root-growth parameters, and crop cover growth parameters. If it is assumed that no crops are present, a control variable allows bypass of the plant-related subroutines.
5. Other constants used in determining bottom boundary conditions, time steps, diffusion coefficients, as specified in the data files. Some of these constants rarely require alteration, but are listed in the data file to define their value for the user and provide the opportunity for change.

In general terms, LEACHM operates in the following manner (Fig. 10–2):

1. Input data are read and initial values are assigned to all variables. The total pesticide present at time zero is partitioned between solution-, sorbed-, and gas-phases according to input values of $K_d$, pesticide solubility, pesticide-saturated vapor density, and the total pesticide mass in the system.
2. Subroutine WATDAT calculates the relationship between soil hydraulic conductivity, soil-water content, and matric potential from input relationships.
3. The simulation begins by selecting a minimum time step entered in (1) above.
4. Subroutine GROWTH simulates canopy and root growth, and provides an estimate of the root-density distribution. If plants are not present, or if insufficient time has elapsed from germination, this subroutine is bypassed.
5. Subroutine POTET calculates daily potential evaporation and transpiration from pan evaporation data (input) and crop cover (GROWTH).
6. Boundary conditions of pesticide application in dry form are updated. If no dry pesticide is applied, these steps are bypassed.
7. Subroutine TSTEP updates the time step size (time increment).
8. Subroutine ETRANS estimates potential transpiration and evaporation during the time increment.
9. Subroutine WUPTAK calculates water uptake by the plant in the time increment as a function of root distribution and matric potential. If no plants are present, this subroutine is bypassed.
10. Subroutine WATFLO simulates soil-water flux density, soil-water content, and matric potential changes during the time increment.

11. Subroutine RRATE calculates new values for the first-order transformation rate constants as a function of environmental variables simulated in LEACHM. While RRATE is available, its use depends upon knowledge of the specific functional relationships between degradation kinetics and environmental variables.

12. Subroutine PSINK calculates pesticide transformations, volatile losses, and plant uptake of pesticide if sufficient information on pesticide uptake functions is available.

13. Subroutine SOLUTE simulates transport of both adsorbed and nonadsorbed pesticide during the time increment. This subroutine includes the simulation of a noninteracting nondegrading (chloride)-type salt for comparison with simulated pesticide species.

14. If the cumulative time does not exceed the stopping time (determined as input), mass balances are calculated and a new time step is determined.

Output of the model at predetermined (in the input data) print intervals includes the following four tables:

1. Hydraulic conductivity and water content for each layer of the soil at the matric potential values of 0, −3, −10, −30, −100, and −500 kPa.
2. Cumulative totals and mass balances of water and pesticide. This includes the amount of pesticide initially in the soil profile, currently in the profile, the simulated change, additions, losses, and a composite mass error. Units in this table are millimeters for water and milligrams per square meter per depth increment for pesticide.
3. A summary by depth of soil-water content, matric potential, soil-water flux between layers, evapotranspiration, pesticide mass, pesticide concentration for both individual species and total pesticide, and the chloride concentration.
4. A summary by depth of root density, water uptake, and pesticide uptake. This table presents the information as a change since the last print and as a cumulative total from time zero.

In summary, the LEACHM model employs the classical Richards equation for water flow to calculate soil-water contents and soil-water fluxes. These values are then used in the convective-dispersive transport equations for predicting solute transport. Thus, transient flow of water and solutes are explicitly linked. Volatilization losses of the pesticide are also calculated in LEACHM. Another unique feature of this model is that is provides predictions of the transport and transformations of the parent chemical as well as up to two daughter products. This is essential for describing the environmental behavior of nematicides such as aldicarb and nemacur, whose oxidative metabolites also possess pesticidal properties. Appropriate specification of surface and bottom boundary conditions will permit simulation of upward movement of soil water as a result of evaporation or because of the presence of shallow water table. Note that LEACHM does not account for preferrential flow of water and solutes in macropores (e.g., root channels and cracks) or for the presence of mobile and immobile soil-water domains, as would be the case in aggregated soils. Although several approaches have been proposed to deal with preferential water and solute flow, none have been incorporated into research models for pesticide fate in soils.

## 10-7 MANAGEMENT MODELS

It is often neither possible nor convenient to use a comprehensive research model to guide pesticide management in a particular situation. This may be because of lack of characterizing soil data or in experience with computer models. Yet, it is still useful to employ a computer model for the purpose of estimating the integrated effect of the various processes that determine pesticide fate under a specific management scenario.

One model that has been developed for such purposes is PRZM. This model (Carsel et al., 1984) is derived from the conceptual, compartmentalized representation of the soil profile shown in Fig. 10-3. This model includes some processes important in management applications, so the mass balance considerations are commensurately more comprehensive. As shall be seen below, this does not (in this case) lead to a model that is computationally more demanding.

The mass balance equations for both the surface and subsurface conceptual horizons (Fig. 10-3) used as the basis for PRZM are as follows. For the surface zone, the mass balance is:

$$\frac{\partial(\theta c_L + \rho c_S)}{\partial t} = -\frac{\partial}{\partial z}[J_{DL} + J_{CL} + J_{TL} + J_{TS} + J_U + J_R + J_{ES} - J_A - J_F] \quad [38]$$

where $J_{DL}$ and $J_{CL}$ are as previously defined, and the other fluxes are defined by subscripts denoting transformation in the liquid phase ($TL$), transformation in the sorbed phase ($TS$), plant uptake ($U$), runoff of dissolved pesticide ($R$), loss of pesticide sorbed to eroded material ($ES$), pesticide applied directly to the soil surface ($A$) and washoff of foliar-applied pesticide ($F$). The mass balance equations for the subsurface zones are identical to Eq. [38], except that the terms $J_R$, $J_F$, and $J_{ES}$ are dropped. Note also that the term $J_A$ applies to subsurface zones only when pesticides are incorporated into the soil.

PRZM does not consider the volatile flux of the applied pesticide, but does include several processes (surface runoff, foliar washoff, and erosion), not included in LEACHM. In PRZM, degradation in the sorbed phase, included conceptually in Eq. [38], is lumped with solution-phase degradation. The functional forms used to represent each process are similar to those in LEACHM, e.g., first-order kinetics for degradation and linear,instantaneous, and reversible equilibrium sorption. Other processes are modeled in PRZM at conceptually consistent levels of resolution. For example, loss of pesticide because of runoff is given by,

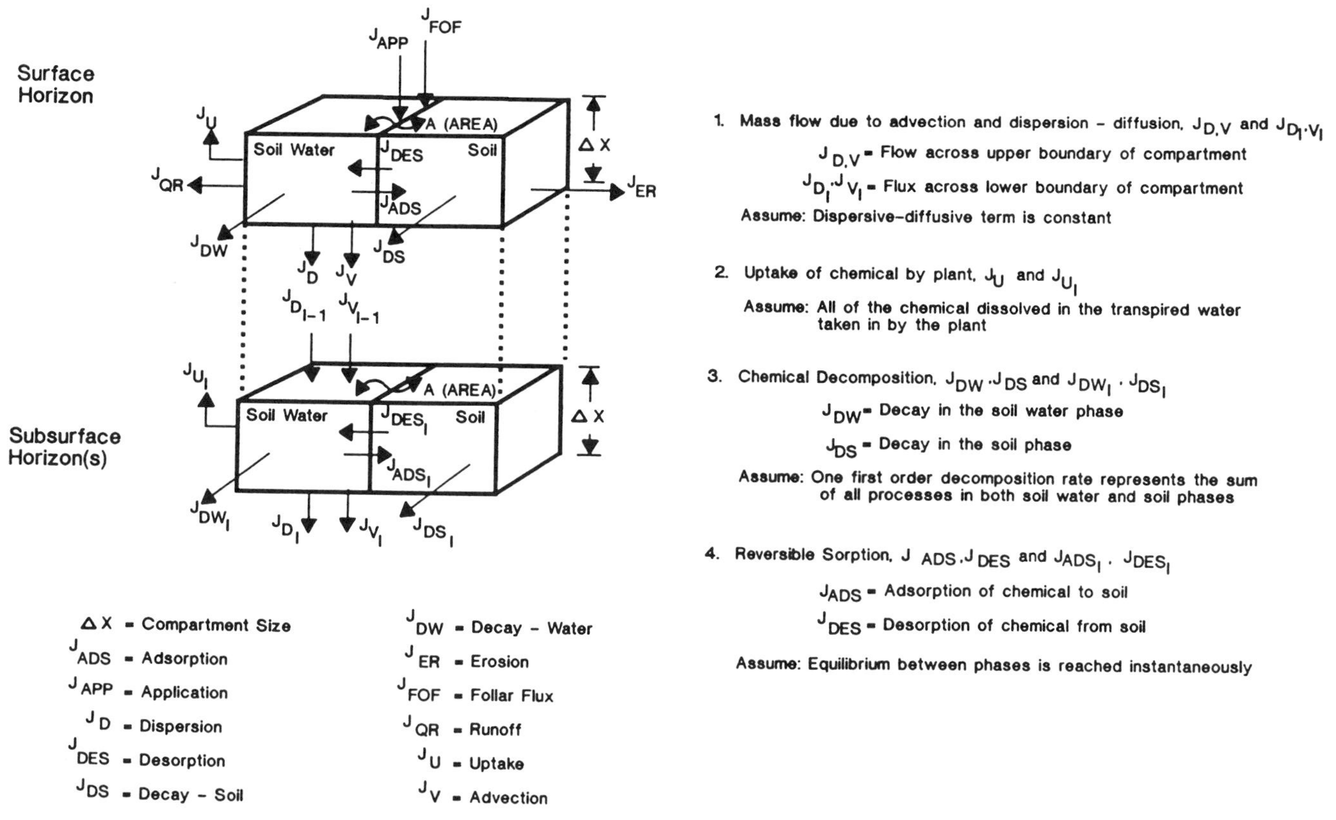


Fig. 10–3. Schematic representation of the processes included in the model PRZM. Taken from Carsel et al. (1984).

$$J_{\mathrm{R}} = \frac{Q}{A_{\mathrm{W}}} c_{\mathrm{L1}} \qquad [39]$$

and pesticide loss because of erosion is calculated by,

$$J_{\mathrm{ES}} = \frac{a\, X_{\mathrm{e}}\, r_{\mathrm{OM}}\, K_{\mathrm{d}}\, c_{\mathrm{L1}}}{A_{\mathrm{W}}} \qquad [40]$$

where $Q$ is the daily runoff depth ($L$), $A_{\mathrm{W}}$ is the watershed area ($L^2$), $c_{\mathrm{L1}}$ the dissolved pesticide concentration at the surface ($M/L^3$), $X_{\mathrm{e}}$ is the erosion sediment loss ($M$), $r_{\mathrm{OM}}$ the enrichment ratio for organic matter, $a$ is a unit conversion factor, and $K_{\mathrm{d}}$ has been defined earlier. These factors, and others included in other defined fluxes, must be either entered as input or calculated using internal routines. Alternatively, default values available in the PRZM library can be used. For management purposes, in which only the approximate fate of the pesticide is to be estimated, the default values may provide useful estimates of pesticide behavior. If, in fact, the particular case being studied represents substantially different conditions than those used in developing the default values, inaccuracies may result.

Equation [38] requires information on soil-water content ($\theta$) and soil-water flux ($q$) in the same manner as required above in Eq. [15]. Although recognizing the importance of accurate description of soil-water flux to accurately estimate pesticide displacement using Eq. [38], PRZM does not use numerical methods to provide this estimate through solution of Eq. [36]. Rather, a much more simplified and approximate representation of water flow is employed. Water balance equations are separately developed for: (i) the surface zone, (ii) horizons comprising the active root zones, and (iii) the remaining lower horizons within the unsaturated zone.

The equations employed are as follows. For the surface zone,

$$(SW)_1^{t+1} = (SW)_1^t + P + SM - I_1 - Q - E_1. \qquad [41]$$

For the root zone,

$$(SW)_1^{t+1} = (SW)_1^t + I_{i-1}. \qquad [42]$$

For the unsaturated zone beyond the maximum rooting depth,

$$(SW)_1^{t+1} = (SW) + I_{i-2} - I_i \qquad [43]$$

where $(SW)_i^t$ is the soil water in layer "i" of the noted zone on day "t", P is precipitation as rainfall minus crop interception, $SM$ is snowmelt, $Q$ is runoff, $E_i$ is evaporation, $U_i$ is transpiration, and $I_i$ is percolation out of zone i.

The USDA-SCS curve number approach (Haith & Loehr, 1979) is used in PRZM for estimating runoff; this precludes the use of the Richards equation for calculating the soil-water percolation term, $I_i$. Instead, two options based on empirical drainage rules are used in PRZM. The first option uses the soil-water contents at field capacity and permanent wilting point as the upper and lower limits, respectively, for specifying the water-storage capacity for each soil layer. If water infiltrating into a given zone is more than that needed to increase the soil-water content to the field-capacity value, the excess water is allowed to percolate into the next zone. Soil-water drainage and redistribution throughout the soil profile are assumed to be completed within 1 d. Plant uptake of water is not permitted to decrease soil-water content below the permanent wilting point value. The drainage rule employed in the second option allows for soil-water redistribution to occur over several days. Each soil zone is presumed to have a maximum soil-water content (e.g., saturated water content) during water infiltration, which decreases in an exponential manner to the field-capacity value over several days. The rate at which redistribution takes place is then specified by a drainage rate parameter.

Daily updating of soil water in the soil profile using the above equations requires additional calculations for runoff, snowmelt, evaporation, and transpiration. Precipitation, pan evaporation, or air temperature are inputs providing the potential energy from which evapotranspiration is estimated. Carsel et al. (1984) described additional details on these calculations.

Although comprehensive in the number of processes considered, PRZM is built upon two simplifications of transport processes that may limit its usefulness for quantitative estimation of pesticide fate. First, the use of a water balance approach and empirical drainage rules to estimate water fluxes in the soil profile will provide only qualitative estimates of water flux. This is particularly true when the soil profile is separated into only three zones (Eq. [41]–[43]). This approach greatly simplifies the water-flow submodel in PRZM, as it is unnecessary to use any finite difference solution to Eq. [36] or to provide any estimate of K-$\theta$-h relations. The model thereby executes rapidly and with a minimum of input data. Such an approach applied to water flow in sandy soil represents a more reasonable approximation of basic process than when applied to a more fine-textured soil. However, the ability of PRZM to describe actual field-measured water contents or water fluxes in any of these cases is not yet well established. It is also unclear what relative error in estimated pesticide movement is produced by the use in PRZM of these approximate estimates of $q$ and $\theta$. They are key variables in the subsequent description of pesticide transport via the use of a finite-difference form of Eq. [38].

A second simplification of transport processes used in PRZM focuses on the use of numerical dispersion to simulate the effects of hydrodynamic dispersion. Numerical dispersion refers to the artifactual transport of chemical to deeper depths than are real because of averaging procedures employed to calculate solute distribution between adjacent nodal positions in a finite difference network. This neglect (or adoption) of numerical dispersion somewhat simplifies the finite difference solution of Eq. [38], as additional expansion of the second-order derivative term, as Bresler (1973) suggested, need not be included to correct for the numerical dispersion. However, by using such dispersion, estimates produced by PRZM of pesticide dispersion during transport are greatly a function of the node spacing (size of depth increment). This could result in an estimate of more pesticide dispersion in the profile than would actually be observed. The consequence is that PRZM would overestimate the leaching of an applied pesticide, and predict movement of pesticide to groundwater when, in fact, no such movement was real according to physical/chemical processes operative in the soil. From the perspective of environmental protection, this may not be entirely bad, as PRZM thereby provides conservative estimates in terms of its evaluation of potential hazards posed by pesticides. It can also be argued, however, that certain pesticides might be judged to pose hazards when no such hazards in fact exist.

Dean and Atwood (1987) have recently incorporated several modifications in the PRZM model to include algorithms for: (i) the fate of daughter products, (ii) simulating application of irrigation, and (iii) allowing lateral drainage of water when perched water tables may be present as a result of impeding layers. Some other changes in PRZM are being made by the model's original developers (R.F. Carsel, 1988, personal communication) to overcome many of the limitations discussed in the foregoing paragraphs. For example, these modifications may include replacement of the empirical drainage rules with an algorithm based on Darcy's law; the use of an improved numerical scheme to correct for numerical dispersion; and inclusion of algorithms for daughter products. A new version of PRZM containing these improvements is expected to be released some time during late 1989 under the name RUSTIC.

## 10–8 MODELS FOR INSTRUCTIONAL PURPOSES

Rao et al. (1976) proposed a management model, based on the concept of piston displacement of water and solutes in soils, to describe the episodic nature of adsorbed solute movement in a uniform soil profile resulting from several rainfall/irrigation events. In this approach, the spreading of the solute pulse resulting from hydrodynamic dispersion is ignored, and the following equations were proposed for estimating the *position* of the solute pulse (either the peak or the leading front):

$$Z_i = Z_{i-1} + [(I_i - I_d)/(\theta_{fc} R)]; \qquad I_d < I_i \tag{44}$$

$$Z_i = Z_{i-1}; \qquad I_d \geq I_i \tag{45}$$

where

$$I_d = \int_0^{Z_{i-1}} [\theta_{fc} - \theta(z)]\, dz, \tag{46}$$

$Z_i$ and $Z_{i-1}$ are the depths ($L$) at which the solute pulse is located after the $i$th and $(i-1)$th events; $I_i$ is the amount ($L$) of water *infiltrating* into the soil profile for the $i$th event; $I_d$ is the amount ($L$) of soil-water deficit resulting from water extraction by plant roots located above the depth $Z_i$; $\theta(z)$ is the volumetric soil-water content ($L^3/L^3$) distribution within the root zone; $\theta_{fc}$ is the field-capacity soil-water content ($L^3/L^3$), which is usually taken to be the volumetric water content at a soil-water potential of $-10$ kPa ($-0.1$ bar); $z$ is the soil depth ($L$); and $R$ is the retardation factor to account for sorption effects as defined earlier.

In the above equations, the values for soil-water deficit, $I_d$, can be estimated by an appropriate submodel that accounts for the daily variations in potential evapotranspirationd emand (PET), root-density distribution $[r(z, t)]$ changes, and soil-water uptake by plant roots. The computational aspects of this submodel can, however, be considerably simplified if it is assumed that root distribution is uniform with depth and time, and by inputting the PET values.

Note from Eq. [44] and [45] that a given rainfall/irrigation causes the solute pulse to be leached only when the amount of water infiltrating into the profile ($I_i$) is larger than the soil-water deficit ($I_d$) created above the solute peak as a result of soil-water uptake by plant roots. Thus, as the solute pulse leaches to lower depths or with increased time elapsed between successive rain/irrigation events, the amount of water infiltrating into the soil, $I_e = (I_i - I_d)$, needs to be larger in order for the solute pulse to be leached deeper than its current position. This modeling approach, based on simplified concepts, provides a means of evaluating the interacting effects of the patterns of rainfall/irrigation, evapotranspiration, sorption,and water-holding capacity of the soil on solute leaching.

Nofziger and Hornsby (1985) used the piston displacement concept, as described above, in developing an interactive microcomputer software package, CMIS, for calculating pesticide movement in soils. In addition to the above equations, degradation of the pesticide was accounted for by assuming first-order kinetics:

$$M(t) = M_o \exp(-\mu t); \; \mu = (0.693/t_{0.5}) \qquad [46]$$

where $M_o$ is the initial amount of pesticide applied ($M$); $M(t)$ is the amount ($M$) remaining at time $t$; $\mu$ is the first-order rate constant for degradation ($1/T$); and $t_{0.5}$ is the degradation half-life ($T$). CMIS does not consider vapor-liquid partitioning of the pesticide; thus, volatilization losses and vapor transport are not calculated by this model.

CMIS contains menu-driven modules for storing, retrieving, and editing files for soils data (soil-water content at $-10$ and 1500 kPa [$-0.1$ and $-15$ bars, respectively]), weather records (rainfall and evapotranspiration data), and pesticide data (sorption coefficient, $K_{oc}$, and degradation half-life, $t_{0.5}$) needed to calculate pesticide leaching and degradation. This software provides both graphical and tabular outputs of the solute pulse depth

($Z_i$) and the fraction of the amount of the applied pesticide, $M(t)/M_o$, still remaining. The interactive and graphic features of this software, and the availability of microcomputers has made CMIS an attractive and popular instructional tool both in the classroom environment and for training extension personnel.

CMIS assumes that the soil properties and root density are uniform with depth. To overcome this limitation, Nofziger and Hornsby (1986, 1987) modified CMIS to permit calculation of pesticide movement in a *layered* soil profile. This software, Chemical Movement in Layered Soil (CMLS), allows the specification of soil properties for up to 25 layers or horizons of varying thickness; soil properties within each layer are assumed to be uniform. In the most recent version of CMLS (Nofziger & Hornsby, 1987), different values for the degradation rate constant can be specified for each soil layer. The fraction of the applied pesticide degrading within each layer is then calculated by assuming first-order kinetics and the residence time within each layer.

Nofziger and Hornsby (1986) discuss the limitations of the simplifying assumptions made in developing CMLS algorithms. The assumptions they made in computing water flow and solute leaching are likely to be better suited for coarse-textured soils that do not contain any layers that impede water flow and, as a result, water redistributes rapidly ($<1$ d). Since the dynamics of soil-water redistribution following an infiltration event are not explicitly accounted for, in finer-textured soils that drain more slowly, the calculated depths would be associated with elapsed time several days longer than those estimated by CMLS. Thus, pesticide leaching in soil profiles with argillic or spodic horizons cannot be described using CMLS. Similarly, upward movement of water is not considered in CMLS and, as a result, simulation of the possible upward migration of a pesticide pulse is neglected. Finally, the root density distribution, $r(z, t)$, is assumed to be uniform with depth and constant over time. Neither of these assumptions is valid for a growing crop.

As is evident from the foregoing discussion, several assumptions simplify the equations for water flow and solute transport to algebraic expressions, reducing the computational aspects to fairly straight forward bookkeeping chores. Thus, for example, $<15$ s are needed for CMLS to compute the progressive leaching of a pesticide pulse during the course of a year. (This time is for CMLS running on an IBM/PC-AT equipped with a math co-processor.) In contrast, LEACHM and PRZM both require longer elapsed time on a microcomputer to complete the simulation of pesticide leaching over 1-yr. Neither of the models provides graphical outputs of the model simulations in an interactive mode. While some guidance is provided in the user's manuals for estimation of model input parameter values, on-line editing and user interaction are not as easy with either LEACHM or PRZM. Thus, CMLS is best suited for *interactive* simulations so essential for instructional purposes. The simplistic representations of the flow processes, however, may not be appropriate for *predictive* purposes, except perhaps in well-drained sandy soils.

## 10–9 MODEL APPLICATIONS

### 10–9.1 Research Models

The LEACHM model has recently been used to evaluate several cases of pesticide fate in field soils. One of these, an assessment of transport and transformations of the nematicide aldicarb (Wagenet & Hutson, 1986), provides a good ilustration of the usefulness of such a research model. A synopsis of the application and interpretation of results follows.

A field study of aldicarb movement was conducted on Palmyra soil (sandy skeletal, mixed mesic, Glossoboric Hapludalfs), at a site with <1% slope. There was no artificial drainage, no water table within 10 m of the surface, and no history of aldicarb application to the field. Soil samples were collected from the field site on 26 May prior to planting and any application of aldicarb. On 7 June (Day 0), the field was planted to potato (*Solanum tuberosum* L.) and aldicarb applied in granular form as a sidedressing at 3.36 kg of active ingredient (a.i.) per hectare. The potato rows were spaced 0.75 m (30 in.) apart, and normal growing practices were followed in all respects during the study. Although it was dry during most of the summer, the potato plants were not irrigated, but depended upon occasional rainfall. Water movement resulting from this rainfall was the driving force for aldicarb leaching during the study period. Potato crops were harvested in late October. Soil samples were collected at approximately 29 d (1 month), 64 d (2 months), and 124 d (4 months) after planting. A bucket auger was used to collect soil samples from within the potato row, equidistant between adjacent plants to a maximum depth of 1.5 m or a shallower depth when sampling was limited by stones. Soil samples were separated into five increments of 0.0 to 0.3, 0.3 to 0.6, 0.6 to 0.9, 0.9 to 1.2, and 1.2 to 1.5 m.

The LEACHM model was applied to the results of this study. Sequential, first-order transformations of aldicarb and its oxidative metabolites have been demonstrated by Zhong et al. (1986), hence, Eq. [15] with $\phi$ appropriately modified in each case, was used to individually simulate the simultaneous transport and transformations of aldicarb ($c_1$), aldicarb sulfoxide ($c_2$), and aldicarb sulfone ($c_3$). Field-measured soil-water contents ($\theta$) and aldicarb residues concentrations were spatially variable. The mean (m) and standard deviation (SD) of measured values (Table 10–4) indicated that variability in $\theta$ increased with depth, and that measured total carbamate residue concentrations ($c_T = c_1 + c_2 + c_3$) were quite variable, even though these residues were almost completely confined to the surface layer (0–0.3 m). In an attempt to accommodate this variability, two simulations were performed, referred to as Case 1 and 2, which differed only in the soil hydraulic properties and which were designed to approximate the water-flow variability that existed in the field. Recognizing that three-textural layers (0–0.6, 0.6–1.2, and 1.2–1.5 m) existed, three different K-$\theta$-h relationships were used in each simulation with an example of the generated K-$\theta$-h relationships shown in Fig. 10–4.

Table 10–4. Volumetric water content and total aldicarb measured and estimated by LEACHM as a function of depth and time. From Wagenet and Hutson (1986).

| | Day 29 | | | | Day 64 | | | | Day 124 | | | |
|---|---|---|---|---|---|---|---|---|---|---|---|---|
| | Measured | | Calculated | | Measured | | Calculated | | Measured | | Calculated | |
| Depth, m | $\mu$ | $\sigma$ | Case 1 | Case 2 | $\mu$ | $\sigma$ | Case 1 | Case 2 | $\mu$ | $\sigma$ | Case 1 | Case 2 |
| | | | | | Water content, $m^2$ $m^{-3}$ | | | | | | | |
| 0.0–0.3 | 0.13 | 0.01 | 0.16 | 0.11 | 0.10 | 0.01 | 0.13 | 0.09 | 0.14 | 0.01 | 0.14 | 0.12 |
| 0.3–0.6 | 0.11 | 0.01 | 0.15 | 0.13 | 0.09 | 0.01 | 0.13 | 0.09 | 0.11 | 0.01 | 0.15 | 0.09 |
| 0.6–0.9 | 0.14 | 0.03 | 0.20 | 0.20 | 0.14 | 0.03 | 0.19 | 0.16 | 0.13 | 0.03 | 0.16 | 0.11 |
| 0.9–1.2 | 0.21 | 0.06 | 0.19 | 0.19 | 0.20 | 0.04 | 0.19 | 0.18 | 0.16 | 0.06 | 0.20 | 0.18 |
| 1.2–1.5 | 0.26 | 0.06 | 0.24 | 0.27 | 0.26 | 0.05 | 0.25 | 0.29 | 0.22 | 0.05 | 0.25 | 0.25 |
| | | | | | Total aldicarb, $\mu g$ $kg^{-1}$ | | | | | | | |
| 0.0–0.3 | 502 | 196 | 445 | 478 | 424 | 296 | 218 | 270 | 25 | 4 | 37 | 41 |
| 0.3–0.6 | ND‡ | ND | 1 | 4 | ND | ND | 1 | 3 | 11 | † | 16 | 51 |
| 0.6–0.9 | -- | -- | 0 | 0 | -- | -- | 0 | 0 | -- | -- | 0 | 1 |
| 0.9–1.2 | -- | -- | 0 | 0 | -- | -- | 0 | 0 | -- | -- | 0 | 0 |
| 1.2–1.5 | -- | -- | 0 | 0 | -- | -- | 0 | 0 | -- | -- | 0 | 0 |

† One value only, the other three replicates were nondetected. ‡ ND = nondetected.

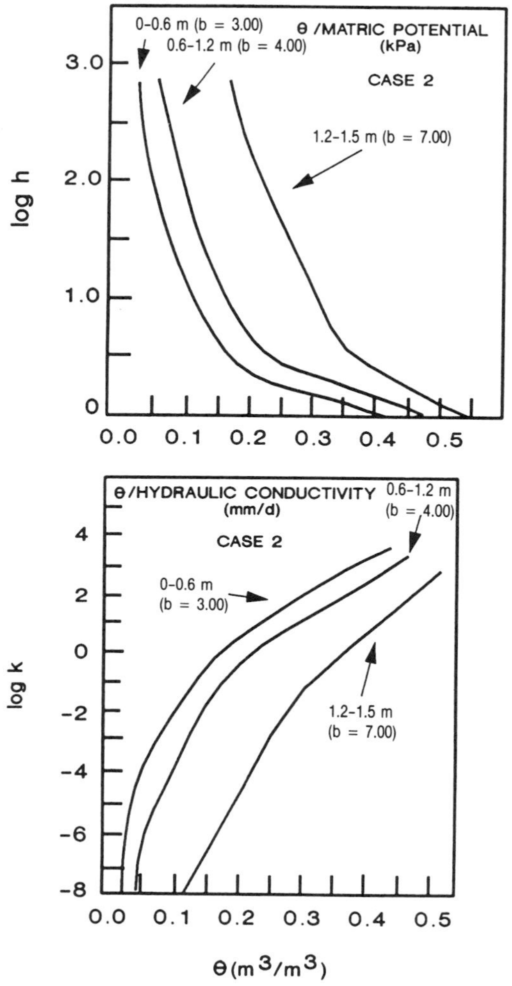


Fig. 10–4. Soil hydraulic properties used in the sensitivity analysis of LEACHM for simulation of aldicarb leaching at the New York field site. Taken from Wagenet and Hutson (1986).

Comparison of measured and simulated values (Fig. 10–5) of both $\theta$ and $c_T$ illustrate generally good agreement, particularly when variability in the measurements is considered. Lack of better agreement could be the result of several inaccuracies in the model, including: (i) characterizationof K-$\theta$-h relationships with depth, (ii) description of plant upake of water, and (iii) values of rate coefficients used in the hydrolysis and oxidation transformation processes. Considering the general lack of information for this study on the first two of these important mediating factors, the relatively good agreement presented here is encouraging. It is interesting to note that Cases 1 and 2 provide generally similar estimation of aldicarb fate, even though

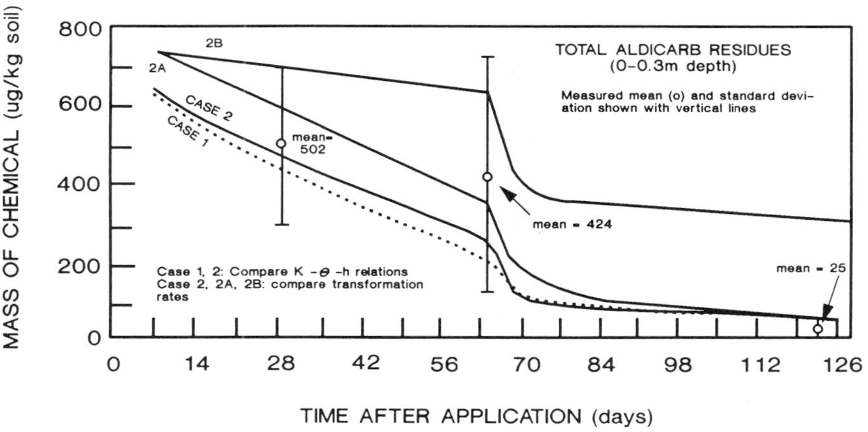


Fig. 10–5. Simulated and measured values of aldicarb residues at the New York field site. LEACHM simulations representing variability in soil hydraulic properties (Cases 1 and 2) and in pesticide degradation rate constants (Cases 2, 2A, and 2B) are shown as solid and dashed lines. Taken from Wagenet and Hutson (1986).

these two cases were different in soil hydraulic properties. It appears that the presence of an actively transpiring root system and the occurrence of transient upward and downward fluxes of water (both of which are considered in the model), may result in simulated pesticide concentration profiles that do not differ markedly. The sources of variability in the measured aldicarb residue concentrations include: (i) cross contamination of soil strata during the sampling operation, (ii) water and solute movement via mechanisms not considered in the models (e.g., macropores), and (iii) extreme relationships between K-$\theta$-h not included in Cases 1 and 2.

Values of $\mu_1$, $\mu_2$, and $\mu_3$ (1/d) used here were taken from a laboratory study of aldicarb transformations in Palmyra soil (Zhong et al., 1986), and for Case 1 were (1/d) $\mu_1 = 0.3840$, $\mu_2 = 0.0528$, and $\mu_3 = 0.0432$. Values of $\mu_1'$ and $\mu_2'$ reported by Zhong et al. ($\mu_1' = 0.2160$; $\mu_2' = 0.019$/d) were decreased for Case 2 by approximately 50% ($\mu_1' = 0.1160$; $\mu_2' = 0.0092$/d) to obtain simulations reported, as greater values underestimated measured aldicarb residues. The degradation pattern produced by the combined effect of these five rate constants is evident by considering the presence of total aldicarb in the 0- to 0.3-m depth (Fig. 10–5), as almost no leaching was predicted to occur beyond that point. The sharp discontinuity at time 63 d is because of two rainfall events on Days 66 and 67, each totaling 39 mm of water.

Cases 2, 2A, and 2B illustrate the sensitivity of LEACHM simulations to the values used for $\mu_i$, and demonstrate the importance of accurate representation of pesticide transformation processes. Case 2A used values of $\mu_i$ identical to Case 2, except $\mu_1'$ was decreased to 0.9116/d. Case 2B used values of $\mu_i$ that were one-tenth those used in Case 2. It was possible, using these values of $\mu_i$, to describe much of the variability in measured total aldicarb residues, although no single set of $\mu_i$ values was entirely accurate.

These simulations infer that the microbial degradation process was both spatially and temporally variable in the field, making it necessary to vary $\mu_i$ with time, as well as over space (represented here by the three simulated cases). A small value of $\mu_i$ early in the season, and a larger value later, perhaps because of increased temperature or microbial populations, would have resulted in fairly accurate simulation of measured values. Information is currently lacking on such temporal behavior of $\mu_i$, but the relative importance of variability in hydraulic properties and microbial degradation (Fig. 10–5) is unresolved. The relationships present in Fig. 10–5 also indicate that laboratory-derived values of $\mu_i$ may not always represent the variation in degradation rates to be expected in the field, but can provide a reasonable first estimate of such processes.

### 10–9.2 Management Models

Carsel et al. (1987) have recently used the management model PRZM to characterize the uncertainty of pesticide leaching in agricultural soils. The usefulness of this model for such purposes was demonstrated for aldicarb using a Monte Carlo simulation technique. A review of this application will illustrate one use of this modeling approach.

Carsel et al. (1987) selected a site located in the eastern Corn Belt (Ohio) for evaluation. This site was selected because field studies (Baker, 1983, 1985) conducted within the basin provided hydrologic (surface runoff) calibration data needed for providing independent estimates of PRZM parameters. Parameter requirements for PRZM fall into four broad categories: hydrology, crop, soil, and chemical. Generation of input for each of these categories is next explained.

Pesticide movement will be largely dependent upon the volume of water recharged and, as a result, will vary from site to site and from year to year within a given site. A population of rainfall totals is required to account for such variations in the rainfall/recharge potential. PRZM requires daily rainfall and pan evaporation totals (centimeters) and average daily temperature (degrees Celsius) as meteorologic input. A 29-yr precipitation record available for the Ohio site was incorporated into the Monte Carlo simulations. Runoff was estimated with the USDA-SCS curve number method (Haith & Loehr, 1979) with the resulting infiltrating water defined rather simply in the context of two commonly reported bulk soil-water-holding characteristics for agricultural soils: field capacity and wilting point ($m^3/m^3$). Field capacity and wilting point are used operationally to define two reference states in each soil layer for predicting soil-water percolation. Although simplistic and lacking in theoretical and physical rigor, this concept has been used in some water-flow models (Baes & Sharp, 1983; Haith & Loehr, 1979; Stewart et al., 1976). The actual values of field capacity, wilting point, and USDA-SCS curve numbers used were calculated from estimated soil properties.

A total 2942 reported soil series in 40 states were evaluated for several soil characteristics required by PRZM for estimating water and pesticide velocity (Carsel et al., 1984) including soil bulk density, percentage of soil organic matter, and soil-water contents at field capacity and wilting point. The soil series were first sorted according to their hydrologic class, with a total of 186, 1216, 870, and 679 series found in USDA-SCS classes A, B, C, and D, respectively. The soils were further evaluated by strata (0–30, 30–60, 60–90, and 90–120 cm). A final sorting by hydrological class and depth for each characteristic indicated that considerable variation existed between and within the hydrological classes.

The value of $K_{oc} = 20 \pm 6$ L/kg for aldicarb was used based upon work of Jones (1986). Using this $K_{oc}$ value, $K_d$ was calculated from randomly selected organic matter contents provided by the Monte Carlo methods. Degradation of aldicarb was recognized to be probably spatially variable, with a range of transformation rates (Jones, 1986) reported to be from 0.012 to 0.023/d, with a mean of 0.015/d. A triangular distribution was assumed for the Monte Carlo simulations using the mean and range of these field-measured values for the degradation rate constants.

The crop-planting window may have an impact on pesticide movement because of variances in water recharge during early to late spring. To examine this, Carsel et al. (1988b) assumed that corn (*Zea mays* L.) was to have been planted at the Ohio site between 20 April and 30 May; model sensitivity tests were conducted with PRZM using an early and late planting window. No significant differences in pesticide leaching were found. Conditions for 5 May, the planting date most commonly reported, were incorporated into the simulations. An aldicarb application rate of 1.68 kg/ha at an incorporation depth of 5 cm was simulated.

A Monte Carlo numerical analysis of the variation in pesticide movement requires the probability density function (PDF) for parameters that affect water and solute movement in soils. These PDFs were obtained by fitting the empirical distributions for each soil characteristic for each hydrologic group and depth used in the simulation (for details, see Carsel & Parrish, 1988). A sample population sufficient to establish inferential statements is required as output from the Monte Carlo exercise. Persuad et al. (1985) determined that a sample population of 200 was adequate to describe noninteracting solute movement in a heterogeneous soil as affected by hydrodynamic dispersion and pore-fluid velocity using a steady-state model. An initial sample size of 2000 was selected to provide a comparison between levels of inference and sample size because the objective of this study was to establish statements of inference.

The PRZM-simulated results of aldicarb leaching (kilogram per hectare) for the 2000 cases were presented by Carsel et al. (1987) as a cumulative probability distribution (Fig. 10–6). The simulated data show that, for greater than 80% of the simulated scenarios, <0.01 kg/ha leached below the 150-cm depth. Similarly, in <5% of the simulated cases did 0.01 kg/ha reach the 200-cm depth. Conversely, the simulation data also indicate that >0.01 kg/ha leach below the 150-cm depth for at least 15% of the scenarios and at least 3% for the 300-cm depth. Unpublished sensitivity analyses conducted by Carsel et al. with a groundwater model using data from recent field evaluations (Rothschild et al., 1982) have demonstrated that aldicarb loadings of approx-

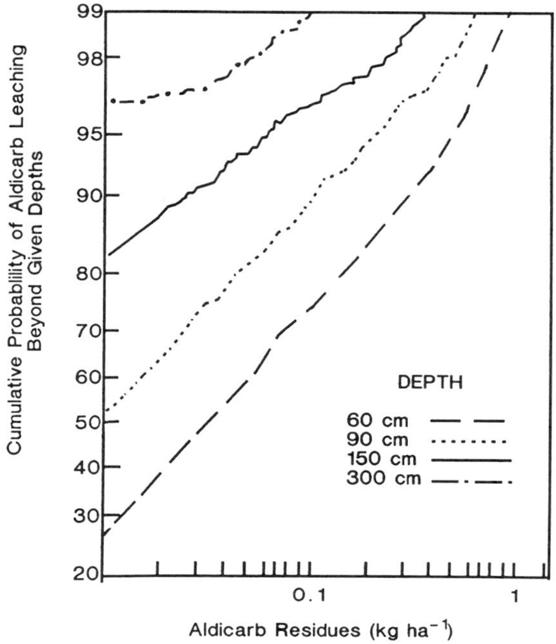


Fig. 10–6. Cumulative probability distributions for aldicarb residues leaching beyond selected profile depths. Taken from Carsel et al. (1988).

imately 0.01 kg/ha or less, will represent concentrations at or below the level of detection (1 $\mu$g/L) in groundwater. If an assumption is made that all of the predicted annual load enters the groundwater at a unit time, 0.1 and 1.0 kg/ha would result in approximately 15 and 150 $\mu$g/L, respectively, in the groundwater. The probability of an annual pesticide loading being >0.1 kg/ha for the 150-cm depth is approximately 5%; for the 300-cm depth, the probability is <1%.

Monte Carlo analyses will provide the associated uncertainty and relative risk of pesticide leaching. One disadvantage, however, is that they do not provide the cause-effect relationships. Because national means and variances for field capacity, wilting point, and organic matter were used in this analyses and may not reflect regional distributions, and because the data base contained soils that may not be used for agriculture (e.g., high clay content or total sands), the individual combinations simulated and PRZM predictions were tabulated and examined. Simple correlations of the amount leached were found for certain rainfall patterns, slower transformation rates, low organic matter, and lower field capacity. The highest correlation was with field capacity and is displayed in Fig. 10–5. In soils with field capacities from approximately 0.08 to 0.09 $m^3/m^3$, the amount of pesticide leached below the 150-cm depth exceeded 0.1 kg/ha for 50 to 90% of the scenarios. These simulation results reported by Carsel et al. (1988b) indicate that in the eastern

Corn Belt soils with field capacities generally $<0.10\ m^3/m^3$ and water tables generally at depths $<300$ cm have the most potential for the appearance of aldicarb residues in groundwater.

Additional investigations were undertaken by Carsel et al. (1988b) using a Soils Information Retrieval System (SIRS) data base (Goran, 1983) and from site-specific hydrologic investigations (Baker, 1983; 1985). These studies showed that approximately 1% of the land in Ohio is classified as sands, loamy sands, and sandy loams. Further analysis indicated that corn is grown on these soil types with Oshtemo sandy loam (coarse-loamy, mixed, mesic Typic Hapludalfs), Spinks loamy sand (sandy, mixed, mesic Psammentic Hapludalfs), and Gilman loam [coarse-loamy, mixed (calcareous) hyperthermic Typic Torrifluvents] being typical for Ohio. The investigations also revealed that depth to groundwater for these soils is $>180$ cm. These data were then used to conduct a sensitivity analysis using the Oshtemo sandy loam (because it had the lowest reported values for subsurface water-holding capacity and organic matter content). A water table depth of 180 cm was used. Simulations were made using an annual application rate (1.68 kg/ha) for the 29-yr rainfall record. The simulations showed that the annual loadings predicted for aldicarb leaching past the 180-cm depth were $<0.01$ kg/ha.

The foregoing simulation studies show that aldicarb would not be expected to appear in groundwater in the Ohio corn belt. If simulations had predicted higher pesticide loading rates to groundwater, additional evaluations would have been required. These evaluations would include linking PRZM to a groundwater contaminant transport model (e.g., R.L. Jones, 1983; unpublished data; Jones et al., 1983) to evaluate the magnitude and frequency of the pesticide residues and distances of migration from treated areas. If these calculations were to indicate levels of concern (e.g., an exceedance of guideline standards), appropriate field evaluations, monitoring or restrictions could be initiated. For example, management practices such as the timing of pesticide applications and restrictions to prevent usage near potable wells have been used in Florida and the northeastern USA.

### 10–9.3 Coupling of Models for the Unsaturated and Saturated Zones

Although there are some sophisticated and simplified models available for describing solute transport in the saturated zone, only a few attempts have been made to couple them to the unsaturated zone pesticide models discussed in the previous sections. While the research models (e.g., LEACHM) have not been coupled to saturated zone models, PRZM outputs (i.e., time series of pesticide loadings at some specified depth) have been used as inputs for groundwater models (Jones et al., 1987; Dean & Atwood, 1987; Carsel et al., 1988a).

Field data collected over a 3-yr period to monitor aldicarb residues in shallow groundwater ($\leq 7$ m below ground surface) beneath a citrus grove were used by Jones et al. (1987) to evaluate simple one- and two-dimensional groundwater models based on a finite-element numerical solution to the advective-dispersive transport equation (Wang & Anderson, 1982). The transport and transformations of aldicarb in the unsaturated zone were also monitored, and these data were used to evaluate PRZM simulations (Hornsby et al., 1983; Jones et al., 1983, 1988). Pesticide loadings to the water table were simulated using PRZM and used as inputs to the groundwater model. Results of the unsaturated zone simulations indicated that about 20 to 30% of the surface-applied aldicarb had entered the shallow groundwater (water table at 3–7 m beneath the 1.7-ha treated area). A range of values for the aldicarb degradation rate constant were used in the two-dimensional version of the saturated-zone model, and the simulations were compared with monitoring data (more than 2000 samples from 174 monitoring wells). Jones et al. (1987) also used the one-dimensional simulation model to perform simulations using a range of degradation rates and water table depths to estimate how far aldiarb residues would have to travel in the saturated zone before concentrations decrease to $\leq 10$ $\mu$g/L. Results of their simulations for 12 scenarios, summarized in Table 10–5, showed that aldicarb residues do not exceed a concentration of 10 $\mu$g/L in shallow groundwater at distances >300 m from the treated area even at sites where aldicarb is applied annually. Carsel et al. (1988a) have used the same modeling approach as Jones et al. (1987) and employed Monte Carlo simulation technique to evaluate the potential threats of groundwater contamination from aldicarb use in North Carolina; this work is an extension to their unsaturated-zone modeling efforts of Carsel et al. (1987) discussed in section 10–9.2.

While Jones et al. (1987) examined aldicarb residues in groundwater at a single field site, Dean and Atwood (1987) examined the same problem for the whole state of Florida. For this purpose, they used PRZM outputs as inputs into a three-dimensional groundwater model, CFEST (Gupta et al., 1982); a more sophisticated groundwater model was required to accommodate a wide range in hydrogeologic conditions found in Florida. For each unsaturated-zone scenario simulation, 14 yr of meteorologic records were

Table 10–5. Predicted movement of aldicarb residues as a function of depth to groundwater and degradation rate.† Adapted from Jones et al. (1987).

| Depth to groundwater | Applied pesticide entering saturated zone | Degradation rate | Maximum distance (m) where residue concentration exceeds 10 $\mu$g/L: Single application | Yearly applications |
|---|---|---|---|---|
| m | % | (half-life in days) | | |
| 1.5 | 38 | 182 | 120 | 132 |
| | | 365 | 224 | 288 |
| 3.0 | 31 | 182 | 108 | 120 |
| | | 365 | 216 | 270 |
| 7.2 | 20 | 182 | 78 | 102 |
| | | 365 | 156 | 240 |

† Note: Parameters used in these calculations include a groundwater velocity of 0.15 m/d and an application rate of 5.6 kg/ha (made on 15 Feb. in each of the 5 yr in the simulation period for the cases with yearly applications, and in only the first year in single application cases.

used with PRZM to generate a time series and frequency distributions of pesticide loadings to groundwater. These loadings were then used in CFEST model to simulate the impact of various factors (e.g., distance from source area to a well; size of the source area; and pesticide loadings as determined by variations in soils and weather) on well-water concentrations of aldicarb residues.

Dean and Atwood's (1987) resuls for PRZM simulations are summarized in Fig. 10–7, where probability plots are shown that allow an estimation of the likelihood that a given value for annual loading of the pesticide to groundwater would be exceeded. These plots suggest that the highest pesticide loadings to groundwater would emanate from citrus groves planted on Ultisols and Entisols with a thin unsaturated zone; it was estimated that there is a 10% chance that loadings to groundwater will exceed 0.3 kg/ha. Pesticide loadings to groundwater from citrus groves planted on Alfisols and Spodosols as well as on Entisols and Ultisols with a thick unsaturated zone are likely to be small; under these conditions, it was estimated that there is only a 10% probability that loadings will exceed 0.01 kg/ha.

A summary of the Dean and Atwood's (1987) simulation results for the saturated zone scenarios is presented in Table 10–6, where the highest calculated pesticide concentrations in well-waters are shown for several combinations of unsaturated and saturated zones found in Florida's citrus-growing areas. The highest calculated concentration is 20.3 $\mu$g/L for the case of a

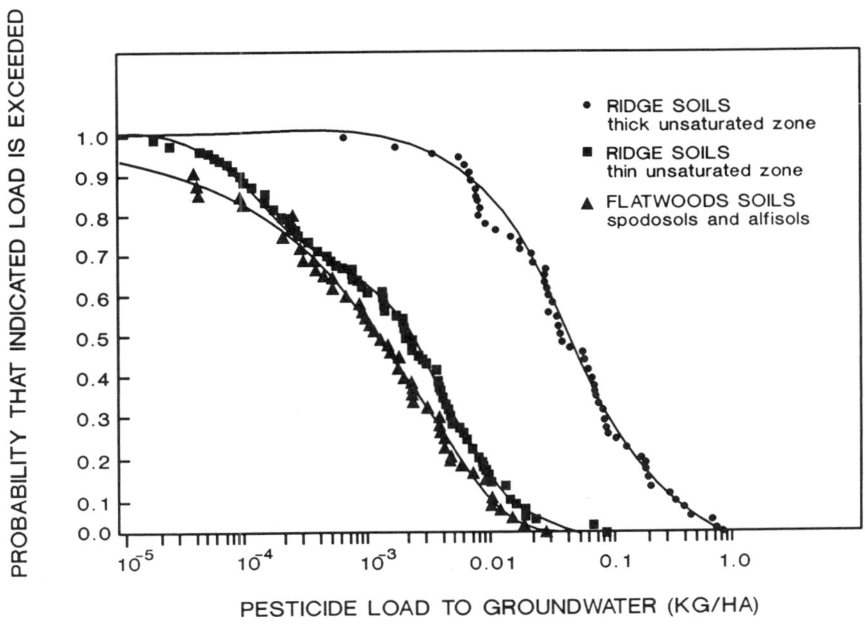


Fig. 10–7. Cumulative probability distributions for aldicarb loadings to groundwater for selected scenarios of use in Florida. Taken from Dean and Atwood (1987).

Table 10–6. Highest calculated aldicarb concentrations (in $\mu$g/L) in the given combined unsaturated/saturated categories.

| Category | | Flatwoods soils | Thick ridge soils | Thin ridge soils |
|---|---|---|---|---|
| | | Floridan worst cases | | |
| 9 144 cm | No decay | No overlap | 7.2E-2 | 2.2 |
| | Decay | No overlap | 1.2E-2 | 3.8E-1 |
| 30 480 cm | No decay | No overlap | 5.6E-2 | 1.7 |
| | Decay | No overlap | 3.1E-3 | 9.1E-2 |
| | | Floridan | | |
| 9 144 cm | No decay | No overlap | 6.9E-2 | 2.0 |
| | Decay | No overlap | 4.1E-4 | 1.2E-2 |
| 30 480 cm | No decay | No overlap | 2.8E-2 | 8.1E-1 |
| | Decay | No overlap | 4.4E-7 | 1.2E-5 |
| | | Surficial worst cases | | |
| 9 144 cm | No decay | 7.5E-1 | 6.9E-1 | 20.3 |
| | Decay | 1.5E-2 | 1.3E-2 | 4.1E-1 |
| 30 480 cm | No decay | 4.5E-1 | 4.1E-1 | 12.2 |
| | Decay | 1.4E-4 | 1.3E-4 | 4.1E-3 |
| | | Two-aquifer worst cases | | |
| 9 144 cm | No decay | 1.4E-2 | 1.2E-2 | 3.8E-1 |
| | Decay | 1.8E-3 | 1.6E-3 | 4.7E-2 |

surficial aquifer, with the well located at 91 m from the source area, and pesticide degradation in the saturated zone ignored. The regulatory standard in Florida for aldicarb in drinking water is 10 $\mu$g/L. Note that when pesticide degradation was accounted for, the concentrations decrease to 0.41 and $\leq$0.004 $\mu$g/L, respectively, in wells located 91 and 300 m away from the treated area. Based on these coupled simulations obtained using PRZM and CFEST models, Dean and Atwood (1987) proposed several recommendations for managing aldicard applications in Florida that would reduce the risk of groundwater concentrations exceeding the regulator standards. Results from field studies, groundwater monitoring programs, and model simulations as discussed above formed the basis for developing a regulatory policy (e.g., well-setback limits, timing of application, and depth of well casing) for aldicarb use in Florida.

## 10–10 CONCLUSION

The development and use of simulation models for describing pesticide fate in soils and groundwater has gained an increasing popularity in recent years. This is because of, in part, the need to have reliable tools for providing management and regulatory guidance in the efficacious, but environmentally safe, use of pesticides. Although site-specific investigations might be preferred, the economic and scientific resources required to conduct such studies are unrealistically high given the vast number of combinations of pesticide-soil-crop-weather-management scenarios that might need to be investigated. Models also permit an assessment of the complex interrelationships among the environmental processes that affect pesticide fate in soils and groundwater.

Some models with varying levels of sophistications have been developed for a variety of applications. Limited availability of data sets from field studies, however, continues to be a major constraint in evaluating the validity of these models before they can be used with confidence for the purposes for which they were developed, i.e., management and regulatory guidance. In only a few cases have predictions of *different models* been compared with a single data set to examine the adequacy of the simplifying assumptions in the models developed for management and instructional purposes. The need is evident for conducting coordinated field studies to collect the required data, to compile data bases and estimation methods for the model input parameters, and to develop objective criteria for assessing model validity. Much of the current research and scientific debate focuses on these important issues.

## REFERENCES

Addiscott, T.M., and R.J. Wagenet. 1985. Concepts of solute leaching in soils: A review of modelling approaches. J. Soil. Sci. 36:411–424.

Baes, C.F., and R.D. Sharp. 1983. A proposal for estimation of soil leaching and leaching constants for use in assessment models. J. Environ. Qual. 12:17–28.

Baker, D.B. 1983. Fluvial transport and processing of sediments and nutrients in large agricultural river basins. USEPA EPA-600/3-83-054. U.S. Gov. Print. Office, Washington, DC.

Baker, D.B. 1985. Regional water quality impacts of intensive row-crop agriculture: A Lake Erie Basin case study. J. Soil Water Conserv. 40:125–132.

Bonazountas, M., and J. Wagner. 1984. SESOIL: A seasonal soil compartment model. A.D. Little and Co., Cambridge, MA.

Bresler, E. 1973. Simultaneous transport of solutes and water under transient unsaturated flow conditions. Water Resour. Res. 9:975–986.

Bresler, E., and G. Dagan. 1981. Convective and pore scale dispersive solute transport in unsaturated heterogeneous fields. Water Resour. Res. 17:1683–1693.

Bresler, E., H. Bieloria, and A. Laufer. 1979. Field test of solution flow models in a heterogenous irrigated cropped soil. Water Resour. Res. 15:645–652.

Briggs, G.G., R.H. Bromilow, and A.A. Evans. 1982. Relationship between lipophilicity and root uptake and translocation of nonionized chemicals by barley. Pestic. Sci. 13:495–504.

Brusseau, M.L., and P.S.C. Rao. 1989. Sorption nonideality during organic contaminant transport in porous media. CRC Crit. Rev. Environ. Control 19:33–99.

Cameron, D.A., A. Klute. 1977. Convective-dispersive solute transport with a combined equilibrium and kinetic adsorption model. Water Resour. Res. 13:183–188.

Carsel, R.F., R.L. Jones, J.H. Hansen, R.L. Lamb, and M.P. Anderson. 1988a. A simulation procedure for groundwater quality assessment of pesticides. J. Contam. Hydrol. 2:125–138.

Carsel, R.F., and R.S. Parrish. 1988. Developing joint probability distributions of soil water retention characteristics. Water Resour. Res. 24:755–769.

Carsel, R.F., R.S. Parrish, R.L. Jones, J.L. Hansen, and R.L. Lamb. 1988b. Characterizing the uncertainty of pesticide leaching in agricultural soils. J. Contam. Hydrol. 2:111–124.

Carsel, R.F., C.N. Smith, L.A. Mulkey, J.D. Dean, and P.P. Jowise. 1984. User's manual for the pesticide root zone model (PRZM): Release 1. USEPA EPA-600/3-84-109. U.S. Gov. Print. Office, Washington, DC.

Childs, S.W., and R.J. Hanks. 1975. Model of soil salinity effects on crop growth. Soil Sci. Soc. Am. Proc. 39:617–622.

Dagan, G., and E. Bresler. 1979. Solute dispersion in unsaturated heterogeneous soil at field scale. I. Theory. Soil Sci. Soc. Am. J. 43:461–466.

Davidson, J.M., and J.R. McDougal. 1973. Experimental and predicted movement of three herbicide in water-saturated soil. J. Environ. Qual. 2:428–433.

Davidson, J.M., C.M. Rieck, and P.W. Santelmann. 1968. Influence of water flux and porous material on the movement of selected herbicides. Soil Sci. Soc. Am. Proc. 32:629–633.

Dean, J.D., and D.F. Atwood. 1987. Exposure assessment modeling of aldicarb in Florida. USEPA EPA-600/3-85/051. U.S. Gov. Print. Office, Washington, DC.

Enfield, C.G., R.F. Carsel, S.Z. Cohen, T. Phan., and D.M. Walters. 1982. Approximate pollutant transport to groundwater. Ground Water 20:711–722.

Gamerdinger, A.P., R.J. Wagenet, and M.Th. van Genuchten. 1990. Application of two-site/two-region models for studying simultaneous nonequilibrium transport and degradation of pesticides. Soil Sci. Soc. Am. J. (in press).

Garbarini, D.R., and L.W. Lion. 1986. Influence of the nature of soil organic carbon on the sorption of toluene and trichloroethylene. Environ. Sci. Technol. 20:1263–1269.

Gelhar, L.W., A. Mantoglu, C. Welty, and K.R. Rehfeldt. 1985. A review of field-scale physical solute transport processes in saturated and unsaturated porous media. EPRI EA-4190. Electric Power Res. Inst., Palo Alto, CA.

Goran, W.D. 1983. An interactive soils information retrieval system users manual. CERL Tech. Rep. N-163. U.S. Army Corps of Engineers, Construction Eng. Res. Lab., Champaign, IL.

Goring, C.A.I., D.A. Laskowski, J.W. Hamker, and R.W. Meikle. 1975. Principles of pesticide degradation in soil. p. 135–172. *In* R. Haque and V.H. Freed (ed.) Environmental dynamics of pesticides. Plenum Press, New York.

Green, R.E., J.M. Davidson, and J.W. Biggar. 1980. An assessment of methods for determining adsorption-desorption of organic chemicals. p. 73–82. *In* A. Banin and U. Kafkafi (ed.) Agrochemicals in soils. Pergamon Press, New York.

Gumbel, E.J. 1958. Statistics of extremes. Columbia Univ. Press, New York.

Gupta, S.K., C.T. Kinkaid, P.R. Meyer, C.A. Newbill, and C.R. Cole. 1982. A multi-dimensional finite-element code for the analysis of coupled fluid, energy and solute transport (CFEST). PNL-4260. Battelle Pacific Northwest Labs., Richland, WA.

Haith, D.A., and R.C. Loehr. 1979. Effectiveness of soil and water conservation practices for pollution control. USEPA EPA-600/3-79-106. U.S. Gov. Print. Office, Washington, DC.

Hamaker, J.W. 1972. Decomposition: Quantitative aspects. p. 253–340. *In* C.A.I. Goring and J.W. Hamaker (ed.) Organic chemicals in the environment. Marcel Dekker, New York.

Hamaker, J.W., and J.M. Thompson. 1972. Adsorption. p. 49–143. *In* C.A.I. Goring and J.W. Hamaker (ed.) Organic chemicals in the environment. Marcel Dekker, New York.

Helling, C.S., and J. Dragun. 1981. Soil leaching tests for toxic organic chemicals. p. 43–88. *In* Protocols for environmental fate and movement of toxicants. Assoc. Official Anal. Chem., Washington, DC.

Hornsby, A.G., and J.M. Davidson. 1973. Solution and adsorbed fluometuron concentration distribution in a water-saturated soil: Experimental and predicted evaluation. Soil Sci. Soc. Am. Proc. 37:823–828.

Hornsby, A.G., P.S.C. Rao, W.B. Wheeler, P. Nkedi-Kizza, and R.L. Jones. 1983. Fate of aldicarb in Florida citrus soils: 1. Field and laboratory studies. p. 936–958. *In* D.M. Nielson and M. Curl (ed.) Characterization and monitoring of the vadose (unsaturated) zone. Natl. Water Well Assoc., Worthington, OH.

Howard, P.H., A.E. Hueber, and R.S. Boethling. 1987. Biodegradation data evaluated for structure/biodegradability relations. Environ. Toxicol. Chem. 6-1-10.

Iskander, I., and H.M. Selim. 1981. Modeling nitrogen transport and transformations in soils. 2. Validation. Soil Sci. 131:303–312.

Jones, R.L. 1986. Field, laboratory, and modeling studies on the degradation and transport of aldicarb residues in soil and groundwater. p. 197–218. *In* W.L. Gardner et al., (ed.) Evaluation of pesticides in groundwater. Symp. Ser. 315. Am. Chem. Soc., Washington, DC.

Jones, R.L., A.G. Hornsby, and P.S.C. Rao. 1988. Degradation and movement of aldicarb residues in Florida citrus soils. Pestic. Sci. 23:307–325.

Jones, R.L., P.S.C. Rao, and A.G. Hornsby. 1983. Fate of aldicarb in Florida citrus soil: 2. Model evaluation. p. 959–978. *In* D. Nielson and M. Curl (ed.) Characterization and monitoring of the vadose (unsaturated) zone. Natl. Water Well Assoc., Worthington, OH.

Jones, R.L., A.G. Hornsby, P.S.C. Rao, and M.P. Anderson. 1987. Movement and degradation of aldicarb residues in the saturated zone under citrus groves on the Florida ridge. J. Contam. Hydrol. 1:265–285.

Jury, W.A., W.F. Spencer, and W.J. Farmer. 1983. Behavior assessment model for trace organics in soil. I. Model description. J. Environ. Qual. 12:558–564.

Jury, W.A., L.A. Stolzy, and P. Shouse. 1982. A field test of the transfer function model for predicting solute transport. Water Resour. Res. 18:369–375.

Jury, W.A., W.F. Spencer, and W.J. Farmer. 1987a. (Errata) Behavior assessment model for trace organics in soil: I. Model description. J. Environ. Qual. 16:448.

Jury, W.A., D.D. Focht, and W.J. Farmer. 1987b. Evaluation of pesticide groundwater pollution potential from standard indices of soil-chemical adsorption and biodegradation. J. Environ. Qual. 16:422–428.

Karickhoff, S.W. 1980. Sorption kinetics of hydrophobic pollutants in natural sediments. p. 193–205. *In* R.A. Baker (ed.) Contaminants and sediments. Vol. 2. Ann Arbor Sci. Publ. Co., Ann Arbor, MI.

Karickhoff, S.W. 1981. Semi-empirical estimation of sorption of hydrophobic pollutants on natural sediments and soils. Chemosphere 10:833–846.

Kay, B.D., and D.E. Elrick. 1967. Adsorption and movement of lindane in soils. Soil Sci. 104:314–322.

Kemper, W.D., and J.C. Van Schaik. 1966. Diffusion of salts in clay-water systems. Soil Sci. Soc. Am. Proc. 30:534–540.

Kenaga, E.E., and C.A.I. Goring. 1980. Relationship between water solubility, soil sorption, octanol-water partitioning, and concentration of chemicals in biota. p. 78–115. *In* J.G.A. Eaton et al. (ed.) Aquatic toxicology. Spec. Tech. Publ. 707. Am. Soc. Test. and Materials, Philadelphia, PA.

Kirkham, D., and W.L. Powers. 1972. Advanced soil physics. John Wiley and Sons, New York.

Knighton, R.E., and R.J. Wagenet. 1987a. Simulation of solute transport using a continuous time Markov process: I. Theory and steady-state application. Water Resour. Res. 23:1911–1916.

Knighton, R.E., and R.J. Wagenet. 1987b. Simulation of solute transport using a continuous time Markov process: II. Application to transient field conditions. Water Resour. Res. 23:1917–1925.

Lee, L.S., P.S.C. Rao, M.L. Brusseau, and R.A. Ogwada. 1988. Nonequilibrium sorption of organic contaminants during flow through columns of aquifer materials. Environ. Toxicol. Chem. 7:779–793.

Leistra, M. 1978. Computed redistribution of pesticides in the root zone of an arable crop. Plant Soil 49:569–580.

Leistra, M. 1979. Computing the movement of ethoprophos in soil after application in spring. Soil Sci. 128:303–311.

Lindstrom, F.T., L. Boersma, and H. Gardiner. 1968. 2,4-D Diffusion in saturated soils: A mathematical theory. Soil Sci. 106:107–113.

Lemley, A.T., R.J. Wagenet, and W.Z. Zhong. 1988. Sorption and degradation of aldicarb and its oxidative products in a soil-water flow system as function of pH and temperature. J. Environ. Qual. 17:408–414.

Lyman, W.J. 1982. Adsorption coefficients for soils and sediments. p. 4.1–4.32. *In* W.J. Lyman et al. (ed.) Handbook of chemical property estimation methods: Environmental behavior of organic compounds. McGraw-Hill Book Co., New York.

Mansell, R.S., H.M. Selim, P. Kanchanasut, J.M. Davidson, and J.G.A. Fiskell. 1977. Experimental and simulated transport of phosphorus through sandy soils. Water Resour. Res. 13:189–194.

McCuen, R.H. 1982. A guide to hydrologic analysis using SCS methods. Prentice-Hall, Englewood Cliffs, NJ.

Nash, R.G. 1988. Dissipation in soil. p. 131–170. *In* R. Grover (ed.) Environmental chemistry of herbicides. Vol. 1. CRC Press, Boca Raton, FL.

Nash, R.G., and E.M. Osborne. 1980. Environmental chemical dissipation (ECD) file. *In* Abstracts of the First Annual Meeting. 23–25 Nov. Soc. Environ. Toxicol. Chem., Washington, DC.

Nielsen, D.R., and J.W. Biggar. 1962. Miscible displacement. III. Theoretical considerations. Soil Sci. Soc. Am. Proc. 26:216–221.

Nimah, M.N., and R.J. Hanks. 1973. Model for estimation of soil water, plant, and atmospheric interrelations: I. Description and sensitivity. Soil Sci. Am. Proc. 37:522–527.

Nofziger, D.L., and A.G. Hornsby. 1985. Chemical movement in soil: IBM PC user's guide. Inst. Food Agric. Sci., Univ. of Florida Circular 654.

Nofziger. D.L., and A.G. Hornsby. 1986. A microcomputer-based management tool for chemical movement in soil. Appl. Agric. Res. 1:50–56.

Nofziger, D.L., and A.G. Hornsby. 1987. CMLS: Interactive simulation of chemical movement in layered soils. Inst. Food Agric. Sci., Univ. of Florida Circular 780.

Ogata, A. 1970. Theory of dispersion in a granular medium. U.S. Geol. Surv. Prof. Paper 411-I. U.S. Geol. Surv., Reston, VA.

Olsen, S.R., and W.D. Kemper. 1968. Movement of nutrients to plant roots. Adv. Agron. 20:91–151.

Ou, L.T., D.H. Gancarz, W.B. Wheeler, P.S.C. Rao, and J.M. Davidson. 1982. Influence of soil temperature and soil moisture on degradation and metabolism of carbofuran in soils. J. Environ. Qual. 11:293–298.

Pacenka, S., and T. Steenhuis. 1984. User's guide for the MOUSE computer program. Cornell Univ., Ithaca, NY.

Parker, J.C., and M.Th. van Genuchten. 1984. Determining transport parameters from laboratory and field tracer experiments. Virginia Agric. Exp. Stn. Bull. 84-3.

Persaud, N., J.V. Giraldex, and A.C. Chang. 1985. Monte-carlo simulation of noninteracting solute transport in a spatially heterogenous soil. Soil Sci. Soc. Am. J. 49:562–568.

Rao, P.S.C., and J.M. Davidson. 1980. Estimation of pesticide retention and tranformation parameters required in nonpoint source pollution models. p. 23–67. *In* M.R. Overcash and J.M. Davidson (ed.) Environmental impact of nonpoint source pollution. Ann Arbor Sci. Publ., Ann Arbor, MI.

Rao, P.S.C, and P. Nkedi-Kizza. 1984. Pesticide sorption on whole soils and soil particle-size separates. p. 5–47. *In* P.S.C. Rao et al. (ed.) Estimation of parameters for modeling the behavior of selected pesticides and orthophosphates. USEPA EPA-600/3-84-019. U.S. Gov. Print. Office, Washington, DC.

Rao, P.S.C., J.M. Davidson, and L.C. Hammond. 1976. Estimation of nonreactive and reactive solute front locations in soils. p. 235–241. *In* Proc. Conf. Hazardous Wastes Res. Symp. USEPA EPA-600/9-76-015. U.S. Gov. Print. Office, Washington, DC.

Rao, P.S.C., J.M. Davidson, R.E. Jessup, and H.M. Selim. 1979. Evaluation of conceptual models for describing kinetics of adsorption-desorption of pesticides during steady flow in soils. Soil Sci. Soc. Am. J. 43:22–28.

Rao, P.S.C., K.S.V. Edvardsson, L.T. Ou, R.E. Jessup, P. Nkedi-Kizza, and A.G. Hornsby. 1986. Spatial variability of pesticide sorption and degradation parameters. p. 100–115. *In* W.L. Gardner et al. (ed.) Evaluation of pesticides in groundwater, Symp. Ser. 315. Am. Chem. Soc., Washington, DC.

Rao, P.S.C., P. Nkedi-Kizza, J.M. Davidson, and L.T. Ou. 1983. Retention and transformation of pesticides in relation to nonpoint source pollution from croplands. p. 126–140. *In* F. Schaller and G. Bailey (ed.) Agricultural management and water quality. Iowa State Univ. Press, Ames.

Robbins, C.W., R.J. Wagenet, and J.J. Jurinak. 1980. A combined salt transport-chemical equilibrium model for calcareous and gypsiferous soils. Soil Sci. Soc. Am. J. 44:1191–1194.

Rothschild, E.R., R.J. Manser, and M.P. Anderson. 1982. Investigation of aldicarb in groundwater in selected areas in the central sand plains of Wisconsin. Ground Water 4:437–445.

Santodonto, J., C. Bush, P.H. Howard, K. Howard, S. DelFavero, P.C. Miles, E.T. Merrick, L.K. Smith, and L.A. Tavares. 1987. TSCATS: A database for chemicaland subject indexing of health and environmental studied submitted under the Toxic Substances Control Act. Environ. Toxicol. Chem. 6:921–927.

Scheidigger, A.E. 1960. The physics of flow through porous media. Univ. of Toronto Press, Toronto, ON.

Selim, H.M., and R.S. Mansell. 1987. Analytical solution for the transport of reactive solute through soils. Water Resour. Res. 12:528–532.

Selim, H.M., J.M. Davidson, and R.S. Mansell. 1976. Evaluation of a two-site adsorption-desorption model for describing solute transport in soils. p. 444–448. *In* Proc. Summer Simul. Conf., Washington, DC. 12–14 July.

Shearer, R.C., J. Letey, W.J. Farmer, and A. Klute. 1973. Lindane diffusion in soil. Soil Sci. Soc. Am. Proc. 37:189–193.

Sharma, M.L., and A.S. Rogowski. 1985. Hydrological characterization of watersheds. p. 291–295. *In* D.G. DeCoursey (ed.) Proc. of the Natural Resour. Modeling Symp., Pingree Park, CO.

Silliman, S.E., and E.S. Simpson. 1987. Laboratory evidence of the scale effect in dispersion of solutes in porous media. Water Resour. Res. 23:1667–1673.

Stewart, B.A., D.A. Wollhiser, W.H. Wischmeier, J.H. Caro, and M.H. Fere. 1976. Control of water pollution from cropland volume II: An overview. USEPA EPA-600/2-75-026b. U.S. Gov. Print. Office, Washington, DC.

Tillotson, W.R., and R.J. Wagenet. 1982. Simulation of fertilizer nitrogen under cropped situations. Soil Sci. 133:133–143.

U.S. Department of Agriculture. 1980. CREAMS: A field-scale model for chemicals, runoff, and erosion from agricultural management systems. *In* W.G. Knisel (ed.) Conservation research report 26. Washington, DC.

van Genchten, M.Th. 1981. Non-equilibrium transport parameters from miscible displacement experiments. Res. Rep. 119. U.S. Salinity Lab., USDA-ARS, Riverside, CA.

van Genuchten, M.Th., and W.J. Alves. 1982. Analytical solutions of the convective-dispersive solute transport equation. Tech. Bull. 1601. U.S. Salinity Lab., Riverside, CA.

van Genuchten, M.Th., and R.W. Cleary. 1979. Movement of solutes in soil: Computer-simulated and laboratory results. p. 349–386. *In* G.H. Bolt (ed.) Soil chemistry: B. Physico-chemical models. Elsevier Publ. Co., Amsterdam, Netherlands.

van Genuchten, M.Th., J.M. Davidson, and P.J. Wierenga. 1974. An evaluation of kinetic and equilibrium equations for the prediction of pesticide movement in porous media. Soil Sci. Soc. Am. Proc. 38:29–35.

van Genuchten, M.Th., and R.J. Wagenet. 1989. Two-site/two-region models for pesticide transport and degradation. Theoretical. Soil Sci. Soc. Am. J. 53:1303–1310.

van Genuchten, M.Th., and P.J. Wierenga. 1976. Mass transfer studies in sorbing porous media. I. Analytical solutions. Soil Sci. Soc. Am. J. 40:473–480.

Wagenet, R.J. 1984. Principles of salt movement in soils. p. 123–140. *In* D.W. Nelson et al. (ed.) Chemical mobility and reactivity in soil systems. SSSA Spec. Publ. 11. ASA and SSSA, Madison, WI.

Wagenet, R.J., and J.L. Hutson. 1986. Predicting the fate of nonvolatile pesticides in the unsaturated zone. J. Environ. Qual. 15:315–322.

Wagenet, R.J., and J.L. Hutson. 1989. LEACHM: A finite difference model for simulating water, salt and pesticide movement in the plant root zone. Continuum. Vol. 2. Version 2.0. New York State Water Resour. Inst., Cornell Univ., Ithaca, NY.

Wagenet, R.J., J.L. Hutson, and J.W. Biggar. 1989. Simulating the fate of a volatile pesticide in unsaturated soil: A case study with DBCP. J. Environ. Qual. 18:78–84.

Wagenet, R.J., and B.K. Rao. 1983. Description of nitrogen movement in the presence of spatially variable soil hydraulic properties. Agric. Water Manage. 6:227–242.

Walker, A. 1976a. Simulation of herbicide persistence in soil: I. Simazine and prometryne. Pestic. Sci. 7:41–49.

Walker, A. 1976b. Simulation of herbicide persistence in soil: II. Simazine and linuron in long-term experiments. Pestic. Sci. 7:50–58.

Walker, A., and P.A. Brown. 1983. Spatial variability in herbicide degradation rates and residues in soil. Crop Prot. 2:17–25.

Wang, H.F., and M.P. Anderson. 1982. Introduction to groundwater modeling. W.H. Freeman, San Francisco.

Warrick, A.W., and D.R. Nielsen. 1980. Spatial variability of soil physical propreties in the field. p. 319–344. *In* D. Hillel (ed.) Applications of soil physics. Academic Press, New York.

Watts, D.G., and R.J. Hanks. 1978. A soil-water-nitrogen model for irrigated corn on sandy soils. Soil Sci. Soc. Am. J. 42:492–499.

Zhong, W.Z., A.T. Lemley, and R.J. Wagenet. 1986. Quantifying pesticide adsorption and degradation during transport through soil to groundwater. p. 61–77. *In* W.J. Garner et al. (ed.) Evaluation of pesticides in groundwater. Symp. Ser. 315. Am. Chem. Soc., Washington, DC.

# Chapter 11

# Efficacy of Soil-Applied Pesticides

**M. LEISTRA,** *Institute for Pesticide Research, Wageningen, Netherlands*

**R. E. GREEN,** *University of Hawaii, Honolulu, Hawaii*

Agricultural crops are continuously threatened by various harmful soil-borne organisms, including fungi, bacteria, nematodes, insects, and weeds. Pests and pathogens may reduce crop yield, and can also be detrimental to crop quality. The spread of harmful organisms, both within and between countries, is also important. In international trade, the occurrence of harmful organisms in trade lots can lead to large financial losses.

The development and spread of pests and diseases can be suppressed by careful farm management, especially by using phytosanitary measures. Crop rotation has been practiced for many centuries. Attempts are being made to breed varieties of crops that are resistant to the most dangerous pests and diseases. In some cases, these attempts have been successful. In some agricultural systems, however, pesticides are indispensable. In many countries, governments aim at systems of integrated pest management by combining good farm management with methods of biological and chemical control. Environmental quality problems may arise when a disproportionate emphasis is put on the use of pesticides.

In this chapter, we use the term *pesticide* in the generic sense, such that it includes all agricultural biocides (e.g., insecticides, nematicides, fungicides, and herbicides). While requirements for effective use of the different specific pesticide types differ, many of the principles are the same. Efficacy (the capacity to produce the desired effect) of pesticides cannot be generalized to encompass all pest-pesticide combinations, hence we will focus our attention in this chapter on some crop-nematode-nematicide combinations that illustrate key principles. Efficacy considerations for other pesticide groups are addressed in detail elsewhere, for example, in the excellent reviews on herbicides by Koskinen and Harper (1987), Hance (1988), and Yaron et al. (1985).

Soil treatments with pesticides are often quite expensive, so it is desirable to protect the crop with doses that are as low as possible. Furthermore, the risk of undesirable side effects is diminished as the dosage decreases. The

 *Pesticides in the Soil Environment*—SSSA Book Series, no. 2.

efficacy of soil-acting pesticides is dependent on many factors: pesticide properties, formulation, application technique, soil properties, climatic conditions, agricultural management, the characteristics of the crop, and the nature and behavior of the harmful organism (Fig. 11-1). Some of the factors can be manipulated while other factors which cannot be controlled must be considered in developing management strategies. Achieving effective control in large soil volumes with a small amount of pesticide is often a difficult task.

Many field studies have assessed the efficacy of soil-applied pesticides. While such studies are valuable in providing a basis for management decisions for specific conditions, they do not usually provide basic information on soil-pesticide interactions that can be readily extended to other crop-soil-management situations. However, there is substantial information available on the separate processes to which pesticides in soil are subjected. This information can be integrated in models that attempt to explain and predict the efficacy of soil-acting pesticides in the field. Simulation models are useful tools to integrate diverse information on pesticides in a quantitative way. Such models should simulate practical conditions and they should be evaluated in well-defined field experiments. They may provide quantitative insight into key factors determining efficacy and highlight approaches for enhancing efficacy.

The behavior of pesticides in soil has already been the subject of some computer-simulation studies. Simulating the temporal and spatial dynamics of pesticide concentration constitutes a prediction of the dosage to which pests and pathogens are exposed. Relationships between pesticide dosage and organismal response obtained in laboratory or greenhouse studies can then be used to estimate the efficacy of a pesticide in field soils. Few studies, however, have integrated the physical-chemical and biological aspects of efficacy in a quantitative way.

The soil-acting pesticides comprise many compounds with divergent biological activities. In this chapter, we illustrate key ideas, principally by reference to research results for the widely used soil-acting nematicides aldicarb,

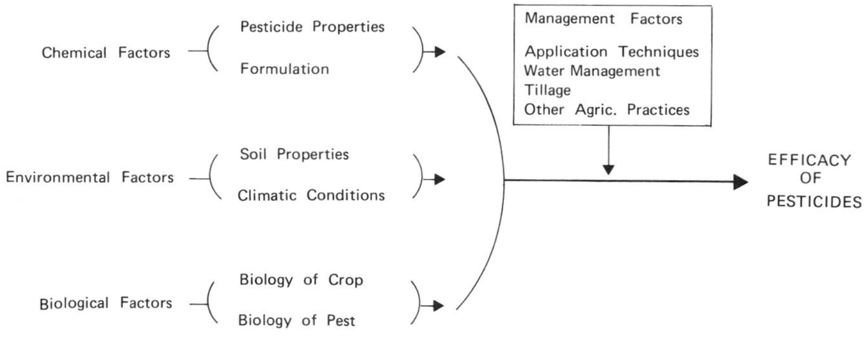


Fig. 11-1. Factors that impact on pesticide efficacy and possible intervention by management practices.

oxamyl, and ethoprophos. The general research approach, however, is similar for many other pesticides and pest control problems. Our purpose is to consider how a quantitative knowledge of physical-chemical-biological interactions of pesticides in soils can be used to improve the effectiveness of pesticides, so that the desired control of crop pests can best be achieved without insult to the environment. Controlled experiments in the field and laboratory, combined with rapidly developing mathematical modeling techniques, have played a key role in elucidating the soil-pesticide interactions that impact pesticide efficacy.

## 11-1 EFFECTS OF FORMULATION AND MODE OF APPLICATION ON EFFICACY

Pesticides need to be formulated to assure ease of application. For spraying, it is necessary to mix the products quickly throughout a comparatively large volume of water. Granular pesticides are often combined with a surplus of filling material to facilitate uniform spreading of the pesticide. Granular formulations are usually preferred for highly toxic pesticides (e.g., aldicarb) because applicator exposure is lower. The following types of formulation are common for soil-acting pesticides:

1. Sprayable formulations, such as emulsifiable concentrates and wettable powders.
2. Granular formulations.
3. Liquid mixtures and concentrated solutions (soil fumigants).

Wilkins (1983) and Furmidge (1984) have given extensive surveys of the many types of formulation. A variety of spraying machines are available that mix sprayable formulations with water and apply them to the soil surface. More recently, various types of granulate distributors have been developed and made available. Subsurface-blade and shank injectors are most often used to introduce liquid soil fumigants at various depths in the soil.

After a product is sprayed on the soil surface, the pesticide is exposed to soil and climatic conditions. When the pesticide is left on the soil surface, the rate of loss may be high. Volatilization from the soil surface depends on the: (i) vapor pressure of the compound, (ii) interactions of the pesticide with soil material, and (iii) climatic conditions. Sunlight may cause substantial photochemical transformation, be it directly or mediated by photosensitizers. Sometimes a granular formulation is used to reduce such losses at the soil surface (Furmidge, 1984). Another remedy is incorporation of the product in soil shortly after application to the surface.

Presumably, the comparatively small masses of formulating materials in sprays do not have much influence on the leachability of the pesticide with water. Gradual release of pesticide from granules may reduce leachability, although the various granular formulations behave differently (Furmidge, 1984).

### 11-1.1 Slow-Release Formulations

Only a few controlled-release formulations for soil application are commercially available. The overall persistence of a pesticide in soil is often increased by such a formulation, since much of the pesticide remains within the formulation. As a result of transformation and slow diffusion, the concentration of active ingredient in the soil around the granules may remain too low to be effective. In fact, soil itself acts as a slow-release formulation in view of the sorptive capacity of soil (Furmidge, 1984). Under dry conditions, the movement of many pesticides in soil is slow and a slow-release formulation may counteract the spread further, thus reducing efficacy (Batterby et al., 1980).

Coppedge et al. (1975) have studied the effect of soil water content on the release of aldicarb from various formulations. The rate of release was lowest at low water contents. The active ingredient must diffuse from the granule via the liquid phase; this is promoted by a high soil water content. Both corncob (*Zea mays* L.) and gypsum ($CaSO_4 \cdot 2H_2O$) formulations of aldicarb were found to be quick-release formulations under moist and wet conditions (Stokes et al., 1970; Coppedge et al., 1975). A slow release of such a nematicide will often be unfavorable because effective nematode control requires fairly high concentrations in the soil soon after sowing or planting of the crop (Batterby et al., 1980). Stokes et al. (1970) has suggested the combination of a quick-release formulation for good initial effect with a slow-release formulation for prolonged effect.

The efficacy of controlled-release formulations of soil-applied nematicides has been tested many times, but only occasionally has a distinct increase in performance been established (Bromilow, 1988). Producing such formulations is rather expensive (Morton, 1986). For these reasons, few controlled-release formulations of soil-acting nematicides are available.

The use of a controlled-release formulation can be advantageous when a low concentration of pesticide should be maintained in soil over a long growing season. Attack of vegetables by insect larvae late in the growing season still might occur with such formulations, even with moderately persistent insecticides (Suett & Thompson, 1985). The efficacy of slow-release formulations on a wide variety of soils throughout the entire cropping season has not yet been adequately researched for most crops.

### 11-1.2 Pesticide Incorporation in the Soil

The distribution of pesticides in soil should be matched to the spatial pattern of the pests and diseases to be controlled, and to that of the plant parts to be protected. This is illustrated in Fig. 11-2. Some insect species occur in the soil at shallow depths, attacking the crop plants just below the soil surface. Effective control of such insects may be achieved by shallow incorporation of insecticide. Soil-applied herbicides are usually applied at the soil surface or are incorporated to shallow depths. On the other hand, parasitic nematodes and fungi attack the whole root system of the crop, and

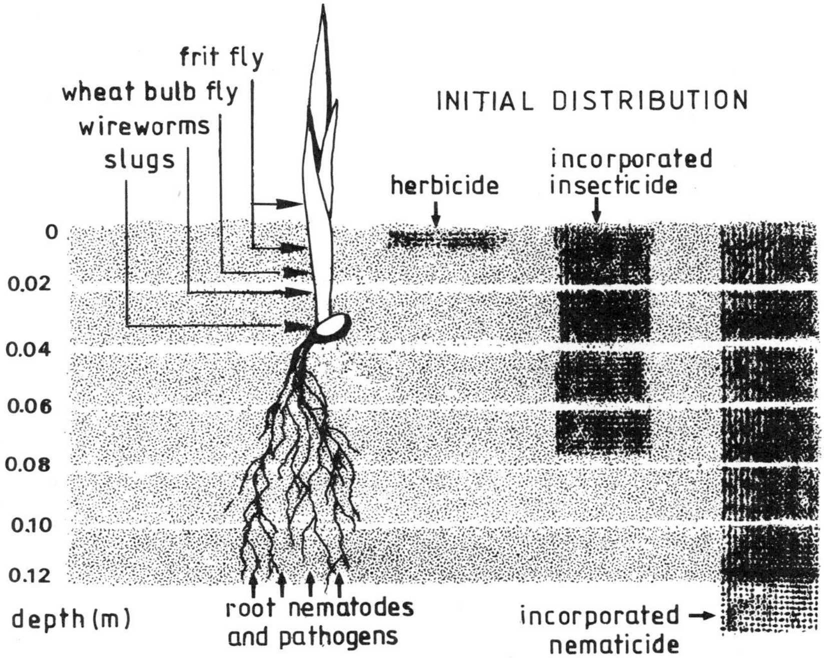


Fig. 11-2. Distribution of target organisms and pesticides in soil. Modified after Briggs (1984).

the density of these organisms may be fairly high throughout the intensely rooted plow layer. Accordingly, nematicides and fungicides should be incorporated homogeneously and deeply (e.g., 0.15–0.30 m) for effective protection of the crop.

Soil-acting nematicides such as aldicarb, oxamyl and ethoprophos are usually applied in granular formulation. Three main types of machines are used for their application: (i) special granulate distributors, (ii) pneumatic fertilizer distributors, and (iii) adapted sowing machines (Andringa, 1981). These nematicides are used in various crops to protect the crop roots against attack by plant-parasitic nematodes. They exert most of their effect on the nematodes via the soil solution and they protect the root systems of the crop only where the concentration in the soil solution is sufficiently high.

Broadcast application of nematicide granules followed by shallow incorporation, for example with a rotary-harrow, has often given suboptimal results (Whitehead et al., 1975; Moss et al., 1976; Smith & Bromilow, 1977; Whitehead et al., 1979). As soon as roots grow out of the thin layer that has been treated, they may be subjected to high nematode population densities in the lower plow layer. Control after shallow incorporation is especially poor when dry conditions prevail in the first months after application. Substantial amounts of rainfall or sprinkler irrigation induce a distinct redistribution of weakly sorbed compounds, and thus can offset the limited effectiveness of shallow incorporation. However, moderately to strongly sorbed nematicides are not easily moved through the soil profile.

The soil-incorporation patterns of various implements have been studied by soil sampling and chemical analysis. Deep cultivation with a rotary cultivator usually resulted in a fairly homogeneous distribution of pesticide in the top 0.15 to 0.20 m (Smelt et al., 1976; Smith & Bromilow, 1977; Whitehead et al., 1979; Bromilow & Lord, 1979; Whitehead et al., 1981). In other instances of rotary tillage, however, the concentrations in the top 0.10 m remained highest. Cultivation with a rotary-harrow or reciprocating harrow resulted in a mixing to only 0.05 to 0.10 m depth (Smelt et al., 1976; Whitehead et al., 1979; Bromilow & Lord, 1979). With other cultivations by rotary-harrow and reciprocating harrow, the main fraction of the granules (about three-quarters) remained in the top 0.05 m (Whitehead et al., 1975; Smith & Bromilow, 1977). A spring-tine cultivator incorporated the granules to a depth of only 0.05 to 0.10 m in soil. The differences between samples were great, indicating a heterogeneous distribution in soil (Whitehead et al., 1975; Smelt et al., 1976). Some of the distributions Smelt et al. (1976) obtained are shown in Fig. 11–3.

Deep and intensive rotary tillage of soils for incorporation of granular nematicides may have serious drawbacks. It is expensive because of the time needed (low driving speeds), the high fuel costs, and equipment wear. The structure of some soils may deteriorate after rainfall leading to puddling on the soil surface and poor trafficability. In other cases, stiff and cloddy subsoil from below may be moved into the seed bed, or the top layer of sandy and peaty soils may become so loose and dry that wind erosion occurs. Thus, it is important to determine which implements can incorporate the granules thoroughly without such drawbacks.

Some interesting attempts have been made to obtain a deep and homogeneous distribution of nematicide in soil without using a rotary cultivator. Whitehead et al. (1975) developed an implement with which granules could be spread horizontally at shallow depths in the soil. In one of their experiments, one-third of the dose of aldicarb was spread at 0.05 m and one-third at 0.10 m, whereas the remainder was broadcast on the soil surface. After this, the soil was cultivated with a rotary harrow set to a working depth of 0.20 m. However, most of the nematicide remained in the top 0.10 m, indicating that vertical mixing by the rotary harrow was minimal (Whitehead et al., 1981). Nematode control with this experimental method was usually poorer than with deep incorporation by rotary cultivation. Granular nematicides were also applied in vertical bands in the soil with an implement described by Whitehead et al. (1987). These applications were followed by an intensive cultivation with a reciprocating harrow. With these methods too, a thick top layer of the soil had to be prepared intensively beforehand. These are interesting first steps to the development of special soil-incorporation implements, and this type of work should be actively continued.

Some crops are especially sensitive to damage by soil-borne parasites in the seedling stage. Pesticide application in the seed furrow or in a narrow band close to the seeds may then be efficacious. The same approach can be used if the compound should be taken up by the crop in an early stage to protect leaves and stems. With such site-specific applications, smaller amounts

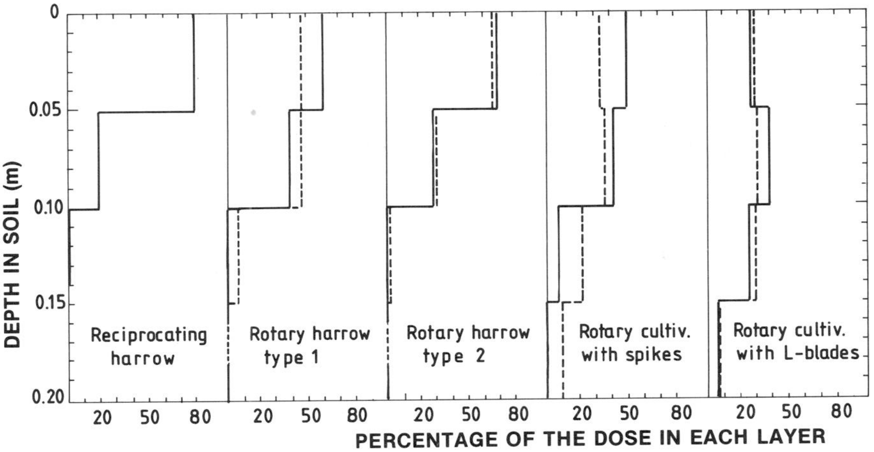


Fig. 11-3. Distribution of ethoprophos in soil after incorporation of surface-applied granules with different implements; one or two experimental fields. After Smelt et al. (1976).

of pesticide are needed than with broadcast applications, and application costs are decreased. However, when roots grow out of the treated zone, they may be attacked by the unaffected nematodes and crop damage may occur. The development of the nematode population during the growing season is hardly influenced by such local application (Whitehead et al., 1975; Whitehead et al., 1979).

### 11-1.3 Application by Irrigation

In many cropping systems, irrigation is an integral part of crop management, especially in low rainfall areas. Chemical application through irrigation systems has increased rapidly in recent years (Thomason, 1987), principally because of the economy of application and the flexibility in timing. Both sprinkler and trickle (drip) irrigation have unique advantages for pesticide application, with sprinklers providing more complete coverage of the soil surface (important for herbicides, for example), while trickle systems deliver the chemicals to specific points where emitters are located. The relative merits of various methods of pesticide application by irrigation will depend on the specific requirements for placement and timing of the treatment as dictated by the synchrony of pest and crop development.

The need to place the pesticide in the zone where protection is required is illustrated in Fig. 11-2. Perennial crops may require chemical control of root pathogens over long periods. Side-dress applications to plant rows are often ineffective (Bromilow, 1988), thus application of pesticide with irrigation water may be a good alternative. However, deep penetration of a chemical at concentrations toxic to the pest in the entire root zone may not be achieved with irrigation. Johnson et al.(1982) found that fenamiphos and its nematicidal oxidation products were most concentrated in the top

0.08 m in layer after application of the nematicide by sprinkler irrigation followed by 326 mm of rainfall and irrigation over 45 d. In this experiment, fenamiphos applied by broadcasting with subsequent rotary cultivation before planting resulted in more uniform incorporation in the top 0.15 m, thus providing better control of root-knot nematodes.

Trickle irrigation appears to be especially well adapted to application of nematicides (Palti & Shoham, 1983; Apt & Caswell, 1988). The combination of precise positioning of the pesticide in the root zone and repeated application at the required rate is especially effective for chemicals that are relatively mobile in the advancing wet front. Effective exploitation of such a system requires a good understanding of the dynamics of both root system and pest population development as well as soil-chemical interactions and soil-water flow. Models of chemical movement may be especially useful for integration of processes involved in pesticide application by trickle irrigation. Complex models that can predict chemical concentration distributions in two dimensions over time are already available (Bresler & Green, 1987; Omary, 1988), but the required input data may often be lacking. Management-oriented models that are based on water flow theory, but simplified for practical application, have been demonstrated to be useful in determining chemical application quantities and times (Keng & Vander Gulik, 1985; Keng et al., 1985). Much remains to be done to incorporate pesticide sorption and degradation processes into simplified two-dimensional management models which will aid in trickle-irrigation system design and operation for effective pesticide application. Simulation of pesticide movement in soils is discussed in more detail later in this chapter (section 11-2.3).

## 11-2 EFFECTS OF SORPTION AND MOVEMENT IN SOIL ON EFFICACY

The sorption of a pesticide on soil reduces its availability to the target organism, both in the soil solution and the soil-gas phase. Weakly sorbed compounds usually exhibit little effect of sorption on efficacy, thus they are usable for a wide range of soils (Whitehead, 1978). Various soil-acting insecticides and nematicides, or their active transformation products, are only weakly sorbed so they are amply available in the soil solution (Briggs, 1984). Moderately or strongly sorbed pesticides will often suffer from reduced efficacy, especially in soils with comparatively high organic matter content (Harris, 1972; Whitehead, 1978). The effect of sorption can sometimes be offset by applying a higher dose to the soil. However, the dose required should not be estimated on the basis of sorption alone, because other factors may also have substantial effects.

### 11-2.1 Sorption and Water Regime Impacts

The availability and movement of the insecticide/nematicide ethoprophos in soil under spring field conditions in the Netherlands were simulated in a computation model by Leistra (1979). Rainfall, water evaporation from the soil surface and water flow in the soil profile were simulated. The rainfall used for the simulation was slightly higher than actual average rainfall. Values of the sorption coefficient measured for two different soils were introduced into the computations. Simulated concentrations of ethoprophos in the liquid phase of the two soils are presented in Fig. 11-4. The effect of sorption on the availability of pesticide in the soil solution is clearly shown. In the sandy loam soil with low sorption, the pesticide was substantially redistributed below the depth of incorporation, providing protection to crop roots below that depth. The uptake of pesticide by the crop will often be increased by such redistribution, thus improving the systemic activity of insecticides and nematicides in plants.

In various field crops, pesticide redistribution in soil is dependent on the rainfall in excess of evapotranspiration. Surplus rainfall during the first few weeks after pesticide application may be decisive for efficacy. Rainfall statistics of an area will indicate the frequency of the years in which the amount of rainfall is inadequate to induce substantial redistribution. Effective pesticide incorporation into a sufficiently deep-tilled layer can lessen the requirement for rainfall to distribute a pesticide, which illustrates how management can overcome environmental constraints that might otherwise limit pesticide efficacy.

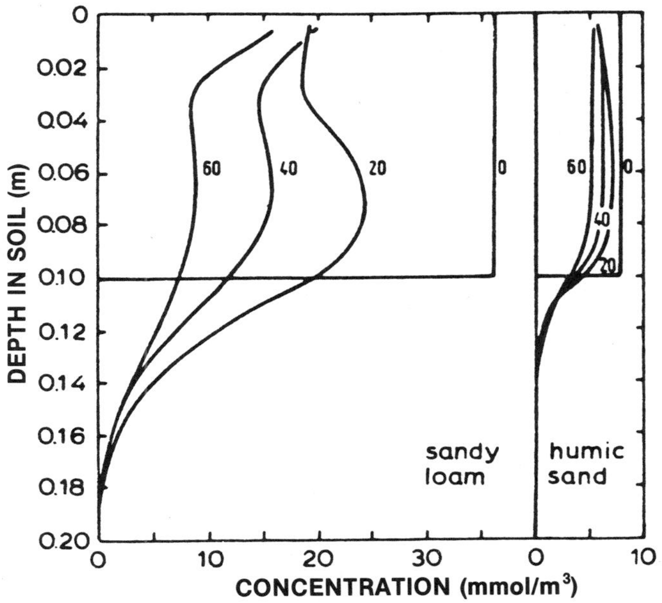


Fig. 11-4. Computed concentrations of ethoprophos in the liquid phase of a sandy loam and of a humic sand soil. Period: April and May 1977. Time: 0, 20, 40, and 60 d after application. After Leistra (1979).

When post-plant nematicide application is necessary, nematicides can be leached into the soil by irrigation water. After distribution of the granules on the soil surface, leaching by sprinkler irrigation may result in a relatively uniform distribution of nematicide in soil. However, this can only occur with compounds that are weakly sorbed on soil, and the practice is only suited to crop-nematicide combinations that do not exhibit phytotoxicity. The proper amount and timing of irrigation subsequent to nematicide application should result in pesticide distribution such that a large fraction of the root system is protected against parasitic nematodes. In practice, this protection is seldom achieved. The use of irrigation to distribute pesticides and increase pesticide efficacy could perhaps be improved through the use of chemical distribution models.

When a crop is seriously attacked by pathogenic soil fungi, attempts are made to distribute a fungicide in the root zone. Unfortunately, most fungicides are moderately to highly sorbed on soils so that distribution by leaching with irrigation water is not successful. A suitably formulated fungicide may be applied as a soil drench in sufficient water to wet the root zone. However, the distribution of the compound in soil will often be irregular because water flow under such conditions is non-uniform. The effect of such drenches is often limited to decreasing the rate of disease spread within the crop.

### 11–2.2 Vapor Movement

Some pesticides can, to a degree, move in soil by vapor diffusion, and thus spread through soil in the gas phase from the initial point of application. A substantial concentration of active ingredient in the gas phase is considered important for the efficacy of soil-acting insecticides. Penetration of the vapor of these compounds into insects via the respiratory system and via the cuticle can be effective (Harris, 1972). The extent of vapor transfer depends primarily on the vapor pressure of the pesticide, but also on the solubility in water, the rate and extent of sorption onto soil material, and environmental conditions (most importantly, soil temperature and moisture content) at the time of application. Various insecticides showing moderate to strong sorption are sufficiently active in soil because of their high vapor pressure (Briggs, 1984).

### 11–2.3 Simulation of Pesticide Movement

Simulation models can be useful tools in the development of recommendations to obtain an optimum distribution of a pesticide in soil by leaching or vapor movement. The movement and transformation of oxamyl in fallow sandy loam soils under field conditions was simulated with a computation model by Leistra et al. (1980). Laboratory data on sorption and transformation rate were introduced as parameters in model equations. The results of computations and field measurements were compared (Fig. 11–5) for well-defined conditions. The model provided a reasonable description

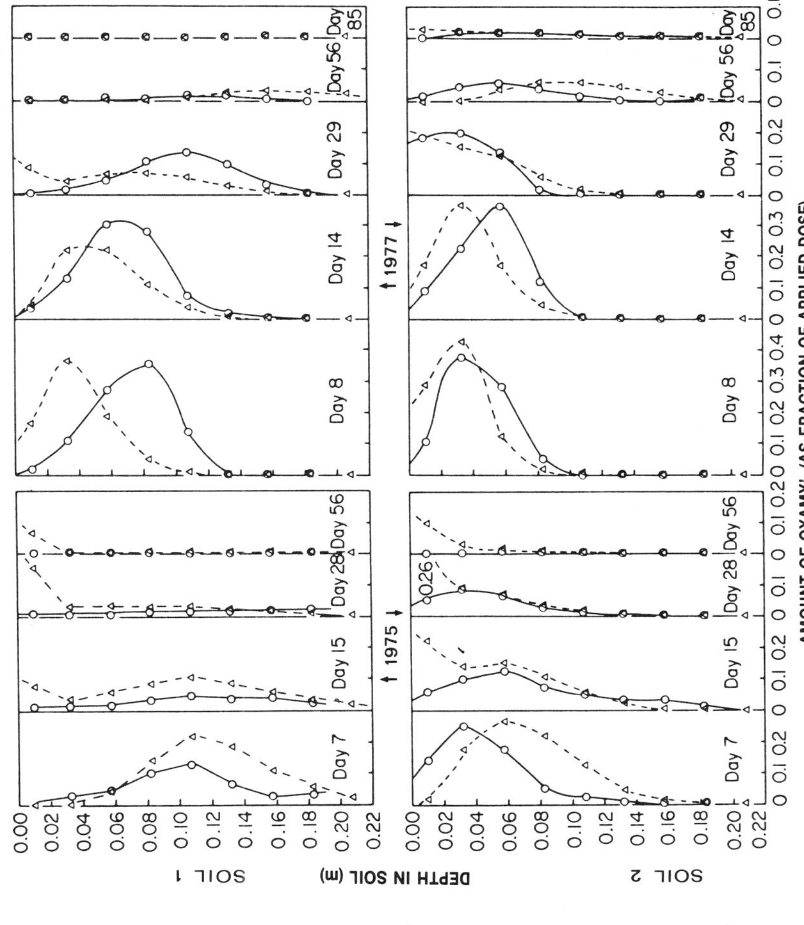


Fig. 11–5. Simulated and measured vertical distribution of oxamyl in two soils in 1975 and 1977. Dashed line = simulated; solid line = measured. After Leistra et al. (1980).

of the movement of oxamyl in soil, although it underestimated movement in the first month and tended to overestimate later movement (1977 data, both soils). Inaccurate description of water flow in soil and of water evaporation at the soil surface presented some problems. The variation in soil characteristics in space and time may be another obstacle for accurate simulations. Evaluation of spatial and temporal variability in soil characteristics and the consequent development of appropriate management practices for pesticides are worthy of simulation studies in the future.

Sorption coefficients and transformation rates of ethoprophos, as measured in the laboratory (Smelt et al., 1981), were introduced into a computer model to simulate ethoprophos movement and transformation in soils under field conditions in winter (Leistra & Smelt, 1981). The movement simulated by the model was somewhat greater than that measured in an experiment with soil columns in the field (Fig. 11–6). Even so, the different behavior of the pesticide in three soils, which varied considerably in chemical and physical properties, was clearly demonstrated. The Wierum, Middenmeer, and Rholde soils are silt loam, sandy loam and loamy sand textures, respectively, and have organic matter contents of 1.6, 1.8, and 6.1%, respectively. Such models are useful in that they reduce the need for carrying out extensive field trials under a wide variety of conditions. The potential for simulating pesticide distribution in the soil with various management alternatives and specific environmental conditions is attractive as a preliminary step in applied research which can lead to more effective design of field experiments.

The impact of chemical transformation on mobility of bioactive transformation products can also be evaluated with computer simulation. Movement and transformation of fenamiphos and its sulfoxide in irrigated soil were computer-simulated by Lee et al. (1986). The input data for sorption and transformation rates for the parent chemical and two oxidation products were measured in the laboratory. The results illustrated restricted movement of the more strongly sorbed fenamiphos and substantial movement of the weakly sorbed sulfoxide with repeated application of fenamiphos through a high-frequency irrigation system. Such a model can be used to estimate the effect of a wide range of irrigation regimes on the distribution of the parent pesticide and its bioactive transformation products in soils that differ in chemical, biological, and physical properties. These models can be effective tools for evaluation of pesticide efficacy, provided the relationship between occurrence and distribution of biotransformation products and subsequent pest control is established.

Acceptable simulation of pesticide movement in soils may be hindered by soil heterogeneity and consequent irregularity of water flow. Clay soils may crack when drying and they often have a cloddy structure. Nicholls et al. (1982) studied the movement of two pesticides and chloride-ion in a strongly structured clay loam soil during a winter period. Movement was simulated with two models: one was based on Darcy's law and the second used the concept of mobile and stagnant water phases in soil. Both models overestimated movement of aldicarb-sulfone and of the herbicide fluometuron in soil. The model with mobile and stagnant water phases provided the best description, possibly because this concept was most applicable to the situation studied.

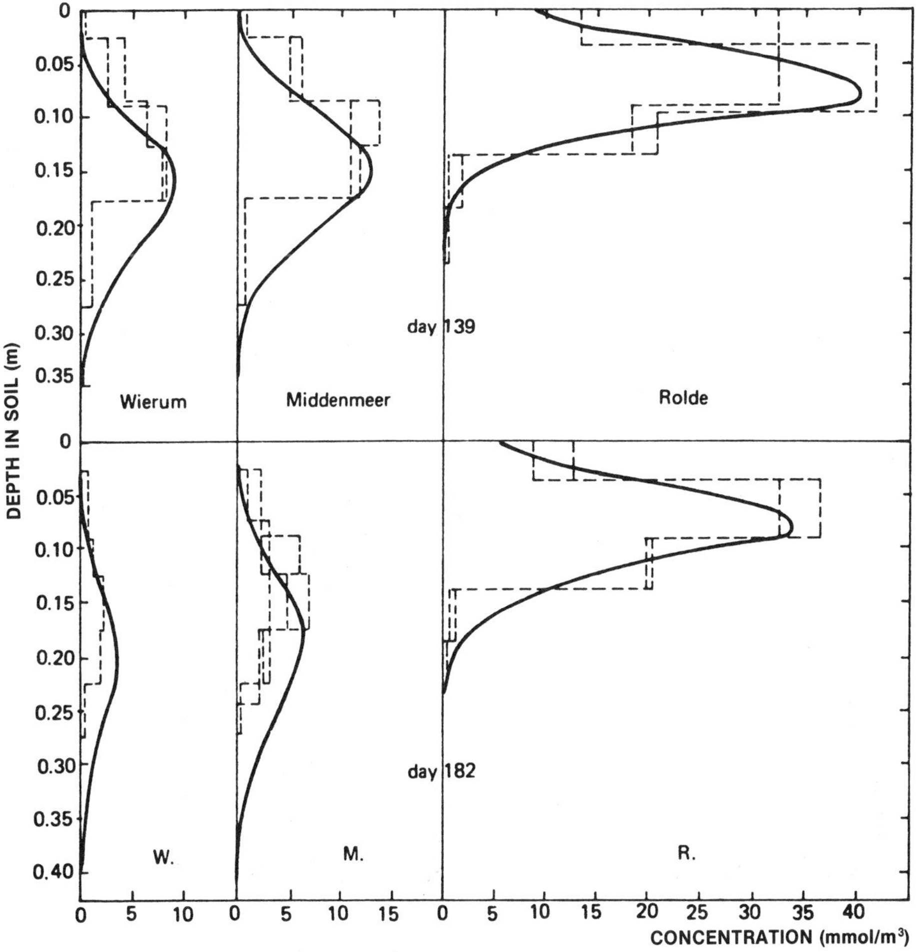


Fig. 11-6. Simulated and measured concentrations (mmol/m$^3$, soil volume basis) of ethoprophos in three soils at Day 139 (12 March) and at Day 182 (25 April) after application in autumn. Dashed line = measured; solid line = simulated. After Leistra and Smelt (1981).

The micro-relief of a tilled field and crop-canopy characteristics can also complicate modeling efforts. The water flow pattern in a system of ridges and furrows in a potato (*Solanum tuberosum* L.) field (loamy sand soil) was studied using a dye (Saffigna et al., 1976). Much of the water from rainfall or irrigation was intercepted by the plants and flowed along defined pathways (e.g., along stems) through the canopy. One fraction of the water infiltrated on some spots on the ridges, whereas another fraction flowed from the ridges to the furrows and infiltrated there. The result was an irregular distribution of solute in the soil system. In situations such as this, the assumption of uniform water infiltration and conduction in soil is inappropriate. Simulation of pesticide movement in soil is most useful for conditions that satisfy the assumptions associated with a specific model's governing equations and boundary conditions.

## 11–3 EFFECT OF PERSISTENCE IN SOIL ON EFFICACY

Moderate persistence of soil-acting pesticides is desirable for long-term protection of crops against parasitic soil organisms. The desirable period of protection may range from weeks to several months, especially when the pesticides are applied at seeding or planting.

In many countries, persistent pesticides are no longer registered for use because of crop and environmental residues. Most of the newer soil-acting pesticides are rapidly transformed, with half-lives of a few weeks up to a month, so they are active during only a fraction of the growing season. This limited persistence makes effective control of soil-borne pathogens quite difficult.

There is an inherent variation in the persistence of a pesticide in various soils. Sometimes there is a correlation between rate of transformation and some soil characteristic. For example, organophosphate and carbamate pesticides are often transformed comparatively fast in soils with high pH. The rate of transformation of a pesticide in some soils may be too high for effective control of soil pests and diseases. Knowledge of the rate of transformation in several soils of differing properties is desirable to allow modification of the recommendations for various soil types.

### 11–3.1 Persistence in Relation to Formulation and Application Method

Sometimes the persistence of a pesticide in soil is increased by applying it as a granular formulation. However, it is questionable whether the exposure of soil pests to a pesticide is improved by such a formulation, because much of the pesticide may be retained in the granules. In a comparison of formulations, the insecticide carbofuran was incubated in a silty clay loam soil in both the technical form and 10% granular formulation (Ahmad et al., 1979). Carbofuran applied in technical form had a half-life of 11 to 13 d, while 63 to 75% of the granular-formulation remained after 35 d. When a pesticide is slowly released from granules and the rate of transformation in the surrounding soil is relatively fast, the exposure level will remain low. Efficacy of a pesticide in relation to exposure of pests to the pesticide over time is discussed in detail in section 11–4.

In some pesticide-soil combinations the type of formulation may not be critical. Little difference was measured in the rate of transformation of oxamyl and ethoprophos when applied in aqueous solution or in granules (Vydate® 10G, Mocap® 20G) to moist soil in incubation studies (Smelt et al., 1987). This can be expected for granules which quickly release the nematicide into the surrounding soil.

### 11–3.2 Accelerated Degradation with Repeated Application

In recent years, there has been an increasing number of reports concerning problem soils in which a specific soil-acting pesticide is no longer active. This phenomenon may occur after repeated application of the pesticide at intervals ranging from a few weeks to years. The failures are usually attributed to rapid transformation of the pesticide in soil. The following examples of such accelerated degradation illustrate that the phenomenon is not limited to a single class of chemicals.

The rate of transformation of diazinon in a soil was studied because of the pesticide's failure to control root aphids in lettuce (*Lactuca sativa* L.) (Forrest et al., 1981). When diazinon was repeatedly applied to soil (14 times) over 3 yr, degradation half-life for subsequent doses was $<2$ d, in contrast to half-lives of 20 to 30 d in three nonadapted soils.

The effect of repeated applications of the insecticide/nematicide carbofuran on the transformation rate in a sandy loam soil (no pesticide application history) was studied by Harris et al. (1984). After the second and third treatment of the soil with carbofuran in the lab, the transformation was much faster than after the first application. Upon repeated application, $<10\%$ of the dosage was left as carbofuran after 1 d. Field-treated soils have shown similar behavior. Carbofuran became ineffective in the soils of various fields with a history of several applications in the course of years (Kaufman et al., 1984); soils exhibiting this response have been designated problem soils with respect to efficacy of the chemical. When carbofuran was incubated in these soils, it was found that the rate of transformation was high, as is shown in Fig. 11–7. When labelled with $^{14}C$ in the carbonyl group, a large fraction of the label was released as $^{14}CO_2$ in a short period. Sterilization of the soils by autoclaving or by gamma-irradiation lowered the rate of transformation (release of $^{14}CO_2$), thus corroborating the role of microorganisms.

The rate of transformation of the fungicide iprodione was measured in a few soils that had been treated with the compound at least three times in recent years, as well as in nearby soils that had not been treated before (Walker et al., 1984). The rates of transformation in the soils treated previously were much higher than those in the soils treated for the first time (Fig. 11–8). The two most important transformation products were the same for the two groups of soils. Treatment of both soil groups with the sterilant sodium azide impeded the transformation of iprodione, indicating that it is a microbial

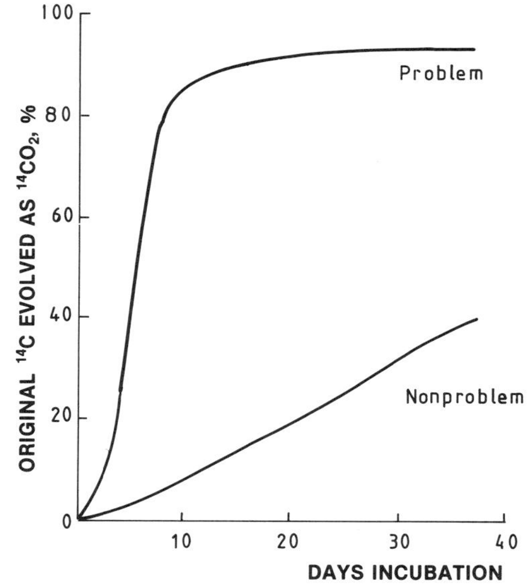


Fig. 11–7. The degradation of $^{14}$C-carbofuran in carbofuran-problem and nonproblem soils. After Kaufman et al. (1984). A problem soil is one in which pesticide efficacy is poor after repeated application of the pesticide.

process. This suggests that the previous applications of iprodione had induced changes in the microbial populations leading to the accelerated transformation.

Rates of transformation of ethoprophos, oxamyl, and aldicarb were measured at 15 °C in loamy soils from plots with different nematicide application histories at two experimental fields (Smelt et al., 1987). In a soil that had been treated annually (four times) with ethoprophos, transformation of this nematicide was much faster than in previously nontreated soil. Accelerated transformation of oxamyl also occurred after its application annually in the field. Aldicarb in nonpretreated soil showed the expected transformation pattern, but in soil that had been treated annually, the amount of aldicarb sulfoxide remained comparatively low and aldicarb sulfone could not be detected at all. Similar effects were measured after renewed applications of the nematicides to the experimental fields. This phenomenon was accompanied with poor nematode control resulting in visible damage to the potato crop.

Several studies have shown that the adaptation of soil microorganisms to a certain pesticide may also accelerate degradation of other pesticides with a closely related chemical structure (Kaufman et al., 1984; Kearney & Kellogg, 1985). In a soil sample that had been treated four times with carbofuran, resulting in accelerated transformation of this compound, various other methyl carbamate pesticides were also transformed at an enhanced rate, while the transformation of pesticides belonging to other chemical groups was not affected (Harris et al., 1984).

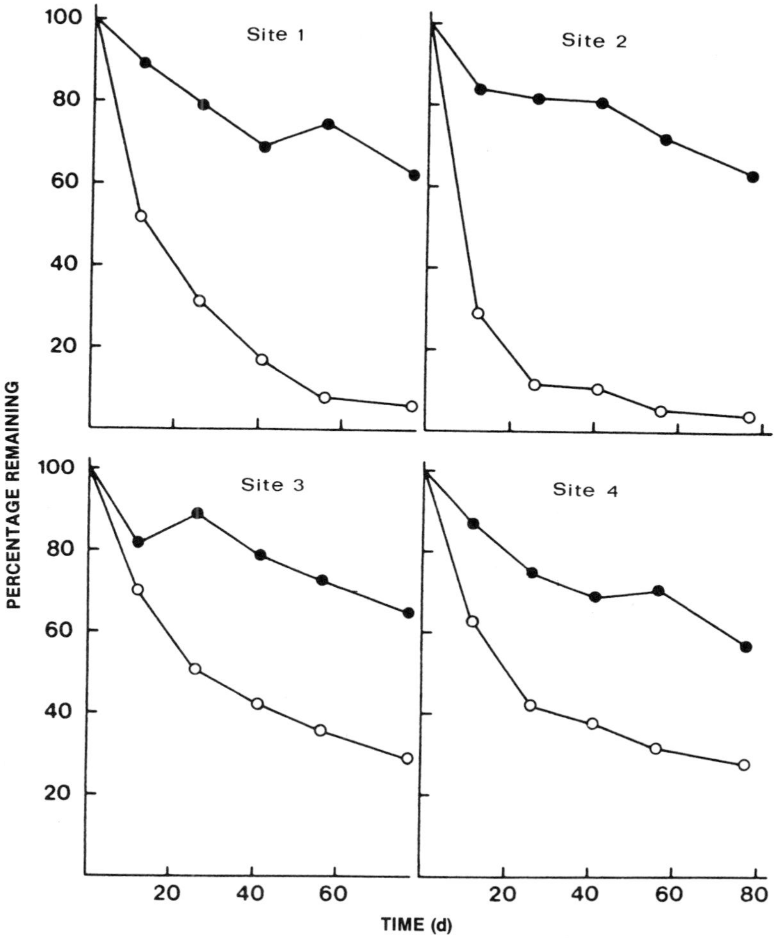


Fig. 11-8. Degradation of iprodione in soil from four sites. Dots = previously untreated soil; circles = soil treated previously with iprodione in the field. After Walker et al. (1984).

This behavior was demonstrated also by Smelt et al. (1987). Previous applications of aldicarb in the field induced accelerated transformation of oxamyl in the incubation study and vice versa; both chemicals are carbamoyl oximes. In soil from plots treated previously with ethoprophos (a phosphorodithioate), the transformation of aldicarb and oxamyl was not accelerated. Similarly, the ethoprophos transformation rate was not affected by earlier oxamyl applications.

Cross adaptation does not seem to occur within large groups of related pesticides with a greater variation in chemical structure. Six organophosphorus insecticides were incubated in a soil that had been adapted to isofenphos (Racke & Coats, 1988). None of these insecticides showed accelerated transformation. Similarly, in a soil adapted to fonofos, the transformation of various other organophosphorus insecticides was not enhanced. In a soil adapted to carbofuran, the transformation rates of aldicarb and its oxidation products were not affected (Suett, 1987). Thus, in these experiments, accelerated transformation in a soil was limited to compounds of similar chemical structure.

The capability of accelerated transformation can develop quickly in a soil, sometimes after a single application. However, in other pesticide-soil combinations this phenomenon does not occur, even after many treatments (Kaufman, 1987). Essential information required for many soil-acting pesticides includes the extent to which accelerated transformation has advanced in practice, and how quickly this phenomenon expands. In most studies on this subject, the researchers used soil from fields where crop protection by a pesticide failed, or soil from experimental fields with abnormally high frequency of application. Under normal agricultural practice, the development of accelerated transformation will proceed more slowly, especially when closely related pesticides are applied at ample intervals. The situation will often be more serious with frequent application, for example in monocultures. Further research is needed to explain why this potency develops so quickly in some soils and so slowly in others. This could be related to the presence of certain microorganisms. Microorganisms capable of fast transformation of a pesticide have been isolated from adapted soil (Kaufman, 1987). Presumably, soil-applied pesticides that are vulnerable to accelerated transformation will vary in their susceptibility to diminished effectiveness under a range of conditions.

Long-term maintenance of pesticide efficacy requires prevention or retardation of accelerated transformation in soil. Long time intervals between pesticide application will be helpful and can be realized by integration of crop rotation and nonchemical methods into pest management systems. Alternating pesticides of different chemical structure may be useful, provided there is no cross adaptation. However, the number of alternative chemicals for a given crop is small and will remain relatively static in the near future. It is uncertain whether a controlled-release formulation can extend performance in an adapted soil. The transformation of carbofuran was slowed by applying it as a granulate (Suett, 1987), presumably because the pesticide was only slowly released into the soil. Although such formulation of the pesticide may retard degradation of the total amount of pesticide in the soil, the concentration in the soil solution may not be adequate to provide control of the pest.

It is not yet known how long the capacity for accelerated transformation is retained by a soil. The phenomenon has been observed after annual or biannual applications of a pesticide, so it may be retained at least a few years. Possible mechanisms for the induction of accelerated transformation of a pesticide in microorganisms were described by Kearney and Kellogg (1985). They speculated that the production of the relevant enzymes may be coded in genes and this genetic information may be retained in a population for a long time. Mixing of a small mass of adapted soil with a large mass of non-adapted soil leads to a fast development of the potency for accelerated transformation (Yarden et al., 1987).

In principle, pesticide formulation could be altered to slow down the transformation of a pesticide in soil (Kaufman & Edwards, 1983). Although such inhibition of organisms or enzymes has been demonstrated, e.g., for some organophosphorus insecticides, it will be difficult to find and develop a substance that works adequately under a wide range of field conditions. Elimination of the adapted microorganisms in field soil will require drastic soil treatments that may not be acceptable. Also, increased persistence of pesticides may not only enhance pest control, but may also result in groundwater contamination.

### 11-3.3 Mathematical Simulation of Pesticide Transformation

Attempts have been made to develop simulation models that depict pesticide persistence in soils under field conditions, based on laboratory measurements of transformation pathways and rates. The rate of transformation of oxamyl in two sandy loam soils was measured in the laboratory at various temperatures and soil water contents. These data were introduced into a simulation model depicting oxamyl behavior in the field (Leistra et al., 1980). The computed and measured concentrations are given in Fig. 11-5. The rate of transformation of oxamyl in soil was simulated quite well. However, the computed accumulation at the soil surface in dry periods (1975 especially) did not actually occur, which indicates that one or more dissipation processes near the surface were missed.

Aldicarb is partly oxidized in soil to its sulfoxide, which in turn is partly oxidized to the sulfone; both products are highly bioactive. Other reactions occur also, including hydrolysis, leading to compounds of comparatively low toxicity. Rates of transformation have been measured for two soils in the laboratory at different temperatures and soil water contents. These relationships were introduced into computer simulations of the behavior of aldicarb and its oxidation products in the field (Bromilow & Leistra, 1980). The concentrations were fairly well simulated with the model, although the production of aldicarb-sulfone under field conditions was somewhat overestimated. A similar sequence of transformations exists for fenamiphos; the greater persistence and lower sorption of fenamiphos-sulfoxide, relative to the parent compound, was shown by simulation to be responsible for the effective distribution of this nematicide in a tropical soil (Lee et al., 1986).

The rate of transformation of ethoprophos under winter field conditions in the Netherlands was simulated by Leistra and Smelt (1981). The relationship between transformation rate and temperature, as measured in the laboratory, was introduced into the computations. The results of the computations were compared with those of measurements in the field (Fig. 11-9). The effect of soil differences and temperature in the field were well represented by simulated transformation of ethoprophos. The rate of transformation in the first month was comparatively fast because of relatively high soil temperatures. The transformation was slower in mid-winter and speeded up in spring. Despite the fairly low temperatures in winter, the transformation in the loam soils (Middenmeer and Wierum) continued at a substantial

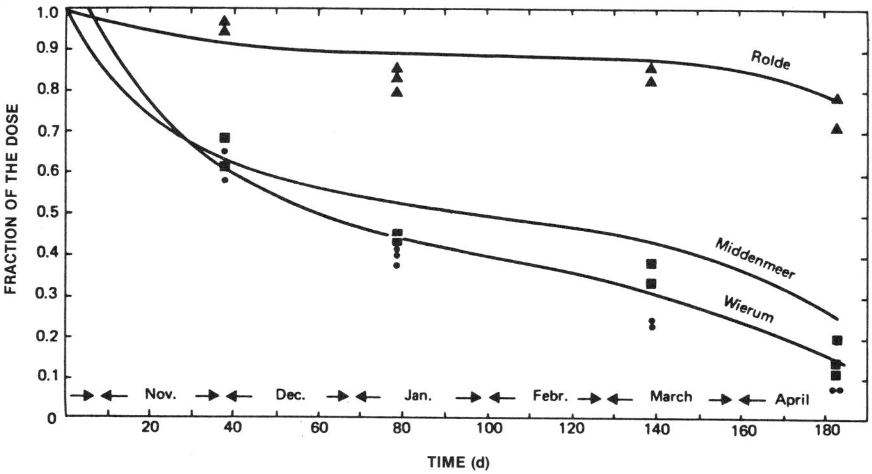


Fig. 11-9. Ethoprophos remaining in soil during the winter half-year of 1978 to 1979. Dots, squares, and triangles: measured amounts for the Wierum, Middenmeer, and Rolde soils. Lines: simulated with a computer model from laboratory data. After Leistra and Smelt (1981).

rate. These results encourage the use of simulation modeling, with appropriate input data from laboratory measurements, to predict transformation of pesticides in the field.

## 11-4 EXPOSURE AND EFFECTS ON ORGANISMS

Knowledge of the concentration distribution of a pesticide in the soil, both in space and time, is extremely useful, whether it is obtained by actual measurement or by computer simulation. This information alone is insufficient to determine the impact on a pest organism unless the dose-response relationship for a given pest-pesticide combination is also known for various stages of the pest's life cycle. In this section, we will address this subject through the example of cyst nematode control by pesticides. Similar illustrations could be given for the control of many other pests, but the case of cyst nematode control will suffice in the interest of brevity.

### 11-4.1 Nematicide Dose-Response Relationships

Organophosphate and carbamate nematicides inhibit the enzyme acetylcholinesterase that functions in the transmission of impulses in the nematode nervous system (Wright, 1980, 1981). Because the transmission is disturbed, changes occur in the behavior and vital abilities of the nematodes and the development and reproduction of the animals is impaired. The effects of low concentrations of these nematicides are largely reversible; after decline of the pesticide concentration over time, many nematodes can resume normal activity. These compounds can be characterized as nematistatic. A fraction of the treated population may die, however, possibly through starvation (Wright, 1980).

The effect of known initial concentrations of nematicides on nematodes in their various life stages has been observed or measured in a series of studies. Studies with two parasitic species of cyst-forming nematodes will be discussed here, those with a potato cyst nematode (*Globodera rostochiensis*) and those with the sugarbeet nematode (*Heterodera schachtii*). The host plants of these nematodes excrete substances from their roots to the soil solution that induce hatching of the larvae and their emergence from the cysts. Subsequently, the larvae move through the soil solution to the roots by orienting on concentration gradients of the hatching agent. They invade the roots by using their stylet and start feeding, thus damaging the root tissues, withdrawing essential substances from the plants, and interfering with water uptake.

#### 11–4.1.1 Effect on Hatching of Nematode Larvae

In a pot experiment with potato, aldicarb was applied at 5 mg/kg in soil (higher than in normal practice). This resulted in only a slight delay in the emergence of the larvae of the potato cyst nematode from their cysts (Hague & Pain, 1973). Osborne (1973) studied the effect of aldicarb on the emergence of larvae from cysts submerged in potato root diffusate. Emergence was completely inhibited for several weeks by exposure to concentrations of 2.5 mg/L (within the practical range) and higher. At a concentration of 1.0 mg/L of aldicarb, hatching was incompletely suppressed. When the cysts were transferred to solutions of root diffusate after 4 wk of exposure to the nematicide, many of the larvae were capable of hatching. However, this capability gradually decreased with time of exposure to aldicarb plus root diffusate, until few viable larvae remained after 12 wk of exposure.

Steudel (1972) studied hatching of larvae of the sugarbeet nematode from their cysts submerged in aldicarb solutions. An almost complete inhibition was caused by an exposure of 5 mg/L (high in the practical range) for some weeks. Even at 1 mg/L of aldicarb in combination with a hatching agent, hatching was substantially inhibited. Aldicarb at a concentration of 4.8 mg/L completely inhibited hatching of larvae of the sugarbeet nematode from the cysts (Hough & Thomason, 1975), but a concentration of 0.48 mg/L only slightly reduced hatching. Steele and Hodges (1975) found that aldicarb concentrations of 1 mg/L (within the practical range) and higher inhibited hatching and emergence of the sugarbeet cyst nematode from cysts in a solution of a hatching agent. However, low concentrations of aldicarb (up to about 0.1 mg/L) stimulated hatching.

In a pot experiment, oxamyl was applied to soil containing potato cyst nematodes in which potato plants were grown (Hague, 1979). At contents of 4 and 8 mg/kg in soil (above the practical range) emergence of larvae from the cysts was delayed for about 2 wk, after which it resumed. An oxamyl content of 2 mg/kg in soil (high in the practical range) resulted in only a

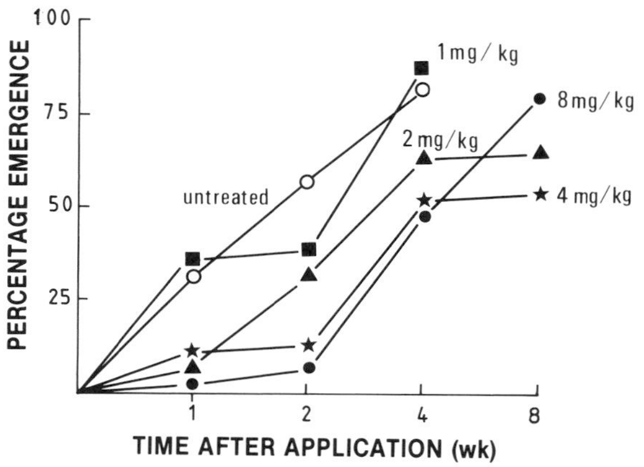


Fig. 11-10. Percentage emergence of second stage larvae of *Globodera rostochiensis* from cysts in soil treated with different quantities of oxamyl. After Hague (1979).

short delay in the emergence of the larvae (Fig. 11-10). When cysts of the potato cyst nematode were submerged in a solution of potato root diffusate plus oxamyl for 1 wk, an oxamyl concentration of 4 mg/L (within the practical range) strongly inhibited emergence of larvae, but at 1 mg/L there was much less inhibition (Evans & Wright, 1982). After transfer of the cysts to a solution of root diffusate alone, the capability of the larvae to emerge was completely restored.

The variable impact of nematicides, applied at different concentrations, on nematode emergence from cysts emphasizes the importance of maintaining an adequate level of nematicide in the soil solution over relatively long periods of time to achieve satisfactory control.

### 11-4.1.2 Effect on Nematode Movement

Transformation products of some pesticides (e.g., aldicarb and fenamiphos) are also active against pests, thus the toxicity of both the parent chemical and its products is important. The effect of aldicarb and its oxidation products on the body movement of second stage larvae of potato cyst nematode has been studied by Nelmes (1970). Aldicarb itself was most active; after 1 d at 1 mg/L the percentage of moving larvae was strongly reduced. Aldicarb sulfoxide at 5 to 10 mg/L (high in the practical range) had a distinct effect on movement after 1 d, but aldicarb sulfone had less effect. At 1 d after transfer to clean water, many larvae had resumed normal movement, except after exposure to aldicarb at 10 mg/L. Stylet movement of the larvae was also affected, mostly by aldicarb itself. The higher activity of aldicarb on the short term could be related to a comparatively quick penetration of this compound into the nematodes (Nelmes, 1970).

Batterby (1979) studied the toxic effects of aldicarb and its two oxidation products on second stage larvae of the sugarbeet cyst nematode. Within 1 d, aldicarb itself was most active; the oxidation products at 2 mg/L had

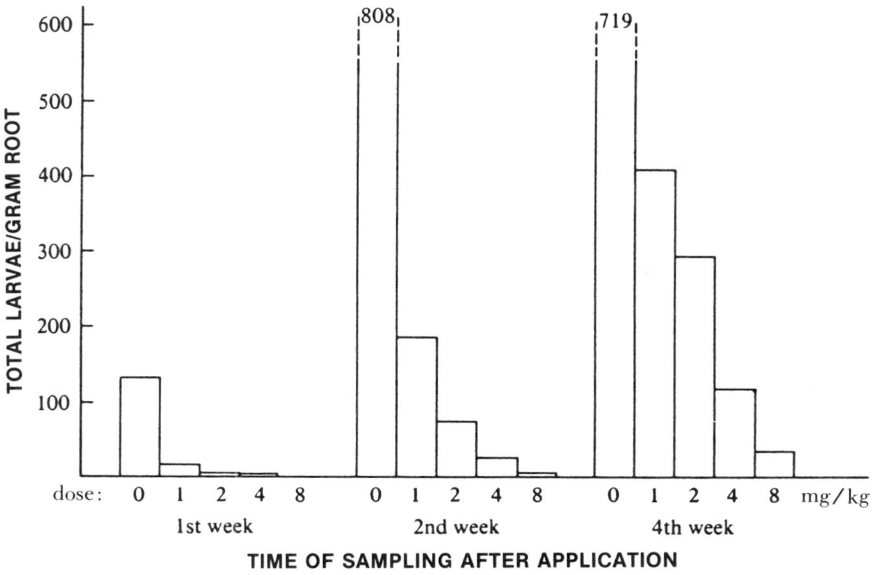


Fig. 11-11. Number of larvae of *Globodera rostochiensis* in roots of potato plants after application of oxamyl. After Hague (1979).

little effect on the percentage of active larvae, although movement was clearly affected. The differences in effect of sulfoxide and sulfone were not great at this concentration. The effects of these compounds became more distinct with increase in time of exposure. Movement of the larvae in moist sand was strongly impaired by concentrations of aldicarb of 1 mg/L and higher, but 0.5 mg/L had little effect (Hough & Thomason, 1975). Further, invasion of larvae into the roots of sugarbeet (*Beta vulgaris* L.) was strongly inhibited at 1 mg/L of aldicarb, but 0.1 mg/L had no clear effect.

### 11-4.1.3 Effect on Nematode Orientation and Penetration of Roots

In some pot experiments with potato plants, Hague and Pain (1973) counted the second stage larvae of the potato cyst nematode that were present in soil outside cysts and plant roots. At aldicarb contents of 5 mg/kg (higher than in practice), many such larvae were found, whereas the number of larvae in the roots was small. Many larvae had emerged from the cysts, but few were able to orient on the plant roots and invade them. It was observed that little food was stored in the digestive tract of these larvae. Similar results were obtained with oxamyl at contents of 4 and 8 mg/kg in soil (Hague, 1979). The number of larvae that had penetrated the roots for each treatment is shown in Fig. 11-11.

The location of nematodes in the soil-root environment can affect the efficacy of applied nematicides. The larvae of the sugarbeet nematode that already had penetrated beet roots were little affected by submergence of the roots in solutions of aldicarb or its oxidation products (Steele, 1976). Only a few individuals left the roots. However, the development of the larvae in the roots was retarded. When the nematicides were applied at 15 d after penetration of the larvae into the roots, the effect on the nematodes was minimal. Development of the potato cyst nematodes that had penetrated the roots was repressed when oxamyl contents in soil were 4 and 8 mg/kg (Hague, 1979).

### 11-4.2 Significance of Effects for Cyst Nematode Control

The general picture that comes forward from the various studies on the effect of practical concentrations of aldicarb and oxamyl on cyst nematodes is:

1. The emergence of larvae from the cysts is inhibited incompletely by the nematicides and for only a short period.
2. Orientation of larvae to host plant roots and the invasion of roots are sensitive to nematicides.
3. Nematodes that are established in plant roots are affected little by nematicides. Thus, the preferred application time is just before or at seeding or planting of the crop.
4. The effect of the nematicides on the various life stages is largely reversible, so many nematodes will resume normal activity after the concentration has fallen sufficiently.
5. The dose of a few kilograms per hectare, as usually applied in normal practice, can protect the crop in an initial period, but the final density of the nematodes may be expected to be higher than the initial one. Only at fairly high doses, for example around 10 kg/ha, can a decrease in the number of nematodes in soil be demonstrated (Hague & Pain, 1973; Whitehead, 1980).

This example of cyst nematode control serves to illustrate the complexity and dynamics of a crop-pest-soil system into which a pesticide is introduced. Effective pest control is not solely a function of pesticide-soil-water interactions, but also depends on the biology of the organism and the manner in which the pest life cycle is impacted by the chemical.

## 11-5 GENERAL DISCUSSION

An impressive body of information is available on the separate processes that play a part in the efficacy of soil-acting pesticides. Various properties of the pesticides are known, the effect of soil and climatic factors on their behavior can be estimated, and their effect on plant-parasitic organisms has been studied. Little has been done, however, to integrate knowledge regarding the physical-chemical and the biological aspects of pesticide efficacy. Knowledge of this area should become less fragmentary by intensifying cooperative research of workers in various disciplines.

The behavior of pesticides in soil under field conditions has been simulated with computer models, and these models have been evaluated in some studies. In systems without great complications, especially with respect to water flow, movement and transformation of pesticides in soil can be realistically simulated. Thus, the exposure of harmful soil organisms can be described quantitatively, so that simulation models can be used increasingly to aid in optimization of nematicide application. The development of computational models for pesticide behavior in soil should be continued. There is a special need for models that can simulate the complicated situations that often occur in practice. Also, modeling efforts must be accompanied by the development and testing of laboratory and field measurement methods that provide key parameters for models; useful models must have appropriate input data.

The persistence of pesticides in soils of widely differing properties should be known. Pesticide transformation rate in some soils may be too high for efficient control of pests. Development of soil microorganisms with increased ability to degrade soil-acting pesticides is a serious problem. Research is needed to find ways to prevent or delay such development. Soil microbiologists should play a key role in this research. For the time being, pesticide application with ample time intervals and alternation with non-allied pesticides are viable remedies.

The population dynamics of harmful soil organisms is an important area of research for the optimal use of various control measures. Simulation of population dynamics with computer models could enhance progress. Undoubtedly, a lot of quantitative biological research will also be needed because of missing and conflicting basic information. Knowledge of the behavior of harmful organisms and the effect of pesticides in various life stages should be extended considerably. The interactions of pests and pesticides with soil and crop should be included in such studies.

In the future, soil scientists can play an important part in the improvement of the efficacy of soil-applied pesticides. Neglecting knowledge on the behavior of pesticides in soil will often lead to pesticide waste (too high doses), to failure of the treatment, or environmental contamination problems. Various new technical developments in agriculture present challenges for further research and they require intensified pesticide research. Application of pesticides by different irrigation practices is an example of rapidly changing technology. As new developments are adapted to different regional conditions, a great variety of agricultural systems may develop. Many of the new application methods should be optimized on the basis of a detailed knowledge of the behavior of pesticides in soil. Such knowledge can lead to improved efficacy of pesticides in agricultural crops and a concomitant reduction in environmental contamination.

## REFERENCES

Ahmad, N., D.D. Walgenbach, and G.R. Sutter. 1979. Degradation rates of technical carbofuran and a granular formulation in four soils with known insecticide use history. Bull. Environ. Contam. Toxic. 23:572–574.

Andringa, J.T. 1981. Strooien en inwerken van granulaten (Distributing and incorporating granulars). Landbouwmechanisatie 32:471–475.

Apt, W.J., and E.P. Caswell. 1988. Application of nematicides via drip irrigation. Ann. Appl. Nematol. (J. Nematol. 20, Supplement) 2:1-13.

Batterby, S. 1979. Toxic effects of aldicarb and its metabolites on second stage larvae of *Heterodera schachtii*. Nematologica 25:377-385.

Batterby, S., G.N.J. Le Patourel, and D.J. Wright. 1980. Effects of aldicarb persistence on field populations of *Heterodera schachtii*. Ann. Appl. biol. 95:105-113.

Bresler, E., and R.E. Green. 1987. Transport of a degradable substance and its metabolites under drip irrigation. Agric. Water Manage. 12:195-206.

Briggs, G.G. 1984. Factors affecting the uptake of soil-applied chemicals by plants and other organisms. p. 35-47. *In* R.J. Hance (ed.) Soils and crop protection chemicals. Monogr. No. 27. Br. Crop Protection Counc., Croydon, England.

Bromilow, R.H. 1988. Physico-chemical properties and pesticide placement. p. 295-308. *In* T. Martin (ed.) Application to seeds and soils. Monogr. 39. Br. Crop Protection Counc. Publ., Thornton Heath, England.

Bromilow, R.H., and M. Leistra. 1980. Measured and simulated behavior of aldicarb and its oxidation products in fallow soils. Pestic. Sci. 11:389-395.

Bromilow, R.H., and K.A. Lord. 1979. Distribution of nematicides in soil and its influence on control of cyst-nematodes (*Globodera* and *Heterodera* spp.). Ann. Appl. Biol. 92:93-104.

Coppedge, J.R., R.A. Stokes, R.E. Kinzer, and R.L. Ridgway. 1975. Effect of soil moisture and soil temperature on the release of aldicarb from granular formulations. J. Econ. Entomol. 68:209-210.

Evans, S.G., and D.J. Wright. 1982. Effects of the nematicide oxamyl on life cycle stages of *Globodera rostochiensis*. Ann. Appl. Biol. 100:511-519.

Forrest, M., K.A. Lord, N. Walker, and H.C. Woodville.1981. The influence of soil treatments on the bacterial degradation of diazinon and other organophosphorus insecticides. Environ. Pollut. A24:93-104.

Furmidge, C.G.L. 1984. Formulation and application factors involved in the performance of soil-applied pesticides. p. 49-64. *In* R.J. Hance (ed.) Soils and crop protection chemicals. Monogr. 27. Br. Crop Protection Counc., Croydon, England.

Hague, N.G.M. 1979. A technique to assess the efficacy of non-volatile nematicides against the potato cyst nematode *Globodera rostochiensis*. Ann. Appl. Biol. 93:205-211.

Hague, N.G.M., and B.F. Pain. 1973. The effect of organophosphorus compounds and oxime carbamates on the potato cyst nematode *Heterodera rostochiensis* Woll. Pestic. Sci. 4:459-465.

Hance, R.J. 1988. Adsorption and bioavailability. p. 1-19. *In* R. Grover (ed.) Environmental chemistry of herbicides. Vol. 1. CRC Press, Boca Raton, FL.

Harris, C.R. 1972. Factors influencing the effectiveness of soil insecticides. Ann. Rev. Entomol. 17:177-198.

Harris, C.R., R.A. Chapman, C. Harris, and C.M. Tu. 1984. Biodegradation of pesticides in soil: Rapid induction of carbamate degrading factors after carbofuran treatment. J. Environ. Sci. Health B19:1-11.

Hough, A., and I.J. Thomason. 1975. Effects of aldicarb on the behavior of *Heterodera schachtii* and *Meloidogyne javanica*. J. Nematol. 7:211-229.

Johnson, A.W., W.A. Rohde, and W.C. Wright. 1982. Soil distribution of fenamiphos applied by overhead sprinkler irrigation to control *Meloidogene incognita* on vegetables. Plant Dis. 66:489-491.

Kaufman, D.D. 1987. Accelerated biodegradation of pesticides in soil and its effect on pesticide efficacy. Proc. 1987 Br. Crop Prodection Conf.—Weeds 2:515-522.

Kaufman, D.D., and D.F. Edwards. 1983. Pesticide/microbe interaction effects on persistence of pesticides in soil. p. 177-182. *In* J. Miyamoto and P.C. Kearney (ed.) Proc. 5th Int. Congr. Pestic. Chem., Vol. 4. Pergamon, Oxford, England.

Kaufman, D.D., Y. Katan, D.F. Edwards, and E.G. Jordan. 1984. Microbial adaptation and metabolism of pesticides. *In* J.L. Hilton (ed.) Agricultural chemicals of the future. Beltsville Symp. Agric. Res. 8:437-451.

Kearney, P.C., and S.T. Kellogg. 1985. Microbial adaptation to pesticides. Pure Appl. Chem. 57:389-403.

Keng, J.C.W., and T. Vander Gulik. 1985. Application of chemicals through a trickle system for soil-borne pest control. I. Derivation of basic physical theory for practical use. Can. Agric. Eng. 27:31-33.

Keng, J.C.W., T.C. Vrain, and J.A. Freeman. 1985. Application of chemicals through a trickle irrigation system for soil-borne pest control. II. The design of a prototype system and test results. Can. Agric. Eng. 27:35-37.

Koskinen, W.C., and S.S. Harper. 1987. Herbicide properties and processes affecting application. p. 9–18. *In* C.G. McWhorter and M.R. Gebhardt (ed.) Methods of applying herbicides. Monogr. 4. Weed Sci. Soc. of Am., Champaign, IL.

Lee, C.-C., R.E. Green, and W.J. Apt. 1986. Transformation and adsorption of fenamiphos, F. sulfoxide and F. sulfone in Molokai soil and simulated movement with irrigation. J. Contaminant Hydrol. 1:211–225.

Leistra, M. 1979. Computing the movement of ethoprophos in soil after application in spring. Soil Sci. 128:303–311.

Leistra, M., R.H. Bromilow, and J.J.T.I. Boesten. 1980. Measured and simulated behaviour of oxamyl in fallow soils. Pestic. Sci. 11:379–388.

Leistra, M., and J.H. Smelt. 1981. Movement and conversion of ethoprophos in soil in winter. 2. Computer simulation. Soil Sci. 131:296–302.

Morton, H.V. 1986. Modification of proprietary chemicals for increasing efficacy. J. Nematol. 18:123–128.

Moss, S.R., D. Crump, and A.G. Whitehead. 1976. Control of potato cyst nematodes, *Globodera rostochiensis* and *G. pallida*, in different soils by small amounts of oxamyl or aldicarb. Ann. Appl. biol. 84:355–359.

Nelmes, A.J. 1970. Behavioral responses of *Heterodera rostochiensis* larvae to aldicarb and its sulfoxide and sulfone. J. Nematol. 2:223–227.

Nicholls, P.H., R.H. Bromilow, and T.M. Addiscott. 1982. Measured and simulated behaviour of fluometuron, aldoxycarb and chloride ion in a fallow structured soil. Pestic. Sci. 13:475–483.

Omary, M.S. 1988. Distribution and concentration in the soil of Nemacur applied through a trickle irrigation system. Ph.D. diss. Clemson Univ., Clemson (Diss. Abstr. 89-15241).

Osborne, P. 1973. The effect of aldicarb on the hatching of *Heterodera rostochiensis* larvae. Nematologica 19:7–14.

Palti, L., and H. Shoham. 1983. Trickle irrigation and crop disease management. Plant Dis. 67:703–705.

Racke, K.D., and J.R. Coats. 1988. Comparative degradation of organophosphorus insecticides in soil: Specificity of enhanced microbial degradation. J. Agric. Food Chem. 36:193–199.

Saffigna, P.G., C.B. Tanner, and D.R. Keeney. 1976. Non-uniform infiltration under potato canopies caused by interception, stemflow and hilling. Agron. J. 68:337–342.

Smelt, J.H., M. Leistra, A. Dekker, and C.J. Schut. 1981. Movement and conversion of ethoprophos in soil in winters: 1. Measured concentration patterns and conversion rates. Soil Sci. 131:242–248.

Smelt, J.H., S.J.H. Crum, W. Teunissen, and M. Leistra. 1987. Accelerated transformation of aldicarb, oxamyl and ethoprophos after repeated soil treatments. Crop Prot. 6:295–303.

Smelt, J.H., S. Voerman, and M. Leistra. 1976. Mechanical incorporation in soil of surface-applied pesticide granules. Neth. J. Plant Pathol. 82:89–94.

Smith, J., and R.H. Bromilow. 1977. Incorporation of granular nematicides in peat soils for control of potato cyst-nematode (*Heterodera rostochiensis* Woll.). Exp. Husb. 33:98–111.

Steele, A.E. 1976. Effects of oxime carbamate nematicides on development of *Heterodera schachtii* on sugar beet. J. Nematol. 8:137–140.

Steele, A.E., and L.R. Hodges. 1975. In-vitro and in-vivo effects of aldicarb on survival and development of *Heterodera schachtii*. J. Nematol. 7:305–312.

Steudel, W. 1972. Versuche zum einfluss von aldicarb auf den schlupfvorgang bei zysten von *Heterodera schachtii* nach langerer einwirkung. Nematologica 18:270–274.

Stokes, R.A., J.R. Coppedge, and R.L. Ridgway. 1970. Chemical and biological evaluation of the release of aldicarb from granular formulations. J. Agric. Food Chem. 18:195–198.

Suett, D.L. 1987. Influence of treatment of soil with carbofuran on the subsequent performance of insecticides against cabbage root fly (*Delia radicum*) and carrot fly (*Psila rosae*). Crop Prot. 6:371–378.

Suett, D.L., and A.R. Thompson. 1985. The development of localized insecticide placement methods in soil. p. 65–74. *In* E.S.E. Southcombe (ed.) Symposium on application and biology. Monogr. 28. Br. Crop Protection Counc. Publ., Thornton Heath, England.

Thomason, I.J. 1987. Challenges facing nematology: Environmental risks with nematicides and the need for new approaches. p. 469–476. *In* J.A. Veech and D.W. Dickson (ed.) Vistas on nematology. Soc. Nematologists, Hyattsville, MD.

Walker, A., A.R. Entwistle, and N.J. Dearnaley. 1984. Evidence for enhanced degradation of iprodione in soils treated previously with this fungicide. p. 117–123. *In* R.J. Hance (ed.) Soils and crop protection chemicals. Monogr. 27. Br. Crop Protection Counc., Croydon, England.

Whitehead, A.G. 1978. Chemical control. Soil treatments. p. 283–296. *In* J.F. Southey (ed.) Plant nematology. 3rd ed. Her Majesty's Stationary Office, London, England.

Whitehead, A.G. 1980. Nematode control with special reference to non-fumigant nematicides. p. 1–17. *In* Nematicides, manual of a workshop. Assoc. of Appl. Biol., Rothamsted, Harpenden, England.

Whitehead, A.G., R.H. Bromilow, K.A. Lord, S.R. Moss, and J. Smith. 1975. Incorporating granular nematicides in soil. p. 133–144. *In* Proc. 8th Br. Insectic. Fungic. Conf. Vol. 1. Br. Crop Protection Counc., Brighton, England.

Whitehead, A.G., R.H. Bromilow, D.J. Tite, P.H. Finch, J.E. Fraser, and E.M. French. 1979. Incorporation of granular nematicides in soil to control pea cyst nematode, *Heterodera goettingiana*. Ann. Appli. biol. 92:81–91.

Whitehead, A.G., D.J. Tite, and R.H. Bromilow. 1981. Techniques for distributing non-fumigant nematicides in soil to control protato cyst nematodes, *Globodera rostochiensis* and *G. pallida*. Ann. Appl. Biol. 97:311–321.

Whitehead, A.G., D.J. Tite, D.J. Fraser, and A.J.F. Nicholls. 1987. Control of beet cyst nematode, *Heterodera schachtii*, by aldicarb applied to the soil by a vertical band-reciprocation harrow technique or by rotary cultivation. Ann. Appl. Biol. 110:127–133.

Wilkins, R.M. 1983. Controlled release—Present and future. Proc. 10th Int. Congr. Plant Prot. 2:554–559.

Wright, D.J. 1980. Mode of action of nematicides, their movement in plants and development of resistance. p. 39–77. *In* Nematicides, manual of a workshop. Assoc. Appl. Biol., Rothamsted, Harpenden, England.

Wright, D.J. 1981. Nematicides: Mode of action and new approaches to chemical control. p. 421–449. *In* B.M. Zuckerman and R.A. Rohde (ed.) Plant parasitic nematodes 3. Academic Press, New York.

Yarden, O., N. Aharonson, and J. Katan. 1987. Accelerated microbial degradation of methyl benzimidazol-2-yl carbamate in soil and its control. Soil Biol. Biochem. 19:735–739.

Yaron, B., Z. Gerstl, and W.F. Spencer. 1985. Behavior of herbicides in irrigated soils. Adv. Soil Sci. 3:121–211.

# Chapter 12

# Impact of Pesticides on the Environment

**Y. A. MADHUN AND V. H. FREED,** *Oregon State University, Corvallis, Oregon*

For more than 1500 yr, attempts to control pests with chemicals achieved only limited and often disappointing success. However, Mueller's discovery of DDT in 1939 and the subsequent development of this potent insecticide in the 1940s led to its wide-scale use in agriculture and public health. Parallel to this development was the discovery of the highly effective growth-regulating herbicide such as 2,4-D and MCP (Davies & Edmundson, 1972; Cremlyn, 1978). For the first time, these new chemicals gave prompt effective control of pest problems that here-to-fore had proven difficult or impossible to control. For example, in 1944 when an outbreak of typhus occurred during World War II, DDT was used to delouse a large civilian population and stop the epidemic. Subsequently, it was used widely to control malaria carrying mosquitoes and other insect vectors (Hayes, 1982).

Dramatic as was the use of these chemicals in protecting human health and increasing agricultural production, they were not without adverse side effects. Well before Rachel Carson's apocalyptic *Silent Spring* was published, (Carson, 1962) scientists had studied some of these side effects (Loosanoff, 1965; Cope, 1965). Residues of the chlorinated hydrocarbon insecticides getting into bodies of water were shown to be directly toxic to aquatic organisms by these and other authors. Indirect effects were also noted, particularly in aquatic environments when herbicides were used to control undesired vegetation. Other problems encountered included persistent residues in soil, air, and food as well as water (Gould, 1966; ACS, 1978). Another concern, especially with chlorinated hydrocarbon insecticides but also true of some organophosphates, is that of "bioaccumulation"—that is deposit and retention of residues ingested through food and water, inhaled or absorbed through dermal exposure (USHEW, 1969; NRC-NAS, 1969; Davies et al., 1975).

It was inevitable that the concerns over the impacts of pesticides expressed so vigorously by Rachel Carson would stir up a lively controversy in view of the benefits accruing to those who used the chemicals. Many defended the view point expressed in the book (Graham, 1970; Vandenbosch,

1977; Weir & Shapiro, 1981; Hay, 1982). These and many other authors expressed fears not only of serious adverse impacts on the environment but on human health as well. But others were equally articulate in defense of the use of pesticides (Beatty, 1973; Maddox, 1972; Barrons, 1981; Claus & Bolander, 1977) pointing to the benefits in increased food production and control of insect vectors. Both sides could agree, however, on the need to know more about the chemicals and their judicious use. There was, and still is, a quantitative difference on the extent to which pesticide should be used and the degree of hazard posed by many of the chemicals (Epstein & Swartz, 1988; Ames & Gold, 1988; Efron, 1984).

Concern over possible impacts of pesticides on the environment and humans is not confined just to the highly industrialized nations. Developing coutries are also experiencing the same concern (Bull, 1982). Problems encountered in the developing countries are similar to those encountered in the industrialized nations, but may be exacerbated by lack of infrastructure and resources to deal with them (Davies et al., 1982).

This chapter addresses some of the impacts of pesticides in the environment and attempts to assess the consequence of such impacts. In this connection, it is well to keep in mind that assessment of the consequences is a function of the nature of the chemical (i.e., toxicity and properties) and the exposures received (i.e., amount, frequency, and duration). These will be addressed in more detail later.

## 12–1 NATURE AND SCOPE OF THE IMPACT

Although pesticide use has undoubtedly contributed to increased farm productivity and to human health, it has also created several problems, including widespread accumulation of residues with damage to wildlife, fisheries, beneficial insects, and even humans. The impact of pesticides on the natural environment is not always obvious, but often insidious. It has more serious effects than is apparent, such as adverse changes in environmental quality that may reduce productive potential rather than overt toxicity. The nature and magnitude of the impact are influenced by many factors. These factors may be considered with respect to pesticide, organism, and biological interactions.

### 12–1.1 Pesticide

#### 12–1.1.1 Properties vs. Environmental Transport and Behavior

Each chemical has a unique set of properties, both chemical and biological, that distinguish it from all other chemicals. By measuring some of these properties (e.g., melting point, vapor pressure, water solubility, absorption of light, or other electromagnetic waves), the chemist can uniquely identify the compound. However, these properties also are important in relation to how the chemical will behave when released in the environment either in use or disposal. For example, a liquid with a high vapor pressure,

Table 12-1. Relation of physicochemical properties to environmental behavior.

| Physical chemical data | Related to |
|---|---|
| Solubility in water | Leaching, degree of adsorption, mobility in environment, and uptake by plants |
| Partition coefficient | Bioaccumulation potential, and adsorption by organic matter |
| Hydrolysis | Persistence in environment or biota |
| Ionization | Route and mechanism of adsorption or uptake, persistence, and interaction with other molecular species |
| Vapor pressure | Atmospheric mobility, and rate of vaporization |
| Reactivity | Metabolism, microbiological, and photochemical and autochemical degradation |

such as benzene or gasoline, is immediately recognized as a volatile compound and one that will evaporate rapidly. In contrast, the high boiling liquids and solids will have lower vapor pressures and thus be less volatile as is the case of lubricating oil.

It has been found that each chemical behaves a little differently in the environment and this difference in behavior is because of variation in the properties of chemicals and how they interact with the environment. Knowing the relationship between properties and behavior permits a prediction of how the chemical will behave in terms of sorption to soil and other solid material, rates of evaporation, water solubility, and bioaccumulations.

Energy (sunshine) flux, as well as wind and water movement, are the driving forces for movement of chemicals. Moisture, soil microorganisms, and temperature as well as light intensity influence the degradation in the environment. The rate and extent of the processes are related to the physical and chemical properties of the compound. Table 12-1 indicates the physical chemical characteristics and their relationship to such fundamental behavior in the environment as adsorption, leaching, vaporization, and breakdown.

Of the various processes in the environment, those particularly related to overall environmental behavior include, sorption, leaching, vaporization, degradation, and in terms of biological effect, bioaccumulation. The individual processes have been discussed in previous chapters. For an assessment of environmental impact, you must examine the interrelationship among the processes. For example, if you are concerned with the uptake and accumulation of a chemical by an organism, remember that adsorption on a particle soil or sediment will substantially reduce the amount of chemical available for uptake. Thus, if the accumulation is by a fish, the same concentration of chemical in water free of sediment will result in a greater uptake than if there is a heavy silt load in water.

**12-1.1.1.1 Solution Behavior.** All of the known pesticides have some measurable solubility in water. The solubility may be low, being a few tenths of a mg $kg^{-1}$ in the case of some compounds and ranging up to quite high water solubilities in the case of others. The solubility in part depends upon the polarity or electrical properties of the compound as well as its intrinsic elemental makeup. The solubility and behavior of compounds in water is of considerable importance in considering the environmental impact of chemicals.

If the pesticide in question is an ionizable one, such as an acid, base, or salt of either of these, the solubility is likely to be greater than that of the less polar materials such as the chlorinated hydrocarbons, organophosphorus compounds, or the carbamates. Of these latter, less polar compounds tend to concentrate at surfaces in water that may be either at the air-water interface or at the surface of colloidal particles in water. Where the surface or interface is with another liquid immiscible with water such as a fatty solvent, the tendency of the chemical to escape into this solvent from water is called *partitioning.*

Where the compound is one of the less soluble and less polar chemicals, the sorption onto the suspended matter in water or partitioning into oil layers on the water surface tends to take the chemical out of the bulk of the water. With many substances, it is found that filtering the suspended matter from water removes the greater part of the pesticide from the water. It is probable, therefore, that much of the pesticide carried by water is in fact carried by particulate matter suspended in water. Further confirmation of this comes from studies of transport of chemical in flowing streams. Here it has been noted that there is "chromatographing" tendency of the material that is felt because of the solid material alternately being deposited on the bottom of the stream and picked up again in the turbulent motion of the water. Ultimately, the pesticide-carrying particles do settle out and remain on the bottom.

It may be noted in passing that as a salt concentration of water increases, the solubility of most compounds tends to decrease.

**12-1.1.1.2 Uptake and Accumulation.** The accumulation of pesticides by organisms is frequently observed. This accumulation results in residues that may be of concern. The mechanism by which materials are accumulated is complex and varied. Many compounds are accumulated through the metabolic activity of the organism, direct ingestion, or it may be through a physical mechanism or all of these processes. With many compounds used as pesticides, the uptake can be related to a physico-chemical property called the *partition coefficient.* If the partition coefficient is high and the degradation rate low, the compound will accumulate in organisms of the food chain with successive increases at each step. This partition coefficient is a measure of the distribution of the chemical between a lipophilic and hydrophilic or aqueous state. The following equation is the simplest form indicating this relationship.

$$K_{ow} = C_{octanol}/C_{water}$$

The partition coefficient, $K_{ow}$, is characteristic for the compound and is dependent on a variety of molecular features of the chemical. However, many studies have demonstrated the value of the partition coefficient in estimating the ease with which a chemical, particularly a nonpolar or nonionized chemical, will be accumulated by a living organism exposed to it.

Partition coefficient is also a good index of the possible adsorption of a compound by soil organic matter. This, together with molar refraction or latent heat of solution, gives an indication of the adsorbability of compounds.

### 12–1.1.2 Rate and Frequency of Pesticide Application

Pesticides by virtue of their nature, as poisons, would be expected to have dose-related effects on target as well as nontarget organisms. For a given compound, the higher the application rate the greater the impact. The relationship between dose and effect, nonetheless, is not linear but rather logarithmic; that is, a dose 10 times higher kills only twice as many organisms (Edward & Thompson, 1973). The implication of this relationship is that it would take extremely large quantities of even a toxic pesticide to bring about complete elimination of an organism.

The rate and frequency of pesticide applications for pest management, whether in agriculture and forestry or in public health, varies widely with the type of pest, crop or situation, climatic conditions, and the chemical. For example, in orchard crops, row crops, and cotton (*Gossypium hirsutum* L.) multiple applications (often of different chemicals) may be needed to control the insect complex. On the other hand, in control of locusts and grasshoppers one or two applications followed by several years of no application may be the norm (Green et al., 1977; Matthews, 1979; Ware, 1978). Similarly, a single application of herbicide may give season-long control of weeds, e.g., atrazine in corn (*Zea mays* L.).

Whether a single or multiple application(s) of one or more pesticide(s) at a given rate is determined by the number of different organisms involved, their relative resistance and the number of generations per year (Martin, 1973) and the persistence of the biological activity of the chemical (Plapp, 1981). The rate of application for a given chemical and a particular organism may vary from a few grams per hectare with some of the pyrethroids and sulfonylurea herbicides to 1 kg or more per hectare with some of the copper fungicides, nematicides, and older herbicides and insecticides (Worthing & Walker, 1983; Royal Society of Chemistry, 1987; Weed Science Society of America, 1983). A problem resulting from frequent applications often is the development of resistance of the organism to the chemical or the supplanting of a susceptible species by a much more tolerant species (Georghiou & Mellon, 1983).

## 12–1.2 Organisms

### 12–1.2.1 Age-Susceptibility

The relative susceptibility of young and mature organisms is an important consideration in assessing the impact of pesticides on aquatic and terrestrial biota. Often the young of a species is more sensitive to the toxic action of pesticides than the old. The evidence for this has been recorded for many compounds. Larvae of dungeness crab (*Cancer magister*) were found to be much more sensitive to carbofuran than adults (Caldwell, 1977). Similarly, younger mallard (*Anas platyrhynchos*) and Japanese quail (*Coturnix corturnix japonica*) were substantially more sensitive to carbofuran than the older birds (Hill & Camardese, 1982; Tucker & Crabtree, 1970; Hudson et al., 1972). This relation seems to hold true for other bird species for which data are available (Eisler, 1985).

There is indication, however, that young animals are not always more susceptible to pesticides than adults of a species. For example, Hudson et al. (1972) found 14 common pesticides to manifest different effects on mallard ducks of several ages. The youngest ducks were more sensitive than the oldest birds to six pesticides and less sensitive to eight. Endrin was almost eight times less toxic to 36-h-old than to 30-d-old mallard ducks.

### 12–1.2.2 Nature of Habitats

The habitat of the organism is an important consideration in assessing the impact of pesticides. In lakes, for instance, toxicity of pesticides to fish was significantly related to depth, stratification, and biological activity (Eisler & Jacknow, 1985). When two mountain lakes in Oregon were treated with toxaphene, the deeper and biologically sparse lake, Miller Lake, remained toxic to fish for several years. Rainbow trout (*Salmo gairdneri*) could not be restocked in the lake for 6 yr after treatment. Whereas, the shallow lake rich in aquatic life, Davis Lake, trout could be restocked within 1 yr after treatments when toxaphene levels were of approximately 1 $\mu g\ L^{-1}$ (Terrierre et al., 1966). Wide variation in fish mortality in different streams was reported following forest spraying with DDT in British Columbia, Canada (Crouter & Vernon, 1959). Fish mortality was much higher in large streams flowing through steep-walled valleys, than in streams situated in well differentiated but not particularly steep-walled valleys.

### 12–1.2.3 Sublethal Effects

Apart from direct lethal exposure, sublethal effects have been recorded in most species of organisms whenever insufficient concentration of a pesticide is consumed to cause mortality. Depending on the type of organism involved, exposure to sublethal concentrations may cause genetic, physiologic, or behavioral changes in the target and nontarget species. Cases of increased tolerance or resistance, reproduction impairment, inhibition of brain enzyme activities, growth reduction or inhibition and spinal deformities were among the more insidious effects documented (Table 12–2). Other adverse effects reported include loss in body weight, reduced hatching success, eggshell thinning, embryo mortality, reduced survival of ducklings, altered feeding habits, impaired incubation behavior, and avoidance response.

Assessment of such effects is becoming more important as residues of persistent pesticides are widely spread in biota even in remote places where no record of pesticide use can be found (Freed, 1970).

### 12–1.2.4 Resistance

Continued or repeated exposure of an organism to a particular pesticide may result in the organism becoming resistant to the chemical. This phenomenon, which is genetic in nature, has been demonstrated in many species of animals including insects and fish (Boyd & Ferguson, 1964; Ferguson, 1970; Vinson et al., 1963; Harris, 1972; Milio et al., 1987; Collins,

Table 12-2. Sublethal effects of pesticides to nontarget organism.

| Pesticide | Organism (species) | Effect | Reference |
|---|---|---|---|
| Toxaphene | Brook trout (*Salvelinus fontinalis*) | Reduced reproduction | Mayer et al., 1975 |
| Diazinon | Fathead minnow (*Pimephales promelas*) | Reduced hatching success | Goodmanet al., 1979 |
| Diazinon | Shrimp (*Mysidopsis bahia*) | Reduced growth and reproduction | Nimmo et al., 1981 |
| Toxaphene | Channel catfish (*Ictalurs punctatus*) | Backbone abnormalities | Mayer et al. 1977 |
| Parathion | Bobwhite quail (*Colinus virginianus*) | Reduced egg production, Reduced body weight | Rattner et al., 1982b |
| Krenite | Bobwhite quail | Impaired embryonic growth | Hoffman, 1988 |
| Krenite | Mallard (*Anas platyrhyunchos*) | Reduced hatching success, Embryo abnormalities | Hoffman, 1988 |
| DDE | Barn owl (*Tyto alba*) | Eggshell thinning, Embryo mortality | Mendenhall et al., 1983 |
| Dieldrin | Barn owl | Eggshell thinning | Mendenhall et al., 1983 |
| Heptachlor | American kastrel (*Falco sparverius*) | Reduced reproductivity | Henny et al., 1983 |
| Parathion | Laughing gull (*Larus africilla*) | Altered incubation behavior, Reduced hatching success, Respiratory failure | Coon, 1983; King et al., 1984; White et al., 1983a |
| Temephose | White fowl | Reduced survival of ducklings | Coon, 1983 |
| Dicrotaphos | Starling (*Sturnus vulgaris*) | Reduced body weight | Grue et al., 1982 |
| Orthene | Meadow vole (*Microtus pennsylvanicus*) | Cholinesterase inhibition | Jett, 1986 |
| Mirex | Freshwater algae | Reduced photosynthesis | Hollister et al., 1975 |
| DDT, Toxaphene | Mosquito fish (*Gambusia affinis*) | Developed resistant strains | Vinson et al., 1963; Boyd & Ferguson, 1964 |
| Bromide | Fathead minnow | Reduced growth | Call et al., 1987 |
| Diuron | Fathead minnow | Reduced survival of fry | Call et al., 1987 |
| Endrin | Yellow bullhead (*Ictalurus natalis*) | Developed resistance | Ferguson & Bingham, 1966 |
| DDT | Magfly (*Heptagenia hebe*) | Developed resistance | Grant & Brown, 197 |
| Chlorpyrifos | German cockroach (*Blatella germanica* L.) | Developed resistance | Milio et al., 1987 |
| Chlorpyrifos | Sawtoothed grain beetle (*Oryzaephilus surinamensis* L.) | Developed resistant strains | Collins, 1985 |

1985; Odenkirchen & Eisler, 1988). Certain cases of insect resistance became so widely spread that alternative methods of control were necessary for the housefly (*Musca domestica* L.), the cotton boll weevil (*Anthonomus grandis*), alfalfa weevil [*Hypera postica* (Gyllenhal)], the tobacco hornworm [*Manduca sexta* (L.)], and the Colorado beetle [*Leptinotarsa decemlineata* (Say)] (Edwards, 1973). Toxaphene-resistant strains of cotton pests appeared in Claifornia, Texas, Egypt, and India (Eisler & Jacknow, 1985).

The mosquito fish (*Gambusia affinis*) was found to have developed resistance to a number of insecticides including endrin, aldrin, toxaphene, dieldrin, and DDT (Vinson et al., 1963; Boyd & Ferguson, 1964). Other fish species that acquired resistance to pesticides are the bluegill sunfish (*Lepomis macrochirus*), the green sunfish (*L. cyanellus*), and the golden Shiner (*Notemigonus chrysoleucas*) (Ferguson et al., 1964).

As long as persistent pesticides continue to be used against pests, resistance will continue to develop and spread; that may pose serious problems to pest control programs and hazards to nontarget organisms. However, the resistance of organisms to chemicals is not an exclusive phenomena. Resistance can develop to pathogenic agents as evidence in breeding of plant varieties that are much more tolerant to a given pathogen than the original population. This may or may not persist because the pathogen itself may undergo mutation to reestablish its pathogenicity. Development of resistance does not necessarily mean the emergence of a "super organism," but merely one whose genetic makeup enables it to tolerate the particular chemical or pathogen that afflicts it. In fact, entomologists studying resistance development in insects have found instances where the altered organism may have lost some other trait beneficial to its survival once the pressure of the selecting agent has been removed.

## 12–1.3 Biological Interactions

### 12–1.3.1 Biomagnification

Organisms have three routes of exposure to pesticides in the environment, namely through ingestion of food and water, respiration, and exterior contact with skin or exo-skeleton. Depending on the chemical, more or less of the pesticide will cross the different barriers encountered, whether intestinal mucosa, lung, or other covering of the respiratory tract, or be absorbed through the skin or external covering of the organism. The exposure may arise through deliberate application of the chemical as to control pests in a crop situation or it may be inadvertent where the organism is exposed to low-level residues remaining after an application or erosion into a nontarget area.

The chemical that crosses the various barriers of the body reaches the metabolizing tissue or a storage depot. If the excretion rate is slow or the chemical refractory to metabolism will accumulate approaching a concentration near an equilibrium level depending upon external concentration to which the organism is exposed. If the rate of excretion or metabolism is slow or (Ware,

1980; Tulp & Hutzinger, 1978) with more fat-soluble chemicals or those that are strongly sorbed to other body constituents the ultimate concentration of chemical in the organism will be higher than the concentration of the chemical in the media to which it was exposed. This is termed *bioaccumulation*. In the case of the fat-soluble chemicals, the accumulation process has many similarities to partitioning, and is related to the octanol water partition coefficient of the chemical (Chiou et al., 1977; Tulp & Hutzinger, 1978; Chiou, 1985). In the case of food chains, an organism at a higher tropic level may accumulate a concentration much greater than would have been found in the agent air, water, food to which the base of the food chain organism was exposed. Thus, particularly in aquatic systems, organisms while at the food chain may have accumulated concentrations of a persistent organochlorine pesticide several thousandfold greater than the water concentration (Hunt & Bischoff, 1960). This magnification of concentration sometimes referred to as *biomagnification* and also *biological concentration factor* is related to the partition coefficient. When the whole fresh tissue of an organism is assayed for the concentration, it will be found that the amount of the chemical is a function of the percent of fat in the organism. However, even with organisms that may have a low percentage of fat if the concentration is determined on the fat basis the value in an organism with a high fat content will be comparable.

The term biomagnification has led some to develop an erroneous concept of what is happening in the environment. They may feel that a great deal or most of the chemical in an environmental compartment is being taken up by a particular organism when, in fact, this is not the case. A simple comparison of the size of the environmental compartment to that of the organism reveals, in fact, that the percentage of chemical in that compartment taken up by the organism is really quite small. By way of example, consider a 20-L aquarium in which there will be 50 g of fish. Now if DDT is introduced into the aquarium at 1 $\mu g\ L^{-1}$ or approximately 50% saturation, the bioconcentration factor or ecological magnification is 40 000, 50 g of fish will have taken up 2 $\mu g$ of chemical. The total amount of chemical in the 20 L, however, will be 20 $\mu g$, hence the fish will have taken up only 10% of the total chemical in their particular ecosystem. This does not mean that the fish did not take up a dangerously high level of the chemicals, but it does illustrate that not all of the chemical available in the ecosystem has been accumulated by the organisms. Another consideration to help keep the matter in perspective is the fact the level of accumulation depends upon the concentration of chemical available. Several studies have shown that, despite a rather high bioconcentration factor, when the concentration in the environment is low, the corresponding concentration in the accumulating organisms will be at a lower level than if the environmental concentration is high.

### 12–1.3.2 Synergism

Another impact of chemicals is the possible interaction of two or more chemicals. The enhancement of the activity of a particular chemical is by now a well-known phenomenon in medicine and was discovered many years ago with organophosphate and pyrethroids where the addition of a second chemical would greatly enhance the biological activity of the primary toxicant. This has been termed *synergism,* though at times it is used as a more generic term to indicate either enhancement or inhibition of the fact. Actually, the latter phenomenon should be referred to as antagonism rather than synergism. The use of piperonyl butoxide with the pyrethroid insecticides is a good example of synergistic action.

## 12–2 IMPACT ON SPECIFIC ORGANISMS

### 12–2.1 Microorganisms

The biological population is an important and fairly large component of the soil ecosystem. It consists of several species of bacteria, fungi, algae, protozoa, earthworms, and insects among other invertebrate organisms (Russell, 1973). As would be expected, there is both beneficial and competitive relationships among these organisms and, of course, plants growing on the soil. Thus, bacteria such as the *Rhizobium* spp. are especially beneficial to leguminous plants whereas certain fungi are pathogenic. Similarly, some of the nematodes can be a limiting factor in plant growth.

Much of the pesticide use is directed to the control of pests on aerial portions of plants or in animals; none-the-less, a goodly proportion of the pesticide reaches the soil (Brown, 1978; Parr, 1974) and may be carried in the soil by water or tillage. Yet other pesticides will be applied directly on or in the soil for control of weeds, nematodes, insects, and fungi. Two different questions arise in using pesticides. First, if the pesticide is being applied to control an organism and the persistence is a factor in the control, whether the soil organisms may metabolize the chemical too rapidly. The second question from almost the diametric view point is whether the pesticide may be so toxic to the soil organisms as to imbalance or seriously reduce the population.

Many studies have been conducted both to ascertain the rate and manner different species of soil organisms metabolize various pesticides and to determine the effect pesticides have on organisms (Khan, 1980; Anderson, 1978; Alexander & Aleem, 1961; Audus, 1964; Bollen et al., 1968; Brown, 1978; Kaufman, 1974). The methods employed to study the effect of pesticides on soil organisms include measurement of soil respiration, nitrification, ATP formation, and many other techniques. Degradation of pesticides by organisms has been studied by comparing rates of disappearance of the chemical in normal and sterilized soil, under different temperature and moisture regimes, as well as following the appearance of a metabolite. Enriched cultures and isolates of specific organisms have also been used.

For the most part, those pesticides applied to plants, i.e., soil surface at moderate rates have had at most a transitory effect on the activity or numbers of one or more species (Anderson, 1978). At high concentration, the

Table 12-3. Some effects of pesticides on microorganisms in soil.

| Pesticides | Effects | Reference |
|---|---|---|
| Dalapon | Inhibited denitrifying activity | Grant & Payne, 1982 |
| Simazine | Increased populations of aerobic $N_2$-fixing bacteria; the anaerobic bacteria decreased | Kaiser et al., 1970 |
| Pyrazon, atrazine, and linuron | Decreased bacteria and algae populations and activity | Hauke-Pacewiczowa, 1971 |
| Trifluralin | Nodulation of legumes reduced but not population of Rhizobia | Brock, 1972 |
| Chlorpropham | Inhibited growth of actinomycetes | Taha et al., 1972 |
| PCP | Decreased populations of fungi and actinomycetes in six soil types | Au, 1968 |
| Dalapon | Decreased bacteria population | Sharma & Saxena, 1972 |
| TCA | Low rates slightly increased populations. High rates decreased spore-forming bacteria | Simon, 1971 |
| Diquat | Inhibited $N_2$-fixation by free-living soil bacteria | Atkinson, 1973 |
| Captan | Fungi decreased but actinomycetes increased | Wainwright & Pugh, 1975 |
| Dieldrin | Population of nitrifying bacteria severely depressed | Mahmoud et al., 1970 |

effect may be quite pronounced especially with the chemicals used as soil fumigants (EDB, DD, and methyl isothiocyanate). Some species may have their numbers markedly reduced, but others may actually be stimulated by low levels of certain chemicals (Anderson, 1978; Audus, 1964). Table 12–3 summarizes the effect of pesticides on different organisms.

In the metabolism of pesticides, soil organisms use several biochemical reactions (Matsumura, 1982). In some instances, a molecule of similar structure—natural or synthetic—may induce or at least stimulate the organisms ability to metabolize a particular compound.

### 12–2.2 Plants

The impact of pesticides on plants is often as dramatic and devastating as the impact on insects and animals. Exposure of a sensitive plant to a growth-regulating chemical will produce remarkable growth distortions, yield loss, and outright death depending on the dose received. With other types of pesticides, the effect may be much more subtle and less obvious. The impact may be to reduce the productivity or produce a residue that renders a crop as unsalable by virtue of the residue.

Plants, like other organisms, e.g., insects or bacteria, may develop resistance to a given pesticide. Different plant species have been found to have developed resistance to the phenoxy, triazine, and the chloralkyl acid herbicides. The result of this resistance may be the development of a population of plants that cannot be controlled even with much higher rates of application of the chemical.

Conversely, the impact may be to eliminate the more susceptible species and elevate what had been a minor but tolerant species to dominance (Haas & Streibig, 1982). This occurs where the pesticide has been applied frequently over a period. This has occurred in wheat (*Triticum aestivum* L.) growing areas where 2,4-D has been applied on an annual basis for many years, but has also been encountered with the bipyridilium compounds triazine and other herbicides in other crops and situations.

During application, pesticides may move from the treatment area to deposit a residue affecting a nontarget species. The herbicide 2,4-D frequently caused symptoms on and damage to such susceptible crops as cotton and grapes (*Vitus* spp.) (Brown et al., 1948; Clore & Bruns, 1953; Goodman, 1953). Similarly, other herbicides have been found to injure sensitive plants by moving from the treated area, among these chemicals are propanil and some of the newer phenoxy aromatic derivatives. Following application, the movement may occur by re-entertainment of vapor or particles in the air.

But herbicides are not the only class of pesticides whose residues can be troublesome either on the target or with movement to another site (Westlake & Gunther, 1966). Organochlorine insecticides as residues in soil from a previous treatment will appear as a residue in root crops or may transfer by vaporization to leaves (Brown, 1978). Also, as a soil contaminant, a pesticide may be transferred to water by soil erosion or leaching as discussed in previous chapters. In water, depending on the activity of the pesticide, it may exert an effect on some of the aquatic plants (Butler, 1977). With some of the pesticides, the amount remaining after a prior treatment can be a limiting factor as to what, if any, crop may be planted. Even though the quantity is small, it will be sufficient to limit the growth of a sensitive species.

Table 12–4 contains a general tabulation of possible impacts of various classes of pesticides. It must be noted that among the members of a given class, the biological activity will vary widely and hence not all members will exert the same impact.

### 12–2.3 Insects

Insects of both aquatic and terrestrial forms exist in large numbers constituting about 75% of the totally described arthropoda, the largest phylum of the animal kingdom (Mulla et al., 1981). Of all insect species on record, only about 5% are known for their undesirable effects on humans or the environment. For many centuries, attempts to control noxious insect pests and vectors with chemicals achieved only limited success. With the advent of synthetic organic chemicals in the 1940s, a new era of pest control unfolded. The use of the highly persistent organochlorine insecticides as insect pest control agents in agriculture, forestry, and public health, had produced a number of deleterious effects including widespread pollution of the natural environment.

The vast majority of insecticides are relatively nonselective, with a broad-spectrum of effects, and usage therefore is hazardous to target species as well as beneficial competitors, predators, and parasites of the target pest

Table 12-4. Pesticides effects on nontarget plants.

| Type of pesticide | Probable impact |
|---|---|
| | Insecticides |
| Organochlorine insecticide: | |
| In soil | Residue in and on plants may effect growth of some crops, erode to surface waters, may effect some aquatic plants. |
| In water | Contaminant, may transfer to plants through irrigation. |
| Organophosphate, carbamate Pyrethroid insecticides: | |
| In soil | Generally short lives, little impact on most plants. |
| In water | Some toxic to certain algae |
| | Herbicides |
| Aromatic acid type: | |
| In soil | Residues effect subsequent crop. Some move readily with water and injure adjacent plants, e.g., picloram. Surface deposits subject to wind and water transport that can cause symptoms or injury. |
| In water | Inhibit growth or kill some aquatic plants. |
| Amides, anilides, nitriles ethers, and carbamates: | |
| In soil | Variable persistence among these herbicide classes. Residues of some may carry over to limit subsequent crops. Some may reach groundwater where heavy deposits and chance of percolation or wash off to open system. |

insects (Ware, 1980). The effects of pesticides on target and nontarget species of terrestrial and aquatic insects have been discussed in many good reviews (Muirhead-Thompson, 1987; Brown, 1978; Edwards, 1973; Mulla et al., 1981) to which interested readers are referred. In what follows, a brief evaluation of the impact of pesticides on nontarget beneficial insects, notably pollinators, is provided.

Honeybees (*Apis mellifera*) apart from this economic value to beekeepers as the main source of honey production, are also the principal pollinators of many imporant fruit trees and seed crops including pears (*Pyrus communis* L.), apples (*Malus domestica* Borkh.), apricots (*Prunus armeniaca* L.), almonds, and alfalfa (*Medicago sativa* L.). Toxicity of pesticides to bees varies greatly, depending on the chemical nature of the compound, formulation, and time of application (Ware, 1980). Atkins et al. (1973) reported that only 20% of the 399 pesticides tested were highly toxic, 15% moderately toxic, and 65% relatively nontoxic to honeybees. As expected, insecticides are more toxic to bees than any other group of pesticides, with the carbamate and organophosphorus insecticides the most lethal (Atkins, 1981). Use of pesticides in the fight against noxious pest insects led to widespread mortality of domestic and wild bees. In California alone (Table 12-5), the loss was as high as 89 000 bee colonies in 1 yr (Atkins, 1981). Application of chlorpyrifos, an organophosphate insecticide, for mosquito control in rice (*Oryza sativa* L.) fields, killed all honeybees within 400 m downwind, 95% at 800 m and 89% at 1200 m in the drift area (Atkins, 1972).

Table 12-5. Bee colonies poisoned by pesticides.

| State | Year | Colonies destroyed | Reference |
|---|---|---|---|
| Connecticut | 1981-1982 | 268 | Anderson & Glowa, 1984 |
| Arizona | 1965-1971 | 61 000 | Ware, 1980 |
| California | 1969 | 82 000 | Ware, 1980 |
| California | 1970 | 89 000 | Ware, 1980 |
| California | 1971 | 76 000 | Ware, 1980 |
| California | 1972 | 40 000 | Ware, 1980 |
| California | 1973 | 36 000 | Ware, 1980 |
| California | 1974 | 54 000 | Ware, 1980 |

Anderson and Glowa (1984), reported extensive poisoning of honeybees following ground spraying of trees against gypsy moth. The bees were poisoned by eight insecticides; with carbaryl as the most frequent contaminant and methyl parathion the most devastating. Other residues in bees included diazinon, acephate, endosulfan, chlordane, methoxychlor, and malathion.

Bees are extremely susceptible to carbofuran. Spray and dust formulations of this insecticide severely damaged honeybees and other airborne pollinators (Finlayson et al., 1979). Application of several insecticides to blooming alfalfa caused heavy mortality of field force of honeybees (Lieberman et al., 1954). When applied in the morning, mortality was highest with TEPP (63%) and lowest with aldrin (10%).

Pesticide residues have been found in bees, pollen, and honey. Residues of the carbamate insecticide Sevin were found in bees and pollen collected at the entrances of bee hives in treated areas (Morse et al., 1963). In another study (Johansen et al., 1957), residue of parathion in honey was measured, but with no apparent effect on the flavor of the honey produced by the bees.

Formulations and time of application are detrimental factors in consideration of pesticide effects to pollinators. Johansen et al. (1957) reported that when equal dosages of parathion dusts and sprays were compared, dusts killed twice as many honeybees as did sprays. Ultra-low-volume (ULV) application of malathion killed large number of honeybees (Hill et al., 1971; Caron, 1979). The ULV formulations are more toxic than diluted sprays (Ware, 1980). Daytime applications are more harmful to honeybees than night sprays. Morning applications of TEPP caused 63% mortality, whereas the same dosage applied in the evening resulted in only 6% kill of honeybees (Lieberman et al., 1954).

The impact of pesticides on wild species of bees that are as effective pollinators as honeybees has not been greatly evaluated. Although wild bees are affected by the same factors as honeybees, the results reported for honeybees may or may not shed much light on the damage to the native species (Ware, 1980; Lieberman et al., 1954).

### 12-2.4 Fish

The hazards of pesticides to fish have been widely discussed (Johnson, 1968; Brown, 1978; Cope & Springer, 1958; Holden, 1973). The large-scale application of pesticides to fields, forests, rivers, and marshlands has an impact on fisheries resulting in mortality or subtle effects on fish reproduction and behavior. Acute toxicity, which, results from short-term exposure to lethal concentration of pesticides, has caused dramatic losses of fishery resources in the USA.

The classic example of pesticide-related fish mortality has been the massive fish kills in the Mississippi and Atchafalaya rivers and associated bayous in Louisiana between 1960 and 1963 (Mount & Putnicki, 1966). Rough estimates of fish losses were listed at 3.5 million in 1960, close to 1 million each in 1961, and again in 1962 and from 5 to 10 million in 1963 (Cottam, 1965). Extracts of mud from the water where fish were dying killed healthy fish, as did tissue extracts from contaminated fish. The highly toxic insecticide, endrin, was singled out as the major cause of this tragedy that affected more than nine species of fish (Mount & Putnicki, 1966).

Forest spraying with pesticides to control insect infestations in the USA and Canada have caused widespread damage to important fish populations. The spruce budworm (*Chloristoneura fumiferana*) spray program with DDT, in the Yellowstone River system in 1955 and on the forests of Maine in 1958 resulted in the death of thousands of fish of many species (Warner & Fenderson, 1962; Cope, 1961). In the Miramichi River, in Brunswick, Canada, the spray program killed many young Atlantic salmon (*Salmo salar* L.), stickleback (*Gasterosteus aculeatus*), and sculpin (*Cottus cognatus*) (Kerwill & Edwards, 1967). Downstream transport of DDT-in-oil mixture was found to be an important factor in extending the harmful affects 48 km or more below spray zones (Elson, 1967). When DDT, at 0.6 kg $ha^{-1}$, was applied, against the black-headed budworm (*Acleris variana* Fern.), on the timberlands of British Columbia, Canada, it almost eliminated Coho salmon (*Oncorhynchus kisutch* W.), young rainbow and Cutthroat trout (*Salmo gairdneri* R. and *S. clarki* R.) and steelhead (Crouter & Vernon, 1959). The reduction in aquatic insects was found to parallel the loss of fish in the streams affected.

In a treatment, of a tidal marsh ditch in Florida, with DDT at 0.2 lb $ha^{-1}$, several thousand of fish were killed including sheepshead minnow (*Cyprinodon variegatus* L.) and striped mullet (*Mugil cephatus* L.) (Crocker & Wilson, 1965). The fish in the ditch accumulated up to 90 mg $kg^{-1}$ of residue within 5 wk after DDT application. In another case involving DDT, a single application of the insecticide in oil to a section of the Blue Nile River in Sudan, for control of the insect (*Tanytarus lewsii*), caused the death of hundreds of fish of many species (Burden, 1956).

In California, pesticides were the confirmed cause of 48 fish kills during 1965 to 1969 (Hunt & Linn, 1970). The total number of fish killed in these incidents was more than 400 000, including 264 160 deaths in 1966 alone. The organochlorine insecticides were implicated in most of the losses; however, the organophosphate insecticides caused two losses and the pentachlorophenol were involved in several other kills.

Over the years, careless handling of pesticides or spillage in streams and rivers have resulted in a number of severe kills. In 1967, the death of about 40 million fishes in the Rhine was attributed to the emptying of a drum of endosulfan into the river near Bonn, West Germany (Brown, 1978). Extensive mortality among brook trout (*Salvelinus fontinalis* M.) and juvenile Atlantic salmon occurred when a spray mixture containing the fungicide nabam and endrin, an insecticide, was flushed from a potato sprayer into the Mill River in Prince Edward Island, Canada, in 1962 (Saunders, 1969).

Acute toxicities of pesticides on fish populations are drastic, on the other hand, chronic toxicities resulting from the exposure to sublethal concentrations of pesticide are much more insidious and difficult to identify. The two types of toxicities for a given pesticide vary, however, with water temperature, water chemistry, and biological factors such as age, sex, size, and health conditions as well as species (Johnson, 1968).

Sublethal exposure to pesticides is known to cause suppression in reproduction success and development of resistant strains in many fish species (Ferguson, 1970; Butler, 1970; Burdick et al., 1964). Populations of the mosquito fish having a past exposure to insecticides showed resistance to DDT, endrin, aldrin, dieldrin, toxaphene, heptachlor, and lindane (Boyd & Ferguson, 1964; Vinson et al., 1963). The level of resistance of up to 2000-fold in resistant strains of mosquito fish remained unchanged after growing for several generations in pesticide-free environments showing that the phenomenon is genetic (Ferguson, 1970).

The phenomenon of sac-fry mortality has been observed in both hatchery and field situations, with several different fish species and more than one pesticide. It was first noticed in lake trout fry (*Salvelinus namaycush* W.) from areas of high DDT application on Lake George watershed in New York (Burdick et al., 1964). Mortality of the trout fry developed when DDT concentration in the eggs exceeded 2.9 mg $kg^{-1}$. The mortality of Coho salmon fry from Lake Michigan ranged from 15 to 73% and occurred during the last stage of yolk sac absorption (Johnson & Pecor, 1969). DDT residues in the salmon eggs, which ranged from 1.1 to 2.8 mg $kg^{-1}$, were three to five times higher than in eggs from Lake Superior and approximately 60 times higher than in eggs from Oregon.

The widespread accumulation of residue in fish from all over the USA has been well demonstrated (Henderson et al., 1969; Johnson & Lew, 1970; Stucky, 1970; Hunter et al., 1980; Schmitt et al., 1981). Few illustrative examples are given in Table 12-6, for pesticide residues in representative fisheries worldiwide. Even in the absence of obvious subtle effects, these residues may increase fish mortality through reproductive failures and susceptibility to increased predication resulting from alteration of fry mobility and behavior (Johnson, 1973).

### 12-2.5 Birds

Soon after the beginning of the modern pesticide era, studies demonstrated the toxicity of these chemicals to animals as were some deaths of free-living wildlife from application of pesticides (Stewart et al., 1946; Hotchkiss & Pough, 1946; Coburn & Treichler, 1946). By the early 1950s, it was well established that dead birds were commonplace in fields sprayed with DDT

Table 12-6. Pesticide residues in representative fish species.

| Fish (species) | Location | Residue ($mg\ kg^{-1}$, wet wt.) DDE | EDDT | Dieldrin | Reference |
|---|---|---|---|---|---|
| Shad (*Dorosoma cepedianumo*) | Mississippi River | 13.00 | | 7.50 | Fleming et al., 1983 |
| (*Sargus vulgarius*) | Alexandria (Bay) Egypt | 0.05 | 0.150 | | El Nabawi et al., 1987 |
| (*Sparus aurutus*) | Mediterranean | 0.04 | 0.07 | | El-Dib & Badawy, 1985 |
| (*Seriolella violacea*) | Chile | ND† | ND | 3.00 | Ober et al., 1987 |
| Trout (*Salmo trutta fario*, Sierra, L.) | Spain (River) | 0.44 | 1.29 | 0.69 | Teran & Sierra, 1987 |
| Mosquito fish (*Gambusia affinis*) | Lake Providence, Louisiana | 2.07 | 2.89 | 0.03 | Niethammer et al., 1984 |
| Blue catfish (*Ictalurus furcatus*) | San Juan, TX | 11.20 | | | White et al., 1983c |
| Rainbow trout (*Salmo gairdneri*) | Snake River, Idaho | 0.62 | 0.68 | 0.01 | Schmitt et al., 1985 |
| Cod, (*Gadus morhua* L.) | Norway | 0.70 | 1.11 | | Skare et al., 1985 |
| Slimy sculpin (*Cottus cognatus* R.) | Lake Superior, Minnesota | 0.15 | 0.24 | | Veith et al., 1977 |

† Not detected/Not determined.

(Hall, 1987). The application of DDT to American elm shade tree (*Ulmus americana*) to control the insect vectors of the Dutch elm disease, caused by a fungus (*Ceratostomella ulmi*) killed large numbers of a variety of birds in many communities in the Atlantic Coast and Midwest states (Hunt & Bischoff, 1960). The use of DDT for tree spraying, was accompanied by large reductions in populations of robins and other insectivorous birds. At least 94 species of birds were found dead or dying in areas treated with DDT to control the disease (Brown, 1978).

The campaign in the 1950s, to eradicate the imported fire ant (*Solenopsis saevissima* Richterring Forel), by using the organochlorine insecticides heptachlor and dieldrin, led to massive mortality of a variety of birds (Smith & Glasgow, 1963). On four farms in Louisiana, 222 dead birds from 28 species were found after the spraying.

The practice in Great Britian of treating seeds with organochlorine insecticides aldrin, dieldrin, and heptachlor to protect spring-sown cereals against insects, resulted in extensive mortality of many seed-feeding birds and other wildlife (Brown, 1978; Dempster, 1975). Birds were reported to drop dead to the ground while in flight (Turtle et al., 1963). Feeding experiments provided evidence to confirm that the cyclodiene-type insecticides aldrin, dieldrin, and heptachlor seed dressings were responsible for the bird mortalities (Turtle et al., 1963). A voluntary ban on the use of these chemicals as cereal seed-dressings during the spring season has resulted in a dramatic decline in the reports of incidents of this nature (Robinson, 1968). Heptachlor seed treatment in Oregon in 1976 to 1987, had also led to the mortality of several species of birds (Fleming et al., 1983).

Grue et al. (1983) found reports of 31 confirmed incidents of wildlife mortality because of organophosphate poisonings in North America and more than 387 in other parts of the world with estimated number of unintentional poisoning ranging from a few individuals to almost three million. Parathion, an organophosphate insecticide, was responsible for the 1981 deaths of 72 wild geese, mostly Canadian geese (*Branta canadensis*), in Texas (White et al., 1982). The dead birds were found in winter wheat fields sprayed with parathion for the control of greenbugs (*Schizaphis graminum*). The same year in Louisiana, about 100 birds, mostly ducks and geese were found dead or dying in rice fields (White et al., 1983d). Azodrin, an organophosphate compound was identified as the cause of death of these waterfowls. Not <1000 birds of 12 species died from organophosphate poisoning in Texas in 1982. The birds died from feeding on rice seed that was treated with dicrotophos or monocrotophos (Flickinger et al., 1984).

Numerous bird mortalities have resulted from applications of the insecticide (carbofuran) to agricultural crops. Carbofuran killed more than 100 birds, mostly dickcissel (*Spiza americana*) and savannah sparrow (*Passerculus sandwichensis*), as well as nine other species of songbirds and sandpipers (Flickinger et al., 1986). The birds died from feeding on planted rice seed treated illegally with carbofuran.

Pesticide properties, species of bird, and other factors make diagnosis of the cause of death of birds a difficult task (Stickel, 1973). Residue levels found in pesticide-poisoned dead birds varied greatly, even when the birds were killed by the same chemical under similar conditions. For the organochlorine insecticides, laboratory studies indicated that residues of DDT in the brain were reliably related to lethality (Stickel et al., 1966). Such studies have shown that lethal residue of DDT (and metabolites) in the brain is in excess of 30 mg $kg^{-1}$.

Sublethal effects occur when even insufficient quantity of pesticide is consumed to cause mortality. One result of DDT contamination in avian species is eggshell thinning, which contributes to lowered population recruitment. The increased frequency of egg breakage in nests of wild birds observed in the 1940s in Britain was shown to be caused by pesticides (Ratcliffe, 1967, 1970). The effect often is assessed by evaluating the relationship between eggshell thickness and pesticide concentration in eggs (Blus et al., 1974). The decrease in eggshell thickness was shown by correlative data to be a major mechanism of pesticide impact on avian populations (Hall, 1987).

In Great Britain, eggshell thickness decreased significantly in 9 out of 17 species; the decrease was from 5 to 19% (Ratcliffe, 1970). These decreases were 19% in peregrine falcon (*Falco peregrinus*), 17% in the sparrow hawk (*Accipiter nisus* L.); 13% in the merlin (*Falco columbarius*); 12% in the shag (*Phalacrocorax aristotelis*), and 10% in the golden eagle (*Aquila chrysaetos*). Hickey and Anderson (1968) reported eggshell thickness declines of 18 to 26% in three species of raptorial birds in the USA. Nearly all brown pelican (*Pelecanus occidentalis*) eggs collected from South Carolina, Florida, and California in 1969 to 1970 exhibited eggshell thinning (Blus et al., 1974).

All eggs analyzed in the study contained measurable quantities of DDE. The relationship between DDE concentration and eggshell thinning in the pelican was essentially linear.

DDE has become widely known to cause eggshell thinning and reproductive impairment particularly in predatory birds that feed on fish or other birds (Stickel, 1973; Ware, 1980). These birds include the bald eagle (*Haliaetus leucocephalus*), osprey (*Pandion halieaetus*), peregrine falcon, sparrow hawk, and brown pelican.

Pesticide residues in eggs reflect concentration in tissues of birds at a given time in the annual cycle and also measure the exposure of embryo (Stickel, 1973). Thereby, these residues provide a useful means in comparing the impact of pesticides on birds at different geographical locations. A list of DDE residues in eggs of wild birds is given in Table 12–7 to illustrate the impact.

Table 12–7. Residues of DDE in eggs of wild birds.

| Bird (species) | Location | Year | DDE ($mg\ kg^{-1}$, wet wt.) | Reference |
|---|---|---|---|---|
| Cooper's Hawk (*Accipiter cooperi*) | Connecticut | 1980 | 17.0 | Pattee et al., 1985 |
| Brown pelican (*Pelecanus occidentalis*) | California | 1969 | 71.0 | Blus et al., 1974 |
| | S. Carolina | 1967 | 4.45 | Blus et al., 1974 |
| | S. Carolina | 1970 | 2.83 | Blus et al., 1974 |
| Black-crowned night-heron (*Nycticorax nycticorax*) | Colorado-Wyoming | 1979 | 3.10 | McEwen et al., 1984 |
| | Massachusetts | 1973 | 4.43 | Custer et al., 1983a |
| | Massachusetts | 1979 | 2.12 | Custer et al., 1983a |
| Common tern (*Sterna hirundo*) | Rhode Island | 1980 | 0.67 | Custer et al., 1983b |
| | Virginia | 1980 | 0.78 | Custer et al., 1983b |
| | New York | 1980 | 0.58 | Custer et al., 1983b |
| Western grebe (*Achmophorus occidentalis*) | California | 1981 | 1.40 | Boellstroff et al., 1985 |
| American white pelican (*Pelecanus erythrorhynchos*) | California | 1969 | 2.34 | Boellstroff et al., 1985 |
| | California | 1981 | 2.38 | Boellstroff et al., 1985 |
| American robin (*Turdus migratorius*) | Washington | 1983 | 11.10 | Blus et al., 1987 |
| Mallard (*Anas platyrhynchos*) | Washington | 1982 | 6.90 | Blus et al., 1987 |
| Turkey vulture (*Cathortes aura*) | California | 1981 | 6.90 | Wiemeyer et al., 1981b |
| Common raven (*Corvus corax*) | California | 1981 | 0.29 | Wiemeyer et al., 1981b |
| Leach's storm petrel (*Oceanodroma leucorhoa*) | Oregon | 1979 | 2.50 | Henny et al., 1982 |
| Herring gull (*Lorus argentatus*) | Ontario, Canada | | 10.40† | Frank & Holdrint, 1975 |
| Black Skimmer (*Rynchops niger*) | Texas | 1979 | 5.6 | White et al., 1984 |
| | Texas | 1980 | 9.4 | White et al., 1984 |
| | Texas | 1981 | 9.8 | White et al., 1984 |
| | Texas | 1984 | 3.2 | Custer & Mitchell, 1987 |

† Sum of DDE, TDE, and DDT.

Although the use of organochlorine pesticides has been banned or severely restricted in he USA, evidence exists for the accumulation and persistence of these chemicals in wildlife. All bald eagle samples collected in a national survey in 1978 to 1981 contained DDE (Reichel et al., 1984). Of the wildlife samples colleced from Washington fruit orchards in 1979 to 1983, DDE was found in 96% and DDT in 46% (Blus et al., 1987).

Although pesticide residue concentrations in wildlife have declined since the 1960s, certain regions have recently had high concentrations or repeatedly have had high levels of residue, particularly DDE (Cain, 1981; Bunck et al., 1987; White et al., 1983b; Henny et al., 1984; Fleming & Cain, 1985; White & Krynitsky, 1986; Niemi et al., 1986; DeWeese et al., 1986). A few examples of residue data are presented in Table 12–8.

### 12–2.6 Mammals

Mammals, like birds, are rarely exposed to direct contact with pesticides, except in cases of intentional sprays or accidents. Death of mammals is usually because of feeding on contaminated sources. Predatory mammals accumulate higher residues than herbivores and other mammals. Feeding experiments indicate the existence of wide variation between individuals and between species of animals in susceptibility to pesticides. Residues of organochlorine insecticides or their metabolites predominate in wild mammals reflecting both distribution and persistence of these chemicals. As in birds, residues of organochlorine pesticides in the brain provide the best diagnostic criteria of lethality (Stickel, 1973).

Widespread mortality of wild mammals in association with major pest control programs in the USA and elsewhere has been reported (Taylor & Blackmore, 1961; Benton, 1951; Smith & Glasgow, 1963; Scott et al., 1959; Clark, 1981). In a treatment program employing dieldrin, one of the highly toxic chlorinated hydrocarbon insecticides, at a rate of 3.4 kg $ha^{-1}$, against Japanese beetles (*Popillia japonica*) in Illinois, ground squirrels (*Citellus tridecemlineatus* and *C. franklinii*) muskrats (*Ondatra zibethica*), and cottontail rabbits (*Sylvilagus floridanus*) were virtually eliminated (Scott et al., 1959). Short-tailed shrews (*Blarina brevicauda*), fox squirrels (*Sciurus niger*), woodchucks (*Marmota monax*), and meadow mice (*Microtus ochrogaster*) appeared to have sustained heavy losses. Other mammals found dead during inspection of the area treated with dieldrin included opossums (*Cidelphis marsupialis*) and moles (*Scalopus aquaticus*). Domestic cats (*Felis catus*) have taken heavy losses also. It was estimated that 90% of the farm cats died following the application of the insecticide.

A high rate of wildlife mortality was observed in the Gulf States treated with heptachlor or dieldrin, for eradication of Argentine fire ants (*Solenopsis saevissima Richtering* Forel) (Smith & Glasgow, 1963). Mammals found dead included the cotton rat (*Sigmodou hispidus*), raccoon (*Procyon lotor*), red fox (*Vulpes fulva*), cottontail rabbit, white-footed mouse (*Peromyscus* sp.), armadillo (*Dasypus novemcinctus*), and opossum. Analysis of dead wildlife from treated areas showed that most of the animals had high residues

Table 12-8. Residue of organochlorine pesticides in birds.

| Bird (species) | Tissue | Location | Year | Residues (mg $kg^{-1}$, wet wt.) DDE | DDT | Dieldrin | Reference |
|---|---|---|---|---|---|---|---|
| Great egret heron (*Costnerodius albus*) | Brain | California | 1978 | 22.00 | 0.30 | 3.70 | Ohlendorf et al., 1981 |
| Great blue heron (*Ardea herodias*) | Brain | N.Carolina | 1976 | 62.00 | 20.00 | 0.83 | Ohlendorf et al., 1981 |
| Herring gull (*Larus argentatus*) | Carcass | Texas | 1978 | 5.00 | ND† | ND | White et al., 1983c |
| Mallard (*Anas platyrhynocos*) | Wings | Maine | 1979–80 | 0.21 | 0.02 | ND | Cain, 1981 |
| | Wings | Alabama | 1979–80 | 0.85 | 0.13 | 0.31 | Cain, 1981 |
| | Wings | Oregon | 1979–80 | 0.57 | 0.02 | ND | Cain, 1981 |
| Pintail (*Anas acuta*) | Breast | Arizona | 1982 | 0.78 | ND | ND | Clark & Krynitsky, 1983 |
| Bald eagle (*Haliaetus leucocephalus*) | Carcass | 32 states | 1978 | 2.50 | 0.12 | 0.17 | Reichel et al., 1984 |
| | Carcass | 32 states | 1981 | 3.00 | 0.23 | 0.20 | Reichel et al., 1984 |
| Sparrow hawk (*Accipiter nisus* L.) | Brain | Spain | | 3.39 | 0.06 | 0.13 | Sierra et al., 1987 |
| Red kite (*Milvus milvus* L.) | Brain | Spain | | 0.06 | 1.16 | 0.02 | Sierra et al., 1987 |
| Barn owl (*Tyto alba*, Scope) | Brain | Spain | | 0.04 | 0.04 | 0.22 | Sierra & Santiago, 1987 |
| Barn swallow (*Hirundo rustica*) | Carcass | Western USA | 1980 | 4.00 | ND | 0.02 | DeWeese et al., 1986 |
| American robin (*Turdus migratorus*) | Carcass | Western USA | 1980 | 0.67 | ND | ND | DeWeese et al., 1986 |
| Brown pelican (*Pelecanus occidentalis*) | Carcass | California | 1980 | 0.93 | ND | ND | Ohlendorf et al., 1985 |
| Black skimmer (*Rynochops niger*) | Carcass | Texas | 1983 | 2.50 | ND | ND | White et al., 1985 |
| Black skimmer | Carcass | Mexico | 1983 | 2.00 | ND | ND | White et al., 1985 |
| Long-billed curlew (*Numenius americanus*) | Brain | Oregon | 1981 | 7.70 | ND | 5.90 | Blus & Henny, 1985 |
| | Brain | Oregon | 1982 | 0.41 | 0.04 | ND | Blus & Henny, 1985 |
| | Brain | Oregon | 1983 | 3.40 | ND | 0.07 | Blus & Henny, 1985 |

† Not detected/Not determined.

of heptachlor and dieldrin. Feeding experiments with dogs indicated that the lethal level of dieldrin in the grain was about 5 mg $kg^{-1}$ (Harrison et al., 1963).

The use of heptachlor as a seed-dressing agent against bulb-fly and wireworm has caused the death of several wild foxes in Great Britain between 1959 and 1960 (Taylor & Blackmore, 1961). Approximately 1300 wild foxes were found dead apparently because of secondary poisoning through eating birds that had died after feeding on grain coated with the chemical.

Massive mortality and population decline of several bat species, including the endangered species the grey bat (*Myctis grisescens*), have been related to pesticide use (Clark et al., 1988; Clark et al., 1978; Clark, 1981; Clark et al., 1983; Geluso et al., 1976; Reidinger, Jr., 1976). Bats, either because of behavior or physiology, seem to accumulate more organochlorine residues than most other mammals and birds at equivalent trophic levels (Reidinger, Jr., 1976). Also, bats are probably the most susceptible mammal to DDT poisoning (Ware, 1980). Residual sprays of DDT wettable powder were used to eliminate bat colonies carrying rabies infections (Brown, 1978).

Residue accumulation and persistence vary widely in mammals. Raccoon (*Procyon lotor*) and river otter (*Lutra canadensis*) accumulated the highest residue at 1 yr after mirex application; by contrast, bobcat (*Lynx rufus*), foxes (*Urocyon* or *Vulpes*), opossums, and skunks (*Spilogale* or *Mephitis* spp.) contained the highest residue 24 wk after treatment (Hill & Dent, 1985). In the year of treatment, shrews (*Blarina brevicanda, Microsorex hoyi,* and *Sorex* sp.) accumulated an average of 15 mg $kg^{-1}$ of DDT and metabolites, while, mice (*Peroomyscus* sp.) and vole (*Clethrionomys gapperi*) contained an average of 1 mg $kg^{-1}$ of residues (Dimond & Sherburne, 1969). The differences prevailed even 9 yr after DDT applications; in shrews, residues were still well above pretreatment levels after 9 yr and showed only little decline after 5 yr. Widespread distributions of pesticide residues in aquatic and terrestrial mammals including game animals, mountain goat (*Oreamnos americanus*), whale (*Delphinapterus leucas*), waterdog (*Nectureus lewisi*), mink (*Mustela vison*), and otter (*Lutra canadensis*) has been reported (Foley et al. 1988; Pillmore & Finley, Jr., 1963; Frank et al., 1979; Culley & Applegate, 1967; Hall et al., 1985; Martineau et al., 1987; Boddicker et al., 1971).

### 12–2.7 Humans

Pesticides are different from most chemicals that humans encounter because they are (i) selected for their biological activity and (ii) deliberately released into the environment. If they were highly specific for one genera of organisms or even one class with little or no activity toward others, there would be less concern over their presence in the environment. While, to be sure, there are wide differences in susceptibility among species and classes of organisms, the activity toward most of them may produce effects at sufficiently high exposure (Hayes, 1982). For example, many of the organophosphate insecticides are quite toxic to mammals.

The impacts of these chemicals, as far as humans are concerned, may be either beneficial or adverse depending on the circumstances. When correctly used for control of a pest, the chemical can ameliorate a vector-borne disease oubreak (e.g., malaria and typhus); assure and increase food production; or achieve environmental improvement for the benefit of wildlife (Brown, 1978; Sheail, 1985). That these are beneficial impacts is widely accepted, though perhaps not by those who long for a pesticide-free world. Their documentation is to be found in the scientific literature of the pest control and public health scientists (e.g., *Journal of Economic Entomology, Weed Science, Residue Reviews,* and WHO Bulletin).

While the benefits accruing from use of pesticides is generally acknowledged, it is the possible adverse effects of these chemicals in the environment that commands much attention. Many have viewed these possible adverse impacts in the most dramatic light beginning with Rachel Carson (1962) and continuing with unabated passion to the present time. The toxicity of high doses of many of the pesticides to humans and other organisms is indeed unquestionable (Hayes, 1982; Murphy, 1986). It is not surprising, therefore, that those who handle and apply the concentrated forms of the chemical—often neglecting the proper precautions and protective clothing—may experience toxic reactions (Davies, 1982) because of overexposure. However, at the much lower doses (or concentrations) encountered in food and water residues or in the environment, the risk of an adverse effect on humans becomes much more a tenuous issue (Murphy, 1986).

Among the toxic effects shown to be produced in laboratory studies with animals at relatively high doses with various pesticides include cancer, mutagenesis, teratogenesis, neuropathy, and, in a few instances, effect on the immune system (Wagner, 1981; Hayes, 1982; Murphy, 1986). The dosages used in laboratory studies are purposefully set relatively high to determine with some accuracy whether or not an effect can be produced (Ottoboni, 1984; Calabrese, 1983). The extrapolation from laboratory studies with relatively high doses to possible impact at the much lower dose (concentration) found in the environment or on food becomes a formidable task. Tolerances or standards for residue levels and concentrations are set on this basis, factoring possible species and age-related susceptibility with another factor for safety. But most of the limited epidemiological studies designed to determine whether or not pesticides in the environment are causing some chromic or acute effect on humans have been at best ambiguous and most quite inconclusive. For example, the study of Hoar et al. (1986) to assess the risk of certain types of cancer among farmers in Kansas appeared to find a positive association (risk) with use of 2,4-D and cancer. However, the findings of the study have met with serious challenge based on several other studies finding no carcinogenesis with 2,4-D. The USEPA, relying on the analysis of the data by scientific panels, took the position that the weight of evidence did not warrant classifying 2,4-D as a carcinogen at the time.

A large segment of the populace evidently has received an apparent low-level exposure not only to pesticides (at least the more persistent ones) as well as many other chemicals as shown by USEPAs "Broad Scan Analysis of FY82 National Adipose Tissue Survey Specimens'' (Stanley et al., 1986). Those handling and applying pesticides clearly have a heavier exposure to the chemical (Richter et al., 1986) than does the general public. The exposure of the general public appears to be sufficiently low that few, if any, instances of measurable health impact have been recorded (Hayes, 1982; Murphy, 1986). Any exception to this usually entails some gross contamination of food, water, or clothing.

Despite the paucity of epidemiological information that would indicate a cause/relation to the level of legal residues in food or water and other sources of low-level exposure of the public, there is a perception that pesticides constitute a larger health threat that seems to be the case. In fact, as Ames et al. (1987) pointed out, pesticides (and many other synthetic chemicals) afford a much smaller impact or health threat (cancer) than do many naturally occurring products or the life style followed by the individual. However, this contention has been vehemently contested by Epstein and Swartz (1988). Nonetheless, the perception is widespread and leads to a serious concern on the part of many. So much so that it must be recognized as one of the ''impacts of pesticides in the environment'' and be addressed in a serious and responsible manner.''

## 12-3 MONITORING

One of the impacts of pesticides in the environment as far as humans are concerned is the need to know the concentration and spatial distribuion of the chemical. Thus, what is desired is to determine the amount of chemicals in or on food, in water, soil, and air to avoid any adverse consequence. It is automatic that a pesticide has biological activity, otherwise it would be ineffective. The fact that in studying the compound in laboratory animals, the high concentrations used to show an effect leads some to conclude that even small amounts ingested over long periods may be hazardous. However, it must be recognized that at some concentrations a ''toxic'' level will be reached. By monitoring, it is possible to take steps to avoid this level.

The basis of concern for escape of chemicals whether it be in the work environment or the more general, ambient environment, is the exposure of people and other organisms and the potential effects of this exposure. If exposure to the multitude of chemicals existing in the environment were innocuous, there would be but small cause for concern. However, exposure is not always innocuous, particularly at the relatively higher levels of exposure encountered in occupational settings or with prolonged (chronic) exposure from general environmental contamination.

### 12-3.1 Exposure

There are three usual routes of exposure experienced by humans: respiratory exposure, dermal exposure, and ingestion (Freed et al., 1980). Occupational settings may result in high-respiratory exposure and if the worker is careless through ingestion. However, the major route of exposure for workers is dermal from spillage, splashes, or contamination of clothing. Of course, there are instances where any combination of the three exposure routes may be experienced. For example, an individual mixing a chemical for use may be exposed by both respiration and dermal routes.

It is known that the intensity of effects and time required to produce effecs from a chemical are dependent upon the dose or amount of chemical entering the body (Freed et al., 1980). To assess the risk of effects to be encountered under any given set of circumstances, the route and rate of chemical reaching the organism must be known. For this reason, it is of interest and importance to trace the pathway of the chemical in the environment. This requires a knowledge of the properties and behavior of the chemical (Gillett et al., 1974; Mackay & Wolkoff, 1973).

### 12–3.2 Dynamics of Chemicals in the Environment

The pathway of a chemical through the environment following release from some source may be circuitous and have many feedback loops (Freed et al., 1977). On a meso- or global-scale, the pathways are dependent upon major factors of the ecological/climatic systems. These factors are known to include solar energy flux and gravitational fields as well as atmospheric and aquatic gradients. The interaction of the physicochemical properties of a substance plays a significant role in its transport and in its availability for transport even on the global scale. At lower levels of environmental resolution, it is probable that physical properties such as vapor pressure, water solubility, and partition coefficient are of major importance (Brown, 1978; Chiou et al., 1977; Gillette et al., 1974; Haque et al., 1974; Kenaga, 1975). Wide differences in the environmental behavior of chemicals and minerals are attributable, in large measure, to their physical properties.

### 12–3.3 Types of Monitoring

Monitoring is a complex of activities having a hierarchial structure. Four different types of monitoring practices have been used to increase knowledge of the location and form of chemical substances in the environment. They are:

1. *Reconnaissance monitoring:* The conduct of periodic observation to disclose changes or trends.
2. *Surveillance monitoring:* Conduct of periodic observations made to support an enforcement program and to ensure compliance.
3. *Subjective monitoring:* Spot-checking for broad or open-ended exploration of a problem.
4. *Objective monitoring:* The conduct of periodic observations with the expressed purpose of providing data for use in developing or confirming the results of quantitative models used to simulate the transport of a substance.

An essential component of any monitoring program is an inventory of either the sources or the agent of pollution. Inventories attempt to identify and to quantify the agent of its source in relation to environmental media as a rational basis for surveillance monitoring. Inventory data may, however, have use beyond that of providing a basis for surveillance monitoring. For example, these data may be useful in establishing parameters for further research or in developing new measurement techniques.

### 12–3.4 Monitoring Options

Concern for human-induced environmental change arises out of a desire to protect humans and other biota. Protection may be brought about either by avoidance or by protective reactions. Knowledge of changes in environmental quality gained through monitoring provides a basis for assessing and anticipating probable effects. The question then arises as to what to monitor. Should it be the biota, including humans, or should it be the environment of the biota? The most feasible course is to monitor the environment; but, on occasion, the biota may be used for monitoring purposes.

The environment consists of three media: air, water, and soil. The quality of these media is commonly monitored for changes because of human activity. Sources of pollutants in air, water, and soil are nearly as varied as human activity and may be categorized in three groups: stationary sources (such as manufacturing or power plants), nonpoint sources (such as an agricultural area), and mobile sources (such as motor vehicles). In point sources, the inventory is rather specific in that it deals with the transport, storage, use, and disposal of pesticides or the containers. Probably, primary attention would be focused in any monitoring on human exposures or environmental contaminations, that would occur during the mixing and application of the chemical, but spills and leakage in transport and storage would certainly be of concern.

### 12–3.5 Monitoring of Humans

Individuals most likely to receive exposure to the chemical are those handling, mixing and loading, or applying the chemical (Tordoir & VanHeenstra, 1980; VanHeenstra & Tordoir, 1982). These are the individuals that not only would most likely be exposed, but also would be most likely to receive the greatest exposure. The bystander observing the application at a distance may receive an exposure of drift, spray or dust, but would experience substantially less of an exposure than a worker unless directly in the path of the application. The general public, which would be the third group of persons who may receive an exposure, would most likely receive it through water or food that had been contaminated either through spills or drift and overspraying of the water supply or food crop (Swan, 1985).

For the public, assessment of exposure is probably most conveniently-done by determination of residues in environmental samples such as water, air, and food supplies. To be sure a nearby population should be observed for overt manifestations of effect, but in well-performed spray operations the exposure should not result in such manifestations. The effect produced by any chemical on an organism is dependent upon the dose or amount of chemical *inside* the organism at any given time. In the case of animals, the dose is expressed in terms of milligrams per kilogram of body weight. Dose is distinguished from the exposure, hence, the skin particularly is something of a barrier to entry of chemicals, though some pass through much more readily and rapidly than others (Freed et al.,1980; Lilis et al., 1978; Swan, 1985).

Membranes of the respiratory tract or the alimentary tract provide less of a barrier to entry, thus inhalation of a chemical or consuming it increases the likelihood of receiving a biologically significant dose. However, dermal absorption represents the major entry route.

Assessment of human exposure to pesticides or other chemicals may be quantitative by either direct measurement methods or the more indirect methods that rely on evaluating the activity of a particular enzyme (e.g., cholinesterase) or analyses of tissues for the presence of the chemical, a metabolite, for observation of particular manifestations.

Among the first direct methods of measuring exposure to pesticides is the use of the alpha cellulose patches attached to clothing or taped to the skin. Shirts have also been used to measure exposure and, of course, even the clothing itself may be analyzed to determine exposure. However, where t-shirts or other items of clothing may be used as the trap for the chemical, care must be taken in the analyses to avoid problems with possible interferences that are extracted from the clothing. Another technique used in direct measurement of exposure is that of skin washes with an appropriate solvent. This technique has been used particularly in the case of hands and forearms where a bag rinse technique has been used.

For respiratory exposure, respirators with special filter pads and various types of air samplers have been used. The air samplers may be a battery-operated dynamic sampler with appropriate absorbent for the chemical, or the use of a passive sampler that depends on diffusion. These direct methods have been used to measure the exposure of workers particularly to a wide variety of pesticides, among them such things as carbaryl, organophosphate pesticides, and at an earlier date the organochlorine pesticides.

### 12–3.6 Environmental Monitoring

Chemicals in the environment may produce adverse effects on biota, water quality, productivity of the soil, or air quality. For these reasons, it is often desirable to monitor during the application or release of chemicals to assure that such adverse affects are not being produced or if they are, that appropriate legislative actions can be taken. In the application of pesticides, the monitoring can be helpful in modifying the application techniques or use of a particular chemical to avoid adverse effects (Neely, 1980; NRC/NAS, 1977).

Among the adverse effects that can be produced by the chemicals is, of course, the direct effect on nontarget species, such as killing of crops, bees, or birds, poisoning of fish and other organisms of the environment by direct contact of the spray. The more persistent of chemicals may produce a subtle effect at lower concentrations such as yield reduction of crops or reducing the reproductive capacityof a species (e.g., thinning of bird egg shells by virtue of bioaccumulation). Another factor is the possibility of residues in soils to be picked up by subsequent crops, or through drift of the chemical to afford residues on the current crops (Brown, 1978).

### 12–3.7 What Samples Should Be Monitored?

In monitoring the environment, the question is, what samples to take to get an accurate picture of the distribution of the residues and make an assessment of possible impact beneficial or adverse. One may take samples of water, soil, some of the available biota and on occasions, air, for determining the distribution and fate of the residues. Of particular value, when using organophosphates, carbamates, or synthetic pyrethroids with a somewhat short persistence, would be water sampling to assure that fish and other aquatic organisms are not adversely effected. Sampling of plants and other organisms in the treated area or just adjacent to determine levels of residues and ensures against excessive drift. They also ensure against possible accumulation that may have an adverse effect on the biota. With some of the persistent chemicals, such as the chlorinated hydrocarbons (e.g., the lindane and dieldrin that have been used extensively in the past) one may find residues in the soil. Additionally, one would want to take samples of water, plants, and resident animals or other organisms to determine the level of exposure occurring during and shortly after the application. For following the residue levels in the environment for a longer period, you may want to look for an accumulator species such as earthworms or fish and relate back either the effect on these species or the residue level found by chemical analyses to the residue in the environment (Blau et al., 1975; Neely, 1980).

## 12–4 PERSPECTIVES ON IMPACTS

The biological activity or toxicity of pesticides, their use on food crops, effect on nontarget species, and occasional occurrence in groundwater has served to focus much public attention on these chemicals. Since 1962 when Rachel Carsons' *Silent Spring* was published, there has been widespread concern that pesticide use might result in irreversible damage to some populations of organisms, produce large numbers of excess cancers in humans, and result in many undiagnosed maladies. It is interesting that a wide segment of the population rates pesticides among some of the top hazards. Studies have shown where morbidity and mortality data were used for a variety of human activities and agents, that the danger attributable to pesticides was quite low compared to other reported injuries and death. While indeed some pesticides have been shown to produce cancer in laboratory animals at somewhat elevated doses, present data shows that, at least, their residues in food comprises a small carcinogenic hazard to humans.

It has been a concern on the part of some that repeated pesticide applications to the environment may cause the chemical to accumulate to the point of creating a nonproductive biological desert. With some of the more persistent material, it is true the chemical remains in the soil for some period, but during the course of that time various biological, chemical, and physical processes are at work bringing about the degradation of the chemical or its loss to the atmosphere where photochemical degradation occurs (Freed et al., 1977). Certainly, in some instances of more persistent or mobile materials repeated applications can result in residues remaining in the soil or leaching into groundwater. Similarly, for chemicals that may persist even over a relatively short time, chemical erosion at the surface may carry the chemical into water courses or if the chemical volatilizes it may be transported by air. However, in the temperate zones and tropical regions substantial degradation begins shortly after the chemical reaches the environment and continues to the ultimate mineralization of the chemical. The rate at which this occurs depends upon temperature, moisture, fertility of the soil, or the ultraviolet radiation intensity to which the chemical may be exposed. The rates of application usually employed in agriculture, forestry, and public health pest control most chemicals will fade over a reasonable time. A few such chemicals such as DDT, whose initial degradation product DDE tends to have a long persistence, particularly in soil.

Early in this chapter, a few remarks were made about the development of resistance to chemicals. Again, particular impact on biological populations has been a matter of speculation and concern that the resulting modified organism would be a worst pest than the original wild population. True that the organism may be more difficult to control, particularly if the chemicals available for its control are limited. This is especially so if those chemicals have a similar mode of action. With other chemicals, however, or better yet the use of integrated pest management (IPM) strategies that employ not only chemical but a variety of other tactics, pests can usually be controlled. We overlooked in such a scenario, however, that the continued pressure of even a biological agent may cause a mutation in a pest that results in resistance to that particular pathogen or disease. One of the authors (V.H. Freed) has seen the development of resistance by mosquitos to a pathogenic nematode that at one time was considered for control of the malaria vector. This occurred in a country in Central America where the nematode was being intensively studied. Pesticides in the environment have an impact on the biological population, whether target or nontarget, on water, and air quality. They also leave residues in food, directly or indirectly. However, judicious and careful use of the pesticides, particularly when incorporated into an IPM program, will not afford a particular risk to nontarget species or to the environment. Further, biological, regulatory, and economic forces operative driving pesticide development toward safer, more selective, and less persistent materials. Part of this is attributable to the fact that pesticides usually have a finite useful life due to development of resistance or the appearance of a more effective and better-adapted chemical or control measure. It is reasonable, therefore, that the impact of chemicals in the environment, specifically the pesicides, be continually ameliorated through the development of new materials and increasing scientific knowledge on the appropriate and safe use of such materials.

## REFERENCES


Alexander, M., and M.I.H. Aleem. 1961. Effect of chemical structure on microbial degradation of aromatic herbicides. J. Agric. Food Chem. 9:44–47.

American Chemical Society. 1978. Cleaning our environment. Am. Chem. Soc., Washington, DC.

Ames, B.N., and L.S. Gold. 1988. Response. Science 240:1045–1047.

Ames, B.N., R. Magaw, and L.S. Gold. 1987. Ranking possible carcinogenic hazards. Science 236:271–280.

Anderson, J.F., and W. Glowa. 1984. Insecticidal poisoning of honey bees in Connecticut. Environ. Entomol. 13:70–74.

Anderson, J.R. 1978. Pesticide effects on non-target soil microorganisms. p. 313–533. *In* I.R. Hill and S.J.L. Wright (ed.) Pesticide microbiology. Academic Press, London.

Atkins, E.L. 1972. Rice field mosquito control studies with low volume dursban sprays in Colusa County, California. V. Effects upon honey bees. Mosquito News 32:538–541.

Atkins, E.L. 1981. Reducing pesticide hazards to honey bees: Mortality prediction techniques and integrated management strategies. Univ. of California Riverside Div. Agric. Sci. Leafl. 2883.

Atkins, E.L., E.A. Graywood, and R.L. MacDonald. 1975. Toxicity of pesticides and other agricultural chemicals to honey bees. Univ. of California Div. Agric. Sci. Leafl. 2287.

Au, F.H.F. 1968. Effect of endothal and certain other selective herbicides on microbial activity in six different soils. Ph.D. diss., Oregon State Univ., Corvallis.

Audus, L.J. 1964. Herbicide behaviour in he soil. p. 163–206. *In* L.J. Audus (ed.) The physiology and biochemistry of herbicides. Academic Press, London.

Barrons, K.C. 1981. Are pesticides really necessary? Regency Gateway, Chicago, IL.

Beatty, R.G. 1973. The DDT myth. The John Day Co., New York.

Benton, A.H. 1951. Effects on wildlife of DDT used for control of Dutch elm disease. J. Wildl. Manage. 15:20–27.

Blau, G.E., W.B. Neely, and D.R. Branson. 1975. Ecokinetics: A study of the fate and distribution of chemicals in laboratory ecosystems. Am. Inst. Chem. Eng. J. 21:854–861.

Blus, L.J., A.B. Belisle, and R.M. Prouty. 1974. Relations of the brown pelican to certain environmental pollutants. Pestic. Monit. J. 7:181–194.

Blus, L.J., and C.J. Henny. 1985. Organochlorine induced mortality and residues in long-billed Curlews from Oregon. Condor 87:563–565.

Blus, L.J., C.J. Henny, C.J. Stafford, and R.A. Grove. 1987. Persistence of DDT and metabolites in wildlife from Washington State orchards. Arch. Environ. Contam. Toxicol. 16:467–476.

Boddicker, M.L., E.J. Hugghins, and A.H. Richardson. 1971. Parasites and pesticide residues of mountain goats in South Dakota. J. Wildl. Manage. 35:94–103.

Boellstroff, D.E., H.M. Ohlendorf, D.W. Anderson, E.J. O'Neill, J.O. Keith, and R.M. Prouty. 1985. Organochlorine chemical residues in white pelicans and Western grebes from the Klamath basin, California. Arch. Environ. Contam. Toxicol. 14:485–493.

Bollen, W.B., J.E. Roberts, and H.E. Morison. 1968. Soil properties and factors influencing aldrin-dieldrin recovery and transformation. J. Econ. Entomol. 51:214–219.

Boyd, C.E., and D.E. Ferguson. 1964. Susceptibility and resistance of mosquito fish to several insecticides. J. Econ. Entomol. 57:430–431.

Brock, J.L. 1972. Effects of the herbicides trifluralin and carbetamide on nodulation and growth of legume seedlings. Weed Res. 12:150–154.

Brown, A.W.A. 1978. Ecology of pesticides. John Wiley and Sons, New York.

Bull, D. 1982. A growing problem: Pesticides and the third world poor. Oxford, England.

Bunck, C.M., R.M. Prouty, and A.J. Krynitsky. 1987. Residues of organochlorine pesticides and polychlorobiphenyls in starlings (*Sturnus vulgaris*), from the continental United States. 1982. Environ. Monit. Assess. 8:59–75.

Burden, E.H.W.J. 1956. A case of DDT poisoning in fish. Nature (London) 178:546–547.

Burdick, G.E., E.J. Harris, H.J. Dean, T.M. Walker, J. Skea, and D. Colby. 1964. The accumulation of DDT inlake trout and the effect on reproduction. Trans. Am. Fish. Soc. 93:127–136.

Butler, G.L. 1977. Algae and pesticides. Residue Rev. 66:19–62.

Butler, P.A. 1970. The sub-lethal effects of pesticide pollution. p. 87–89. *In* J.W. Gillett (ed.) The biological impact of pesticides in the environment. Oregon State Univ. Press, Corvallis.

Cain, B.W. 1981. Nationwide residues of organochlorine compounds in wings of adult mallards and black ducks, 1979–80. Pestic Monit. J. 15:128–134.

Calabrese, E.J. 1983. Principles of animal extrapolation. John Wiley and Sons, New York.

Caldwell, R.S. 1977. Biological effects of pesticides on dungeness Crab. USEPA EPA-600/3-77-131. U.S. Gov. Print. Office, Washington, DC.

Call, D.J., L.T. Brooke, R.J. Kent, M.L. Knuth, S.H. Poirier, J.M. Huot, and A.R. Lima. 1987. Bromacil and diuron herbicides: Toxicity, uptake and elimination in freshwater fish. Arch. Environ. Contam. Toxicol. 16:607–613.

Caron, D.M. 1979. Effects of some ULV mosquito abatement insecticides on honey bees. J. Econ. Entomol. 72:148–151.

Carson, R.L. 1962. Silent spring. Riverside Press, Cambridge, MA.

Chiou, C.T. 1985. Partition coefficients of organic compounds in lipid-water systems and correlation with fish bioconcentration factor. Environ. Sci. Technol. 19:57–62.

Chiou, C.T., V.H. Freed, D.W. Schmedding, and R.L. Kohnert. 1977. Partition coefficient and bioaccumulation of selected organic chemicals. Environ. Sci. Technol. 11:475–478.

Chiou, C.T., and D.W. Schmedding. 1981. Measurement and inter-relations of octanol water partition coefficient and water solubiliy of organic chemicals. p. 28. *In* Test protocols for environmental fate and movement of toxicants. Proc. Symp. Assoc. Off. Anal. Chem., Arlington, VA.

Clark, D.R., Jr. 1981. Bats and environmental contaminants: A review. U.S. Fish Wildl. Serv. Spec. Sci. Rep. Wildl. 235:1–27.

Clark, D.R., Jr., F.M. Bagley, and W.W. Johnson. 1988. Northern Alabama colonies of the endangered gray bat (*Myotis grisescens*): Organochlorine contamination and mortality. Biol. Conserv. 43:213–225.

Clark, D.R., Jr., R.L. Clawson, and C.J. Stafford. 1983. Gray bats killed by dieldrin at two additional Missouri caves: Aquatic macroinvertebrates found dead. Bull. Environ. Contam. Toxicol. 30:214–218.

Clark, D.R., Jr., and A.J. Krynitsky. 1983. DDT: Recent contamination in New Mexico and Arizona? Environment 25:27–31.

Clark, D.R., Jr., R.K. LaVal, and D.M. Swineford. 1978. Dieldrin-induced mortality in an endangered species, the gray bat (*Myotis grisescens*). Science 199:1357–1359.

Claus, C., and K. Bolander. 1977. Ecological sanity. David McKay Co., New York.

Clore, W.J., and V.F. Bruns. 1953. The sensitivity of concord grapes to 2,4-D. Proc. Am. Soc. Hortic. 61:125–134.

Coburn, D.R., and R. Treichler. 1946. Experiments on toxicity of DDT to wildlife. J. Wildl. Manage. 10:208–216.

Collins, P.J. 1985. Resistance to grain protectants in field populations in sawtoothed grain beetle in Southern Queensland. Aust. J. Exp. Agric. 25:683–686.

Coon, N.C. 1983. Summaries of selected studies on wildlife pollution. U.S. Fish Wildl. Serv., Patuxent Wildl. Res. Ctr., Laurel, MD.

Cope, O.B. 1961. Effecs of DDT spraying for spruce budworm on fish in the Yellowstone river system. Trans. Am. Fish. Soc. 90:239–251.

Cope, O.B. 1965. Agricultural chemicals and freshwater ecological systems. p. 115–128. *In* C. Chichester (ed.) Research in pesticides. Academic Press, New York.

Cope, O.B., and P.F. Springer. 1958. Mass control of insects: The effects on fish and wildlife. Bull. Environ. Soc. A 4:52–56.

Cottam, C. 1965. The ecologists' role in problems of pesticide pollution. BioScience 15:457–463.

Cremlyn, R. 1978. Pesticides-preparation and mode of action. John Wiley and Sons, New York.

Crocker, R.A., and A.J. Wilson. 1965. Kinetics and effects of DT in a tidal march ditch. Trans. Am. Fish. Soc. 94:152–159.

Crouter, R.A., and E.H. Vernon. 1959. Effects of black-headed budworm control on salmon and trout in British Columbia. Can. Fish. Cult. 24:23–40.

Culley, D.D., and H.G. Applegate. 1967. Residues in fish, wildlife, and estuaries. Pestic. Monitor. J. 1:21–28.

Custer, T.W., C.M. Bunck, and T.E. Kaiser. 1983a. Organochlorine residues in Atlantic Coast black-crowned night-haron eggs. 1979. Colon. Waterbirds 6:160–167.

Custer, T.W., R.M. Erwin, and C. Stafford. 1983b. Organochlorine residues in common ferm eggs from nine Atlantic Coast Colonies, 1980. Colon. Waterbirds 6:197–204.

Custer, T.W., and C.A. Mitchell. 1987. Organochlorine contaminants and reproductive success of black skimmers in South Texas. 1984. J. Field Ornithol. 58:480–489.

Davies, J.E., A. Barquet, V.H., Freed, R. Haque, C. Morgade, R. Sonneborn, and C. Vaclavek. 1975. Human pesticide poisoning by a fat soluble organophosphate insecticide. Arch. Environ. Health. 30:608–613.

Davies, J.E., and W.F. Edmundson (ed.). 1972. Epidemiology of DDT. Futura Publ. Co., Mount Kisco, NY.

Davies, J.E., V.H. Freed, and F. Whittemore (ed.). 1982. An agromedical approach to pesticide management: Some health and environmenal consideration. Univ. of Miami, Miami.

Dempster, J.P. 1975. Effects of organochlorine insecticides on animal populations. p. 231–248. *In* F. Moriarty (ed.) Organochlorine insecticides: Persistent organic pollutants. Academic Press, New York.

DeWeese, L.R., L.C. McEwen, G.L. Hensler, and B.E. Petersen. 1986. Organochlorine contaminants in passeriformes and other avian prey of the peregrine falcon in the western United States. Environ. Toxicol. Chem. 5:675–693.

Dimond, J.B., and J.A. Sherburne. 1969. Persistence of DDT in wild populations of small mammals. Nature (London) 221:486–487.

Edwards, C.A. 1973. Persistent pesticides in the environment. CRC Press, Cleveland.

Edwards, C.A., and A.R. Thompson. 1973. Pesticides and the soil fauna. Residue Rev. 45:1–79.

Efron, E. 1984. The apocalyptics. Simon and Schuster, New York.

Eisler, R. 1985. Carbofuran hazards to fish, wildlife, and invertegrate: A synoptic review. U.S. Fish Wildl. Serv. Biol. Rep. 85. U.S. Gov. Print. Office, Washington, DC.

Eisler, R., and J. Jacknow. 1985. Toxaphene hazards to fish, wildlife, and invertebrate: A synoptic review. U.S. Fish Wildl. Serv. Biol. Rep. 85. U.S. Gov. Print. Office, Washington, DC.

El-Dib, M.A., and M.I. Badawy. 1985. Organochlorine insecticides and PCBs in water, sediment, and fish from the Mediterranean sea. Bull. Environ. Contam. Toxicol. 34:216–227.

El Nabawi, A., B. Heinzow, and H. Kruse. 1987. Residue levels of organochlorine chemicals and polychlorinated biphenyls in fish from the Alexandria region, Egypt. Arch. Environ. Contam. Toxicol. 16:689–696.

Elson, P.F. 1967. Effects of wild young salmon of spraying DDT over New Brunswick forests. J. Fish Res. Board Can. 24:731–767.

Epstein, S.S., and J.B. Swartz. 1988. Carcinogenic risk estimation. Science 240:1043–1045.

Ferguson, D.E. 1970. The effects of pesticides on fish: Changing patterns of speciation and distribution. p. 83–86. *In* J.W. Gillett (ed.) The biological impact of pesticides in the environment. Oregon State Univ. Press, Corvallis, OR.

Ferguson, D.E., and C.R. Bingham. 1966. Endrin resistance in the yellow bullhead (*Ictalurus natalis*). Trans. Am. Fish. Soc. 95:325–326.

Ferguson, D.E., D.D. Culley, W.D. Cotton, and R.P. Dodds. 1964. Resistance to chlorinated hydrocarbon insecticides in three species of freshwater fish. BioScience 14:43–44.

Finlayson, D.G., J.R. Grahmam, R. Greenhalgh, J.R. Roberts, E.A.H. Smith, P. Whitehead, R.F. Willes, and I. Williams. 1979. Carbofuran: Criteria for interpreting the effects of its use in environmental quality. Publ. NRCC 16740. Natural Resourc. Counc., Quebec.

Fleming, W.J., and B.W. Cain. 1985. Areas of localized organochlorine contamination in Arizona and New Mexico. Southwest. Nat. 30:269–277.

Fleming, W.J., D.R. Clark, Jr., and C.J. Henny. 1983. Organochlorine pesticides and PCB's: A continuing problem for the 1980. p. 186–199. *In* Trans. 48th N. Am. Wildl. Natural Resour. Conf. Wildlife Manage. Inst., Washington, DC.

Flickinger, E.L., C.A. Mitchell, D.H. White, and E.J. Kolbe. 1986. Bird poisoning from misuse of the carbamate furadan in Texas rice field. Wildl. Soc. Bull. 14:59–62.

Flickinger, E.L., D.H. White, C.A. Mitchell, and T.G. Lamont. 1984. J. Assoc. Off. Anal. Chem. 67:827–828.

Foley, R.E., S.J. Jackling, R.J. Sloan, and M.K. Brown. 1988. Organochlorine and mercury residues in wild mink and otter: Comparison with fish. Environ. Toxicol. Chem. 7:363–374.

Frank, R., and M.V.H. Holdrint. 1975. Residue of organochlorine compounds and mercury in birds' eggs from the Niagara peninsula, Ontario. Arch. Environ. Contam. Toxicol. 3:205–218.

Frank, R., M.V.H. Holdrinet, and P. Suda. 1979. Organochlorine and mercury residues in wild mammals in Southern Ontario, Canada 1973–74. Bull. Environ. Contam. Toxicol. 22:500–507.

Freed, V.H. 1970. Global distribution of pesticides p. 1–10. *In* J.W. Gillett (ed.) The biological impact of pesticides in the environment. Oregon State Univ. Press, Corvallis, OR.

Freed, V.H., C.T. Chiou, and R. Haque. 1977. Chemodynamics: Transport and behavior of chemicals in the environment—A problem in environmental health. Environ. Health Persp. 20:55.

Freed, V.H., C.T. Chiou, and R. Haque. 1980. Physicochemical factors in routes and rates of human exposure to chemicals. p. 59–84. *In* J.D. McKinney (ed.) Environmental health chemistry. Ann Arbor Sci. Publ., Ann Arbor, MI.

Geluso, K.N., J.S. Altenback, and D.E. Wilson. 1976. Bat mortality: Pesticide poisoning and migratory stress. Science 194:184–186.

Georghiou, G.P., and R. Mellon. 1983. Resistance in time and space. p. 1–46. *In* G.P. Georghiou and T. Saito (ed.) Pest resistance to pesticides: Challenges and prospects. Plenum Press, New York.

Gillett, J.W., J. Hill, IV, A.W. Jarvinen, and W.P. Schoor. 1974. A conceptual model for the movement of pesticides through the environment. USEPA EPA-660/3-74-024. U.S. Gov. Print. Office, Washington, DC.

Goodman, L.R., D.J. Hansen, D.L. Coppage, J.C. Moore, and E. Matthews. 1979. Diazinon: Chronic toxiciy to, and brain acetylcholinesterase inhibition in the sheepshead minnow (*Cyprinodon variegatus*). Trans. Am. Fish. Soc. 108:479–488.

Goodman, V.H. 1953. 2,4-D injury to cotton. Mississippi Agric. Exp. Stn.Cir. 185.

Gould, R.F. (ed.). 1966. Organic pesticides in the environment. Adv. Chem. Ser. 60. Am. Chem. Soc., Washington, DC.

Graham, F. 1970. Since silent spring. Houghton Mifflin Co., Boston.

Grant, C.D., and A.W.A. Brown. 1967. Development of DDT resistance in certain mayflies in New Brunswick. Can. Entomol. 99:1040–1050.

Grant, M.A., and W.J. Payne. 1982. Effects of pesticides on denitrifying activity in salt marsh sediments. J. Environ. Qual. 11:369–372.

Green, M.B., Hartley, G.S., and T.F. West. 1977. Chemicals for crop protection and pest control. Pergamon Press, Oxford, UK.

Grue, C.E., W.J. Fleming, D.G. Busby, and E.F. Hill. 1983. Assessing hazards of organophosphate pesticides to wildlife. p. 200–220. *In* Trans. 48th N. Am. Wildl. Natural Resour. Conf. Wildlife Manage. Inst., Washington, DC.

Grue, C.E., G.V.N. Powell, and M.J. McChesney. 1982. Care of nestling by wild female starlings exposed to an organophosphate pesticide. J. Appl. Ecol. 19:327–335.

Haas, H., and J.C. Streibig. 1982. Changing patterns of weed distribution as a result of herbicide use and other agronomic factors. p. 57–79. *In* H.M.N. LeBaron and J. Gressel (ed.) Herbicide resistance in plants. John Wiley and Sons, New York.

Hall, R.J. 1987. Impact of pesticides on bird populations. p. 85–111. *In* G.J. Marco et al., (ed.) Silent spring revisited. Am. Chem. Soc., Washington, DC.

Hall, R.J., R.E. Ashton, Jr., and R.M. Prouty. 1985. Pesticide and PCB residues in Neuse river waterdog (*Necturus lewisi*). Brimleyana 10:107–109.

Haque, R., D.W. Schmedding, and V.H. Freed. 1974. Aqueous solubility, adsorption, and vapor behavior of polychlorinated biphenyl arochlor 1254. Environ. Sci. Technol. 8:139–142.

Harris, C.R. 1972. Factors influencing the effectiveness of soil insecticides. Ann. Rev. Entomol. 17:177–198.

Harrison, D.L., P.E.G. Maskell, and D.F.L. Money. 1963. Dieldrin poisoning of dogs. N. Z. Vet. J. 11:23–31.

Hauke-Pacewiczowa, T. 1971. The effect of herbicides on the activity of soil microflora. Pam. Pulawski. 46:5–48.

Hay, A. 1982. The chemical scythe. Plenum Press, New York.

Hayes, W.J., Jr. 1982. Pesticides studied in man. Williams and wilkens, Baltimore.

Henderson, C., W.L. Johnson, and A. Inglis. 1969. Organochlorine insecticide residues in fish. Pestic. Monit. J. 3:145–171.

Henny, C.J., L.J. Blus, and R.M. Prouty. 1982. Organochlorine residues and shell thinning in Oregon seabird eggs. Murrelet 63:15–21.

Henny, C.J., L.J. Blus, and C.J. Stafford. 1983. Effects of heptachlor on American kestrels in the Columbia Basin of Oregon. J. Wildl. Manage. 47:1080–1087.

Hickey, J.J., and D.W. Anderson. 1968. Chlorinated hydrocarbons and eggshell changes in raptorial and fish-eating birds. Science 162:271-273.

Hill, E.F., and M.B. Camardese. 1982. Subacute toxicity testing with young birds: Response in relation to age and interest variability of LC-50 estimates. p. 41-65. *In* D.W. Lamb and E.E. Kenaga (ed.) Avian and mammalina wildlife toxicology: Second conference. ASTM STP 757. Am. Soc. Test. Material, Philadelphia.

Hill, E.F., and D.M. Dent. 1985. Mirex residues in seven groups of aquatic and terrestrial mammals. Arch. Environ. Contam. Toxicol. 14:7-12.

Hill, E.F., D.A. Eliason, and J.W. Kilpatrick. 1971. Effects of ultra-low volume applications of malathion in Hale County, Texas. J. Med. Entomol. 8:173-179.

Hoar, S.K., A. Blair, F.F. Holmes, O.D. Boysan, R.J. Robel, R. Hover, and J.F. Fraumeni, Jr. 1986. Agricultural herbicide use and risk of lymphoma and soft-tissue sarcoma. J. Am. Med. Assoc. 256:1141-1147.

Hoffman, D.J. 1988. Effects of krenite brush control agent on embryonic development in mallards and bobwhite. Environ. Toxicol. Chem. 7:69-75.

Holden, A.V. 1973. Effects of pesticides on fish. p. 213-253. *In* C.A. Edwards (ed.) Environmental pollution by pesticides. Plenum Press, New York.

Hollister, T.A., G.E. Walsh, and J. Forester. 1975. Mirex and marine unicellular algae: Accumulation, population growth and oxygen evaluation. Bull. Environ. Contam. Toxicol. 14:753-759.

Hotchkiss, N., and R.H. Pough. 1946. Effect on forest birds of DDT used for gypsy moth control in Pennsylvania. J. Wildl. Manage. 10:202-207.

Hudson, R.H., R.K. Tucker, and M.A. Haegele. 1972. Effect of age on sensitivity: Acute oral toxicity of 14 pesticides to mallard ducks of several ages. Toxicol. Appl. Pharmacol. 22:556-561.

Hunt, E.G., and A.I. Bischoff. 1960. Inimical effects on wildlife of periodic DDD applications to Clear Lake. Calif. Fish Game 46:91-106.

Hunt, E.G., and J.D. Linn. 1970. Fish kills by pesticides. p. 97-103. *In* J.W. Gillett (ed.) The biological impact of pesticides in the environment. Oregon State Univ. Press, Corvallis.

Hunter, R.G., J.H. Carroll, and J.C. Randolph. 1980. Organochlorine residues in fish of lake Texoma, October 1979. Pestic Monit. J. 14:102-107.

Inis, W.B. (ed.). 1979. Production to crop protection. ASA and CSSA, Madison, WI.

Jett, D.A. 1986. Cholinesterase inhibition in meadow voles following field applications of orthene. Environ. Toxicol. Chem. 5:255-259.

Johansen, C.A., M.D. Coffey, and J.A. Quist. 1957. Effect of insecticide treatments to alfalfa on honey bees, including insecticidal residue and honey flavor analysis. J. Econ. Entomol. 50:721-723.

Johnson, D.L. 1968. Pesticides and fishes—a review of selected literature. Trans. Am. Fish. Soc. 97:398-424.

Johnson, D.W. 1973. Pesticide residues in fish. p. 181-212. *In* C.A. Edwards (ed.) Environmental pollution by pesticides. Plenum Press, New York.

Johnson, D.W., and S. Lew. 1970. Chlorinated hydrocarbon residues in representative fishes of southern Arizona. Pestic. Monit. J. 4:57-61.

Johnson, H.E., and C. Pecor. 1969. Coho salmon mortality and DDT in Lake Michigan. N. Am. Wildl. Natural Res. Conf. Trans. 34:159-166.

Khan, S.U. 1980. Pesticides in the soil environment. Elsevier Sci. Publ. Co., Amsterdam.

Kaiser, P., J.J. Pochon, and R. Cassini. 1970. Influence of triazine herbicides on soil microorganisms. Residue Rev. 32:211-233.

Kaufman, D.D. 1974. p. 133-202. *In* W.D. Guenzi (ed.) Pesticides in soil and water. SSSA, Madison, WI.

Keenleyside, M.H.A. 1959. Effects of spruce budworm control on salmon and other fishes in New Brunswick. Can. Fish. Cult. 24:17-22.

Kenaga, E.E. 1975. Partitioning and uptake of pesticides in biological systems. p. 217-273. *In* R. Haque and V.H. Freed (ed.) Environmental dynamics of pesticides. Plenum Publ. Corp., New York.

Kerwill, C.J., and H.E. Edwards. 1967. Fish losses after forest sprayings with insecticides in New Brunswick, 1957-1962. As shown by caged specimens and other observations. J. Fish Res. Board Can. 24:708-729.

King, K.A., D.H. White, and C.A. Mitchell. 1984. Nest defense behavior and reproductive success of laughing gulls sublethally dosed with parathion. Bull. Environ. Contam. Toxicol. 33:499-504.

Lee, R.E., Jr. (ed.). 1976. Air pollution from pesticides and agricultural processes. CRC Press, Cleveland.

Lieberman, F.V., G.E. Bohart, G.F. Knowlton, and W.P. Nye. 1954. Additional studies on the effect of field applications of insecticides on honey bees. J. Econ. Entomol. 47:316–320.

Linder, R.L., R.B. Dahlgren, and Y.A. Greichus. 1970. Residues in the brain of adult pheasants given dieldrin. J. Wildl. Manage. 34:954–956.

Loosanoff, V.L. 1965. Pesticides in seawater and the possibilities of their use in mariculture. p. 135–146. *In* C. Chichester (ed.) Research in pesticides. Academic Press, New York.

Mackay, D., and A.W. Wolkoff. 1973. Rate of evaporation of low-solubility contaminants from water bodies to atmosphere. Environ. Sci. Technol. 7:611.

Maddox, J. 1972. The doomsday syndrome. McGraw-Hill Book Co., New York.

Mahmoud, S.A.Z., K.G. Selim, and T. El-Mokadem. 1970. Effect of dieldren and lindane on soil microorganisms. Zentralbl. Bakteriol. Abt. 125:134–149.

Martin, H. 1973. The scientific principles of crop protection. 6th ed. Edward Arnold Ltd., Publ., London.

Martineau, D., P. Beland, C. Desjardins, and A. Lagace. 1987. Levels of organochlorine chemicals in tissues of beluga whales (*Delphinapterus leucas*) from the St. Lawrence estuary, Quebec, Canada. Arch. Environ. Contam. Toxicol. 16:137–147.

Matsumura, F. 1982. Degradation of pesticides in the environment by microorganisms and sunlight. p. 67–90. *In* F. Matsumura and C.R. Krishna Murti (ed.) Biodegradation of pesticides. Plenum Press, New York.

Matthews, G.A. 1979. Pesticide application methods. Longman, London.

Mayer, F.L., Jr., P.M. Mehrle, Jr., and W.P. Dwyer. 1975. Toxaphene effects on reproduction, growth and mortality of brook trout. USEPA EPA-600/3-75-013. U.S. Gov. Print. Office, Washington, DC.

Mayer, F.L., Jr., P.M. Mehrle, Jr., and W.P. Dwyer. 1977. Toxamphene: Chromic toxicity to fathead minnows and channel catfish. USEPA EPA-600/3-7-069. U.S. Gov. Print. Office, Washington, DC.

McEwen, L.C., C.J. Stafford, and G.L.Hensler. 1984. Organochlorine residues in eggs of black-crowned night herons from Colorado and Wyoming. Environ. Toxicol. Chem. 3:367–376.

Mendenhall, V.M., E.E. Klass, and M.A.R. McLane. 1983. Breeding success of barn owls (*Tyto alba*) fed low levels of DDE and dieldrin. Arch. Environ. Contam. Toxicol. 12:235–240.

Milio, J.F., P.G. Koehler, and R.S. Patterson. 1987. Evaluation of three methods for detecting chloropyrifos resistance in German cockroach (Orthoptern: Blattellidae) populations. J. Econ. Entomol. 30:44–46.

Morse, R.A., L.E. St. John, and D.J. Lisk. 1963. Residue analysis of sevin in bees and pollen. J. Econ. Entomol. 56:415–416.

Mount, D.I., and G.J. Putnicki. 1966. Summary report of the 1963 Mississippi fish kill. Trans. N. Am. Wildl. Conf. 31:177–184.

Muirhead-Thompson, R.C. 1987. Pesticide impact on stream fauna with special reference to macroinvertebrates. Cambridge Univ. Press, Cambridge, England.

Mulla, M.S., L.S. Mian, and J.A. Kawecki. 1981. Distribution, transport, and fate of the insecticides malathion and parathion in the environment. Residue Rev. 81:1–159.

Murphy, S. 1986. Toxic effects of pesticides. p. 519–581. *In* C.D. Klaassen et al. (ed.) Casarett and Doull's toxicology: The basic science of poisons. 3rd ed. Macmillan Publ. Co., New York.

Neely, W.B. 1980. Chemicals in the environment. *In* Proc. USEAP Workshop Transport and Fate of Toxic Chemicals in the Environment, 1981. Marcel Dekker, New York.

Niemi, G.J., T.E. Davis, G.D. Veith, and B. Vieux. 1986. Organochlorine chemical residues in herring gulls, ring-billed gulls, and common terns of western Lake Superior. Arch. Environ. Contam. Toxicol. 15:313–320.

Niethammer, K.R., D.H. White, T.S. Bakett, and M.W. Sayre. 1984. Presence and biomagnification of organochlorine chemical residues in Oxbow lake of northeastern Louisiana. Arch. Environ. Contam. Toxicol. 13:63–74.

Nimmo, D.R., T.L. Hamaker, E. Matthews, and J.C. Moore. 1981. An overview of the acute and chronic effects of first and second generation pesticides on an estuarine mysid. p. 3–19. *In* J. Vernberg et al. (ed.) Biological monitoring of marine pollutants. Academic Press, New York.

National Research Center/National Academy of Sciences. 1977. Analytical studies for the U.S. Environmental Protection Agency. Vol. 6. Environmental Monitoring. NAS, Washington, DC.

Ober, A., M. Valdivia, and I.S. Maria. 1987. Organochlorine pesticide residues in Chileau fish and shellfish species. Bull. Environ. Contam. Toxicol. 38:528–533.

Odenkirchen, E.W., and R. Eisler. 1988. Chloropyrifos hazards to fish, wildlife, and invertebrate: Synoptic review. U.S. Fish Wildl. Serv. Biol. Rep. 85. U.S. Gov. Print. Office, Washington, DC.

Ohlendorf, H.M., D.W. Anderson, D.E. Boellstroff, and B.M. Mulhern. 1985. Tissue distribution of trace elements and DDE in Brown Pelicans. Bull. Environ. Contam. Toxicol. 35:183–192.

Ohlendorf, H.M., F.C. Schaffner, T.W. Custer, and D.J. Stafford. 1985. Reproduction and organochlorine contaminants in terns at San Diego bay. Colonial Waterbirds 8:42–53.

Ohlendorf, H.M., D.M. Swineford, and L.N. Locke. 1981. Organochlorine residues and mortality of herons. Pestic. Monit. J. 14:125–135.

Oppenorth, F.J. 1976. Development of resistance to insecticides. p. 41–64. *In* R.L. Metcalf and J.J. McKelvey, Jr. (ed.) The future for insecticides. . . needs and prospects. John Wiley and Sons, New York.

Ottoboni, M.A. 1984. The dose makes the poison. Vincente Books, Berkeley, CA.

Parr, J.F. 1974. Effect of pesticides on microorganisms in soil and water. p. 315–340. *In* W.D. Guenzi (ed.) Pesticides in soil and water. SSSA, Madison, WI.

Pattee, O.H., M.R. Fuller, and T.E. Kaiser. 1985. Environmental contaminants in Eastern Cooper's hawk eggs. J. Wildl. 49:1040–1044.

Pillmore, R.E., and R.B. Finley, Jr. 1963. Residues in game animals resulting from forest and range insecticide applications. p. 409–422. *In* Trans. 28h N. Am. Wildl. Natural Resourc. Conf., Wildlife Manage. Inst., Washington, DC.

Plapp, F.W., Jr. 1981. The nature, modes of action and toxicity of insecticides. p. 3–16. *In* D. Pimentel (ed.) CRC handbook of pest management in agriculture. Vol. 3. CRC Press, Boca Raton, FL.

Ratcliffe, D.A. 1967. Decrease in eggshell weight in certain birds of prey. Nature (London) 215:208–210.

Ratcliffe, D.A. 1970. Changes attributable to pesticides in egg breakage frequency and eggshell thickness in some British birds. J. Appl. Ecol. 7:67–115.

Rattner, B.A., L. Sileo, and C.G. Scanes. 1982a. Hormonal responses and tolerance to cold of female quail following parathion ingestion. Pestic. Biochem. Physiol. 18:132–138.

Rattner, B.A., L. Sileo, and C.G. Scanes. 1982b. Oviposition and the plasma concentrations of LH, progesterone and corticosterone in bobwhite quail fed parathion. J. Reprod. Fert. 66:147–155.

Reichel, W.L., S.K. Schmeling, E. Cromartie, T.E. Kaiser, A.J. Krynitsky, T.G. Lamont, B.M. Mulhern, R.M. Prouty, C.J. Stafford, and D.M. Swineford. 1984. Pesticide, PCB, and lead residues and necropsy data for bald eagles from 32 states—1978–1981. Environ. Monit. Assess. 4:395–403.

Reidinger, R.F., Jr. 1976. Organochlorine residues in adults of six southwestern bat species. J. Wildl. Manage. 40:677–680.

Ritcher, E.D., Z. Rosenvald, L. Kaspi, S. Levy, and N. Bruener. 1986. Sequential cholinesterase tests and symptoms for monitoring organophosphate absorption in field workers and in persons exposed to pesticide spray drift. *In* E.A.H. VanHeemstra-Lequin and N.J. Van-Sittert (ed.) Toxicology letters: Biological monitoring of workers manufacturing, formulating and applying pesticides. Elsevier, Amsterdam.

Robinson, J. 1968. Organochlorine insecticides and birds. Chem. Br. 4:158–161.

Royal Society of Chemistry. 1987. The agrochemicals handbook. 2nd ed. R. Soc. of Chem., Nottingham, England.

Russell, E.W. 1973. Soil conditions and plant growth. 10th ed. Longman, London.

Sandman, T.M. 1986. Explaining environmental risks. Office of Toxic Substances, USEPA, Washington, DC.

Saunders, J.W. 1969. Mass mortalities and behavior of brook trout and juvenile atlantic Salmon in a stream polluted by agricultural pesticides. J. Fish. Res. Board Can. 26:695–699.

Schmitt, C.J., J.L. Ludke, and D.F. Walsh. 1981. Organochlorine residues in fish: National pesticide monitoring program, 1970–74. Pestic Monit. J. 14:136–155.

Schmitt, C.J., J.L. Zajicek, and M.A. Ribick. 1985. National pesticide monitoring program: Residues of organochlorine chemicals in freshwater fish, 1980–81. Arch. Environ. Contam. Toxicol. 14:225–260.

Scott, T.G., Y.L. Willis, and J.A. Ellis. 1959. Some effects of a field application of dieldrin on wildlife. J. Wildl. Manage. 23:409–427.

Sharma, L.N., and S.N. Saxena. 1972. Influence of dalapon (2,2-dichloropropionic acid) on biotic potential of soil. Andhra Agric. J. 18:74–78.

Sheail, J. 1985. Pesticides and nature conservation. Clarendon Press, Oxford, England.

Shorey, H.H. 1961. Synergism of carbamate insecticides against cabbage looper and beet armyworm larvae. J. Econ. Entomol. 54:1243–1247.

Sierra, M., and D. Santiago. 1987. Organochlorine pesticide levels in barn owls collected in Leon, Spain. Bull. Environ. Contam. Toxicol. 38:261–265.

Sierra, M., M.T. Teran, A. Gallego, M.J. Diez, and D. Santiago. 1987. Organochlorine contamination in three species of diurnal raptors in Leonn, Spain. Bull. Environ. Contam. Toxicol. 38:254–260.

Simon, L. 1971. Effects of herbicides on microorganisms. Folia Microbiol. 16:516.

Skare, J.U., J. Stenersen, N. Kveseth, and A. Polder. 1985. Time trends of organochlorine chemical residues in seven sedimentary marine fish species from a Norwegian fjord during the period 1972–1982. Arch. Environ. Contam. Toxicol. 14:33–41.

Smith, R.D., and L.L. Glasgow. 1963. Effects of heptachlor on wildlife in Louisiana. Proc. 7th Annu. Conf. South. Assoc. Game Fish Commun. 17:140–154.

Stanley, J.S. 1986. Broad scan analysis of the FY 1982 national human adipose tissue survey specimens. Vol. 1. Executive Summary. USEPA EPA-560/5-86-035. U.S. Gov. Print. Office, Washington, DC.

Stanley, J.S., K.E. Boggess, J.E. Going, G.A. Mack, J.C. Remmers, J.J. Breen, F.W. Kutz, J. Carra, and P. Robinson. 1986. Broad scan analysis of human adipose tissue for the EPA FY 1982 NHATS Repository. *In* F.C. Kopfler and G.F. Craun (ed.) Environmental epidemiology. Lewis Publ., Chelsea, MI.

Stewart, R.E., J.B. Cope, C.S. Robbins, and J.W. Brainerd. 1946. Effects of DDT on birds at the Patuxent research refuge. J. Wildl. Manage. 10:195–201.

Stickel, L.F. 1973. Pesticide residues in birds and mammals. p. 254–333. *In* C.A. Edwards (ed.) Environmental pollution by pesticides. Plenum Press, London.

Stickel, L.F., and L.I. Rhodes. 1970. The thin eggshell problem. p. 31–35. *In* J.W. Gillett (ed.) The biological impact of pesticides in the environment. Oregon State Univ. Press, Corvallis.

Stickel, L.F., W.H. Stickel, and R. Christensen. 1966. Residues of DDT in brains and bodies of birds that dead on dosage and in survivors. Science 151:1549–1551.

Stucky, N.P. 1970. Pesticide residues in channel catfish from Nebraska. Pestic. Monit. J. 4:62–66.

Swan, A.V. 1985. *In* W.W. Holland et al. (ed.) Oxford textbook of public health. Vol. 3. Oxford Univ. Press, Oxford, England.

Taha, S.M., S.A.Z. Mahmound. A.M. Abdel-Hafez, and A.S. Hamed. 1972. Effect of some pesticides on rhizosphere microflora of cotton plants. II: Herbicides Egypt. J. Microbiol. 7:53–61.

Tames, B.M., R. Magaw, and L.S. Gold. 1987. Ranking possible carcinogenic hazard. Science 236:271.

Taylor, J.C., and D.K. Blackmore. 1961. A short note on the heavy mortality of foxes during the winter 1959–1960. Vet. Rec. 73:232–233.

Teran, M.T., and M. Sierra. 1987. Organochlorine insecticides in trout (*Salmo trutta fario* L.) taken from four rivers in Leon, Spain. Bull. Environ. Contam. Toxicol. 38:247–253.

Terrierre, L.C., U. Kiigemagi, A.R. Gerlack, and R.L. Borovicka. 1966. The presence of toxaphene in lake water and its uptake in aquatic plants and animals. J. Agric. Food Chem. 14:66–69.

Thompson, A.R., and C.A. Edwards. 1974. Effects of pesticides on nontarget invertebrates in freshwater and soil. p. 341–386. *In* W.D. Guenzi (ed.) Pesticides in soil and water. SSSA, Madison, WI.

Tordoir, W.F., and E.A.H. VanHeemstra (ed.). 1980. 5th Int. Workshop Sci. Comm. Pesticides. Int. Assoc. Occupational Health. Elsevier Sci. Publ. Co., New York.

Tucker, R.K., and D.G. Crabtree. 1970. Handbook of toxicity of pesticides to wildlife. U.S. Fish Wildl. Serv. Resour. Publ. 84. U.S. Gov. Print. Office, Washington, DC.

Tulp, M.The., and O. Hutzinger. 1978. Some thoughts on aqueous solubility and partition coefficients of PCB, and the mathematical correlation between bioaccumulation and physicochemical properties. Chemosphere 7:849–860.

Turtle, E.E., A. Taylor, E.N. Wright, R.J.P. Thearle, H. Egan, W.H. Evans,and N.M. Soutar. 1963. The effects on birds of certain chlorinated insecticides used as seed dressings. J. Sci. Food Agric. 14:567–577.

U.S. Department of Health, Education, and Welfare. 1969. Report of The Secretary's Commission on pesticides and their relationship to environmental health. U.S. Gov. Print. Office, Washington, DC.

Vandenbosch, R. (ed.). 1977. Pesticide conspiracy. Prism Press, New York.

Van Heemstra, E.A.H., and W.F. Tordoir (ed.). 1982. Education and safe handling in pesticide application. *In* Proc. 6th Int. Workshop Sci. Comm. Pesticides. Int. Assoc. Occupational Health, Buenos Aires, Argentine. 12–18 Mar. 1981. Elsevier Sci. Publ. Co., New York.

Veith, G.D., D.W. Kuehl, F.A. Puglisi, G.E. Glass, and J.G. Eaton. 1977. Residues of PCB's and DDT in the Western Lake Superior ecosystem. Arch. Environ. Contam. Toxicol. 5:487–499.

Vinson, S.B., C.E. Boyd, and D.E. Ferguson. 1963. Resistance to DDT in the mosquito fish, *Gambusia affinis.* Science 139:217–218.

Wagner, S.L. 1981. Clinical toxicology of agricultural chemicals. Oregon State Univ., Corvallis.

Wainwright, M., and G.J.F. Pugh. 1975. Effect of fungicides on the numbers of microorganisms and frequencyof cellulolytic fungi in soils. Plant Soil 43:561–572.

Ware, G.W. 1978. Pesticides: Theory and application. W.H. Freeman and Co., San Francisco.

Ware, G.W. 1980. Effects of pesticides on nontarget organisms. Residue Rev. 76:173–201.

Warner, K., and O.C. Fenderson. 1962. Effects of DDT spraying for forest insects on Maine trout streams. J. Wildl. Manage. 26:86–93.

Weir, D., and M. Shapiro. 1981. Circle of poisons: Pesticides in people in a hungry world. Inst. for Food Development, San Francisco.

Westlake, W.E., and F.A. Gunther. 1966. Occurrence and mode of introduction of pesticides in the environment. ACS 60. Am. Chem. Soc., Washington, DC.

White, D.H., and A.J. Krynitsky. 1986. Wildlife in some areas of New Mexico and Texas accumulate elevated DDE residues 1983. Arch. Environ. Contam. Toxicol. 15:149–157.

White, D.H., C.A. Mitchell, and E.F. Hill. 1983a. Parathion alters incubation behavior of laughing gulls. Bull. Environ. Contam. Toxicol. 31:93–97.

White, D.H., C.A. Mitchell, and T.E. Kaiser. 1983b. Temporal accumulation or organochlorine pesticides in shorebirds wintering on the south Texas coast, 1979–1980. Arch. Environ. Contam. Toxicol. 12:241–245.

White, D.H., C.A. Mitchell, H.D. Kennedy, A.J. Krynitsky, and M.A. Ribick. 1983c. Elevated DDE and toxaphene residues in fishes and birds reflect local contamination in the lower Rio Grande Valley, Texas. Southwest. Nat. 28:325–333.

White, D.H., C.A. Mitchell, E.J. Kolbe, and W.H. Ferguson. 1983d. Azodrin poisoning of water fowl in rice fields in Louisiana. J. Wildl. Dis. 19:373–375.

White, D.H., C.A. Mitchell, E.J. Kolbe, and J.M. Williams. 1982. Parathion poisoning of wild geese in Texas. J. Wildl. Disease 18:389–391.

White, D.H., C.A. Mitchell, and D.W. Swineford. 1984. Reproductive success of black Skimmers in Texas relative to environmental pollutants. J. Field Ornithol. 55:18–30.

White, D.H., C.A. Mitchell, and C.J. Stafford. 1985. Organochlorine concentrations, whole body weights, and lipid content of black skimmers wintering in Mexico and in South Texas, 1983. Bull. Environ. Contam. Toxicol. 34:513–517.

Wiemeyer, S.N., R.M. Jurek, and J.F. Moore. 1986. Environmental contaminants in surrogates, food, and feathers of California condors (*Gymnogyps californianus*). Environ. Monit. Assess. 6:91–111.

Wiemeyer, S.N., T.G. Lamont, C.M. Bunck, C.R. Sindelar, F.J. Gramlich, J.D. Fraser, and M.A. Byrd. 1984. Organochlorine pesticide, polychlorobiphenyl and mercury residues in bald eagle eggs—1969–1979 and their relationships to shell thinning and reproduction. Arch. Environ. Contam. Toxicol. 13:529–549.

Worthing, C.R., and B. Waker (ed.). 1983. The pesticide manual. 7th ed. The Br. Crop Protection Counc. Lavenham Press Ltd., Lavenham, England.

Weed Science Society of America. 1983. Herbicide handbook. 5th ed. Weed Sci. Soc. of Am., Champaign, IL.

# Chapter 13

# Risk/Benefit and Regulations

**DAVID J. SEVERN,** *Clement Associates, Inc., Fairfax, Virginia*

**GARY BALLARD,** *U.S. Environmental Protection Agency, Washington, DC*

The Federal Insecticide, Fungicide and Rodenticide Act (FIFRA) is a licensing statute under which the USEPA grants the use of a pesticide according to conditions specified by a product label. Before registration is granted, all pesticides are evaluated under a "no unreasonable adverse effects" standard, taking into account the economic, social, and environmental costs and benefits of use. The purpose of this chapter is to describe the components of risk and benefits assessments that support this process.

Manufacturers are required to submit extensive scientific studies to support the registration of a pesticide. Part 158 of Title 40 of the Code of Federal Regulations establishes the types of data that must be submitted. These requirements include studies on the potential for short-term toxic effects, such as acute poisoning or skin and eye irritation; long-term effects such as tumor formation, birth defects, inheritable genetic damage or other adverse reproductive effects; hazards to wildlife; behavior of the pesticide in the environment after application; and the nature and quantity of pesticide residues likely to occur in food or animal feed as a result of use. Data from these studies are used to evaluate the extent of human and environmental exposure to pesticides and the possible consequences of that exposure.

Evaluations of pesticide data fall into two general categories: (i) those which concern new pesticide chemicals or applications to register new uses of previously registered pesticide and (ii) those which review the data supporting pesticides already on the market. For new pesticides or new uses, if the data are complete and demonstrate that a pesticide will not cause unreasonable adverse effects when used in accordance with widespread and commonly recognized use practices, a full registration is granted under section 3 of FIFRA.

However, data requirements and science assessment methods have changed over time, and the data base supporting an existing pesticide registration may not include acceptable studies for all required tests. Section 4 of FIFRA instructs USEPA to reexamine by current scientific standards the

 *Pesticides in the Soil Environment*—SSSA Book Series, no. 2.

health and environmental safety data base supporting all currently registered pesticides, and to reregister those that meet the no unreasonable adverse effects standard. In practical terms, "current scientific standards" are represented by the data requirements in 40 CFR 158, which are thus brought to bear on the reregistration of older pesticides.

The agency has also established a Special Review process for identifying those pesticides that may be causing unreasonable adverse effects (40 CFR 154). Risk criteria for initiating this process include potential for acute injury, induction of cancer or other chronic or delayed effects, and acute or chronic impacts on nontarget organisms or threats to endangered species. All of these potential risks are to be evaluated in the context of likely exposure to humans or wildlife. The Special Review process is designed to be an in-depth review of the risks and benefits of pesticide use whenever that use appears to pose a significant risk. The process allows for interested parties to submit evidence to modify USEPA's preliminary risk assessment and to evaluate the benefits of continued use, for general public comment, for review of the scientific data and conclusions supporting USEPA's proposed regulatory position by the FIFRA Scientific Advisory Panel and by the USDA, and finally for appeal to the USEPA administrator for an adjudicatory hearing and then to the courts.

## 13-1 HUMAN RISK ASSESSMENT

The risk assessment process employed by USEPA to make regulatory decisions on pesticides is a combination of separate assessments for hazard and for exposure, as shown schematically in Fig. 13-1. Hazard signifies the inherent toxicological properties of a pesticide, while risk is an evaluation of potential impacts of the pesticide on humans or wildlife, taking into account likely levels of exposure. In fact, since the toxicological properties of a pesticide are intrinsic to the chemical and its formulation, regulation of pesticides basically focuses on the determination of use conditions under which exposure is acceptable.

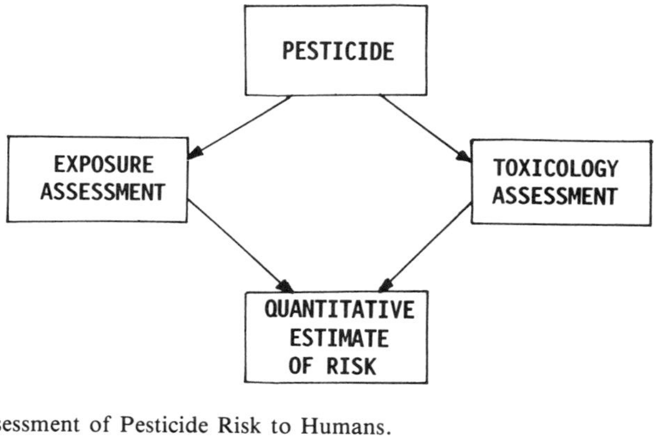


Fig. 13-1. Assessment of Pesticide Risk to Humans.

The toxicology studies normally required to support the registration of a pesticide are summarized in Table 13–1. In the evaluation of these studies, the lowest dose eliciting any type of adverse toxicological response is to be identified. This dose is known as the lowest observed adverse effect level (LOAEL). The dose immediately below the LOAEL in the same study must, by definition, elicit no toxicological response, and is called the *no observed adverse effect level* (NOAEL).

To establish tolerances for pesticide residues in food, the NOAEL is converted into an Acceptable Daily Intake (ADI). The ADI is defined as "the daily exposure level of a pesticide residue which, during the entire lifetime of man, appears to be without appreciable risk on the basis of all facts known at the time. The ADI is expressed in terms of milligrams of the pesticide, as it appears in the diet, per kilogram of body weight per day. "Without appreciable injury" is taken to mean the practical certainty that injury will not result even after a lifetime of exposure. The ADI may be changed as new information becomes available" (Paynter et al., 1975).

The NOAELs are converted into ADIs using a series of numerical factors often called *safety factors*; these factors allow for uncertainties in extrapolation from animals to humans and for the differences in sensitivities that exist within the human population. Examples of safety factors used to establish ADIs for various types of studies are listed in Table 13–2 (Farber, 1985).

More recently, the USEPA has developed the concept of reference dose (RfD) to replace the traditional ADI. The RfD is a benchmark dose opera-

Table 13–1. Toxicology data requirements for pesticide registration.†

| | |
|---|---|
| Acute studies | |
| Acute oral toxicity | Rat |
| Acute dermal toxicity | Rat |
| Acute inhalation toxicity | Rat |
| Primary eye irritation | Rabbit |
| Primary dermal irritation | Rabbit |
| Dermal sensitization | Rabbit |
| Acute delayed neurotoxicity | Hen |
| Subchronic studies | |
| 90-d feeding studies | Rodent and nonrodent |
| Chronic studies | |
| Chronic feeding | Rodent and dog |
| Oncogenicity | Rat and mouse |
| Teratogenicity | Two species |
| Reproduction | Two generation |
| Mutagenicity studies | |
| Gene mutation | |
| Structural chromosome aberration | |
| Other genotoxic effects | |
| Special studies | |
| General metabolism | |

† Adapted from 40 CFR 158.135.

Table 13-2. Standard safety factors for toxicological effects.†

| Effect | Safety factor |
|---|---|
| Cholinesterase inhibition based on 2-yr rodent or dog studies | 10 |
| Cholinesterase inhibition based on human NOAEL | 10 |
| General toxicity based on chronic studies | 100 |
| Teratogenic effects | At least 100 |
| General toxicity based on subchronic studies | 1000 |

† Note: These are minimal safety factors, and actual margins of safety are likely to be higher.

tionally derived from the NOAEL, using order of magnitude uncertainty factors reflecting the quality of the toxicology data used to determine the LOAEL (Dourson & Stara, 1983). The calculation is functionally the same as used to derive ADIs, but the use of uncertainty factor rather than safety factor emphasizes the scientific confidence or uncertainty in the process and avoids the connotation of "safety." A clear conclusion cannot be categorically drawn that all doses below the RfD are acceptable and that all doses above the RfD are unacceptable (USEPA, 1986a).

The direct expression of risk for the types of toxicological effects in Table 13-2 is the margin of safety (MOS) or margin of exposure (MOE), which is the ratio of the NOAEL to the actual or likely exposure encountered. A recent example of this type of risk assessment is the teratogenic risk assessment prepared for the special review of the herbicide cyanazine. Studies carried out in rabbits by the manufacturer and reviewed by USEPA were used to establish a NOAEL of 1 mg/kg per d by oral administration and 573 mg/kg per d by dermal administration. These two NOAELs were compared with numerical estimates of human exposure to cyanazine to calculate margins of safety. Using pesticide applicator exposure as an example, the special review estimated dermal exposure doses of 0.86 mg/kg per d when gloves were worn during application and 80 mg/kg per d when gloves were not worn. Dividing the dermal NOAEL by these exposure levels led to margins of safety of 670 and 7, respectively (USEPA, 1986b).

For those pesticides that are determined on the basis of submitted tests to be carcinogenic in laboratory animals, or for which epidemiologic investigations provide direct evidence of human carcinogenicity, a different procedure for assessing risk is used. First, the weight of evidence of the available carcinogenicity and/or epidemiologic data is reviewed in order to decide how likely a chemical is to be a human carcinogen. The USEPA has established categories of evidence based on human and animal data. For example, chemicals with sufficient human evidence of a causal association between exposure and cancer are classified in category A, while chemicals for which either some or no epidemiologic data are available but for which animal studies are sufficient to show carcinogenicity are classified in category B1 or B2 (USEPA, 1986c).

For chemicals judged to be carcinogenic, an extrapolation from the high doses used in the animal feeding studies to much lower human exposure levels is then carried out to arrive at estimates of risk. Uncertainty is inherent in this process as well, arising from the necessity to extrapolate both from animals to humans and from the high doses used in the animal studies to the much lower likely levels of human exposure. The risks and the associated doses are presented in tabular form to help regulatory authorities to determine the level of risk that can be tolerated after consideration of scientific and other factors. The basic principles for this type of risk assessment have been discussed by the National Academy of Sciences (NAS, 1983) and in EPA's Guidelines for Carcinogenic Risk Assessment (USEPA, 1986c).

A recent example of a regulatory assessment carried out according to these procedures is the estimate of excess cancer risk from exposure to the herbicide alachlor. The USEPA concluded in a special review of alachlor that studies carried out in both rats and mice demonstrated significant oncogenic responses, and therefore classified alachlor as a probable human carcinogen (category B2). A dose-response extrapolation from the observed tumor rates in the animals treated at a range of dose levels yielded a value of $8 \times 10^{-2}$ for the parameter used to calculate an upper bound of the probable risk associated with exposure to alachlor. This value was then used to construct a table of lifetime cancer risks associated with exposure to alachlor in drinking water (USEPA, 1986d). The USEPA's final regulatory decision concluded that the risk associated with drinking water containing alachlor in the 0.2 to 2.0 $\mu$g/L range will generally not exceed a risk of one in one million (USEPA, 1987).

## 13-2 ECOLOGICAL RISK ASSESSMENT

The wildlife toxicity studies normally required to support the registration of a pesticide are summarized in Table 13-3. These studies include acute, subacute, reproduction, and simulated and full field studies arranged in a hierarchial or tier system that progresses from the basic laboratory tests to the applied field tests. The results of each tier are evaluated to determine the potential of the pesticide to cause adverse effects and to determine whether further testing is needed. The short-term test results are used to determine the acute toxicity effect levels and to compare these levels with likely environmental residue levels, to assess potential impacts on aquatic and terrestrial wildlife. The long-term and field studies may be required to estimate the potential for chronic effects, taking into account measured or estimated residues in the field.

An extensive and detailed description of the components of the ecological risk assessment processes used for pesticide registration decisions has recently been published (USEPA, 1986e). It describes the rationale for the particular species to be tested, the tiered relationships among the various studies, and the different exposure assessment approaches needed to evaluate risk to both aquatic and terrestrial populations. For aquatic populations, this risk assessment may take several forms. The simplest form is a ratio between an aquatic toxicity endpoint (either acute or chronic) and the concentration that would be achieved if a pesticide were to be applied uniformly

Table 13-3. Wildlife and aquatic organism data requirements.

| |
|---|
| Avian and mammalian data |
| Avian single-dose oral $LD_{50}$ |
| Avian dietary $LC_{50}$ |
| Wild mammal toxicity |
| Avian reproduction |
| Simulated and actual field testing |
| Honeybee acute contact $LD_{50}$ |
| Honeybee foliar residue toxicity |
| Aquatic organism data |
| Freshwater fish acute toxicity |
| Freshwater invertebrate acute toxicity |
| Estuarine/marine organism acute toxicity |
| Fish early life stage study |
| Aquatic invertebrate life-cycle study |
| Aquatic organism accumulation |
| Simulated or actual field testing |

† From 40 CFR 158.145.

to a pond. If the ratio is greater than unity under this worst-case approach, then the presumption is made that the pesticide will not pose any risks to aquatic organisms.

If this ratio is less than unity, then the assessment would proceed to the next step, in which preliminary estimates of likely surface runoff or spray drift contributions to a pond are used to estimate an environmental concentration for comparison with aquatic toxicity endpoints. The same ratio approach is used; if the comparison still suggests that a potential risk to aquatic organisms might result from the use of a pesticide, then more sophisticated testing for exposure and effects, perhaps including simulated or actual field testing, would be carried out.

Similar approaches are used for the evaluation of risk to terrestrial and avian species, using literature values or actual field studies to estimate likely residues in wildlife diets for comparison with toxicity endpoints. A standard tabulation of residue levels of 28 pesticides on 60 crops (Hoerger & Kenaga, 1972) is used to calculate maximum residues likely to occur immediately after application. Specific residue information on the pesticide of interest is used when available to corroborate this generic approach. Standard dry feed consumption factors are then used to calculate the dietary exposure of species of interest.

## 13-3 ENVIRONMENTAL FATE AND TRANSPORT

A battery of environmental fate studies is required to support the registration of pesticides used outdoors; these studies are summarized in Table 13-4. The purpose of this test battery is to account for the behavior of the pesticide after its release into the environment, by tracing all pathways of degradation and transport. The ultimate goal of the testing scheme is to enable conditions of use to be established such that unacceptable environmental exposure is avoided.

Table 13-4. Environmental fate data requirements.†

| | |
|---|---|
| Degradation data | |
| Hydrolysis | Laboratory test |
| Photodegradation in water | Laboratory test |
| Photodegradation on soil | Laboratory test |
| Photodegradation in air | Laboratory test |
| Metabolism data | |
| Aerobic soil metabolism | Laboratory test |
| Anaerobic soil metabolism | Laboratory test |
| Anaerobic aquatic metabolism | Laboratory test |
| Aerobic aquatic metabolism | Laboratory test |
| Mobility data | |
| Adsorption/desorption data | Laboratory test |
| Volatility | Laboratory test |
| Volatility | Field study |
| Dissipation data | |
| Soil dissipation | Field study |
| Aquatic dissipation | Field study |
| Forestry dissipation | Field study |
| Long-term soil dissipation | Field study |
| Accumulation data | |
| Rotational crop uptake | Laboratory or confined plot |
| Rotational crop uptake | Field study |
| Irrigated crop uptake | Field study |
| Fish accumulation | Laboratory test |
| Aquatic nontarget uptake | Field study |

† From 40 CFR 158.130.

The major pathways of environmental transport are shown schematically in Fig. 13-2. It is to be understood that conversion to daughter products or complete degradation may occur at any point in the scheme. The environmental fate review of a pesticide thus addresses the following questions:

1. Where does the pesticide go?
2. How long does it persist?
3. What is its pattern and rate of degradation?

Environmental fate studies were first required for pesticide registration in 1970 (USDA, 1970). After 10 yr of discussion on the most useful set of data for regulatory purposes and on the best scientific procedures for carrying out the various studies, guidelines (recommended testing protocols) were published in 1982 as a nonregulatory companion to the formal data requirements of Part 158 (USEPA, 1982a). The general strategy of the test battery is that laboratory tests are to be carried out under controlled conditions in order to evaluate individual degradation pathways and identify and characterize degradates. In most cases, radiolabeled pesticide active ingredients are used for these studies. If individual degradates are formed in significant amounts, then separate environmental fate studies using these degradates as test materials may be required. The field studies are then performed, using the knowledge gained from the laboratory studies to guide the detailed protocol design and to select the appropriate degradates for quantitation of environmental residue levels.

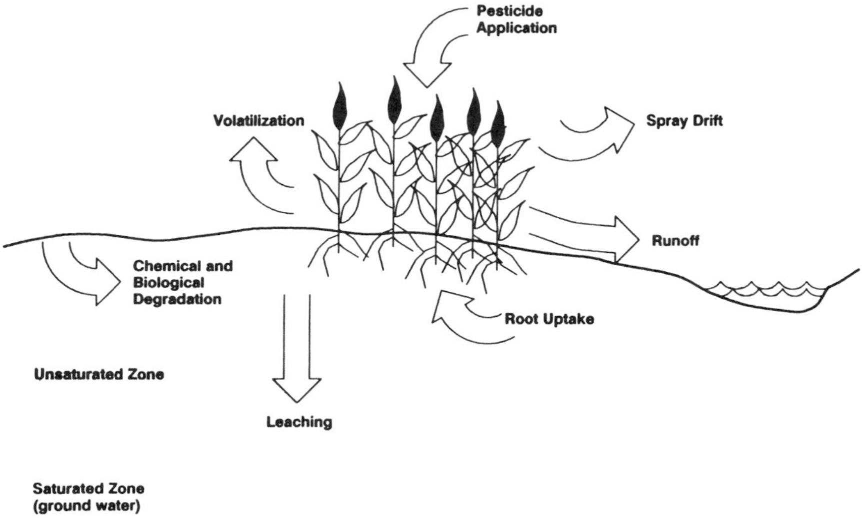


Fig. 13-2. Pathways of pesticide degradation and transport.

Pesticides are released into the environment as formulated products, which are designed to deliver the active ingredient to the target insect, weed, or other organism as efficiently as possible. Major formulation types include emulsifiable concentrates and wettable powders, to be diluted in water or other carriers for spraying, dusts and granular formulations designed for direct application, and encapsulated products designed to maintain a slow steady release of the active ingredient. The type of formulation employed has the ability to modulate the interaction of the active ingredient with the target pest and with the environment. Thus, while the laboratory portion of the environmental fate test battery is to be carried out using the pesticide chemical itself, the field studies use representative formulation types as the materials to be tested in order to investigate the behavior of the pesticide under conditions reflecting actual use.

The hydrolysis and photolysis studies evaluate chemical patterns of degradation. The potential for hydrolysis is a key environmental determinant, since transportation pathways so often lead to water, either through direct application to water bodies, spray drift, overland runoff, or leaching, or through direct contact with soil moisture. Protocols for the hydrolysis study are designed to reflect realistic environmental conditions: temperature of 25 °C and pHs of 5, 7, and 9. Irradiation by sunlight may result in pesticide degradation either in water, on foliar or soil surfaces, or in the air. Either natual sunlight or artificial light sources may be used in the photolysis studies, but if artificial sources are used the emission spectrum and light intensity must reflect natural radiation.

Metabolism studies investigate the pattern of pesticide decay caused by soil and water-borne microorganisms. The enormous diversity of agricultural soils and their microbial populations renders its necessary to conduct degradation studies in soils representative of proposed uses. The test conditions must reflect realistic conditions of soil moisture and temperature, as well as both aerobic and anaerobic conditions. The overall goal of the studies is to identify residues occurring in soil down to 10 parts per billion, although this may not be feasible in all cases. Isotopically labeled pesticides are usually used in these studies, so that all of the chemical applied to the soil may be accounted for, including the evolution of $CO_2$ or other volatile degradates. The usual duration of the soil metabolism studies is 1 yr, to establish patterns of decline of the parent pesticide and formation and decay of degradates. Residues that may persist beyond a crop season and be available for uptake by a subsequently planted crop will thus be identified for quantitation in the rotational crop uptake studies.

For pesticides intended for direct application to water, similar studies are carried out using water and sediment characteristic of proposed use sites.

Mobility studies have as the principal goal the direct evaluation of pesticide transport through soil, so that the likelihood of the pesticide leaching to groundwater after application may be evaluated. The diversity of agricultural soil types requires that several soils be tested for pesticide mobility; four different soils ranging from agricultural sand to clay loam are usually tested. These tests are intended to provide a direct measure of the Freundlich adsorption constant $K_d$; the Pesticide Guidelines, however, allow either a direct adsorption/desorption procedure or the less direct procedures of soil column or soil thin-layer chromatography to be used. In the latter two cases, relationships between $R_f$, $K_{oc}$, and $K_d$ may be used to convert the experimental results into a value for $K_d$ (Hamaker, 1975).

These degradation, metabolism, and mobility studies are all performed in the laboratory. The dissipation studies are then carried out under actual field conditions, using information on identification of metabolites and degradates gained from the laboratory studies. These studies "synthesize" all of the individual transport and dissipation pathways into an overall quantitative description of pesticide environmental behavior. The studies are usually carried out at two agricultural sites representative of the pesticide use pattern; data on overall movement and decline of residues are usually collected for 78 wk.

These studies use conventional chemical residue techniques to analyze the soils. However, for some pesticides whose recommended application rates may be low, such conventional techniques may not be sufficiently powerful to detect residues of concern. A method that can investigate the fate of radiolabeled pesticides under actual field conditions has been developed by Harvey (Harvey, 1980). This method uses stainless steel cylinders driven into the soil at the site of interest to isolate an undisturbed column of soil, to which radiolabeled pesticides may be applied under confined field conditions.

A final series of environmental fate and transport studies examines the potential for pesticide residues to bioaccumulate in subsequent crops planted in fields where pesticides have previously been applied and in fish where the potential for aquatic impacts from pesticide use exists. Laboratory rotational crop accumulation studies using radiolabeled pesticides examine the ability of persistent residues to be transported into various portions of follow crops. If significant residues of parent pesticide and/or its metabolites or degradates (but not radioactivity incorporated into natural plant constituents) are detected in the laboratory studies, then field studies under actual rotational cropping conditions are performed to evaluate the actual extent of this uptake. The field studies may take the same form as residue studies carried out in support of pesticide tolerances, described later in this chapter, if a tolerance to cover residues in the rotated crop is requested.

Fish accumulation studies are typically carried out for pesticides which have the potential because of their use pattern to reach water by direct application or by spray drift or runoff, and which have the environmental properties of persistence in water and high octanol/water partition coefficient. The preferred test method exposes bluegill sunfish or channel catfish to radiolabeled pesticides in a flow-through system for 28 d, with a subsequent depuration period of 14 d. Radioactivity is determined in whole body, edible tissue, and viscera as a function of time for the uptake and depuration phases; and residues occurring at highest levels are chemically identified. The results of the study are used to evaluate the likely extent of pesticide residues in fish used as human food sources.

The above-described studies, along with more specialized designed to evaluate pesticide fate and transport in other environments such as forests or irrigated crops, constitutes the base of environmental data that is assembled in order to evaluate the proposed registration of a pesticide. These data serve as essential input to exposure assessments that may be required for risk assessment activities during the registration, reregistration, or Special Review processes.

## 13–4 ENVIRONMENTAL EXPOSURE ASSESSMENTS

The transport and fate data from the studies described in the previous section, combined with information on pesticide use rates, application practices, and extent of use, form the basis for environmental exposure assessments. These assessments are usually media-specific, since specific populations, such as the population of a small town which obtains its drinking water from a well in an agricultural area or fish in a river receiving agricultural runoff, are commonly identified as target populations for which exposure and risk assessments are needed. Among the most common situations for which an articulation of pesticide exposure is needed are the direct exposure of pesticide applicators and fieldworkers, assessments of spray drift and deposition, levels of residues that foraging birds or freshwater fish are exposed to, and human dietary exposure via the food supply or drinking water (Fig. 13–3).

### 13–4.1 Applicator Exposure

Direct contact with formulated products and spray-diluted mixtures represents the most immediate route of exposure to pesticides. Though limited

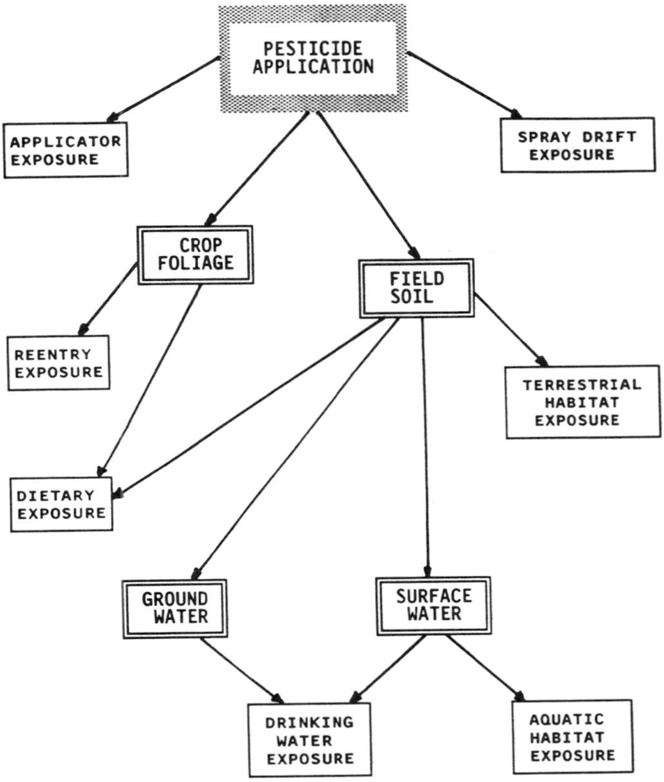


Fig. 13-3. Pathways of pesticide exposure.

to those who mix and load or apply pesticides, this route has high potential for dermal and inhalation exposure. Over approximately the last 10 yr, a fairly standard approach towards assessment of applicator exposure has been used in the regulatory process for pesticides (Reinert & Severn, 1985). This approach may follow one of two basic elements: either direct measurement of applicator exposure through field-monitoring studies, or utilization of extant "surrogate" data obtained from studies with other pesticides or even nonpesticidal chemical exposure situations.

A number of field techniques have been developed to measure dermal and inhalation exposure directly (Davis, 1980). For dermal exposure, the usual procedure is to attach pads of absorbent material such as gauze or cellulose, protected by impervious backing, to the clothes of the persons being monitored. After the exposure period is completed, the center portion of the pads is cut out for residue analysis. For the hands, which usually are the highest exposed part of the body, both cotton gloves (worn under normal work gloves) and periodic handwashing techniques have been used. Inhalation exposure is usually monitored by air samplers which collect air in the breathing zone of the person being monitored and pump it through an absorbent agent such as charcoal or polyurethane foam plugs. These techniques have been reviewed and incorporated into a new section of the Pesticide Assessment Guidelines (USEPA, 1986f).

Many studies of applicator and mixer/loader exposure have been carried out using these techniques (see for example Wolfe et al., 1967; Maddy et al., 1985). In general, the experience gained from these studies shows that dermal exposure is usually much greater than inhalation exposure, and that most of the dermal exposure is to the hands. For application techniques for which a large number of studies have been carried out, a clear central tendency of the dermal deposition data is apparent, independent of the particular pesticides used in the studies. Such a central tendency would be expected in many spraying situations, because the controlling factors are the delivery system and the physical dynamics of the spray mixture rather than the chemical properties of the particular pesticide. For this reason, compilations of exposure data have been used as surrogate data to predict likely exposure in the absence of actual field studies. In the best studied example, an extensive array of data on dermal exposure during application by airblast spray equipment demonstrates a strong correlation between the rate of pesticide application per acre and resulting dermal exposure (Reinert & Severn, 1985).

In contrast to the above "passive dosimetry" techniques that measure external deposition on the skin or air concentrations available for inhalation exposure, biological monitoring of exposed individuals (most commonly by analysis of urinary metabolites) can, in principle, lead directly to the absorbed dose. The great advantage of this approach is that it avoids the problem of estimating the efficacy of clothing as a protective barrier and the extent of absorption through skin or by the lung. On the other hand, interpretation of observed metabolite levels in urine requires knowledge of the pharmacokinetics of the pesticide in question, and the actual route and source of exposure cannot always be determined.

The advantages and disadvantages of the passive dosimetry and biological monitoring approaches have been discussed in detail (USEPA, 1986f). Depending on the exposure situation under consideration, either of these general approaches may provide acceptable data on applicator exposure which can serve the needs of the risk assessment process.

### 13-4.2 Reentry Exposure

A different mode of exposure to pesticides may occur when agricultural workers enter sites where pesticides have been previously applied. In this situation the exposure is not to the pesticide itself or to a dilute spray solution but rather to a weathered and environmentally modified residue present on foliar surfaces or soil. Episodes of fieldworker poisonings arising from organophosphate residues in tree fruit crops were first reported in the early 1950s (Abrams & Leonard, 1950; Ingram, 1951). Since that time the phenomenon has been extensively studied (Federal Working Group on Pest Management, 1974; Gunther et al., 1977; Popendorf, 1980).

For purposes of exposure and risk assessment, the environmental processes of dissipation and human contact need to be evaluated. The residues with highest potential for fieldworker exposure are dislodgeable residues on lead surfaces; methods for measuring these residues involve collecting leaf punches and washing off the residue with detergent for residue analysis (Gunther et al., 1973; Iwata et al., 1977). Surface soil residues can also lead to fieldworker exposure, particularly during harvesting of crops growing close to the ground; procedures for determination of residues available for exposure have also been published (Spencer et al., 1977; Berck et al., 1981).

Some reentry exposure monitoring studies have been carried out to measure the extent of dermal contact with dislodgeable residues (Spear et al., 1977; Gunther et al., 1977; Zweig et al., 1983). The techniques used to measure this exposure are similar to those described above for applicator exposure. Although such monitoring studies indicate a wide variability in dermal exposure, results of these studies allowed Popendorf (1980) to establish a correlation between the level of dislodgeable residues on citrus leaves and the resulting dermal exposure of harvesters.

A combination of studies measuring the temporal decline of dislodgeable residues in field sites and the relationship between this residue level and the corresponding dermal exposure experienced by the harvesters allows an exposure assessment as a function of time to be constructed. Toxicology data for a pesticide are used to determine an Allowable Exposure Level; and comparison of this Level with the time dependence of exposure allows a determination of the elapsed time after pesticide application at which the actual exposure is not likely to exceed this level. This elapsed time is called the Reentry Interval; such intervals are established as a condition of pesticide registration in cases where pesticide use patterns and toxicology properties indicate a potential for reentry hazard. Guidelines for carrying out studies and proposing reentry intervals have recently been published (USEPA, 1984a). Although the process of reentry exposure is complex and influenced by meteorological variables, formulation types, and details of fieldworker activities, this procedure represents a rational method of evaluating reentry exposure and determining safety precautions. A similar method for determining reentry intervals based on direct observation of clinical signs of poisoning leads to reentry intervals of the same magnitude (Iwata et al., 1982).

### 13-4.3 Spray Drift

Spray drift is of concern as a route of both wildlife exposure and direct human exposure, and includes consideration of downwind swath deposition and extended airborne displacement. Two basic types of studies are needed to evaluate the potential for spray drift during the pesticide registration process: droplet spectrum analysis and field deposition studies; these studies are required when humans, plants, or animals may be readily exposed by aerial transport (40 CFR 158). Though aerial application is the most likely source of spray drift, some types of surface application, such as orchard airblast application or spray irrigation, may also lead to off-target drift and deposition (Holst, 1984; Nigg et al., 1984).

Protocols for measuring the droplet size spectrum and for monitoring drift and deposition during actual pesticide application have been published (USEPA, 1984d). The droplet size spectrum study measures the size range of droplets produced by pesticide formulations under specific conditions of nozzle type and wind shear velocity. The field deposition studies typically involve multiple swath passes over a target area to evaluate the effects of weather, topography, and vegetation on the droplet size and deposition. Data on droplet sizes and deposition patterns are used as inputs to mathematical models which predict foliar and aquatic concentrations or direct human exposure (USEPA, 1986e; Ghassemi et al., 1982).

### 13–4.4 Groundwater

One of the more recent issues facing the pesticide regulatory process is the assessment of the potential for agricultural chemicals to leach through soil and appear in groundwater. Roughly half the U.S. population relies on groundwater as a source of drinking water, and in rural areas this proportion may reach 90% or more. In the last few years, several pesticides have been detected in groundwater under conditions where agricultural use could be clearly identified as the source of the contamination (Cohen et al., 1984, 1986).

The data obtained from the environmental fate studies described earlier in this chapter form the basis for the assessment of groundwater contamination potential. Pesticides which are both persistent and mobile in soil have the potential to reach groundwater, though of course both persistence and mobility are influenced by soil type and temperature, rainfall and recharge, crop type, and other factors. Cohen et al. (1984) concluded that pesticides with hydrolysis half-lives of more than about 24 wk, soil half-lives of more than about 2 to 3 wk, and soil absorption constants of less than about five have considerable potential to leach.

Mathematical models are necessary to organize all of the pesticide-specific and site-specific information that contributes to an assessment of leaching potential. The Pesticide Root Zone Model (PRZM) (Carsel et al., 1985) is currently used by EPA to assess leaching potential. The PRZM is a continuous simulation model which uses actual rainfall data on a daily basis; it incorporates the USDA-SCS curve number technique to estimate the amounts of infiltration and runoff. The PRZM has been used in connection with a model describing transport in the saturated zone to predict maximum concentrations of several pesticides in groundwater underneath the site of application (Donigian & Carsel, 1987).

Many different modeling approaches are available to evaluate the downward transport of chemicals from the land surface. A detailed discussion of modeling is presented in chapter 10 by Wagenet and Rao.

Assessments of groundwater contamination potential and resulting human exposure used in the pesticide regulatory process rely on both modeling and actual groundwater monitoring data. A recent example of such an assessment is the Special Review Position Document for alachlor, in which extensive monitoring data and PRZM modeling simulations led USEPA to conclude that alachlor has the potential to leach below the root zone in the case of soybean [*Glycine max* (L.) Merr.], but is likely to be sufficiently retained by the deeper corn root zone to enable microbial degradation to dissipate soil residues (USEPA, 1986d).

### 13-4.5 Surface Water

Runoff of pesticide chemicals from application sites has been studied intensively for many years. Pesticides can travel both in the aqueous phase and while adsorbed to soil particles that are entrained by erosion. As with transport to groundwater, the potential for surface transport by runoff is governed both by persistence and mobility properties of pesticides and by meteorological and soil conditions at the site. No single chemical or physical property of a pesticide uniquely determines its potential for transport via runoff, and there is no single laboratory or field test that pesticide registrants are required to perform to evaluate runoff potential.

The Office of Pesticide Programs relies on both modeling approaches and monitoring studies to evaluate potential human and environmental exposure from runoff and surface water contamination. A recent example of an exposure assessment for fish-eating birds is a simulation of DDT runoff and aquatic concentrations resulting from the application of dicofol containing DDT that was carried out for the Arroyo Colorado waterway in southern Texas (USEPA, 1984b). Runoff losses of DDT from the application of dicofol to citrus groves in the drainage area were estimated to be about 2% of DDT applied, based on an extensive review of the literature on pesticide runoff (Wauchope, 1978). The Exposure Analysis Modeling System (EXAMS) was then used to estimate a steady-state concentration of DDT in the river basin. The application of known fish bioconcentration factors resulted in an estimate of DDT in fish tissue that exceeded dietary levels shown to reduce eggshell thickness.

## 13-5 DIETARY EXPOSURE

Exposure to pesticides in the food supply is regulated by the establishment of tolerances for pesticide residues in foods and animal feeds. A tolerance is the maximum residue concentration legally allowed for a pesticide and its metabolites or degradation products in or on a particular raw agricultural commodity, processes food, or animal feed item. Tolerances are set under the authority of the Federal Food, Drug and Cosmetic Act (FFDCA). Section 408 of FFDCA applies to residues on raw agricultural commodities, and Section 409 applies to processed foods or feeds.

Section 409 contains the well-known Delaney Clause, which specifically prohibits the use of cancer-causing agents as food or feed additives. This prohibition is an absolute risk standard that does not allow consideration of benefits in deciding whether to grant use of a pesticide for food-processing operations. Moreover, some pesticide residues are concentrated above levels allowed in raw agricultural commodities during some food processing operations; these residues are also regulated under Section 409 and thus are subject to the no-risk provision of the Delaney Clause. Recently, the National Academy of Sciences reviewed these issues and concluded that current law and regulations are inconsistent in regulating residues in raw and processed foods (NAS, 1987).

The data needed to obtain a tolerance are listed in Table 13-5.

The first consideration in evaluating dietary exposure is the characterization of the actual residue remaining on the treated crop. The composition of the final residue must be determined before a complete residue detection methodology and quantitative field residue data can be generated. The use of radiolabeled pesticide chemicals is usually the only way to identify the terminal residue. The site of the radiolabel must ensure that no significant unlabeled residue would escape detection; ring labeling is therefore preferred for most aromatic or cyclic compounds, and tritium labeling is usually unsatisfactory.

The results of the radiochemical studies are used to identify the total toxic residue in or on the treated commodity. This term is used to describe the sum of the parent pesticide and its degradation products, metabolites (which may be either free or bound to the substrate), and impurities of toxicological concern. All of these components of the residue, except for plant metabolites, are toxicologically characterized by the set of toxicology studies described earlier in this chapter. If it happens that plant metabolic pathways produce products different from those found in animal metabolism studies (and which would thus not contribute to the spectrum of toxicological response observed in animal feeding studies), then separate feeding studies would have to be carried out using these metabolites. All components of the total toxic residue are included in the tolerance expression for the pesticide, and residue analytical methods must be developed for each component.

Residue characterization studies are carried out in plants to identify the components and understand the distribution of the residue. Residues of a pesticide applied to crop plants may remain largely on plant foliage, or may be absorbed through the foliage or taken up through the root system. The

Table 13-5. Residue chemistry data requirements.

| | |
|---|---|
| Chemical identity of the pesticide | |
| Nature of the residue: | In plants |
| | In livestock |
| Residue analytical method | |
| Magnitude of the residue: | Crop field trials |
| | Processed foods and feeds |
| | Meat/milk/poultry/eggs |
| | Potable water |
| | Irrigated crops |
| | Food handling |
| Residue reduction studies | |

residue studies must thus include analysis of all plant parts used for food or feed. Since different plants have different metabolic processes, metabolism studies are needed for each type of plant for which use is proposed. For example, metabolism studies in bean (*Phaseolus* sp.) plants are considered to be representative of all legumes, but the study results would not be applicable to root crops such as potato (*Solanum tuberosum* L.) or carrot (*Daucus carota*).

Metabolism studies are also carried out for domestic animals, to characterize residues in edible tissue of livestock, milk, and eggs. These studies are carried out whenever a pesticide is registered for direct application to livestock or animal premises, or when residues occur in crops used for animal feed. In general, separate studies are carried out for ruminants and poultry, usually in goats and chickens. Other studies are needed if the pesticide is to be applied directly to livestock, since dermal or inhalation exposures may result in different metabolic patterns than oral dosing. Of primary concern are those items such as muscle, liver, kidney, and fat tissues that are direct human food items or that can be rendered into animal foods and thus reenter the food chain.

Once the nature of the residue has been determined and acceptable methods have been devised to analyze the appropriate tissues, then field studies to quantify the level of residue are carried out. These studies are performed under the actual conditions of the proposed use, using actual formulations and maximum application rate, and reflecting as closely as possible the actual mode of application and agricultural conditions. For example, some crops may be grown with little or no rainfall. In such a case, the field study would reflect these conditions and additional field studies using irrigation would be needed to extend the use of the proposed product to irrigated crop conditions.

Detailed protocols or guidelines describing these studies have been published by USEPA as part of the Pesticide Assessment Guidelines (USEPA, 1982b).

Based on data from the field studies, and perhaps on additional studies on the effect of food processing on the identity and level of residues, a pesticide tolerance is established for a raw agricultural commodity if it can be shown that the maximum residue contribution to the human diet will not exceed the ADI. The dietary exposure assessment thus takes the form of a Theoretical Maximum Residue Contribution (TMRC), in which the tolerance level for an individual food commodity is multiplied by the amount of the commodity that a person would eat. The amounts of individual commodities in the diet are expressed as *food factors*, which are estimates of average per capita food consumption. The food factors are based primarily on a 1965 to 1966 USDA Nationwide Food Consumption Survey, and are calculated on the basis of a 60-kg person who eats 1500 g each day. For cases where several tolerances have been established for different commodities, the TMRC is the sum of the residue contributions for each of these commodities.

Several assumptions are, of necessity, built into this dietary exposure assessment procedure. The TMRC calculation assumes that 100% of the crop is treated, that the residue is not decreased by cooking or other processing, and that pesticide residues exist up to the level of the tolerance. It must also be assumed that everyone eats a standard 1.5 kg diet, of which various food groups comprise constant portions and weighs 60 kg.

Often these assumptions are not representative of actual consumption patterns nor of actual pesticide residue levels in the food as consumed. For these reasons, a new dietary exposure assessment procedure called the Tolerance Assessment System (TAS) is being developed in the USEPA (D.S. Saunders and B.J. Peterson. 1987. Unpublished data.). The central element of this new system is a greatly expanded tabulation of actual food consumption data. These data were collected by the USDA for its 1977 to 1987 Nationwide Food Consumption Survey. In that survey, approximately 30 000 individuals were asked to record what they ate for three consecutive days. Some 370 distinct food items were reported to have been consumed at least once.

To convert these data into a form usable for dietary exposure assessment, each food item was related back to the appropriate raw agricultural commodity. Thus, for example, flour in a baked food item represents an amount of wheat (*Triticum aestivum* L.) consumption, and sugar consumption was translated to the appropriate amounts of beet (*Beta* sp.) and cane. Finally, the consumption of each commodity by each respondent in the survey was divided by the respondent's body weight, to calculate the intake of each commodity in terms of grams of food eaten per kilogram of body weight.

The individuals in the USDA survey were selected to be statistically representative of different age, sex, and ethnic groups, as well as of different geographic regions of the USA. The TAS can provide estimates of consumption for different subpopulations, such as children from the age of 1 to 6 yr.

Actual pesticide dietary intakes are calculated from the TAS data base by multiplying the amount of a commodity consumed by the pesticide residue level in the commodity. The residue levels used as input for TAS may be the tolerance level, the actual observed levels from food-monitoring studies, or either of these levels corrected for the percentage of crops treated with the pesticide.

The TAS can generate assessments of either chronic (yearly) or acute (daily) dietary exposure. These two types of consumption estimates yield extremely different numbers. For example, the total amount of a particular commodity consumed in an entire year, when divided by 365, yields an estimate of chronic exposure, the average daily consumption. On the other hand, on the days that the commodity is actually eaten, the consumption for that day may be much higher, since consumption is not constant throughout the year.

These exposure values may then be used in conjunction with the toxicological evaluation of a pesticide to derive estimates of risk, in the same manner as for other routes of exposure. If the toxicological effect of concern is an acute effect only, or can be engendered as a result of a single instance of exposure, then the TAS daily exposure estimates would be used, while for chronic health effect endpoints the corresponding chronic exposure estimates would be appropriate.

## 13-6 BENEFITS ASSESSMENT

The benefit component of the risk/benefit assessment procedure arises from the criteria for registerability of pesticides under FIFRA. Section 3(c)(5) states that EPA shall register a pesticide when it determines that:

> . . .it will perform its intended function without unreasonable adverse effects on the environment.

Section 2(bb) of FIFRA defines unreasonable adverse effects on the environment to be:

> . . .any unreasonable risk to man or the environment, taking into account the economic, social, and environmental costs and benefits of the use of any pesticide.

With this definition, Congress made it clear that society has a need to trade off risks or costs against the benefits of pesticide use.

The term *benefits* needs to be carefully defined in the context of pesticide regulation. In the normal lexicon of economics, the term is usually encountered in the context of a cost/benefit analysis of a government regulation, and refers to the gain or value to society from the governmental action. For example, the benefit may be a reduction in the number of smoggy days resulting from a limitation in smokestack emissions. However, the term benefits as used in FIFRA refers to the value of the pest control provided by the pesticide use. The benefits are measured in terms of improved agricultural productivity with respect to yields and benefits, reduction in nuisance pests, or other such measures. Unless otherwise stated, the following discussion will use the term within this latter meaning.

In May 1976, the USEPA issued an interim guideline for carrying out economic impact analyses of proposed regulatory decisions on pesticides (USEPA, 1976). The content of the guideline defined the factors to be considered and the procedures to be used in assessing the economic impact resulting from future regulatory actions affecting carcinogenic pesticides. While specifically addressing only pesticides shown to cause cancer in laboratory animals, the guidelines became the basis for all economic impact analyses prepared by USEPA, regardless of the type of hazard assessment under consideration. As discussed earlier in this chapter, unreasonable adverse effects triggering the Special Review process may include a variety of potential acute or chronic human health effects or potential environmental damage, including effects on fish, wildlife, or the physical environment. Economic analyses of proposed regulatory actions under the guideline focused on the anticipated loss in benefits to users from cancelling the registration or otherwise restricting the use of a pesticide. The guideline specifically excluded, however, the consideration of the impacts of regulatory decisions on pesticide manufacturers.

The guideline specifies the content of an economic impact analysis as follows:

1. Identification of the major uses of the pesticide, including estimated quantities used by crop or other application.
2. Preliminary identification of the minor uses of the pesticide, including estimated quantities used by category, such as lawn and garden or household uses.
3. Identification of registered alternative products for these uses, including an estimate of their availability.
4. Determination of the change in costs to the user of providing equivalent pesticide treatment with any available substitute products.
5. Assessment of the impact of a proposed regulatory decision upon user productivity (e.g., yield per acre or total output) from using available substitute pesticides or if no pesticide is used.
6. If the impacts upon either user costs or productivity are significant, then a qualitative assessment of the impact of the regulatory decision on production of major agricultural commodities and retail food prices of these commodities should be carried out.

### 13–6.1 Agricultural Uses

Thus, benefit analyses for pesticides concentrate on the effects of regulatory restrictions on the production costs of the commodities and the value of the commodity produced on a per unit basis (such as per hectare) and a total basis. The effects on production costs for any commodity are expressed by the following relationships:

$$C_{\mathrm{p,r}} + C_{\mathrm{o,r}} - C_{\mathrm{p,n}} - C_{\mathrm{o,n}} \qquad [1]$$

where $C_{\mathrm{p,r}}$ is pest control costs assuming problem pesticide is restricted or not available; $C_{\mathrm{o,r}}$ is other production costs (not including pesticides) assuming problem pesticide is restricted or not available; $C_{\mathrm{p,n}}$ is pest control costs assuming problem pesticide continues to be available; $C_{\mathrm{o,n}}$ is other production costs (not including pesticides) assuming problem pesticide continues to be available.

These relationships may be rearranged to:

$$(C_{\mathrm{p,r}} - C_{\mathrm{p,n}}) + (C_{\mathrm{o,r}} - C_{\mathrm{o,n}}) \qquad [2]$$

which represents the change in production costs from a regulatory restriction as the sum of the changes in pest control costs and the changes in other production costs. It cannot be predicted a priori whether Eq. [2] will be positive or negative, because the components of pest control costs and other production costs may either increase or decrease as a result of regulatory restrictions on a pesticide chemical. Each benefits analysis must be done on a case-by-case basis to arrive at conclusions about the effects on production costs.

The pest control cost component must consider such factors as the differences in prices of pesticide products, differences in application rates or in the number of applications needed during a growing season, and any differences in application methods. It may be the case, for example, that alternative pesticides to a pesticide under Special Review may cost less per pound, but because of reduced efficacy additional applications must be made during the growing season to achieve an equivalent level of pest control. Inputs from biologists are essential to provide comparative relationships among pesticides registered for the same uses.

In the evaluation of the productivity effects on an agricultural commodity from cancelling or otherwise restricting a problem pesticide, the analyst needs to evaluate the following expression:

$$P_n X_n - P_r X_r \qquad [3]$$

where $P_n$ is the price of the commodity $X_n$ with no regulatory action, $X_n$ is the output of the commodity with no regulatory action, and $P_r$ and $X_r$ are the price and output of the same commodity under new regulatory restrictions. In effect, Eq. [3] expresses the difference in the value of commodity $X$ produced before and after regulatory restrictions are placed on a pesticide used in the production of commodity $X$.

Although this expression is conceptually straightforward, its evaluation can encompass significant data and estimation problems that need to be resolved. An important question is what is the effect of a pesticide under analysis on agricultural production relative to alternative pesticides or using no pesticide at all. The question of agricultural productivity revolves around determining whether the pesticide has some effect in providing either higher yields or higher quality or both as compared to using alternative pest control strategies.

A related, but distinct, concept is *pesticidal efficacy*, which describes the impact of a pesticide on a pest population. Obviously, efficacy is a partial determinant of product performance; but other factors must be considered. For example, a pesticide may be efficacious in reducing a particular pest species, but if the population of the pest is at a low level at the time of treatment, then the treatment will contribute little to crop productivity. Thus, knowledge of the extent and severity of pest infestations must be gathered to properly evaluate the benefits of application of an agricultural pesticide.

The decision to apply a pesticide to protect a crop from insect damage or diseases, or to reduce weed competition for light, water, and nutrients is governed by the concept of economic threshold of pest infestation. The value of the protection achieved from application must be considered in relation to the cost of the pesticide and the related cost of application. For example, suppose an insecticide available to control insect pests on soybean crops costs $15 (U.S. dollar) per acre to purchase and apply. If the expected market price for soybean crops is $5 per bushel, treatment with the pesticide would be justified when an insect infestation reaches a level of severity in which a yield loss of more than three bushels per acre might occur. The economic threshold is defined in terms of potential loss of productivity from a pest infestation vs. the cost of the pesticide application.

In addition to impacts on productivity, the benefits analysis must also consider whether any potential changes in output may be great enough to significantly affect the market price of the treated commodities. Available data or estimates on consumer demand and supply elasticities are used to determine the direction and magnitude of price changes. Elasticities are the relationships between quantities and prices of goods bought and sold. Elasticities are presented in terms of the relative change in quantity in proportion to a small change in price. The elasticity of demand is given by the expression:

$$e = (P/q) \times (dq/dP) = [(dq/q)/(dP/P)] \quad [4]$$

where $P$ is the price of a good, $q$ is the quantity demanded, and $dq/dP$ is the derivative of the quantity with respect to the price.

The values of the above expression have been empirically measured for many commodities; they are affected by such factors as consumer tastes and income. The significance of elasticity measures is that they provide the relationship between price levels and levels of production. Estimated prices and production levels can then be used to compute total values of production for an agricultural commodity.

The preceding discussion gives a brief overview of the considerations necessary to assess the benefits of use of an agricultural pesticide. After the economic effects of restricting or cancelling a pesticide are determined, this information can be used along with the risk information to make regulatory decisions.

A recent example of the assessment of benefits of pesticide use is contained in the Special Review of Alachlor (USEPA, 1986d). Of the several crops on which alachlor is used extensively, the largest potential impacts of cancellation of the registration were determined to be in field corn production. The USEPA determined that shifting to alternative herbicides would both increase the cost of weed control and result in lower corn yields. The benefits analysis predicted an increase in the costs of treatment of about $1.50 per acre, and also estimated average yield losses of 4.8 bu of corn per acre. Analysis of market response to those impacts led USEPA to conclude that cancellation of alachlor would result in annual losses to farmers who would normally use alachlor on corn of $318 to $552 million in the first year after cancellation. Similar analyses of other major crop uses of alachlor were also carried out as input to the decision-making process.

### 13-6.2 Nonagricultural Uses

Although agricultural uses of pesticides tend to dominate the market, there are significant pesticide uses besides those for food, feed, and fiber production. Other important uses of pesticides include disease vector and nuisance pest control, commercial and industrial uses, and pest control for aesthetic purposes. Nonagricultural pesticide uses obviously involve a variety of economic considerations. In analyzing the benefits of these uses, however, there are many similarities to the economic calculations performed for agricultural uses. The evaluation of the cost of achieving pest control involves concepts generally parallel to those used in agricultural settings. The impacts of reduced pesticide performance in the nonagricultural sector are more complicated, however, since the desired results may not normally manifest themselves in market outputs. For example, the reduced performance in controlling weeds in turf may have a detrimental effect on the appearance of a public park. Evaluating reductions in benefits is generally imprecise; however, the benefits of nonagricultural pesticide use can be analyzed on a case-by-case basis where data permit.

## 13-7 CONCLUSION

The foregoing summaries of exposure, risk, and benefits assessment procedures demonstrate the broad scope and interdisciplinary reach of the review process that supports pesticide regulatory decision-making. It has not been possible in this review to more than briefly introduce the various subject areas; for each disciplinary area there exists an enormous and constantly expanding technical literature. The intent has been to describe briefly the scope of each disciplinary area of the assessment process, indicate the areas of intersection between them, and insert recent key references to facilitate a more detailed introduction.

In contrast to other environmental statutes that set health-based standards, FIFRA explicitly requires the balancing of the risks of pesticide use to society against the benefits to be derived from use. This chapter has attempted to describe the risk and benefit assessment processes; these processes have as their central mission the assembly of all relevant information and the evaluation of the completeness, scientific validity, and uncertainty of the data.

The National Academy of Sciences (NAS, 1983) has recommended that regulatory agencies maintain a clear conceptual distinction between risk *assessment* and consideration of risk *management* alternatives, where political and social considerations can influence the choice of a regulatory strategy. The combining of the outputs of the risk and benefit assessment processes to make regulatory decisions enters the sphere of risk management. The role of the scientist in pesticide regulation is to ensure that the technical basis underlying the risk and benefits assessments is complete and supports a fair exposition of policy choices (Ruckelshaus, 1984).

This chapter reviews the central concerns of the evolving assessment processes supporting pesticide regulation. It is clear that these processes must continue to evolve, to keep pace with emerging technologies such as new chemical classes of pesticides that are effective at low application rates, with the pest control and environmental impact potentials of genetically engineered organisms, and with the continued need for vigorous monitoring of the environment for both exposure and possibly subtle environmental impacts of pesticides.

## REFERENCES


Abrams, H.K., and A.R. Leonard. 1950. Toxicology of organic phosphate insecticides. Calif. Med. 73:183.

Berck, B., Y. Iwata, and F.A. Gunther. 1981. Worker environment research: Rapid field method for estimation of organophosphate insecticide residues on citrus foliage and in grove soil. J. Agric. Food Chem. 29:209–216.

Carsel, R.F., L.A. Mulkey, M.N. Lorber, and L.B. Baskin. 1985. The pesticide root zone model (PRZM): A procedure for evaluating pesticide leaching threats to groundwater. Ecol. Modell. 30:49–69.

Cohen, S.Z., S.M. Creeger, R.F. Carsel, and C.G. Enfield. 1984. Potential pesticide contamination of groundwater from agricultural uses. *In* R.F. Krueger and J.N. Seiber (ed.) Treatment and disposal of pesticide wastes. ACS Symp. Ser. 259. Am. Chem. Soc., Washington, DC.

Cohen, S.Z., C. Eiden, and M.N. Lorber. 1986. Monitoring ground water for pesticides. *In* W.Y. Garner et al. (ed.) Evaluation of pesticides in ground water. ACS Symp. Ser. 315. Am. Chem. Soc., Washington, DC.

Davis, J.E. 1980. Minimizing occupational exposure to pesticides: Personnel monitoring. Residue Rev. 75:33–50.

Donigian, A.S., Jr., and R.F. Carsel. 1987. Modeling the impact of conservation tillage practices on pesticide concentrations in ground and surface waters. Environ. Toxicol. Chem. 6:241–250.

Dourson, M.L., and J.F. Stara. 1983. Regulatory history and experimental support of uncertainty (safety) factors. Regul. Toxicol. Pharmacol. 3:224–238.

Farber, T.M. 1985. Standard operating procedure 1002: Establishing an ADI for Pesticide Chemicals. USEPA, Office of Pesticide Programs, Washington, DC.

Federal Working Group on Pest Management. 1974. Occupational exposure to pesticides. U.S. Gov. Print. Office, Washington, DC.

Ghassemi, M., P. Painter, M. Powers, N.B. Akesson, and M. Dellarco. 1982. Estimating drift and exposure due to aerial application of insecticides in forests. Environ. Sci. Technol. 16:510–514.

Gunther, F.A., W.E. Westlake, J.H. Barkley, W. Winterlin, and L. Langbehn. 1973. Establishing dislodgeable foliar residues on leaf surfaces. Bull. Environ. Contam. Toxicol. 9:243–251.

Gunther, F.A., Y. Iwata, G.E. Carman, and C.A. Smith. 1977. The citrus reentry problem: Research on its causes and effects, and approaches to its minimization. Residue Rev. 67:1–139.

Hamaker, J.W. 1975. The interpretation of soil leaching experiments. p. 115–133. *In* R. Haque and V. Freed (ed.) Environmental dynamics of pesticides. Plenum Press, New York.

Harvey, J., Jr. 1980. A simple method of evaluating soil breakdown of $^{14}C$-pesticides under field conditions. Residue Rev. 85:149–158.

Hoerger, F.D., and E.E. Kenaga. 1972. Pesticide residues on plants. Correlation of representative data as a basis for estimation of their magnitude in the environment. p. 9–28. *In* Environmental quality. Vol. I. Academic Press, New York.

Holst, R.W. 1984. Federal regulatory aspects of spray drift. Pesticide Drift Management Symp., Sioux Falls, SD. September.

Ingram, F.R. 1951. Health hazards associated with use of airplanes to dusting crops with parathion. Am. Ind. Hyg. Assoc. Q. 12:165.

Iwata, Y., J.B. Knaak, R.C. Spear, and W.J. Foster. 1977. Worker reentry into pesticide treated crops. I. Procedures for the determination of dislodgeable residues on foliage. Bull. Environ. Contam. Toxicol. 18:649.

Iwata, Y., J.B. Knaak, G.E. Carman, M.E. Dusch, and F.A. Gunther. 1982. Fruit residue data and worker reentry research for chlorthiophos applied to California citrus trees. J. Agric. Food Chem. 30:215–222.

Maddy, K.T., R.G. Wang, J.B. Knaak, C.L. Liao, S.C. Edmiston, and C.K. Winter. 1985. Risk assessment of excess pesticide exposure to workers in California. *In* R.C. Honeycutt et al. (ed.) Dermal exposure related to pesticide use. ACS Symp. Ser. 273. Am. Chem. Soc., Washington, DC.

National Academy of Sciences. 1983. Risk assessment in the federal government: Managing the process. Natl. Academy Press, Washington, DC.

National Academy of Sciences. 1987. Regulating pesticides in food: The Delaney paradox. Natl. Academy Press, Washington, DC.

Nigg, H.N., J.H. Stamper, R.M. Queen, and J.L. Knapp. 1984. Fish mortality following application of phenthoate to Florida citrus. Bull. Environ. Contam. Toxicol. 32:587–596.

Paynter, O.E., J.G. Cummings, and M.H. Rogoff. 1975. United States pesticide tolerances system (Internal Draft Doc.). USEPA, Office of Pesticide Programs, Washington, DC.

Popendorf, W. 1980. Exploring citrus harvesters' exposure to pesticide contaminated foliar dust. Am. Ind. Hyg. Assoc. J. 41:652–659.

Reinert, J.C., and D.J. Severn. 1985. Dermal exposure to pesticides. *In* R.C. Honeycutt et al. (ed.) Dermal exposure related to pesticide use. ACS Symp. Ser. 273. Am. Chem. Soc., Washington, DC.

Ruckelshaus, W.D. 1984. Risk in a free society. Risk Anal. 4:157–162.

Spear, R.C., W.J. Popendorf, J.T. Leffingwell, T.H. Milby, J.E. Davies, and W.F. Spencer. 1977. Fieldworkers' response to weathered residues of parathion. J. Occup. Med. 19:406–410.

Spencer, W.F., Y. Iwata, W.W. Kilgore, and J.B. Knaak. 1977. Worker reentry into pesticide treated crops. II. Procedures for the determination of pesticide residues on the soil surface. Bull. Environ. Contam. Toxicol. 18:656–662.

U.S. Department of Agriculture. 1970. Pesticide regulation notice 70-15. USDA, Washington, DC.

U.S. Environmental Protection Agency. 1976. Health risk and economic impact assessments of suspected carcinogens: Interim procedures and guidelines. Fed. Register 40, 21402-21405, May 25. U.S. Gov. Print. Office, Washington, DC.

U.S. Environmental Protection Agency. 1982a. Pesticide assessment guidelines. Subdivision N: Environmental Fate. USEPA EPA-540/9-82-021. U.S. Gov. Print. Office, Washington, DC.

U.S. Environmental Protection Agency. 1982b. Pesticide assessment guidelines. Subdivision O: Residue chemistry. USEPA EPA-540/9-82-023. U.S. Gov. Print. Office, Washington, DC.

U.S. Environmental Protection Agency. 1984a. Pesticide assessment guidelines. Subdivision K: Reentry protection. Office of Pesticide Programs, Washington, DC.

U.S. Environmental Protection Agency. 1984b. Dicofol special review position. Doc. 2/3. Office of Pesticide Programs, Washington, DC.

U.S. Environmental Protection Agency. 1984c. Alachlor special review position. Doc. 1. Office of Pesticide Programs, Washington, DC.

U.S. Environmental Protection Agency. 1984d. Pesticide assessment guidelines, Subdivision R: Pesticide spray drift evaluation. USEPA EPA-540/9-84-002. U.S. Gov. Print. Office, Washington, DC.

U.S. Environmental Protection Agency. 1986a. Reference dose (RfD): Description and use in health risk assessments. Integrated risk information system (IRIS): Appendix A. Office of Health and Environmental Assessment, Washington, DC. USEPA EPA-600/8-86-032a. U.S. Gov. Print. Office, Washington, DC.

U.S. Environmental Protection Agency. 1986b. Cyanazine special review technical support document. Office of Pesticide Programs, Washington, DC.

U.S. Environmental Protection Agency. 1986c. Guidelines for cancer risk assessment. Fed. Register 51:33992–34003. U.S. Gov. Print. Office, Washington, DC.

U.S. Environmental Protection Agency. 1986d. Alachlor special review technical support document. Office of Pesticide Programs, Washington,DC.

U.S. Environmental Protection Agency. 1986e. Standard evaluation procedure for ecological risk assessment. USEPA EPA-540/9-85-001. U.S. Gov. Print. Office, Washington, DC.

U.S. Environmental Protection Agency. 1986f. Pesticide assessment guidelines, Subdivision U: Applicator exposure monitoring. Office of Pesticide Programs, Washington, DC.

U.S. Environmental Protection Agency. 1987. Alachlor: Notice of intent to cancel registration; Conclusion of special review. Fed. Register 52:49480–49503. U.S. Gov. Print. Office, Washington, DC.

Wauchope, R.D. 1978. The pesticide content of surface water draining from agricultural fields—A review. J. Environ. Qual. 7:459–472.

Wolfe, H.R., W.F. Durham, and J.F. Armstrong. 1967. Exposure of workers to pesticides. Arch. Environ. Health 14:622–633.

Zweig, G., R. Gao, and W. Popendorf. 1983. Simultaneous dermal exposure to captan and benomyl by strawberry harvesters. J. Agric. Food Chem. 31:1109–1113.

# Chapter 14

# Epilogue: A Closing Perspective

**GEORGE W. BAILEY,** *U.S. Environmental Protection Agency, Athens, Georgia*

In this book, we have documented the research direction, efforts, and accomplishments in agricultural chemistry over the last 40 yr, particularly with respect to advances made since *Pesticides in Soil and Water* (Guenzi, 1974) was published by the Soil Science Society of America in 1974.

The development of a whole generation of pesticides was telescoped into the few years corresponding with the Second World War. DDT, 2,4-D, and indole acetic acid (IAA) were credited with saving millions of lives. New families of pesticides—the triazines, carbamates, organophosphates, substituted ureas, and bipyridinium compounds—sprang into use in the 1950s, 1960s, and 1970s. Thus, started one side of the chemical revolution in American agriculture. Concomitantly, with the development of new pesticide products was the change in fertilizer management practices from the use of rock phosphates, green manure crops, and animal manures as nutrient sources to the adoption of synthetically produced fertilizers and monoculture agriculture.

Rachel Carson's *Silent Spring* (Carson, 1962) created a national awareness and concern over the adverse effects of pesticides on birds of prey, farmers, other segments of the environment, and the populace. Although this was not the first report of environmental abuse (Young & Nicholson, 1951, had noted fish kills in 15 rural Alabama tributary streams to the Tennessee River when rainfall-generated run-off occurred shortly after pesticides were applied to cotton fields), the book created intense concern at the national and even the international levels. The era of environmental awareness started then and continues to the present.

## 14-1 PESTICIDE FATE, IMPACTS, AND REGULATION—CURRENT STATUS

After more than 40 yr of research, we now understand many of the basic pesticide transport and transformation processes and the causative factors controlling these processes. We also have good insight into the kinetics con-

 *Pesticides in the Soil Environment*—SSSA Book Series, no. 2.

trolling these processes and knowledge of the acute and chronic impacts of many pesticides at high concentration on biological processes in soil. We have modeled these fate processes and incorporated or linked them into comprehensive exposure assessment methodologies. Based upon model assessments and some field evaluations, we have identified and evaluated candidate best management practices to prevent off-site adverse impacts and have promulgated regulatory requirements for pesticide use and disposal.

For many of the authors of this book and their colleagues, this research has constituted all or the majority of professional research careers. In the last 30 yr, we have been involved in the groundswell of public awareness, outcry, and concern over the adverse or perceived adverse effects of pesticides. At our government and university laboratories, we were the recipients of significant levels of research funds generated by this era of environmental awareness and concern. We helped complete the initial qualitative description of the fate processes in soils; identified and defined causative factors, pathways, products, and mechanisms of pesticide transformation or degradation in soils; and identified major transport pathways and the effect of soil properties and management factors on their transport. We answered the cry for predictive tools voiced by those who sought to meet regulatory requirements and contributed to the conceptualization, development, testing, and partial validation of computer simulation models. We are involved in completing the era of model testing, and application for regulatory decision making, and cost-effectiveness assessment.

We have contributed to affecting a change in agriculture's initial philosophy on its role in pesticide pollution of the environment and its prevention from the initial posture that agriculture is not a polluter. Agriculture now recognizes that its responsibility does not stop at the fence line. It recognizes it is part of the problem and will work cooperatively with other segments of society towards resolving and solving its environmental impacts.

We have changed our own personal research philosophy from one in which our responsibility ends when the article is sent off to the editor to one in which we are responsible for interpreting our results to the user community—agribusiness and state and federal regulatory and extension agencies. We have incorporated information into Section 208 plans (P1 92-500) and been involved in identification, evaluation, and implementation of best management practices.

Although much has been done already, many questions still remain. Through a combination of more highly sophisticated and available analytical instrumentation (e.g., atomic force and scanning tunneling microscopy for characterizing environmental surface properties, and more specific and sensitive mass and nuclear magnetic resonance, NMR, spectrophotometers for identifying degradation products), expert systems for pesticide structure-reactivity analysis, more powerful and available computers and software, more advanced techniques in DNA cloning and continued adequate financial support, the challenges and opportunities of the 21st century will be met.

### 14–1.1 System Definition

Prior to identifying research needs, it is prudent to define the current state of the art of the pesticide-soil/porous media-plant system and the nature of these interactions. This can be seen pictorially in Fig. 14–1 and Table 14–1. Land use/cultural practices (deterministic inputs) determine the crop type, the pesticide(s) type and formulation, and the application technology used. Exogenous environmental factors are uncontrolled and determine the stochastic nature of pesticide fate in, through, and from the soil to connected surface and the groundwater systems.

This output or soil/porous media-plant system response can be multimedia in dimension (land to water via overland flow or subsurface/aquifer flow, land to atmosphere via volatilization or aeolian erosion, and atmosphere to land via chemical spray drift). Applied pesticides may stay for various lengths of time in one or more residence zones—plant canopy; crop residue surface; and active surface soil, root, vadose, and saturated zones. Fate processes determine pesticide availability and concentration for transport. Pesticide chemical properties—water solubility, ionic state, vapor pressure, hydrophobic/hydrophilic character, and presence of coordinate covalent bond-forming groups—and the initial pesticide concentration determine how much pesticide moves along each transport route. Soil system properties

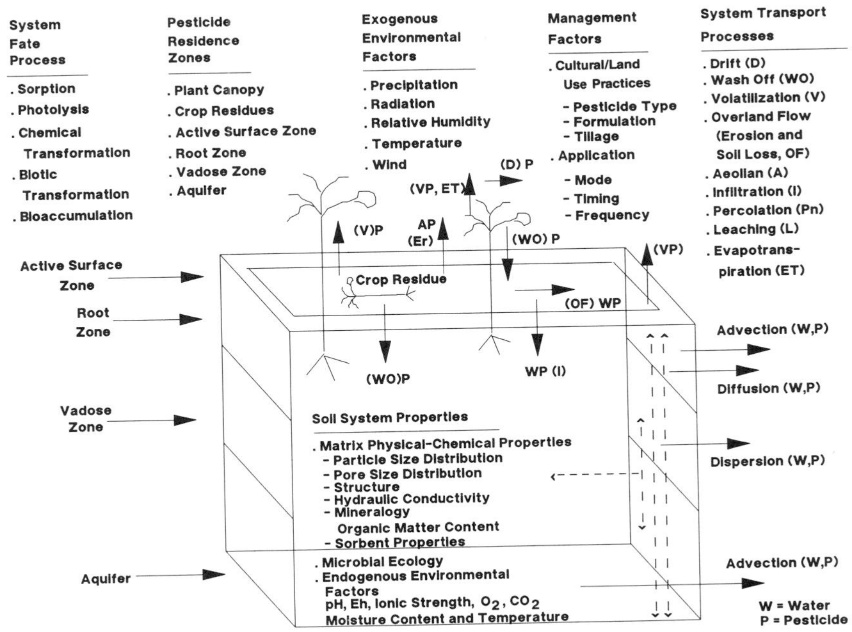


Fig. 14–1. System description of pesticide-soil/porous media-plant interactions.

Table 14-1. Status of knowledge concerning pesticide fate processes.

I. Transformation processes
  A. Sorption
    1. Hydrophobic pesticides
      Causative factors known
      Sorbate—hydrophobic, low water solubility
      Sorbent—organic matter content
      Predictor—$K_{oc}$ or $K_{ow}$ × fraction of organic matter; well-tested relationship
      Isotherm behavior—variable—reversible or irreversible, linear or nonlinear
      Desorption—kinetically controlled
    2. Hydrophilic pesticides
      Causative factors not well known
      Sorbate—acidic in character, pKa
      Basic in character, pKb
      Ionic
      Sorbent—not well known
      Lewis acid-base character or complexation capability
      Predictor—prototype, not tested
      Desorption—not well understood
  B. Microbial transformations
    1. Hydrophobic and hydrophilic pesticides in an oxidizing environment
      Causative factors known—temperature, pH, nutrient, and moisture content
      Mechanisms known for major families
      Transformation products known (their environmental behavior not well characterized)
      Organism specificity—essentially known
      Kinetics—well known
    2. Hydrophobic and hydrophilic pesticides in an anaerobic environment
      Causative factors—Eh, pH, $O_2$ nutrient content—not well known
      Redox chemistry not well understood
      Kinetics—not well defined
      Organism specificity—not well known
  C. Chemical transformations
    1. Hydrophobic and hydrophilic pesticides in an oxidizing environment
      Structure-reactivity relationships fairly well known
      Causative factors—pH, temperature, ionic strength—well known
      Surface catalytic properties of heterogeneous media surfaces—early definitive stage
      Mechanisms—hydrolysis, (acid, base, and neutral) substitution, elimination, and ring cleavage—well known
      Kinetics—generally well known
      Products—generally well known
    2. Hydrophilic and hydrophobic pesticides in an anaerobic environment
      Causative factors not well known
      Mechanisms—early definitive stage
      Kinetics—not well known
      Products—not well known
  D. Photolysis
    1. Water
      Causative factors—fairly well known
      Role of Fe, Mn, and DOC content currently being defined
      Mechanisms—known for major families
      Predictor—prototype available
    2. Soil surface
      Causative factors, pathways, mechanisms, known, research in its infancy

(continued on next page)

Table 14-1. Continued.

| |
|---|
| 3. Foliar or crop residue surface |
| Similar stage as photolysis on soil surface |
| Effect of leaf morphology and biochemical character on process not defined |
| E. Bioaccumulation/biomagnification |
| Causative factors not entirely known |
| Data on hydrophobic pesticide bioaccumulation/magnification known |
| Predictor—first-generation pharmacokinetic prototype available for fish |
| II. Transport processes |
| A. Volatilization |
| 1. Soil surface |
| Causative factors—vapor density, soil matrix properties, and meteorological conditions—well known |
| Predictors—screening level simulation models available |
| 2. Foliar surface |
| Causative factors not well known |
| Predictor—not available |
| B. Drift |
| Causative factors—partially known |
| Predictor—simulation models |
| C. Erosion/pesticide overland flow |
| Causative factors—generally well known for solution and sediment-bound pesticides |
| Predictor—simulation models available |
| D. Washoff |
| Causative factors—partially known |
| Predictor—prototype model available |
| E. Infiltration |
| Causative factors—known |
| Predictor—equations available |
| F. Percolation and leaching in subsoil |
| Causative factors—generally known except for spatial variability of soil properties |
| Predictor—coupled advective/dispersion model with linear or nonlinear/reversible or irreversible sorptive algorithms are available. Generally, not well tested. Transfer function model available but not well tested. |
| G. Groundwater/pesticide transport in aquifer |
| Causative factors not well-defined |
| Predictor—1, 2, and 3D models available, not widely tested |
| H. Aeolian pickup and transport |
| Causative factors generally known |
| Applicable pesticides—sorbed pesticide |
| Predictor—models available, but in question and being re-evaluated |

determine the flow characteristics (of both water and pesticide), sorption determines the availability for transport (vapor vs. aqueous vs. particulate), and the kinetics of transformation/bioaccumulation determine the concentration present to be transported. The hydrologic cycle interacts with chemical properties and soil characteristics to transform and transport pesticides in runoff and leachate to groundwater.

Authors of this book have defined the state of the art in pesticide research under six topics: (i) fate processes—sorption (mechanisms and estimation techniques); chemical, photochemical, and microbial transformation pathways; mechanisms and kinetics; (ii) transport processes—vapor transport and volatilization, aerial transport/drift and canopy deposition, migration to surface water, movement through soils to groundwater and through aquifers, washoff from foliar/crop residue surfaces; (iii) effect of pesticides on biota; (iv) pesticide efficacy; (v) modeling spray drift and fate of pesticides in soils; and (vi) philosophy, procedures, and problems of pesticide registration.

This chapter intends to prognosticate and define needs and priorities for research to be carried out in the 1990s, with results to be implemented early in the 21st century. To accomplish this, what is conventionally done is to: (i) review the state of the art (as described in the previous chapters) and delineate, based upon progress to date, what research needs to be done in the future; (ii) discuss with other knowledgeable colleagues their perception of the research needs and priorities (and filter out, when necessary, disciplinary bias); and (iii) review earlier attempts by advisory panels, symposia participants, to identify and define research needs.

The conventional wisdom and approach will be used; however, in addition, another dimension will be invoked—identification of societal driving forces that will potentially influence the public and private awareness of the need for research, priority setting, and funding levels.

## 14–2 FUTURE RESEARCH AND DEVELOPMENT NEEDS

### 14–2.1 Potential Societal Driving Forces Determining Future Research Directions and Priorities

Influencing and determining pesticide research needs, priorities, funding, and implementation are nine exogenous factors.

#### 14–2.1.1 Increasing Use of Conservation Tillage

Crosson (1983) projects that, by the year 2010, conservation tillage will be used on 50 to 60% of the tillable cropland in the USA. The potential impact of this second major change in U.S. agriculture in the last 40 yr (the first being the switch from rotational agriculture to chemical-based agriculture in the 1950s) is not fully known, but attempts have been made to evaluate it (Unger, 1977, 1978, 1979; Crosson, 1983; Crosson & Brubaker, 1982). Bailey et al. (1985) point out the environmental and health implications as well as emphasize the need to identify and evaluate the agronomic virtues of conservation tillage. Results from their assessment of mainly off-site impacts indicate an increased potential for atmospheric losses of pesticides, a decrease in soil-erosion-related losses runoff, but an increase in plant-residue-associated erosion, as well as increased potential for leaching through the unsaturated zone to groundwater.

#### 14–2.1.2 Increasing Evidence of Pesticide Pollution of Groundwater

The USEPA has initiated a nationwide study, to be completed in 1990, to define the extent of groundwater pollution by pesticides. Seven percent of the nation's rural wells will be sampled and analyzed for pesticides.

### 14-2.1.3 Increasing Use of Supplemental Irrigation

Past research shows that supplemental irrigation will increase in eastern agriculture, particularly in the southeastern states (especially in the Coastal Plain region), Great Lake states, and the Mississippi Delta (Hanson & Pagano, 1982; Shulsted et al., 1982). The environmental threat of pesticides to groundwater quality from the widespread use of this irrigation practice, especially on coarse-textured soils, could be substantial. This trend will be accelerated if global warming projections for these regions become apparent.

### 14-2.1.4 Changes in the Regulatory Philosophy and Recommended Pollution Management Strategies

Changes in regulatory requirements may necessitate the collection of additional information not currently being generated or data with more precision and accuracy than formerly required. If the current approach to registering pesticides as Severn and Ballard discussed (see chapter 13 in this book) is drastically altered, additional research may be needed.

### 14-2.1.5 Changes in Pesticide Structure and Properties

Field and laboratory research on the transport and transformation of pesticides done in the 1960s and 1970s has been conducted mainly on hydrophobic compounds that exhibit low water solubilities, chronic toxicity characteristics, high bioaccumulation/biomagnification tendencies, and strong tendencies to persist for long periods in the environment. The current suite of pesticides developed in the 1980s essentially have the opposite physical, chemical, and biological properties. They are more water soluble, often acutely toxic, less persistent, more hydrophilic, and more mobile.

### 14-2.1.6 Global Climate Changes

The possible effect of widespread changes of monoculture agriculture accelerated by destruction of tropical rain forests and conversion to monoculture agriculture on global climate and the converse—impact of global climate changes on the intensiveness and extensiveness of monoculture agriculture—and the role of pesticides in this mode of agriculture may raise some serious questions.

### 14-2.1.7 Perceived Future Health and Prosperity of American Agriculture

Factors favorable for agri-business—increased export demand; favorable prices and profit margins, credit rates, and debt structures; good climatic conditions; and lower production costs—create a climate for expanded research in agricultural production including formulation and application technology and new pesticides for both major and minor crops.

### 14-2.1.8 Enhanced Residential Pesticide Usage

There is enhanced pressure to develop additional pest-specific pesticides that have even greater efficacy for urban and suburban use.

#### 14-2.1.9 Model-Testing and Sensitivity Analysis

A combination of the identification of critical processes influencing model performance, the level of uncertainty, and changing regulatory procedures may require further process level research.

## 14-3 RESEARCH NEEDS AND PRIORITIES

Based on the driving forces and other factors enumerated above, research activities need to be carried out in six prioritized areas:

### 14-3.1 Model Testing, Refinement, and Application

1. Determine through sensitivity analysis the critical transport or transformation process(es) affecting pesticide migration to surface water, groundwater, and the atmosphere. Deficiencies will be highlighted and requirements for additional process research, and algorithm structure and component changes will be identified. Refinement and testing of surface and subsurface models for exposure assessment or fate evaluation will be completed; testing, refinement, and validation of previously developed exposure assessment/fate surface and subsurface models will be completed.
2. Fully evaluate candidate best management practices in terms of multimedia pollution control effectiveness, that is runoff vs. groundwater vs. terrestrial exposures, using a computer simulation model.
3. Include uncertainty analysis in all assessment evaluations. Develop a better quantitative understanding of uncertainty, particularly of those causative factors controlling uncertainty, and define the precise role of uncertainty in environmental decision-making. Define the uncertainty associated with deterministic vs. stochastic modeling approaches.

### 14-3.2 Transport Processes

1. Quantify the spatial and temporal variability of soil properties by soil series, correlate them to water and solute movement, and incorporate them into appropriate multimedia exposure assessment models and expert systems. Define and test both deterministic and stochastic transport models for predicting water and solute transport for both the unsaturated and the saturated zones and define the degree of uncertainty associated with both approaches.
2. Develop an understanding of the factors controlling pesticide volatilization from foliar and crop residue surfaces; develop models to predict this mode of transport and test them along with those developed to predict the pesticides losses from the soil surface. Incorporate these submodels into comprehensive multimedia exposure assessment packages.
3. Determine the extent and magnitude of overland transport of pesticide granules and pesticides bound on crop residues and the effect of vari-

ous management practices on the onsite losses via this pathway particularly for new chemical-conservation till combinations.

4. Define the factors influencing volatilization from surface water (e.g., in rice culture) including the role of the surface organic monolayer on volatilization loss rates.

### 14–3.3 Pesticide Residence Zone Characterization

1. Characterize the physical, chemical, and biological properties of the 1- to 10-mm thick soil surface active zone created at the time of pesticide application and lasting until the time of the first rainfall event. Correlate these properties with transformation kinetics and volatilization losses. Define and quantify the factors controlling changes in soil moisture content in the pesticide-active surface zone. High temperatures and low moisture content may permit separation of the chemical kinetics from biological transformation processes.

2. Develop an expanded soils data base to include information on soil characteristics below the soil profile that influence pesticide leaching both in the unsaturated and the saturated zones. This includes pH, organic matter content, redox potential, hydraulic conductivity, porosity, particle size, microbial character, and activity and bulk density with depth. Correlate the properties of the C and D horizons of representative soil series and the vadose zone and determine whether such properties correlate well with the physical, chemical, and biological properties of major aquifer materials. Valid correlations would permit extrapolation of existing subsoil data to underlying uncharacterized aquifer materials.

### 14–3.4 Pesticide Delivery Systems and Formulation Technology

Develop, test, and refine cost-effective pesticide delivery systems using formulation technologies that will minimize or eliminate chemical spray drift and optimize effective interaction with plant foliar surfaces, while meeting rural indoor air quality standards.

### 14–3.5 Fate Processes

1. Increase our basic understanding of pesticide anaerobiosis in terms of causative factors—$O_2$ content, pH, $CO_2$ content, redox potential, and nutrient content—and of microbiological characteristics. Define transformation mechanisms in terms of pesticide structure and causative factors and develop predictive kinetic equations.

2. Use the electronic pesticide structure—reactivity relationship approach of Karickhoff et al. (1989) to estimate physical-chemical properties of pesticide transformation or daughter products. Incorporate the predictive method into exposure assessment methodologies for pathway analysis and risk assessment.

3. Define the magnitude and importance of pesticide adsorption, chemical reactions (photolysis and hydrolysis), and plant uptake relative to amounts

lost because of chemical volatilization from foliar surfaces to specify the concentration available for foliar washoff. Relate transformation behavior to leaf surface properties and pesticide properties.

4. Develop and test predictive relationships for the binding of polar organic pesticides to environmental surfaces. Further characterize the chemical composition, functionality and structure of soil organic matter and the nature and extent of organic coatings on mineral surfaces as the basis to better estimate sorption of both polar and nonpolar organic pesticides in porous media of low to very low organic matter content. Use the research results to develop a quantitative method of determining mineral contribution to sorption of hydrophobic and hydrophilic pesticides.

5. Examine microbial transformation kinetics in terms of the true degrader community population, spatial availability of the pesticide on the sorbent surface, redox potentials, and other causative factors.

6. Define the role of chemical equilibrium between solution and sorbed pesticide on the mechanism and rate of pesticide breakdown.

7. Establish the importance of photolysis on the soil surface and the causative factors—pesticide and formulation properties, pH, organic matter and water content, depth of pesticide and radiation penetration into the soil—on transformation mechanisms and kinetics.

8. Define what soil and meteorological factors and cultural practices enhance the accelerated and undesirable loss of pesticide efficacy.

9. Define the processes and kinetics whereby pesticides are polymerized and synthesized into humus compounds of soil. Determine the stability of these products and the ultimate availability and persistence of these adjunct compounds.

10. Develop and test a deterministic, pharmacokinetically based model to predict pesticide bioaccumulation/biomagnification in plants, soil microflora, and fauna.

11. Characterize redox processes as a function of system properties as a basis for understanding anaerobiosis and predicting anaerobic transformation processes and kinetics.

12. Define the significant extent and rate of microbial and chemical transformation processes in the subsurface domain, both in the unsaturated and the saturated zone.

### 14–3.6 Bioengineered Control Technology

Develop through biotechnology techniques (both in vivo and in vitro genetic manipulations), a pesticide-specific microorganism that will effectively degrade a specific pesticide to acceptable daughter or final degradation products. Continue research to evaluate the nature and properties of plasmids and identify the mechanisms for designing and cloning pesticide-specific degradation pathways.

## 14-4 SUMMARY

Challenging new research opportunities for both the 1990s and into the 21st century includes: (i) refining tested/validated first-generation exposure assessment models, including the updating of all transport and transformation process subroutines; (ii) characterizing redox processes as a function of system properties as basis for understanding anaerobiosis and predicting anaerobic transformation processes and kinetics; (iii) modeling photolytic processes on soil and foliar surfaces; (iv) enhancing the precision and lowering the degree of uncertainty of predicting pesticide transformation and transport processes as mandated by changing regulatory needs; (v) predicting pesticide fate in major aquifers based entirely on site characteristics; and (vi) increasing our knowledge of uncertainty and error propagation in the modeling context.

As the foregoing indicates, we have come a long way in these research ventures, but much remains to challenge our efforts. We hope this book sets the stage and points out some of the directions for future endeavors.

## REFERENCES

Bailey, G.W., L.A. Mulkey, and R.R. Swank, Jr. 1985. Environmental implications of conservation tillage: A systems approach. p. 240–265. *In* F.M. D'Itri (ed.) A systems approach to conservation tillage. Lewis Publ., Chelsea, MI.

Carson, R.L. 1962. Silent spring. Riverside Press, Cambridge, MA.

Crosson, P.R. 1983. Trends in agriculture and possible environmental futures. p. 425–452. *In* F.W. Schaller and G.W. Bailey (ed.) Agricultural management and water quality. Iowa State Univ. Press, Ames.

Crosson, P.R., and S. Brubaker. 1982. Resource and environmental effects of U.S. agriculture. Resource. for the Future, Washington, DC.

Guenzi, W.D. (ed.). 1974. Pesticides in soil and water. SSSA, Madison, WI.

Hanson, J., and J. Pagano. 1982. Growth and prospects for irrigation in the eastern United States. *In* Trends in U.S. irrigation: Three regional studies. USEPA EPA-600/3-82/069. U.S. Gov. Print. Office, Washington,DC.

Karickhoff, S.W., L.A. Carreira, C. Melton, V.K. McDaniel, A.N. Vellino, and D.E. Nute. 1989. Predicting chemical reactivity by computer: The ultimate SAR. Environmental research brief. USEPA, Athens, GA. USEPA EPA/600/M-89/017. U.S. Gov. Print. Office, Washington, DC.

Shulstead, R.N., R.D. May, B.D. Herrington, and J.M. Erstine. 1981. The economic potential for the expansion of irrigation in the Mississippi Delta region. *In* Trends in U.S. irrigation: Three regional studies. USEPA EPA-600/3-82/069. U.S. Gov. Print. Office, Washington, DC.

Unger, S.W. 1977. Environmental implications of trends in agriculture and silviculture. Vol. 1. Trend identification and evaluation. USEPA EPA-600/3-77-121. U.S. Gov. Print. Office, Washington, DC.

Unger, S.W. 1978. Environmental implications of trends in agriculture and silviculture. Vol. 2. Environmental effects of trends. USEPA EPA-600/3-78-102. U.S. Gov. Print. Office, Washington, DC.

Unger, S.W. 1979. Environmental implications of trends in agriculture and silviculture. Volume 3. Regional crop production trends. USEPA EPA-600/3-79-047. U.S. Gov. Print. Office, Washington, DC.

Young, L.A., and H.P. Nicholson. 1951. Stream pollution resulting from the use of organic pesticides. Prog. Fish Cult. 13:193–198.

# CHEMICAL INDEX

| Common name | Chemical name |
|---|---|
| Acephate | see Orthene |
| Alachlor | 2-Chloro-2′,6′-diethyl-*N*-(methoxymethyl)-acetanilide |
| Aldicarb | 2-Methyl-2-(methylthio) propionaldehyde *O*(methylcarbamoyl)oxime |
| Aldrin | (1R, 4S, 4aS, 5S, 8R, 8aR) 1,2,3,4,10,10-hexachloro-1,4,4a,5,8,8a-hexahydro-1,4:5,8-dimethanonaphthalene |
| Ametryne | 2-Ethylamino-4-isopropylamino-6-methylthio-1,3,5-triazine |
| Amitrole | 1*H*-1,2,4-Triazol-3-amine |
| Atrazine | 2-Chloro-4-ethylamino-6-isopropylamino-1,3,5-triazine |
| Azinophosmethyl | *O,O*-Dimethyl *S*-[(4-oxo-1,2,3-benzotriazin 3(4H)-yl) methyl] phosphorodithioate |
| Azodrin | Dimethyl (E)-1-methyl-3-(methylamino)-3-oxo-1-propenyl phosphate |
| Bentazon | 3-(1-Methylethyl)-1*H*-2,1,3-benzothiadiazin-4(3*H*)-one 2,2-dioxide |
| γ-BHC | 1,2,3,4,5,6-Hexachloro-cyclohexane, gamma isomer |
| Bromacil | 5-Bromo-3-*sec*-butyl-6-methyluracil |
| Butocarboxim | 3-(Methylthio)-2-butanone *O*-[(methylamino)carbonyl]-oxime |
| Butylate | *S*-Ethyl diisobutylthiocarbamate |
| BT | see Butrizol |
| Butrizol | 4-*N*-Butyl-4*H*-1,2,4-triazole |
| Captan | *cis*-*N*-Trichloromethylthio-4-cyclohexene-1,2-dicarboximide |
| Carbaryl | 1-Naphthyl *N*-methylcarbamate |
| Carbendazim | 2-(Methoxycarbomylamino)-benzimidazole |
| Carbofuran | 2,3-Dihydro-2,2-dimethyl-7-benzofuranyl methylcarbamate |
| Carboxin | 5,6-Dihydro-2-methyl-*N*-phenyl-1,4-oxathiin-3-carboxamide |
| Chlordane | 1,2,4,5,6,7,8,8-Octachlor-2,3,3a,4,7,7a-hexahydro-4,7-methanoindane |
| Chloridazon | 5-Amino-4-chloro-2-phenyl-3-(2H)-pyridazinone |
| Chloropicrin | Trichloronitromethane |
| Chloropropham | Isopropyl 3-chlorophenylcarbamate |
| Chlorpyrifos | *O,O*-Diethyl *O*-(3,5,6-trichloro-2-pyridinyl)-phosphorothioate |
| Chlorsulfuron | 2-Chloro-*N*[(4-methoxy-6-methyl-1,3,5-triazin-2-yl)aminocarbonyl]-benzenesulfonamide |
| Ciodrin | Dimethyl phosphate of α-methylbenzyl-3-hydroxy-*cis*-crofonate |
| CNP | 2,4,6-Trichlorophenyl-4-nitrophenyl ether |
| Crotoxyphos | Dimethyl phosphate of alpha-methylbenzyl 3-hydroxy-*cis*-crotonate |
| Cyanazine | 2-[[4-Chloro-6-(ethylamino)-S-triazin-2-yl]amino]-2-methylpropionitrile |
| Cycloate | *S*-Ethyl-*N*-cyclohexyl-*N*-ethylthiocarbamate |
| 2,4-D | (2,4-Dichlorophenoxy) acetic acid |

(continued on next page)

| Common name | Chemical name |
|---|---|
| Dacthal | see DCPA |
| Dalapon | 2,2-Dichloropropionic acid |
| 2,4-DB | 4-(2,4-Dichlorophenoxy)butyric acid |
| DBCP | 1,2-Dibromo-3-chloropropane |
| 2,4-DCP | 2,4-Dichlorophenol |
| DCPA | Dimethyl tetrachloroterephthalate |
| DD | 1,3-Dichloro-1-propene mixture with 1,2-dichloropropane |
| DDD | 1,1-Dichloro-2,2-bis(*p*-chlorophenyl) ethane |
| DDE | Dichlorodiphenyldichloroethylene |
| DDT | Dichloro diphenyl trichloroethane |
| Diazinon | *O,O*-Diethyl *O*-(2-isopropyl-4-methyl-6-pyrimidinyl) phosphorothiote |
| Dicamba | 3,6-Dichloro-*o*-anisic acid |
| Dichlorvos | 2,2-Dichlorovinyl dimethyl phosphate |
| Dicofol | 4,4-Dichloro-alpha-trichloro-methylbenzhydrol |
| Dicrotophos | (E)-2-Dimethylcarbamoyl-1-methylvinyl dimethyl phosphate |
| Dieldrin | (1R,4S,4aS,5R,6R,7S,8S,8aR)-1,2,3,4,10,10-Hexachloro-1,4,4a,5,6,7,8,8a-octahydro-6,7-epoxy-1,4:5,8-dimethanonaphthalene |
| Dinitramine | $N_4$,$N_4$-Diethyl-$\alpha,\alpha,\alpha$-trifluoro-3,5-dinitrotoluene-2,4-diamine |
| Dioxin | 2,3,7,8-Tetrachlorodibenzo-*p*-dioxin |
| Diphenamid | *N,N*-Dimethyl-2,2-diphenylacetamide |
| Diquat | 1,1′-Ethylene-2,2′-bipyridylium ion |
| Disulfoton | *O,O*-Diethyl *S*-(2-[ethylthio]ethyl) phosphorodithioate |
| Diuron | 3-(3,4-Dichlorophenyl)-1,1-dimethylurea |
| DMDE | 2,2-Bis(*p*-methoxyphenyl)-1,1-dichloroethane |
| EDB | see Ethylene dibromide |
| Endosulfan | 6,7,8,9,10,10-Hexachloro-1,5,5a,6,9,9a-hexahydro-6,9-methano-2,4,3-benzodioxathiepin-3-oxide |
| Endrin | 3,4,5,6,9,9-Hexachloro-1a,2B,2B,3,6,6aB,7B,7a-octahydro-2,7:3,6-dimethanonaphth [2,3] oxirene |
| EPN | *O*-Ethyl *O-p*-nitrophenyl phenylphosphonothioate |
| EPTC | *S*-Ethyl dipropylthiocarbamate |
| Ethoprop | *O*-Ethyl *S,S*-dipropyl phosphorodithioate |
| Ethoprophos | see Ethoprop |
| Ethylene dibromide | 1,2-Dibromoethane |
| Ethyl parathion | see Parathion |
| Ethyl pirimiphos | see Pirimiphos, ethyl |
| Ethylan | 1,1-Dichloro-2,2-bis(4-ethylphenyl) ethane |
| Fenamiphos | Ethyl 3-methyl-4-(methylthio)phenyl(1-methylethyl)phosphoramidate |
| Fenarimol | 3-(2-Chlorophenyl)-3-(4-chlorophenyl)-5-pyrinidinemethane |
| Fenitrothion | *O,O*-Dimethyl-*O*-4-nitro-*m*-tolyl phosphorothioate |

(continued on next page)

| Common name | Chemical name |
|---|---|
| Fenpropathrin | (RS)-α-Cyano-3-phenoxybenzyl 2,2,3,3-tetramethyl-cyclopropanecarboxylate |
| Fensulfothion | Diethyl-4-(methylsulphinyl) phenyl phosphorothionate |
| Fenthion | *O,O*-Dimethyl-*O*-[4-(methylthio)-*m*-tolyl] phorphorothioate |
| Fenvalerate | (RS)-α-Cyano-3-phenoxybenzyl (RS)-2-(4-chlorophenyl)-3-methylbutyrate |
| Fluridone | 1-Methyl-3-phenyl-5-[3-(trifluoromethyl)phenyl]-4(1H)-pyridinone |
| Fluchloralin | *N*-(2-Chloroethyl)-α,α,α-trifluoro-2,6-dinitro-*N*-propyl-*p*-toluidine |
| Fluometuron | 1,1-Dimethyl-3-(α,α,α-trifluoro-*m*-tolyl)urea |
| Fonofos | *O*-Ethyl-*S*-phenylethylphosphonodithioate |
| Glyphosate | Isopropylamine salt of *N*-(phosphono-methyl)glycine |
| Heptachlor | 1,4,5,6,7,8,8-Heptachloro-3a,4,7,7a-tetrahydro-4,7-methanoindene |
| Hexazinone | 3-Cyclohexyl-6-(dimethylamino)-1-methyl-1,3,5-triazine-2,4-(1*H*,3*H*)-dione |
| IPC | see Propham |
| Iprodione | 3-(3,5-Dichlorophenyl)-*N*-(1-methylethyl)2,4-dioxo-1-imida-zolidinecarboxamide |
| Isodrin | 1,2,3,4,10,10-Hexachloro-1,4,4a,5,8,8a-hexahydro-1,4-*endo*,*endo*-5,8-dimethanonaphthalene |
| Isofenphos | 1-Methylethyl-2({ethoxy([1-methylethyl]-amino)phosphino-thioyl}oxy)benzoate |
| Karsil | *N*-(3′4′-Dichlorophenyl)2-methylpentanamide |
| Kelthane | 1,1-Bis(*p*-chlorophenyl)-2,2,2-trichloroethanol |
| Kepone | 1,1a,3,3a,4,5,5a,6-Decachlorooctahydro-1,3,4-methano-2*H*-cyclobuta(c,d)pentalene-2-one |
| Krenite | (Fosamine-Ammonium) ammonium ethyl (aminocarbonyl) phosphonate |
| Lindane | see γ-BHC |
| Linuron | 3-(3,4-Dichlorophenyl)-1-methoxy-1-methylurea |
| Malathion | *O,O*-Dimethyl phosphorodithioate of diethyl mercaptosuc-cinate |
| MCPA | 4-Chloro-2-methylphenoxyacetic acid |
| Metamitron | 3-Methyl-4-amino-6-phenyl-1,2,4-triazin-5(4*H*)-one(I) |
| Methidathion | *O,O*-Dimethyl phosphorodithioate, *S*-ester with 4-(mercapto-methyl-2-methoxy $\Delta^2$-1,3,4-thiadiazolin-5-one |
| Methomyl | *S*-Methyl-*N*-(methylcarbamoyl)oxy)-thioacetimidate |
| Methoxychlor | 2,2-Bis (*p*-methoxyphenyl)-1,1,1-trichloroethane |
| Methyl parathion | see Parathion |
| Metobromuron | 3-(4-Bromophenyl)-1-methoxy-1-methylurea |
| Metolachlor | 2-Chloro-*N*-(2-ethyl-6-methylphenyl)-*N*-(2-methoxy-1-methylethyl) acetamide |
| Metribuzin | 4-Amino-6-(1,1-dimethylethyl)-3-(methylthio)-1,2,4-triazin-5(4*H*)-one |

(continued on next page)

| Common name | Chemical name |
|---|---|
| Mevinphos | Alpha isomer of 2-carbomethoxy-1-methyl-vinyl dimethyl phosphate |
| Mexacarbate | 4-Dimethylamino-3,5-xylylmethylcarbamate |
| Mirex | Dodecachlorooctahydro-1,3,4-metheno-2*H*-cyclobuta(cd) pentalene |
| Monocrotophos | Dimethyl-(E)-1-methyl-2-(methylcarbamoyl)-vinyl phosphate |
| Monuron | 3-(*p*-Chlorophenyl)-1,1-dimethylurea |
| MSMA | Methylarsonic acid, monosodium salt |
| Nabam | Disodium ethylene-1,2-bisdithiocarbamate |
| Napropamide | 2-(*a*-Naphthoxy-*N*,*N*-diethylpropionamide |
| Nemacur | Ethyl-3-methyl-4-(methylthio)-phenyl-(1-methylethyl) phosphoramide |
| Nitrofen | 2,4-Dichlorophenyl-*p*-nitrophenyl ether |
| Orthene | *O*,*S*-Dimethyl acetylphosphoramidothioate |
| Oxamyl | Methyl *N'*,*N'*-dimethyl-*N*-([methylcarbamoyl]oxy)-1-thiooxamimidate |
| Paraoxon | *O*,*O*-Diethyl *O*-*p*-nitrophenyl phosphate |
| Paraquat | 1,1'-Dimethyl-4,4'-bipyridinium ion |
| Parathion | *O*,*O*-Diethyl *O*-4-nitrophenyl phosphorothioate |
| PCP | Pentachlorophenol |
| Pentachlorophenol | see PCP |
| Permethrin | (3-Phenoxyphenyl)methyl(I)*cis*,*trans*-ethenyl-2,2-dimethylcyclopropane-carboxylate |
| Phorate | *O*,*O*-Diethyl *S*-([ethylthio] methyl)-phosphorodithioate |
| Phosalone | *O*,*O*-Diethyl-*S*-[96-chlorobenzoxazalone-3-yl)-methyl]phosphorodithioate |
| Phosmet | *N*-(Mercaptomethyl-phthalimide-*S*-(*O*,*O*-dimethylphorphorodithioate) |
| Picloram | 4-Amino-3,5,6-trichloropicolinic acid |
| Pirimiphos, ethyl | 2-Diethylamino-6-methyl-4-pyrimidinyl diethyl phosphorothioate |
| Prometon | 2,4-Bis(isopropylamino)-6-methoxy-*s*-triazine |
| Prometryn | 2,4-Bis(isopropylamino)-6-(methylthio)-*s*-triazine |
| Propachlor | 2-Chloro-*N*-isopropylacetanilide |
| Propanil | *N*-(3,4-Dichlorophenyl) propionamide |
| Propazine | 2-Chloro-4,6-bis(isopropylamino)-*s*-triazine |
| Propham | Isopropylphenylcarbamate (IPC) |
| Prothiophos | *O*-(2,4-Dichlorophenyl)-*O*-ethyl *S*-propyl phosphoro dithioate |
| Pyrazon | 5-Amino-4-chloro-2-phenyl-3-pyridazinone |
| Resmethrin | 5-Benzyl-3-furylmethyl (1RS)-*cis*,*trans*-chrysanthemate |
| Ronnel | *O*,*O*-Dimethyl *O*-(2,4,5-trichlorophenyl)-phosphorothioate |
| Rotenone | 1,2,12,12a-Tetrahydro-8,9-dimethoxy-2-(1-methylenyl)-(1)benzopyrano(3,4,*b*)furo(2,3,*h*) (1)benzopyran-6(6H)-one |

(continued on next page)

| Common name | Chemical name |
|---|---|
| Sevin | see Carbaryl |
| Silvex | 2-(2,4,5-Trichlorophenoxy)propionic acid |
| Simazine | 2-Chloro-4,6-bis(ethylamino)-*s*-triazine |
| Sulprofos | *O*-Ethyl *O*-(4-[methylthio] phenyl)-*S*-propyl phosphorodithioate |
| Sustar | 3-Trifluoromethylsulfonamido-*p*-acetotoluide |
| Swep | Methyl-3,4-dichlorocarbanilate |
| Syringic acid | 4-Hydroxy-3,5-dimethoxybenzoic acid |
| 2,4,5-T | 2,4,5-Trichlorophenoxyacetic acid |
| TCA | Sodium trichloroacetate |
| 2,3,7,8-TCDD | 2,3,7,8-Tetrachlorobenzodioxin |
| TDE | see DDD |
| Temephos | *O,O*-(Thiodi-4,1-phenylene) bis(*O,O*-dimethylphosphorothioate) |
| TEPP | Tetraethyl diphosphate |
| Terbufos | *S*-{([1.1-Dimethylethyl] thio)methyl} *O,O*-diethyl phosphorodithioate |
| Terbutryn | 2-tert-Butylamino-4-ethylamino-6-methylthio-*s*-triazine |
| Thiabendazole | 2-(4′-Thiazolyl)-benzimidazole |
| Thidiazuron | *N*-Phenyl-*N*-1,2,3-thiadiazol-5-yl urea |
| Thiophanate | 1,2-Bis(3-ethoxycarbonyl-2-thioureido)benzene |
| Thiram | Bis(dimethylthio-carbamoyl)disulfide |
| Toxaphene | Chlorinated camphene (content of combined chlorine, 67–69%) |
| Trichlorfon | Dimethyl(2,2,2-trichloro-1-hydroxyethyl)phosphonate |
| Trifluralin | $\alpha,\alpha,\alpha$-Trifluoro-2,6-dinitro-*N,N*-dipropyl-*p*-toluidine |

# SUBJECT INDEX